Multivariate Polysplines

Dedicated to the memory of Tseni

Multivariate Polysplines: Applications to Numerical and Wavelet Analysis

Ognyan Kounchev

*Institute of Mathematics, Bulgarian Academy of Sciences,
Sofia, Bulgaria*

ACADEMIC PRESS
A Harcourt Science and Technology Company

San Diego San Francisco New York
Boston London Sydney Tokyo

This book is printed on acid-free paper.

Copyright © 2001 by ACADEMIC PRESS

All Rights Reserved.
No part of this publication may be reproduced or transmitted in any form or by any means, electronic or mechanical, including photocopying, recording, or any information storage and retrieval system, without the prior permission in writing from the publisher.

Academic Press
A Harcourt Science and Technology Company
Harcourt Place, 32 Jamestown Road, London NW1 7BY, UK
http://www.academicpress.com

Academic Press
A Harcourt Science and Technology Company
525 B Street, Suite 1900, San Diego, California 92101-4495, USA
http://www.academicpress.com

ISBN 0-12-422490-3

Library of Congress Catalog Number: 2001089852

A catalogue record of this book is available from the British Library

Typeset by Newgen Imaging Systems (P) Ltd., Chennai, India
Printed in Great Britain by MPG Books Ltd, Bodmin.

01 02 03 04 05 06 MP 9 8 7 6 5 4 3 2 1

Contents

Preface		xv
1	**Introduction**	1
1.1	Organization of material	3
	1.1.1 Part I: Introduction of polysplines	3
	1.1.2 Part II: Cardinal polysplines	4
	1.1.3 Part III: Wavelet analysis using polysplines	5
	1.1.4 Part IV: Polysplines on general interfaces	6
1.2	Audience	6
1.3	Statements	7
1.4	Acknowledgements	8
1.5	The polyharmonic paradigm	10
	1.5.1 The operator, object and data concepts of the polyharmonic paradigm	10
	1.5.2 The Taylor formula	11
Part I	**Introduction to polysplines**	15
2	**One-dimensional linear and cubic splines**	19
2.1	Cubic splines	19
2.2	Linear splines	21
2.3	Variational (Holladay) property of the odd-degree splines	23
2.4	Existence and uniqueness of odd-degree splines	26
2.5	The Holladay theorem	27
3	**The two-dimensional case: data and smoothness concepts**	29
3.1	The data concept in two dimensions according to the polyharmonic paradigm	29
	3.1.1 "Parallel lines" or "strips"	32
	3.1.2 "Concentric circles" or "annuli"	33
3.2	The smoothness concept according to the polyharmonic paradigm	34
	3.2.1 The "strips"	34
	3.2.2 The "annuli"	36

4 The objects concept: harmonic and polyharmonic functions in rectangular domains in $\mathbb{R}^2$ — 39
- 4.1 Harmonic functions in strips or rectangles — 40
- 4.2 "Parametrization" of the space of periodic harmonic functions in the strip: the Dirichlet problem — 43
- 4.3 "Parametrization" of the space of periodic polyharmonic functions in the strip: the Dirichlet problem — 46
  - 4.3.1 The biharmonic case — 46
  - 4.3.2 The polyharmonic case — 49
- 4.4 Nonperiodicity in y — 52

5 Polysplines on strips in $\mathbb{R}^2$ — 57
- 5.1 Periodic harmonic polysplines on strips, $p = 1$ — 59
- 5.2 Periodic biharmonic polysplines on strips, $p = 2$ — 60
  - 5.2.1 The smoothness scale of the polysplines — 60
- 5.3 Computing the biharmonic polysplines on strips — 61
- 5.4 Uniqueness of the interpolation polysplines — 64

6 Application of polysplines to magnetism and CAGD — 67
- 6.1 Smoothing airborne magnetic field data — 67
- 6.2 Applications to computer-aided geometric design — 71
  - 6.2.1 Parallel data lines Γ_j — 71
  - 6.2.2 Nonparallel data curves Γ_j — 74
- 6.3 Conclusions — 75

7 The objects concept: harmonic and polyharmonic functions in annuli in $\mathbb{R}^2$ — 77
- 7.1 Harmonic functions in spherical (circular) domains — 77
  - 7.1.1 Harmonic functions in the annulus — 79
  - 7.1.2 "Parametrization" of the space of harmonic functions in the annulus and the ball: the Dirichlet problem — 82
  - 7.1.3 The Dirichlet problem in the ball — 85
  - 7.1.4 An important change of the variable, $v = \log r$ — 86
- 7.2 Biharmonic and polyharmonic functions — 86
  - 7.2.1 Polyharmonic functions in annulus and circle — 87
  - 7.2.2 The set of solutions of $L^p_{(k)} u(r) = 0$ — 89
  - 7.2.3 The operators $L^p_{(k)}(d/dr)$ generate an Extended Complete Chebyshev system — 90
- 7.3 "Parametrization" of the space of polyharmonic functions in the annulus and ball: the Dirichlet problem — 92
  - 7.3.1 The one-dimensional case — 92
  - 7.3.2 The biharmonic case — 92
  - 7.3.3 The polyharmonic case — 95
  - 7.3.4 Another approach to "parametrization": the Almansi representation — 96

	7.3.5	Radially symmetric polyharmonic functions	97
	7.3.6	Another proof of the representation of radially symmetric polyharmonic functions	98
8	**Polysplines on annuli in $\mathbb{R}^2$**		**101**
	8.1	The biharmonic polysplines, $p = 2$	103
	8.2	Radially symmetric interpolation polysplines	104
		8.2.1 Applying the change of variable $v = \log r$	107
		8.2.2 The radially symmetric biharmonic polysplines	108
	8.3	Computing the polysplines for general (nonconstant) data	109
	8.4	The uniqueness of interpolation polysplines on annuli	110
	8.5	The change $v = \log r$ and the operators $M_{k,p}$	111
	8.6	The fundamental set of solutions for the operator $M_{k,p}(d/dv)$	113
9	**Polysplines on strips and annuli in $\mathbb{R}^n$**		**117**
	9.1	Polysplines on strips in $\mathbb{R}^n$	118
		9.1.1 Polysplines on strips with data periodic in y	119
		9.1.2 Polysplines on strips with compact data	121
		9.1.3 The case $p = 2$	122
	9.2	Polysplines on annuli in $\mathbb{R}^n$	122
		9.2.1 Biharmonic polysplines in $\mathbb{R}^3$ and $\mathbb{R}^4$	127
		9.2.2 An "elementary" proof of the existence of interpolation polysplines	128
10	**Compendium on spherical harmonics and polyharmonic functions**		**129**
	10.1	Introduction	129
	10.2	Notations	130
	10.3	Spherical coordinates and the Laplace operator	131
	10.4	Fourier series and basic properties	134
	10.5	Finding the point of view	136
		10.5.1 The functions $r^k \cos k\varphi$ and $r^k \sin k\varphi$ are harmonic for $k \geq 0$	136
		10.5.2 The functions $r^k \cos k\varphi$ and $r^k \sin k\varphi$ are polynomials	136
		10.5.3 The functions $r^k \cos k\varphi$ and $r^k \sin k\varphi$ are homogeneous of degree $k \geq 0$	137
		10.5.4 The functions $r^k \cos k\varphi$ and $r^k \sin k\varphi$ are a basis of the homogeneous harmonic polynomials of degree k	137
		10.5.5 The multidimensional Ansatz	138
	10.6	Homogeneous polynomials in $\mathbb{R}^n$	138
		10.6.1 Examples of homogeneous polynomials	139
	10.7	Gauss representation of homogeneous polynomials	139
		10.7.1 Gauss representation in $\mathbb{R}^2$	140
		10.7.2 Gauss representation in $\mathbb{R}^n$	141

viii Contents

10.8	Gauss representation: analog to the Taylor series, the polyharmonic paradigm	145
	10.8.1 The Almansi representation	146
10.9	The sets $\mathcal{H}_k$ are eigenspaces for the operator Δ_θ	147
10.10	Completeness of the spherical harmonics in $L_2(\mathbb{S}^{n-1})$	149
10.11	Solutions of $\Delta w(x) = 0$ with separated variables	152
10.12	Zonal harmonics $Z_{\theta'}^{(k)}(\theta)$: the functional approach	153
	10.12.1 Estimates of the derivatives of $Y_k(\theta)$: Markov–Bernstein-type inequality	159
10.13	The classical approach to zonal harmonics	159
10.14	The representation of polyharmonic functions using spherical harmonics	164
	10.14.1 Representation of harmonic functions using spherical harmonics	166
	10.14.2 Solutions of the spherical operator $L_{(k)}^p f(r) = 0$	168
	10.14.3 Operator with constant coefficients equivalent to the spherical operator $L_{(k)}^p$	168
	10.14.4 Representation of polyharmonic functions in annulus and ball	173
10.15	The operator $r^{n-1} L_{(k)}^p$ is formally self-adjoint	177
10.16	The Almansi theorem	179
10.17	Bibliographical notes	185
11	**Appendix on Chebyshev splines**	**187**
11.1	Differential operators and Extended Complete Chebyshev systems	187
11.2	Divided differences for Extended Complete Chebyshev systems	191
	11.2.1 The classical polynomial case	191
	11.2.2 Divided difference operators for Chebyshev systems	194
	11.2.3 Lagrange–Hermite interpolation formula for Chebyshev systems	196
11.3	Dual operator and ECT-system	197
	11.3.1 Green's function and Taylor formula	198
11.4	Chebyshev splines and one-sided basis	199
	11.4.1 TB-splines, or the Chebyshev B-splines as a Peano kernel for the divided difference	201
	11.4.2 Dual basis and Riesz basis property for the TB-splines	203
11.5	Natural Chebyshev splines	204
12	**Appendix on Fourier series and Fourier transform**	**209**
12.1	Bibliographical notes	212
Bibliography to Part I		**213**

Part II Cardinal polysplines in $\mathbb{R}^n$ — 217

13 Cardinal L-splines according to Micchelli — 221
13.1 Cardinal L-splines and the interpolation problem — 221
13.2 Differential operators and their solution sets U_{Z+1} — 226
13.3 Variation of the set $U_{Z+1}[\Lambda]$ with Λ and other properties — 228
13.4 The Green function $\phi_Z^+(x)$ of the operator $\mathcal{L}_{Z+1}$ — 229
13.5 The dictionary: L-polynomial case — 232
13.6 The generalized Euler polynomials $A_Z(x;\lambda)$ — 232
13.7 Generalized divided difference operator — 236
13.8 Zeros of the Euler–Frobenius polynomial $\Pi_Z(\lambda)$ — 237
13.9 The cardinal interpolation problem for L-splines — 238
13.10 The cardinal compactly supported L-splines Q_{Z+1} — 239
13.11 Laplace and Fourier transform of the cardinal TB-spline Q_{Z+1} — 241
13.12 Convolution formula for cardinal TB-splines — 243
13.13 Differentiation of cardinal TB-splines — 244
13.14 Hermite–Gennocchi-type formula — 245
13.15 Recurrence relation for the TB-spline — 246
13.16 The adjoint operator $\mathcal{L}_{Z+1}^*$ and the TB-spline $Q_{Z+1}^*(x)$ — 248
13.17 The Euler polynomial $A_Z(x;\lambda)$ and the TB-spline $Q_{Z+1}(x)$ — 250
13.18 The leading coefficient of the Euler–Frobenius polynomial $\Pi_Z(\lambda)$ — 253
13.19 Schoenberg's "exponential" Euler L-spline $\Phi_Z(x;\lambda)$ and $A_Z(x;\lambda)$ — 254
13.20 Marsden's identity for cardinal L-splines — 257
13.21 Peano kernel and the divided difference operator in the cardinal case — 257
13.22 Two-scale relation (refinement equation) for the TB-splines $Q_{Z+1}[\Lambda;h]$ — 259
13.23 Symmetry of the zeros of the Euler–Frobenius polynomial $\Pi_Z(\lambda)$ — 261
13.24 Estimates of the functions $A_Z(x;\lambda)$ and $Q_{Z+1}(x)$ — 264

14 Riesz bounds for the cardinal L-splines Q_{Z+1} — 267
14.1 Summary of necessary results for cardinal L-splines — 270
14.2 Riesz bounds — 271
14.3 The asymptotic of $A_Z(0;\lambda)$ in k — 278
14.4 Asymptotic of the Riesz bounds A, B — 281
    14.4.1 Asymptotic for TB-splines Q_{Z+1} on the mesh $h\mathbb{Z}$ — 282
14.5 Synthesis of compactly supported polysplines on annuli — 283

15 Cardinal interpolation polysplines on annuli — 287
- 15.1 Introduction — 287
- 15.2 Formulation of the cardinal interpolation problem for polysplines — 288
- 15.3 $\alpha = 0$ is good for all L-splines with $L = M_{k,p}$ — 290
- 15.4 Explaining the problem — 293
- 15.5 Schoenberg's results on the fundamental spline $L(X)$ in the polynomial case — 294
- 15.6 Asymptotic of the zeros of $\Pi_Z(\lambda; 0)$ — 298
- 15.7 The fundamental spline function $L(X)$ for the spherical operators $M_{k,p}$ — 300
  - 15.7.1 Estimate of the fundamental spline $L(x)$ — 303
  - 15.7.2 Estimate of the cardinal spline $S(x)$ — 304
- 15.8 Synthesis of the interpolation cardinal polyspline — 305
- 15.9 Bibliographical notes — 306

Bibliography to Part II — 307

Part III Wavelet analysis — 309

16 Chui's cardinal spline wavelet analysis — 313
- 16.1 Cardinal splines and the sets V_j — 313
- 16.2 The wavelet spaces W_j — 315
- 16.3 The mother wavelet ψ — 317
- 16.4 The dual mother wavelet $\widetilde{\psi}$ — 318
- 16.5 The dual scaling function $\widetilde{\phi}$ — 319
- 16.6 Decomposition relations — 319
- 16.7 Decomposition and reconstruction algorithms — 321
- 16.8 Zero moments — 322
- 16.9 Symmetry and asymmetry — 323

17 Cardinal L-spline wavelet analysis — 325
- 17.1 Introduction: the spaces V_j and W_j — 326
- 17.2 Multiresolution analysis using L-splines — 329
- 17.3 The two-scale relation for the TB-splines $Q_{Z+1}(x)$ — 331
- 17.4 Construction of the mother wavelet ψ_h — 333
- 17.5 Some algebra of Laurent polynomials and the mother wavelet ψ_h — 337
- 17.6 Some algebraic identities — 339
- 17.7 The function ψ_h generates a Riesz basis of W_0 — 343
- 17.8 Riesz basis from all wavelet functions $\psi_{2^{-j}h}(x)$ — 345
- 17.9 The decomposition relations for the scaling function Q_{Z+1} — 352
- 17.10 The dual scaling function $\widetilde{\phi}$ and the dual wavelet $\widetilde{\psi}$ — 356
- 17.11 Decomposition and reconstruction by L-spline wavelets and MRA — 362
- 17.12 Discussion of the standard scheme of MRA — 368

**18 Polyharmonic wavelet analysis: scaling and rotationally
invariant spaces** . . . 371
    18.1   The refinement equation for the normed TB-spline $\widetilde{Q}_{Z+1}$ . . . 372
    18.2   Finding the way: some heuristics . . . 373
    18.3   The sets PV_j and isomorphisms . . . 375
    18.4   Spherical Riesz basis and father wavelet . . . 377
    18.5   Polyharmonic MRA . . . 379
    18.6   Decomposition and reconstruction for polyharmonic
wavelets and the mother wavelet . . . 384
    18.7   Zero moments of polyharmonic wavelets . . . 391
    18.8   Bibliographical notes . . . 393

Bibliography to Part III . . . 395

Part IV   Polysplines for general interfaces . . . 397

19 Heuristic arguments . . . 399
    19.1   Introduction . . . 399
    19.2   The setting of the variational problem . . . 401
    19.3   Polysplines of arbitrary order p . . . 403
    19.4   Counting the parameters . . . 404
    19.5   Main results and techniques . . . 405
    19.6   Open problems . . . 406

20 Definition of polysplines and uniqueness for general interfaces . . . 409
    20.1   Introduction . . . 409
    20.2   Definition of polysplines . . . 411
    20.3   Basic identity for polysplines of even order $p = 2q$ . . . 415
        20.3.1   Identity for $L = \Delta^{2q}$ . . . 416
        20.3.2   Identity for the operator $L = L_1^2$ . . . 417
    20.4   Uniqueness of interpolation polysplines and
extremal Holladay-type property . . . 421
        20.4.1   Holladay property . . . 425

21 *A priori* estimates and Fredholm operators . . . 429
    21.1   Basic proposition for interface on the real line . . . 429
    21.2   *A priori* estimates in a bounded domain with interfaces . . . 432
    21.3   Fredholm operator in the space $H^{2p+r}(D\backslash ST)$ for $r \geq 0$ . . . 436
        21.3.1   The space Λ_1 for $L = \Delta^p$ . . . 437
        21.3.2   The case $L = \Delta^2$ . . . 442
        21.3.3   The set Λ_1 for general elliptic operator L . . . 442

22 Existence and convergence of polysplines . . . 445
    22.1   Polysplines of order $2q$ for operator $L = L_1^2$ . . . 445
    22.2   The case of a general operator L . . . 447

xii *Contents*

	22.3	Existence of polysplines on strips with compact data	450
	22.4	Classical smoothness of the interpolation data g_j	451
	22.5	Sobolev embedding in $C^{k,\alpha}$	452
	22.6	Existence for an interface which is not C^∞	453
	22.7	Convergence properties of the polysplines	454
	22.8	Bibliographical notes and remarks	459

23 Appendix on elliptic boundary value problems in Sobolev and Hölder spaces — 461

- 23.1 Sobolev and Hölder spaces — 461
  - 23.1.1 Sobolev spaces on manifolds without boundary — 461
  - 23.1.2 Sobolev spaces on the torus $\mathbb{T}^n$ — 463
  - 23.1.3 Sobolev spaces on the sphere $\mathbb{S}^{n-1}$ — 464
  - 23.1.4 Hölder spaces — 465
  - 23.1.5 Sobolev spaces on manifolds with boundary — 466
  - 23.1.6 Uniform C^m-regularity of $\partial\Omega$ — 467
  - 23.1.7 Trace theorem — 468
  - 23.1.8 The general Sobolev-type embedding theorems — 469
  - 23.1.9 Smoothness across interfaces — 470
- 23.2 Regular elliptic boundary value problems — 472
  - 23.2.1 Regular elliptic boundary value problems in $\mathbb{R}^n_+$ — 474
- 23.3 Boundary operators, adjoint problem and Green formula — 475
  - 23.3.1 Boundary operators in neighboring domains — 477
  - 23.3.2 The Green formula for the operator $L = \Delta^p$ — 479
- 23.4 Elliptic boundary value problems — 479
  - 23.4.1 A priori estimates in Sobolev spaces — 480
  - 23.4.2 Fredholm operator — 480
  - 23.4.3 Elliptic boundary value problems in Hölder spaces — 482
  - 23.4.4 Schauder's continuous parameter method — 483
- 23.5 Bibliographical notes — 484

24 Afterword — 485

Bibliography to Part IV — 487

Index — 491

Preface

In the present the theory of Partial Differential Equations (PDEs) is so overwhelmed by the study of Boundary Value Problems that one can hardly believe that from a global perspective these are no more than a modest part of the properties of the differential equations. Apparently, the Qualitative theory of PDEs is a lot more difficult. This may be understood by using an analogy with the one-dimensional case: the boundary value problems on a compact interval are hardly a topic to discuss for the algebraic polynomials when we consider the last as solutions of ordinary differential equations. Topics of interest are the *Descartes'* rule of signs or the *Budan-Fourier* theorem for the number of sign changes (or zeros) in a compact interval, and other lot deeper properties[1]. On the other hand we are quite far from proving analogs of the Descartes' rule and the Budan-Fourier theorem for polyharmonic functions; even the formulation of the proper analogs is a problem. Similar questions for arbitrary higher-order elliptic equations or for nonlinear equations seem to be rather advanced.

The main message of the present book is that the solutions of higher-order elliptic equations, in particular, the polyharmonic functions, may be used as building blocks of multivariate splines – which we call *polysplines* – in much the same way as the one-dimensional polynomials are used to build the one-dimensional splines. We study *cardinal polysplines* and *polyharmonic wavelets* in a complete analogy with the one-dimensional polynomial *cardinal splines* and *cardinal spline wavelets*. All these results may be considered as a step in the direction of qualitative theory of elliptic PDEs.

The reader should not be scared by the big volume of the present book. It has become bigger for reasons of readability. Another reason for the increase of the volume is that the book is intended for readers with varied backgrounds. The *primary purpose* was to provide readers having a modest (or no) background in PDEs, and more interests in CAGD, spline and wavelet analysis, with an exposition of the theory of polysplines at least in special domains. Thus the biggest Part I has appeared. Once such reader has overcome the initial Chapters of Part I he/she might be willing to see the new developments in *cardinal polysplines* and *polyharmonic wavelet analysis* in Part II and Part III. The *secondary purpose* was to provide readers having more considerable background in PDEs with a proper introduction to the basics of the one-dimensional spline theory and wavelet analysis and the smooth transition to the theory of polysplines in Part IV.

In the present volume we were able to cover only some part of the topics of Numerical Analysis: interpolation by polysplines, cardinal interpolation for special break-surfaces,

---

[1] Consult the first part of the famous book of problems in analysis of Polya and Szegö, or [50, p. 89] (this reference is found at the end of Part I.

convergence of the polyspline interpolation in special cases. The polyharmonic wavelet analysis has outweighed the very interesting topics as

- "Polyharmonic" Euler-Maclaurin formulas and Bernoulli polysplines,
- Optimal recovery and polysplines,
- Peano kernels and mean-value properties for polyharmonic functions, and
- Approximation and interpolation theory by polyharmonic functions and polysplines.

They are left for a next volume.

Sofia – May, 2000
Madison – March, 2001

Ognyan I. Kounchev

Chapter 1

Introduction

The last decade of the twentieth century was marked by the penetration of partial differential equations (PDEs) and the methods used to study them into the multivariate constructive theory of functions, in particular in approximation theory and spline analysis. This trend differs from previous developments when standard objects such as polynomial and rational functions and splines were used to approximate solutions of partial differential equations.

The present book introduces and develops a new type of multivariate spline known as polysplines.[1] Although in the one-dimensional case there is a satisfactory theory of one-dimensional splines, which includes all kinds of generalizations such as Chebyshev splines and L-splines, in the multivariate case there are several alternatives which can be considered to be multivariate splines in their own right, such as box splines, simplex splines, radial basis functions etc. So far there are general principles which come from an intuitive understanding of what a multivariate spline should be and which we wish to discuss.

What is a multivariate spline? We will keep close to the following understanding of a spline: assume that a domain $D \subset \mathbb{R}^n$ be given and a disjoint family of subdomains D_j such that $\cup D_j = D$, and the boundaries $\mathrm{bdry}(D_j)$ are smooth enough, so that the normal n exists almost everywhere on $\mathrm{bdry}(D_j)$. Then a *spline* is a function u defined in D which is assembled of functions u_j defined on D_j. These pieces are of similar nature and match up to a certain degree d of smoothness on the joint boundaries. Imagine for simplicity that $D \subset \mathbb{R}^2$ and $D = D_1 \cup D_2$, and $\overline{D}_1 \cap \overline{D}_2 = \Gamma$ which is a curve (see Figure 1.1).

Throughout this book the joint boundary Γ where two pieces match will be called the **interface** or **break-surface**.

---

[1] The name which completely characterizes the polysplines is "piecewise polyharmonic splines". So far the name "polyharmonic splines" has been used by W. Madych for radial basis functions made of the fundamental solution of the polyharmonic operator.

2   *Multivariate polysplines*

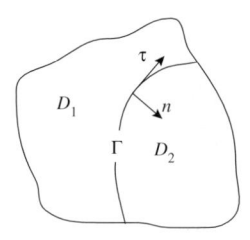

Figure 1.1.

Then we require that

$$u_1 = u_2 \quad \text{on } \Gamma,$$

$$\frac{\partial}{\partial n} u_1 = \frac{\partial}{\partial n} u_2 \quad \text{on } \Gamma,$$

$$\cdots$$

$$\frac{\partial^d}{\partial n^d} u_1 = \frac{\partial^d}{\partial n^d} u_2 \quad \text{on } \Gamma,$$

where $\partial/\partial n$ denotes the normal derivative (one of the two directions) on Γ. If we also require more smoothness of the functions u_1 and u_2 on the joint boundary Γ, say $u_1 \in C^{d_1}(\overline{D_1})$ and $u_2 \in C^{d_1}(\overline{D_2})$, if Γ is also smooth enough, we may differentiate the above equalities in the direction τ tangential to Γ and obtain the equalities of the mixed derivatives up to order d_1, i.e.

$$\frac{\partial^l}{\partial \tau^l} \frac{\partial^k}{\partial n^k} u_1 = \frac{\partial^l}{\partial \tau^l} \frac{\partial^k}{\partial n^k} u_2,$$

where the indices l and k satisfy $l + k \leq d_1$ and $0 \leq k \leq d$. Let us fix a point y on Γ. To write the last equalities in a simpler way, let us introduce a local coordinate system on the surface Γ by putting $y = 0$ and by choosing the normal vector (one of the two directions) to Γ to coincide with the coordinate axis x_2. Then the above equality at the point y will read as follows:

$$\frac{\partial^l}{\partial x_1^l} \frac{\partial^k}{\partial x_2^k} u_1(0) = \frac{\partial^l}{\partial x_1^l} \frac{\partial^k}{\partial x_2^k} u_2(0).$$

The really big questions arise if we are given a "data function" f on the set Γ which has to be *interpolated* by the spline u, i.e. if we would like to have

$$u_1 = u_2 = f \quad \text{on } \Gamma.$$

Then the problem is to find for every d a reasonable *class of functions* u_1 and u_2 for which this interpolation equality can be solved for a large class of data functions f. This problem is a real intellectual challenge. In the present book we provide a solution only for the integers $d = 2p - 2 \geq 0$, where $p \geq 1$ is an integer. The functions u_1 and u_2

then satisfy the equations

$$\Delta^p u_1 = 0 \quad \text{in } D_1,$$
$$\Delta^p u_2 = 0 \quad \text{in } D_2,$$

where Δ^p is the polyharmonic operator.

Our approach to this problem is widely based as will be seen in Chapters 3 and 4.

The theory of polysplines does not appear from nowhere. Even in its simplest cases it relies heavily upon (and even reduces to) Chebyshev's one-dimensional theory and L-splines.

The reader has to be clear about the theory of polysplines. It is a theory which is a genuine synthesis between two important areas in mathematics: approximation theory and elliptic partial differential equations. This has increased the volume of the present book by the inclusion of four appendixes: the *Compendium* on spherical harmonics and polyharmonic functions, and *appendixes* on elliptic boundary value problems (BVPs), Chebyshev splines, and Fourier analysis.

Let us discuss right at the beginning of this book the so-called "data concept" of the polysplines, addressing mainly the readers interested in practical applications: Where do such data come from on a whole curve Γ in $\mathbb{R}^2$ (or a surface Γ in $\mathbb{R}^n$)? Is not this rather restrictive and making the polysplines uninteresting for many applications, since the practical measurements are on a discrete set of points?

The answer to this argument is: In many practical applications (airborne, satellite data, CAGD data) one may find a natural set of curves in $\mathbb{R}^2$ (respectively surfaces in $\mathbb{R}^n$) which contain the discrete data points. Further, in order to obtain data defined on the whole curves one may apply the well-known one-dimensional methods for extending data – one-dimensional splines, etc. In a similar way, and inductively, one proceeds in the case of dimension $n \geq 3$.[2]

1.1 Organization of material

1.1.1 Part I: Introduction of polysplines

In Part I we provide a logical basis for the notion of polysplines. In Chapter 2 we cover some basic results in one-dimensional spline theory. In Chapters 3 and 4 we explain in detail the meaning of the data, smoothness and object concepts of the polyharmonic paradigm, and their implementation in spline analysis. In Chapter 5 we introduce polysplines in the plane $\mathbb{R}^2$ when the interface set is equal to a finite number of parallel straight lines (polysplines on strips). In Chapter 8 we consider the case when the interface set equals a finite number of concentric circles in $\mathbb{R}^2$ (polysplines on annuli). In the adjacent chapters we provide the necessary basics on harmonic and polyharmonic functions on the strip, annulus and ball. The advantage of the two-dimensional case is that the reader has absolutely no need to be familiar with PDEs. All that is necessary is some basic results on the Fourier series which are provided in Chapter 12. The main

---

[2] See also one of the statements in the last Chapter 24, "Afterword".

4   Multivariate polysplines

result is that the computation of the interpolation polysplines in the two cases reduces to a computation of infinitely many proper one-dimensional L-splines, where L are operators with constant coefficients. In the case of polysplines on strips in $\mathbb{R}^2$ these operators are given by

$$L\left(\frac{d}{dt}\right) = \left(\frac{d^2}{dt^2} - \xi^2\right)^p,$$

where ξ is a real number, and in the case of the polysplines on annuli in $\mathbb{R}^2$ the operators are given by

$$L\left(\frac{d}{dt}\right) = \prod_{l=0}^{p-1}\left(\frac{d}{dt} - (k+2l)\right)\prod_{l=1}^{p}\left(\frac{d}{dt} - (-2-k+2l)\right)$$

for all integers $k \geq 0$.

In Chapter 6 we provide experimental proofs for the superiority of the polysplines over well established methods such as kriging, minimum curvature, radial basis functions (RBFs) etc.[3] This may be the most important reason for a reader whose main interest is in numerical analysis to study polysplines. To such a reader we have to say that almost all of Part I is simplified and algorithmic – the results are programmable.

Simple as they are, the polysplines in $\mathbb{R}^2$ have all the main features of the polysplines with arbitrary interfaces and the proof of their existence is not easier. For that reason we leave all the proofs of the existence of the interpolation polysplines to Part IV.

In Chapter 9 we define the polysplines on strips and on annuli in $\mathbb{R}^n$. For the polysplines on annuli we need the structure of a polyharmonic function on the annulus (Almansi-type theorem by I. Vekua in $\mathbb{R}^2$ and by S. L. Sobolev in $\mathbb{R}^n$). This is thoroughly studied in Chapter 10. Matters are definitely simpler for the polysplines on strips since we only need the Fourier transform of functions on strips. All necessary facts about the Fourier transform are provided in Chapter 12. The basic result of Chapter 9 is again as in the two-dimensional case – the polysplines on strips and annuli reduce to some special types of one-dimensional L-splines where L is an operator with constant coefficients (see Theorem 9.3, p. 119, and Theorem 9.7, p. 124).

Part I is logically very self-contained and would meet the interests of a reader occupied with smoothing methods and computer-aided geometric design (CAGD) but not with wavelet analysis.

1.1.2   Part II: Cardinal polysplines

In Part II we concentrate on the polyspline analog to Schoenberg's one-dimensional cardinal splines. This theory will be very important for the wavelet analysis studied in Part III.

As became clear in Part I, we need the theory of the one-dimensional L-splines for the successful study of polysplines on strips and annuli. In Chapter 13 we start with the theory of cardinal L-splines (where L is an operator with constant coefficients) as

---

[3] The comparison has been made with all methods which are implemented in the "Surfer" package (Golden Software, Golden, Colorado).

presented by Ch. Micchelli and I. Schoenberg, and include some results from N. Dyn and A. Ron. The classical polynomial case studied by Schoenberg is a simple special case when the operator $L = (d^{n+1})/(dt^{n+1})$.

The major discovery of Part II is that the *cardinal polysplines on annuli* are polysplines which have as interfaces (break-surfaces) all concentric spheres S_j with radii e^j (or more generally ab^j where the constants $a, b > 0$). They are studied in Chapter 15. We prove that if on every such sphere S_j a function f_j is prescribed such that $\|f_j\|_{L_2}$ has a "power growth" as $|j|^\gamma$ for some $\gamma > 0$ then the interpolation polyspline u exists, i.e.

$$u = f_j \quad \text{on } S_j \quad \text{for all } j \in \mathbb{Z},$$

and satisfies an estimate of the type

$$\|u(r\theta)\|_{L_2(\mathbb{S}_\theta^{n-1})} \leq C|\log r|^\gamma.$$

Here $\mathbb{S}^{n-1}$ is the unit sphere in $\mathbb{R}^n$.

In Chapter 14 we prove that the shifts of the compactly supported L-spline Q (for a fixed operator L) form a Riesz basis. This result is a generalization of the polynomial case considered by Ch. Chui. This is a preparation for the wavelet analysis in Part III.

The case of cardinal polysplines on strips is definitely easier to consider. In that case the cardinal polysplines have as break-surfaces infinitely many parallel hyperplanes which are equidistant. There are again two cases – the periodic polysplines and the polysplines with fast decay. Unfortunately, lack of space prevents us from giving a detailed description of these results.

1.1.3 Part III: Wavelet analysis using polysplines

The *wavelet analysis* in Part III can be read without any preparation in the area, but it would be best if the reader were already familiar with Chui's results on cardinal spline wavelet analysis.

In Chapter 16 we present briefly Chui's results on cardinal spline wavelet analysis so that the reader is familiar with the material that will be generalized.

In Chapter 17 we provide a thorough study of the cardinal L-spline wavelet analysis, i.e. we use cardinal L-splines on refining grids $2^j \mathbb{Z}$ and study the structure of the spaces. The germs of this theory have been laid by C. de Boor, R. DeVore and A. Ron. The case of nonuniform L-spline wavelets (finite case) has been considered by T. Lyche and L. Schumaker. We will prove generalizations of all Chui's basic results for cardinal spline wavelet analysis.

Chapter 18 may be considered as the apex of the pyramid built in the previous chapters. We use all results proved up to this point as bricks for the synthesis of the "polyspline wavelets" on annuli which represent a *spherical polyharmonic multiresolution analysis*. It should be noted that they are compactly supported. It seems that this will be a new framework for further studies in multivariate multiresolution analysis.

On the other hand the *cardinal polysplines on strips* generate a *parallel polyharmonic multiresolution analysis*. Technically its description is easier but no compactly supported "polyspline wavelets" exist!

1.1.4 Part IV: Polysplines on general interfaces

Part IV has a quite different flavour. It is not concrete analysis as the previous parts where we have considered the two special cases of polysplines. It considers the polysplines for general interfaces (break-surfaces). Even more generally, we introduce a very general class of polysplines, which are piecewise solutions of a large class of higher-order elliptic equations. We prove a generalization of the Holladay extremal property as well as the existence of interpolation polysplines for the so-called "even-order polysplines". Unfortunately, there is no simpler setting than Sobolev spaces or Hölder spaces if one wants to obtain solutions. For that reason Part IV may be read without problem only by somebody who is already familiar with the classical theory of elliptic BVPs. Almost all the necessary references are given in Chapter 23. We also provide an extensive introduction to explain the leading ideas of polyspline theory, and references which are classified according to their level and accessibility. Although this is the most abstract part of the book the most technical and deep remains Part III (wavelet analysis).

The following are the surveys and appendixes which we have provided as necessary for our study:

- *Compendium* on spherical harmonics and polyharmonic functions in annular domains following Stein and Weiss, and Sobolev (Chapter 10, p. 129). Not available elsewhere.
- *Survey* of cardinal L-splines (or Chebyshev splines) following Micchelli (Chapter 13, p. 221). Not available elsewhere.
- *Survey* of the sharp estimates and properties of the fundamental function L of the cardinal polynomial splines following several papers by Schoenberg (Section 15.5, p. 294). Not available elsewhere.
- *Survey* of the cardinal polynomial wavelets following Chui (Chapter 16, p. 313).
- *Appendix* on Chebyshev splines following Schumaker (Chapter 11, p. 187).
- *Appendix* on elliptic BVPs in Sobolev and Hölder spaces (Chapter 23, p. 461).
- *Appendix* on Fourier analysis (Chapter 12, p. 209).

1.2 Audience

Part I (except for the existence theorems in Chapter 9), and Parts II and III could be read by anyone who has a good knowledge of classical mathematical analysis, while the *Compendium* on spherical harmonics and polyharmonic functions makes the exposition self-contained. One may say that technically this is nineteenth-century mathematics, wavelets also being included. Part IV is somewhat different. The application of elliptic BVPs is essential and is unavoidable. It would be best if the reader was already familiar with this material. However, we have supplied Part IV with an extended and comprehensive introduction, to at least enable the reader who is not competent in this area to understand that the leading ideas stem from one-dimensional spline theory. This accords with the author's conviction that ideas are more important to the development of mathematics than formulas.

The author has misgivings about not providing elementary proofs to the existence theorems in Chapter 9, but that might have been a lengthy process. For that reason the elementary proofs are left to the reader; we have only supplied the main hints.

The following parts of the book may be used as graduate course texts:

1. Introduction to polysplines (Chapters 2–8). Knowledge of PDEs is not required. Chapter 9, devoted to the polysplines in $\mathbb{R}^n$, may be included with the exception of the existence theorems.
2. Introduction to spherical harmonics and polyharmonic functions (Chapter 10).
3. Introduction to cardinal L-splines (Micchelli's approach) (Chapter 13). Introduction to cardinal polysplines (Chapter 15).
4. Introduction to cardinal L-spline wavelet analysis (approach of Chui and de Boor et al.) (Chapter 17). Introduction to cardinal polyharmonic wavelet analysis (Chapter 18).

1.3 Statements

- The author is convinced that, in the near future, polyharmonic functions will be studied in *multivariate mathematical analysis* in the same way that polynomials are studied in one-dimensional mathematical analysis. From this point of view the constructive theory of functions (including approximation theory) is a qualitative theory of PDEs, whereas the solutions to the elliptic equations form the case of Chebyshev systems.
- Why polyharmonic functions? That is the most natural question which would be asked by any practically oriented mind. "Polynomials are so simple! And *radial basis functions* are simpler to compute! Why should we learn this complicated theory?" First, the answer to this question is far from obvious. There are two ways to answer the above questions. The first is by supplying sufficiently beautiful theorems and other notional arguments which show that the theory of polysplines is *a* natural multivariate spline theory. The second way is simply by providing more experimental material which shows that the polysplines are better than the usual splines, kriging, minimum curvature, RBFs, and other known methods.

  This book is devoted mainly to the first way, the notional one. We show the flexibility of the polysplines concept and provide sufficient argumentation to show that it is a very natural generalization of the notion of one-dimensional splines. We also devote a chapter to showing what the polysplines can do in practice.
- It was the initial intention of the author to scatter the results of this research in several papers. In fact, the most important fundamental results have been published in journals or conference proceedings. But the brevity of such publications does not allow for an extended presentation, explaining the ideas, making comparisons and finishing with a complete description of the illustrative examples.
- One may consider as a proof that the polysplines are a genuine generalization of the one-dimensional splines the fact that all cases of symmetric interfaces are reduced to studying one-dimensional Chebyshev splines. We prove this in Chapter 9 for the case of polysplines on strips, and for the case of polysplines on annuli, all in $\mathbb{R}^n$.

- The Chebyshev systems (see work by Karlin and Studden, or Schumaker) play an important role in univariate approximation theory. In fact, all the beautiful results in the classical and trigonometric moment problem and best approximation theory are also available when the Chebyshev systems are used instead.

  The very deep reason for using pieces of polyharmonic functions as bricks for constructing splines comes from *approximation theory*. It is based on the observation (thus far not rigorously proved) that the polyharmonic functions play the role of Chebyshev systems in several dimensions. We would refer to a series of papers by the author and others establishing theorems in approximation theory through polyharmonic functions which are analogs to one-dimensional approximation theory through polynomials. This area is called by the author the *polyharmonic paradigm*.

- Finally, we have to say that a major motivation to write "yet another theory in analysis" was the experimental success of the polysplines. Experiments with data from magnetism have shown that polysplines give much better results than the Briggs algorithm (minimum curvature), kriging, and RBFs functions which are very popular in that area. The experiments with data in CAGD have shown the superiority of the polysplines even over the usual spline methods, for which CAGD is a priority area.

- The exposition is by no means uniform. Some places are written at an extremely elementary level, things are oversimplified, especially where algorithmic questions are concerned, in order to allow a constructively thinking reader to be able to make the algorithm. Other places, in particular those devoted to the existence theory and applications of Sobolev spaces etc., are intended for the more advanced reader and may be rather tough.

- Many of the results in this book are appearing for the first time. If it is not indicated that the result belongs to someone else, then the result is new.

- We wish to help the reader to find references to a formula, theorem, or chapter–section–subsection, by providing a *page number* to most of these references.

- The name *spline* is associated with smoothness, or smoothly joining pieces of analytic entity. This will be our point of view for the generalization which we plan to do. For that reason it is clear that we have to employ the notions of multivariate smoothness as Sobolev and Hölder spaces. This makes things a lot more complicated than the one-dimensional case but it could not be otherwise if one wants to obtain a genuine multivariate spline theory.

- Many of the books or papers cited here have Russian, French, or other versions. For that reason we quote the number of the theorem, proposition, lemma, section etc., which is independent of the translation.

1.4 Acknowledgements

Many results in the present book were obtained by the author during his stay as a Humboldt Research Fellow at the University of Duisburg in 1992–1994, and later by a grant from the VW-Foundation in 1996–1999. The author wishes to express his gratitude to the Department of Mathematics, University of Duisburg, and especially to Professor Werner Haussmann, for the scientific collaboration and hospitality. The author has lectured over parts of the book as a Visiting Professor at the University of Hamburg in 1997 and 1999, for which he is very grateful to Professor Wolf Hoffmann. The Working Group on Elliptic Boundary Value Problems (formerly at the Max Planck Society) of Professor Bert-Wolfgang Schulze at the University of Potsdam has often played host to the author since 1992, and many solutions and new ideas for the future have taken shape near Sanssouci.

- Special thanks are due to Dr. Hermann Render (Department of Mathematics, University of Duisburg), for many useful remarks on the whole text. Also some basic Lemmata in the Chapter on interpolation cardinal polysplines are due to him, as well as proofs of missing argumentation at several places in Part III on wavelet analysis. Dr. Vladimir Nikiforov has checked carefully the text of Part I and made some useful linguistic remarks.

- At the early stages of the polysplines when they needed to prove practical efficiency, I got a good encouragement from Professor Erik Grafarend (Institute of Geodesy, University of Stuttgart). The first data set, that of airborne data from the Cobb Offset, has been provided to me by Dr. Richard O. Hansen (formerly Professor at Colorado School of Mines, now at Pearson & de Ridder & Johnson in Denver, Colorado), as well as the first "hurrah" confirming that polysplines perform definitely better than Kriging, Minimum Curvature, Radial Basis Functions, etc. I enjoyed also the enthusiasm and the support in working with practical data from Dr. Dimiter Ouzounov (currently at NASA, Goddard Space Flight Research Center), by the time he was at the Institute of Geophysics, Bulgarian Academy of Sciences. I learned from him about the "Surfer" and "Grapher" – both Golden Software products.

- In 1992 Kurt Jetter (then Professor at University of Duisburg, now at University of Hohenheim) invited me to give the first lectures on polysplines at the Seminar of Approximation Theory. In 1993 Kurt was really willing to see the *Polywavelets*. So far the expectation was that there exists a *refinement equation*. Who knew that no such exists? A rather more complicated structure as the *refinement operator* is the reality.

- During the galley-proof period of the present project the author enjoyed the spline-wavelet atmosphere at the University of Wisconsin at Madison thanks to the hospitality of Professors Carl de Boor and Amos Ron. Some improvements in the text are resulting from the experience gained at lecturing about polyharmonic functions and polysplines at the seminar on Approximation theory.

- I used the wonder of Internet by e-mailing from time to time Carl de Boor, Larry Schumaker and Charles Micchelli, and got fast and competent advice on cardinal splines and L-splines.

- I profited many times from the TEX and computer experience of Dr. Hermann Hoch (University of Duisburg) and Dr. Pencho Marinov (Institute of Parallel Computations, Bulgarian Academy of Sciences).

1.5 The polyharmonic paradigm

The polysplines of the present book are a result of the application of the so-called polyharmonic paradigm to spline analysis.

Until recently the *polynomial paradigm* has been a dominant concept in the multivariate constructive theory of functions. The polynomial paradigm comprizes all theories which rely upon multivariate polynomials or "tensor product" constructs.

Let us briefly explain what will be understood by the term *polyharmonic paradigm*. Its aim is to distinguish the polyharmonic functions as a proper multivariate generalization of the *one-dimensional polynomials*. The main thesis is that the polyharmonic functions have to be used as a main building block in the multivariate constructive theory of functions. The polyharmonic paradigm has to be understood as an alternative to the polynomial paradigm.

The polyharmonic paradigm consists of *operator*, *object* and *data concepts* which are related, and which we briefly describe below.

1.5.1 The operator, object and data concepts of the polyharmonic paradigm

The *operator concept* consists of replacing the operator d/dt and its powers in one dimension through the Laplace operator Δ and its powers (called polyharmonic operators), or symbolically written

$$\frac{d}{dt} \longrightarrow \Delta, \tag{1.1}$$

$$\frac{d^p}{dt^p} \longrightarrow \Delta^p.$$

The *objects concept* consists of replacing the one-dimensional polynomials which are solutions to the equation

$$\frac{d^p}{dt^p} f(t) = 0$$

with the solution of the polyharmonic equation

$$\Delta^p f(x) = 0.$$

The function t_+^{p-1} defined by

$$t_+^{p-1} := \begin{cases} t^{p-1} & \text{for } t \geq 0, \\ 0 & \text{for } t \leq 0, \end{cases}$$

satisfies

$$\frac{d^p}{dt^p} t_+^{p-1} = \delta(t)$$

where δ is the *Dirac* delta function in $\mathbb{R}$. We will replace it with the function $R_p(x)$ which satisfies

$$\Delta^p R_p(x) = \delta(x)$$

where δ is the Dirac delta function in $\mathbb{R}^n$. One look at the explicit formulas for R_p (see [3], or the solutions to the operator $L_{(0)}^p$ in Proposition 10.32, p. 168) shows how similar they are to the functions t_+^{p-1}.

The *data concept* consists of replacing the discrete set of data points $\{t_1, t_2, \ldots, t_N\}$ in the one-dimensional case with surfaces $\{\Gamma_1, \Gamma_2, \ldots, \Gamma_N\}$ of dimension $n-1$ in $\mathbb{R}^n$, or

$$t_j \longrightarrow \Gamma_j.$$

In view of the *polyharmonicity paradigm* one may reconsider the basic notions of mathematical analysis in several dimensions, like *finite* (or *divided*) *differences*, *multivariate Taylor formula* etc.

We will illustrate the richness of the above formal side of the polyharmonic paradigm by considering the multivariate analogs of the Taylor formula.

1.5.2 The Taylor formula

In the case of the one-dimensional Taylor formula

$$f(t) = f(t_0) + \frac{t-t_0}{1!} f'(t_0) + \frac{(t-t_0)^2}{2!} f''(t_0) + \cdots + \frac{(t-t_0)^{p-1}}{(p-1)!} f^{(p-1)}(t_0)$$

$$+ \frac{1}{(p-1)!} \int_{t_0}^{t} f^{(p)}(\tau) (t-\tau)_+^{p-1} d\tau$$

we find that the *Almansi, Pizzetti* and *Green* formulas are also *genuine analogs* to it.

The *Almansi formula*

Let $f(x)$ be a polyharmonic function in the ball $B = B(0; R)$. Then the Almansi formula reads as

$$f(x) = f_0(x) + |x|^2 f_1(x) + |x|^4 f_2(x) + \cdots + |x|^{2p-2} f_{p-1}(x) \qquad \text{for } x \in \mathbb{R}^n,$$

where $f_k(x)$ are harmonic functions in the ball B. Compared with the Taylor formula, it is obtained by the scheme

$$\frac{d}{dt} \longrightarrow \Delta, \quad \frac{d^p}{dt^p} \longrightarrow \Delta^p,$$

and since the constants $f^{(k)}(t_0)$ are solutions of

$$\frac{d}{dt}\left(f^{(k)}(t_0)\right) = 0,$$

they are replaced by solutions of the Laplace equation,

$$\Delta f_k(x) = 0.$$

12 Multivariate polysplines

The *Pizzetti formula*

This reads as

$$\int_{S(x_0;R)} f \, d\sigma(y) = f(x_0) + a_{n1} |x - x_0|^2 \Delta f(x_0) + a_{n2} |x - x_0|^4 \Delta^2 f(x_0) + \ldots$$

for all $x \in S(x_0; R)$, where a_{nj} are some coefficients depending only on the dimension of the space n. Due to $|x - x_0| = R$ the replacement scheme is

$$\text{data point } t_0 \longrightarrow \text{data surface } S(x_0; R)$$

$$\text{point value at } t_0 \longrightarrow \int_{S(x_0;R)} f \, d\sigma(y),$$

$$\text{value } \frac{d^k}{dt^k} f(t_0) \longrightarrow \Delta^k f(x_0).$$

The *Green formula*

Let $f(x)$ be a polyharmonic function of order p in the domain D having sufficiently smooth boundary. Assume that f belongs to $C^{2p}(\overline{D})$. Then the Green formula reads as

$$f(x) = \Omega_n \sum_{j=0}^{p-1} \int_{\partial D} \left\{ \Delta^j f(z) \frac{\partial R_{j+1}(z - x)}{\partial \nu_z} - \frac{\partial \Delta^j f(z)}{\partial \nu_z} R_{j+1}(z - x) \right\} d\sigma(z),$$

see *Aronszajn–Creese–Lipkin* [3, p. 9], or formula (23.6) on p. 479. Here R_j is the properly normalized fundamental solution to the equation $\Delta^j R_j(x) = \delta(x)$, and Ω_n is a constant (in (10.3), p. 130); we have denoted by $\frac{\partial}{\partial \nu_z}$ the inward normal derivative.

The above formula is obtained through the rather unusual scheme

$$\text{data point } t \longrightarrow \text{data point } x,$$

$$\text{value } f(t) \longrightarrow \text{value } f(x),$$

$$\text{data point } t_0 \longrightarrow \text{data surface } \partial D,$$

$$\text{value } \frac{d^{2j}}{dt^{2j}} f(t_0) \longrightarrow \int_{\partial D} \Delta^j f(z) \frac{\partial R_{j+1}(z - x)}{\partial \nu_z} d\sigma(z),$$

$$\text{value } \frac{d^{2j+1}}{dt^{2j+1}} f(t_0) \longrightarrow -\int_{\partial D} \frac{\partial \Delta^j f(z)}{\partial \nu_z} R_{j+1}(z - x) \, d\sigma(z).$$

Note that the k'th term

$$u_k(x) = \Omega_n \sum_{j=0}^{k} \int_{\partial D} \left\{ \Delta^j f(z) \frac{\partial R_{j+1}(z - x)}{\partial \nu_z} - \frac{\partial \Delta^j f(z)}{\partial \nu_z} R_{j+1}(z - x) \right\} d\sigma(z)$$

satisfies
$$\Delta^{k+1} u_k(x) = 0 \quad \text{in } D.$$

Thus we see that we have different ways to make a multivariate analog to the Taylor formula. When we generalize univariate results we have to choose a genuine generalization by finding the proper scheme.

Part I

Introduction to polysplines

We have tried to solve the difficult problem of writing introductory chapters on polysplines which are accessible to a wide circle of readers, and which are not overburdened with technicalities. To ease the exposition we have added a special compendium of polyharmonic functions and spherical harmonics (Chapter 10).

Part I of the book is mainly addressed to readers who do not have a specialist background in PDEs. For someone with a background in PDEs, a quick skim through the pages in order to gain familiarity with the notation to be used in the rest of the book will suffice. Such a reader would find interesting results, some of which are not very popular among the PDE community, in the chapters devoted to the *polysplines on annuli* and in Chapter 10. There are also other new results such as:

- Representation of *polyharmonic functions by spherical harmonics*, and the decomposition of the action of the polyharmonic operator Δ^p based on the variable $v = \log r$.
- *Almansi representation* of polyharmonic functions in the *ball* and in the *annulus*.

Part I may also serve as a brief introduction to spline theory for someone who is inexperienced in this area. We start with Chapter 2 where the simplest one-dimensional spline theory is outlined briefly but in sufficient detail, without B-splines and deeper results. Let us emphasize that we make a special point of the case of one-dimensional cubic splines which are one of the most important nontrivial cases.

The ideal reader would be one who has a good background in splines, knows the main results in this area, and also understands the ideas. However, we start our exposition from an elementary level and as we have already said, experience in splines is not necessary.

In order to introduce the polysplines in a way that is easy to comprehend, we start in Chapter 3 with the simplest case, of polysplines in the plane $\mathbb{R}^2$. We will appeal to the complete analogy with the one-dimensional case. It would be advantageous if the reader had some initial familiarity with harmonic functions, since this concerns the core of the nature of the main objects, the polysplines, and this is technically the more complicated part. But such knowledge is not assumed and as we already said we have tried to do the exposition independent of extra competence in PDEs.

Part I contains virtually everything one needs to make a computer implementation of the two-dimensional polysplines when the data is lying on parallel straight lines or on concentric circles. Examples of such applications to smoothing *airborne magnetic data* and applications to *CAGD of surfaces* are given in Chapter 6.

Chapter 2

One-dimensional linear and cubic splines

Let us consider the case of *cubic splines* and *linear splines*, where the basic principles of spline theory are very transparent.

We assume that a finite sequence of points (usually called *knots*) is given in the compact interval $[a, b]$:

$$a = x_1 < x_2 < \cdots < x_{N-1} < x_N = b.$$

2.1 Cubic splines

The *cubic spline* $u(x)$ is a function which coincides with a cubic polynomial $u_j(x)$ (i.e. with a polynomial of degree ≤ 3) on every interval $[x_j, x_{j+1}]$ for $j = 1, \ldots, N-1$, and

$$u(x) \in C^2(a, b).$$

Thus the cubic spline is a sequence of $N - 1$ cubic polynomials

$$u(x) = \begin{cases} u_1(x) = A_1 + B_1 x + C_1 x^2 + D_1 x^3 & \text{in } [x_1, x_2], \\ u_2(x) = A_2 + B_2 x + C_2 x^2 + D_2 x^3 & \text{in } [x_2, x_3], \\ \cdots & \cdots \\ u_{N-1}(x) = A_{N-1} + B_{N-1} x + C_{N-1} x^2 + D_{N-1} x^3 & \text{in } [x_{N-1}, x_N]. \end{cases} \quad (2.1)$$

(See Figure 2.1.)

The number of the coefficients A_j, B_j, C_j is

$$(N - 1) \times 4.$$

20   Multivariate polysplines

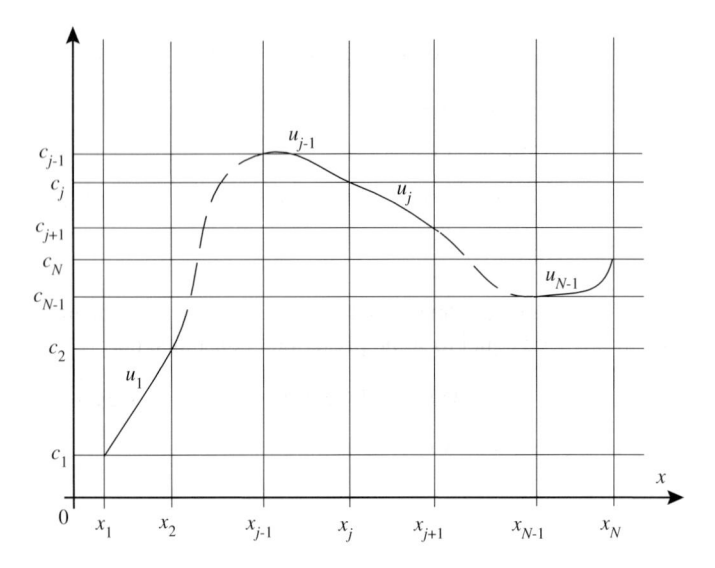

Figure 2.1.   One-dimensional cubic spline.

The smoothness requirement for the cubic spline, $u(x) \in C^2(a, b)$ means that we have the following *matching conditions* between the derivatives:

$$\begin{cases} u_1(x_2) = u_2(x_2), \ u_2(x_3) = u_3(x_3), \ \ldots, \ u_{N-2}(x_{N-1}) = u_{N-1}(x_{N-1}), \\ u_1'(x_2) = u_2'(x_2), \ u_2'(x_3) = u_3'(x_3), \ \ldots, \ u_{N-2}'(x_{N-1}) = u_{N-1}'(x_{N-1}), \\ u_1''(x_2) = u_2''(x_2), \ u_2''(x_3) = u_3''(x_3), \ \ldots, \ u_{N-2}''(x_{N-1}) = u_{N-1}''(x_{N-1}). \end{cases} \quad (2.2)$$

Definition 2.1 *The piecewise cubic function $u(x)$ satisfying $u(x) \in C^2(a, b)$, or equivalently, the matching conditions (2.2), is called a **cubic spline** on the grid $\{x_j\}_{j=1}^N$. More generally, one defines the **polynomial spline of degree** p as a function $u(x)$ which is a piecewise polynomial of degree p on every interval $[x_j, x_{j+1}]$ and which belongs to $C^{p-1}(a, b)$.*

Evidently, the number of conditions (2.2) is equal to

$$(N - 2) \times 3.$$

It follows that $(N - 1) \times 4 - (N - 2) \times 3 = N + 2$ free parameters remain at our disposal.

Cubic splines are used to interpolate some data which are given at the knots $\{x_j\}_j$. This means that for the cubic splines we use the *knot set* simultaneously as a *data set*, i.e.

$$\textbf{data set} = \{x_j\}_{j=1}^N.$$

Additionally, we impose the following conditions:

$$u(x_j) = c_j \quad \text{for} \quad j = 1, \ldots, N,$$

i.e. together with the continuity conditions of the cubic spline we have

$$u(x_j) = u_j(x_j) = c_j \quad \text{for} \quad j = 1, \ldots, N-1, \qquad (2.3)$$
$$u(x_N) = u_{N-1}(x_N) = c_N.$$

Definition 2.2 *The cubic spline $u(x)$ satisfying conditions (2.3) is called **interpolation spline** with data $\{c_j\}_j$. The data set is assumed to coincide with the knot set $\{x_j\}$.*

Since the number of conditions in (2.3) is N, we see that there remain $(N+2)-N = 2$ free parameters. We may impose some "boundary conditions", e.g. the first derivative at the end points will be made equal to some prescribed constants, i.e.

$$u'(x_1) = u_1'(x_1) = d_1, \qquad (2.4)$$
$$u'(x_N) = u_{N-1}'(x_N) = d_2.$$

For simplicity we may put

$$d_1 = d_2 = 0.$$

Let us summarize the above. Assume that the data c_j for $j = 1, \ldots, N$, be given. Then it is clear that conditions (2.2)–(2.4) represent a linear system for finding the unknown coefficients A_j, B_j, C_j, D_j. Our count has shown that the number of unknowns is equal to the number of equations. Hence, one might conclude that there is a unique solution.

We will not go deep into solving the resulting linear system which is usually solved in two steps, with one of the systems having a dominating diagonal. For an illuminating discussion see de Boor [9, Chapter IV, p. 53][1]. A similar discussion is given by Powell [42, Chapter 18]. In the case of polynomial splines of *odd degree* we will provide below an elegant proof of the existence (and uniqueness) of interpolation splines, using the so-called basic identities.

Let us remark that the cases of *linear and cubic splines* and, in general, splines of *odd* polynomial degree are very similar. The most important feature of these odd-order splines is that the data points and the knot points coincide, i.e.

data set = knot set.

The situation with splines of degree 2, the so-called *parabolic* splines, is completely different. De Boor [9, Chapter VI] has thoroughly explained that it is very *inappropriate* to choose knot points which coincide with the data points. In general, all splines of *even* polynomial degree satisfy that principle, namely, that the knot points have to differ from the data points. Otherwise the error available in the data does not decay exponentially [9, Chapter VI, p. 75, 79].

2.2 Linear splines

The linear splines are much simpler and we provide them since they will be considered as the one-dimensional analog to the so-called *harmonic polysplines*.

---

[1] The bibliography is at the end of the present part, p. 211.

22   Multivariate polysplines

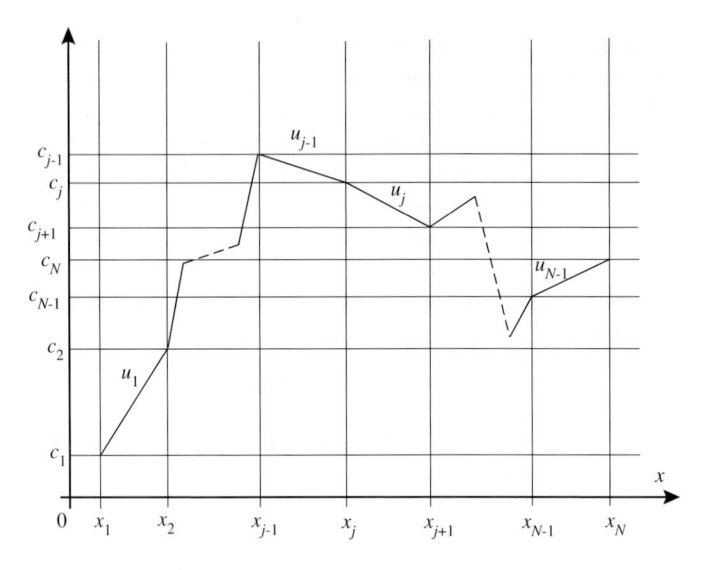

Figure 2.2.   The linear spline consists of pieces of lines which join continuously.

The *linear spline* $u(x)$ is a function which coincides with a linear polynomial $u_j(x)$ (i.e. with a polynomial of degree ≤ 1) on every interval $[x_j, x_{j+1}]$ for $j = 1, \ldots, N-1$, and
$$u(x) \in C(a, b).$$
Thus the linear spline is a sequence of $N - 1$ linear polynomials
$$u(x) = \begin{cases} u_1(x) = A_1 + B_1 x & \text{for } x_1 \leq x \leq x_2, \\ u_2(x) = A_2 + B_2 x & \text{for } x_2 \leq x \leq x_3, \\ \ldots & \ldots \\ u_{N-1}(x) = A_{N-1} + B_{N-1} x & \text{for } x_{N-1} \leq x \leq x_N. \end{cases}$$

(See Figure 2.2.)

The continuity requirement for the linear spline, $u(x) \in C(a, b)$ means that we have only the following *matching conditions*:
$$u_1(x_2) = u_2(x_2), \quad u_2(x_3) = u_3(x_3), \quad \ldots, \quad u_{N-2}(x_{N-1}) = u_{N-1}(x_{N-1}). \tag{2.5}$$

Now the interpolation linear spline satisfies the following conditions:
$$u(x_j) = c_j \quad \text{for } j = 1, \ldots, N,$$
and they reduce to the system
$$A_j + B_j x_j = c_j, \quad \text{for } j = 1, 2, \ldots, N-1,$$
$$A_j + B_j x_{j+1} = c_{j+1}, \quad \text{for } j = 1, 2, \ldots, N-1,$$
which uniquely determines the unknown coefficients A_j, B_j.

Remark 2.3 *Running ahead of events, let us say that one remarkable feature of the multivariate polyspline, which are the main object of the present book, is that a case analogous to the even polynomial degree does not appear at all, and the data set (union of surfaces) for the polysplines always coincides with the knot set (union of surfaces) of the polysplines. In a certain sense their computation, as we will later see, is always stable. A confirmation of this principle is the special case of polysplines which are reduced to computing one-dimensional L-splines which are only of even order. This will become clear in the following chapters. Let us clarify the somewhat misleading terminology: the algebraic polynomial $P(x)$ of odd degree $2q - 1$ is an L-spline of even order $2q$ since it satisfies the equation*

$$\frac{d^{2p}}{dx^{2p}} P(x) = 0,$$

where the operator $L = (d^{2p})/(dx^{2p})$. So the order (or degree) of the polynomial differs from its order as an L-spline.

2.3 Variational (Holladay) property of the odd-degree splines

A wider circle of mathematicians was attracted by the splines in the early 1960s thanks to their beautiful variational property which was discovered by J.C. Holladay. This property is available only in the case of splines of odd polynomial degree and somehow corresponds to the above observation that only in this case may the knot points coincide with the data points. We will state this result in the case of cubic splines.

First, let us introduce some spaces to be used in the sequel. For every integer $s \geq 1$ let us denote by $H^s([a, b])$ the class of functions f defined on the interval $[a, b]$ and having an absolutely continuous $(s - 1)$th derivative there, and $f^{(s)} \in L_2([a, b])$.

Consider the extremal problem

$$\inf \int_a^b |f''(x)|^2 \, dx, \tag{2.6}$$

where the infimum is taken over the set of all functions $f \in H^2([a, b])$ which satisfy the interpolation conditions (2.3), p. 21, i.e.

$$f(x_j) = c_j, \tag{2.7}$$

and the boundary conditions

$$f'(x_1) = d_1, \qquad f'(x_N) = d_2. \tag{2.8}$$

(This is the formulation available in Ahlberg *et al.* [1, Chapter III.4, Theorem 3.4.3], cf. also Laurent [30, p. 162]) (see Figure 2.3). We will prove that this problem has a unique solution and this solution is necessarily a cubic spline $u(x)$.

24   Multivariate polysplines

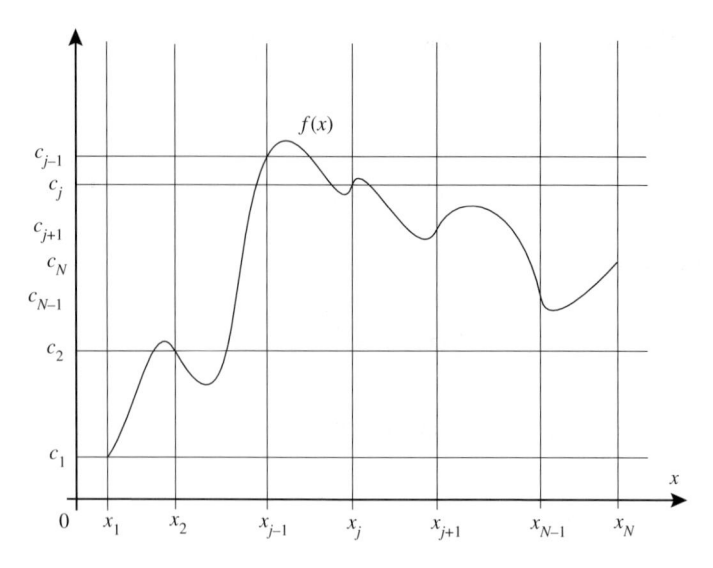

Figure 2.3.   The graph of an arbitrary function $f(x)$ which satisfies the interpolation conditions.

In other words, we will prove that the solution to the extremal problem (2.6), with conditions (2.7) and (2.8), is a function $u \in C^2(a, b)$ such that

$$\frac{d^4}{dx^4} u(x) = 0$$

in every interval (x_j, x_{j+1}) for $j = 1, \ldots, N - 1$.

Since this result and the idea of its proof will be generalized to the multivariate polysplines in Section 20.4, p. 421, we will provide the detailed proof below. We will prove the *basic identity* for odd-degree splines.

Theorem 2.4 *Let $p \geq 1$ be an integer. Let $u(x)$ be a polynomial spline of degree $2p - 1$ and let the function $f \in H^p([a, b])$ be such that*

$$u(x_j) = f(x_j) \quad \text{for} \quad j = 1, \ldots, N,$$
$$u^{(\ell)}(x_1) = f^{(\ell)}(x_1) \quad \text{for} \quad \ell = 0, \ldots, p - 1,$$
$$u^{(\ell)}(x_N) = f^{(\ell)}(x_N) \quad \text{for} \quad \ell = 0, \ldots, p - 1.$$

Then the following identity holds:

$$\int_a^b |f^{(p)}(x)|^2 \, dx = \int_a^b |u^{(p)}(x)|^2 \, dx + \int_a^b |f^{(p)}(x) - u^{(p)}(x)|^2 \, dx.$$

Proof The proof is based on the following equality:

$$\int_a^b |f^{(p)}(x) - u^{(p)}(x)|^2 \, dx$$

$$= \int_a^b |f^{(p)}(x)|^2 \, dx - 2 \int_a^b \{f^{(p)}(x) - u^{(p)}(x)\} u^{(p)}(x) \, dx - \int_a^b |u^{(p)}(x)|^2 \, dx.$$

Indeed, we have the representation

$$I = \int_a^b \{f^{(p)}(x) - u^{(p)}(x)\} u^{(p)}(x) \, dx = \sum_{j=2}^N \int_{x_{j-1}}^{x_j} \{f^{(p)}(x) - u^{(p)}(x)\} u^{(p)}(x) \, dx.$$

After integrating by parts p times all derivatives from $u(x)$ carry over to $\{f^{(p)}(x) - u^{(p)}(x)\}$, and we obtain

$$I_j = \int_{x_{j-1}}^{x_j} \{f^{(p)}(x) - u^{(p)}(x)\} u^{(p)}(x) \, dx$$

$$= \sum_{\ell=1}^p (-1)^{\ell+1} \{f^{(p-\ell)}(x) - u^{(p-\ell)}(x)\} u^{(p+\ell-1)}(x)\big|_{x_{j-1}}^{x_j}$$

$$= \sum_{\ell=1}^p (-1)^{\ell+1} \{f^{(p-\ell)}(x_j) - u^{(p-\ell)}(x_j)\} u^{(p+\ell-1)}(x_j)$$

$$- \sum_{\ell=1}^p (-1)^{\ell+1} \{f^{(p-\ell)}(x_{j-1}) - u^{(p-\ell)}(x_{j-1})\} u^{(p+\ell-1)}(x_{j-1}).$$

Since $u \in C^{2p-2}(a, b)$ and the functions $f, f', \ldots, f^{(p-1)}$ are continuous on $[a, b]$, there remain only some of the terms

$$I = \sum_{j=2}^N I_j$$

$$= \sum_{\ell=1}^{p-1} (-1)^{\ell+1} \{f^{(p-\ell)}(x_N) - u^{(p-\ell)}(x_N)\} u^{(p+\ell-1)}(x_N)$$

$$- \sum_{\ell=1}^{p-1} (-1)^{\ell+1} \{f^{(p-\ell)}(x_1) - u^{(p-\ell)}(x_1)\} u^{(p+\ell-1)}(x_1)$$

$$+ \sum_{j=2}^N (-1)^{N+1} \{f(x) - u(x)\} u^{(2p-1)}(x)\big|_{x_{j-1}}^{x_j}.$$

Finally, we obtain

$$\int_a^b |f^{(p)}(x) - u^{(p)}(x)|^2 \, dx = \int_a^b |f^{(p)}(x)|^2 \, dx - 2I - \int_a^b |u^{(p)}(x)|^2 \, dx. \quad (2.9)$$

The proof of the theorem is obvious, since all terms of I disappear – the first two sums due to the boundary conditions, and the second sum due to the interpolation conditions. This implies the identity

$$\int_a^b |f^{(p)}(x)|^2 \, dx = \int_a^b |u^{(p)}(x)|^2 \, dx + \int_a^b |f^{(p)}(x) - u^{(p)}(x)|^2 \, dx.$$

Within the proof of the above theorem we have also obtained the following lemma.

Lemma 2.5 *Let $p \geq 1$ be an integer, the polynomial spline $u(x)$ be of degree $2p - 1$, and the function $f \in H^p([a, b])$. Then the following identity holds:*

$$\int_a^b f^{(p)}(x) u^{(p)}(x) \, dx$$

$$= \sum_{\ell=1}^{p-1} (-1)^{\ell+1} f^{(p-\ell)}(x_N) u^{(p+\ell-1)}(x_N) - \sum_{\ell=1}^{p-1} (-1)^{\ell+1} f^{(p-\ell)}(x_1) u^{(p+\ell-1)}(x_1)$$

$$+ \sum_{j=2}^{N} (-1)^{N+1} f(x) u^{(2p-1)}(x)|_{x_{j-1}}^{x_j}.$$

The proof is obtained from (2.9), p. 25, by making the substitution

$$f(x) \longrightarrow f(x) - u(x).$$

Remark 2.6 For the curious reader we would recommend jumping far ahead and seeing how the polyharmonic paradigm works by inspecting Theorem 20.7, p. 416, which contains an identity involving the polyharmonic operator Δ^p, which is a direct analog to Lemma 2.5. The generalization is obtained by the replacement scheme (1.1), p. 10, i.e.

$$\frac{d^p}{dt^p} \longrightarrow \Delta^p.$$

2.4 Existence and uniqueness of interpolation odd-degree splines

Now we will see the importance of the above identities by providing an elegant proof for the existence of interpolation splines of odd degree.

Theorem 2.7 *For arbitrary constants $\{c_j\}_j$ and $\{d_\ell^1, d_\ell^2\}_\ell$ there exists a spline $u(x)$ of degree $2p - 1$ such that*

$$u(x_j) = c_j \quad \text{for} \quad j = 1, \ldots, N, \tag{2.10}$$

$$u^{(\ell)}(x_1) = d_\ell^1 \quad \text{for} \quad \ell = 0, \ldots, p - 1, \tag{2.11}$$

$$u^{(\ell)}(x_N) = d_\ell^2 \quad \text{for} \quad \ell = 0, \ldots, p - 1. \tag{2.12}$$

This spline is unique.

Proof As in the case of cubic splines we may formulate the problem of finding the interpolation spline $u(x)$ as a linear system having

$$2p \times (N-1)$$

unknowns which will be the coefficients of the $(N-1)$ polynomials on every interval $[x_j, x_{j+1}]$. The smoothness requirement $u(x) \in C^{2p-2}(a,b)$ and the above interpolation and boundary conditions (2.10)–(2.12), give precisely $2p \times (N-1)$ equalities. The question is whether or not this linear system is solvable with the right sides defined by the constants $\{c_j\}$ and $\{d_\ell^1, d_\ell^2\}$. From linear algebra we know that a linear system with a square matrix is uniquely (!) solvable if and only if its homogeneous system[2] has only the trivial solution. But the homogeneous system will be obtained if all constants are zero, i.e.

$$c_j = 0 \quad \text{for} \quad j = 1, \ldots, N,$$
$$d_\ell^1 = 0 \quad \text{for} \quad \ell = 0, \ldots, p-1,$$
$$d_\ell^2 = 0 \quad \text{for} \quad \ell = 0, \ldots, p-1.$$

Let us assume that we have a polynomial spline $u_0(x)$ which satisfies these zero conditions.

We apply Lemma 2.5, p. 26, for $f = u_0$ and $u = u_0$ and obtain

$$\int_a^b u_0^{(p)}(x) u_0^{(p)}(x)\, dx = 0.$$

Since $u_0 \in C^{2p-2}$, this implies $u_0^{(p)}(x) = 0$, hence, $u_0(x)$ is a polynomial of degree at most $p-1$. But the boundary conditions (2.11) which are now zero imply that $u_0(x) \equiv 0$. This implies the uniqueness of the solution to the linear system. Hence, by the simple argument of linear algebra mentioned above (uniqueness and existence are equivalent for a linear system), we obtain the existence of the solution. This completes the proof of the theorem. ∎

This proof will provide a proper generalization for polysplines, see Exercise 8.8, p. 110.

Exercise 2.8 Prove that the interpolation spline in Theorem 2.7 belongs to the space $H^p([a,b])$.

2.5 The Holladay theorem

Now as a simple consequence of the above theorems we obtain Holladay's famous result.

---

[2] The term "homogeneous system" is often used as a synonym to "system with zero right sides" in linear algebra as well as in PDEs.

Theorem 2.9 (Holladay) *Let $p \geq 1$ be an integer. Consider the minimization problem*

$$\inf \int_a^b |f^{(p)}(x)|^2 \, dx \qquad (2.13)$$

where the function $f \in H^p([a,b])$ and satisfies the following interpolation and boundary conditions:

$$f(x_j) = c_j \quad \text{for} \quad j = 1, \ldots, N, \qquad (2.14)$$

$$f^{(\ell)}(x_1) = d_\ell^1 \quad \text{for} \quad \ell = 0, \ldots, p-1, \qquad (2.15)$$

$$f^{(\ell)}(x_N) = d_\ell^2 \quad \text{for} \quad \ell = 0, \ldots, p-1, \qquad (2.16)$$

where $\{c_j\}_j$ and $\{d_\ell^1, d_\ell^2\}_\ell$ are some given constants. Then the solution of this problem exists and is a polynomial spline $u(x)$ of degree $2p-1$ which satisfies the same interpolation and boundary conditions as f.

Proof We know from Theorem 2.7, p. 26, that an interpolation spline $u(x)$ satisfying conditions (2.14)–(2.16) exists. Substituting it in the identity

$$\int_a^b |f^{(p)}(x)|^2 \, dx = \int_a^b |u^{(p)}(x)|^2 \, dx + \int_a^b |f^{(p)}(x) - u^{(p)}(x)|^2 \, dx$$

of Theorem 2.4, p. 24, we obtain

$$\int_a^b |u^{(p)}(x)|^2 \, dx \leq \int_a^b |f^{(p)}(x)|^2 \, dx$$

which completes the proof since $u(x)$ is in the space $H^p([a,b])$. ∎

Remark 2.10 *It is clear that we may impose other types of boundary conditions different from (2.15) and (2.16) in order to obtain the identity of Theorem 2.4, p. 24. The technique for treating all of them is very similar; it is based on the identities in Theorem 2.4 and Lemma 2.5, p. 26. They do not represent any essential novelty, and for that reason in this book we will confine our considerations mainly to the above type of boundary conditions. More or less this remark will apply later to the polysplines, where we will impose a standard type of boundary conditions similar to the one-dimensional case above.*

Chapter 3

The two-dimensional case: data and smoothness concepts

In searching for the multivariate generalization our ideal will be the one-dimensional splines of odd degree considered in the previous chapter. As we have seen they satisfy the following fundamental property:

data set = knot set

which we wish to generalize to the multivariate case.

First, we have to explain the **data concept** of the polyharmonic paradigm by comparing it with other existing concepts.

3.1 The data concept in two dimensions according to the polyharmonic paradigm

Taking the one-dimensional interval $[a, b]$ and the *data points* x_j on it, namely

$$a = x_1 < x_2 < \cdots < x_N = b$$

there may be different approaches to generalizing the setting in the multivariate case.

The *watershed* dividing the different multivariate generalizations is the way one understands the points in $\mathbb{R}$. The main idea is that one may consider the points x_j as sets of dimension 0 in $\mathbb{R}$

$$\dim(\{x_j\}) = 0,$$

but one may also consider them as sets of dimension $n - 1$ in $\mathbb{R}$

$$\dim(\{x_j\}) = n - 1.$$

30 Multivariate polysplines

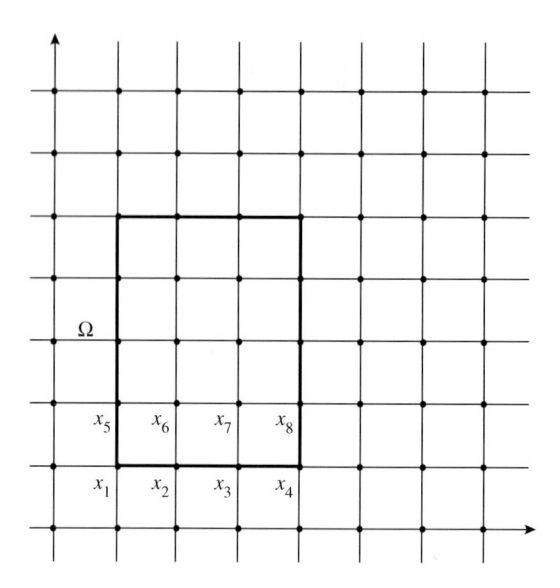

Figure 3.1. The data points of the box splines lie on a regular grid.

1. Now let us adopt the first concept, i.e. $\dim(\{x_j\}) = 0$. Accordingly, we have the first possible generalization by considering a domain Ω in $\mathbb{R}^n$ and a set of discrete points $x_j \in \Omega$, $j = 1, \ldots, N$. This is essentially the data concept of the *box* and *simplex* splines. Let us note that in the case of these splines there is an evident discrepancy between the *data set* and the set where the *singularities* = *knots* are lying. The data set coincides with the points $\{x_j\}$ and is zero-dimensional but the set of singularities occupies some $(n-1)$-dimensional surfaces. This is the opposite of the one-dimensional case where these two sets coincide, or at least have the same dimension (see Figure 3.1). This is also the data concept of the *radial basis functions* (RBFs) (see Figure 3.2).

2. As said above, the second approach is to consider the points x_j in $\mathbb{R}$ as sets of dimension 0 but of codimension 1. Thus the multivariate analog in $\mathbb{R}^n$ to the points in $\mathbb{R}$ would be sets, in particular surfaces, Γ_j, $j = 1, \ldots, N$, which have codimension 1, i.e.

$$\dim \Gamma_j = n - 1.$$

(see Figure 3.3).

The *data concept* of the *polyharmonic paradigm* states that the *data set* and the set of *singularities* coincide, i.e. just as in the one-dimensional case of odd-degree splines we have

data set = knot set = singularity set.

This means that the splines which we will construct will consist of analytic pieces between the surfaces Γ_j and the interpolation and the smoothness conditions will be

The two-dimensional case: data and smoothness 31

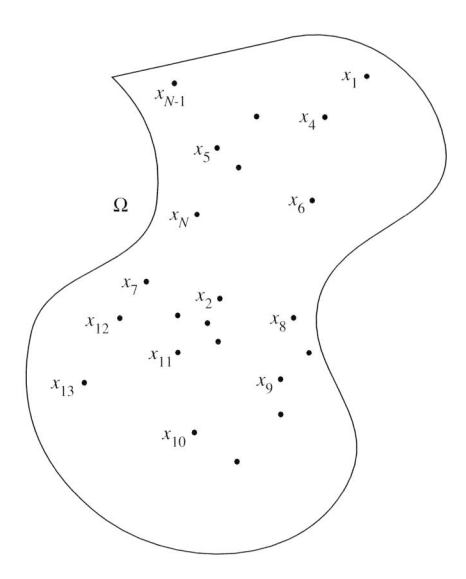

Figure 3.2. Completely scattered data of the radial basis functions.

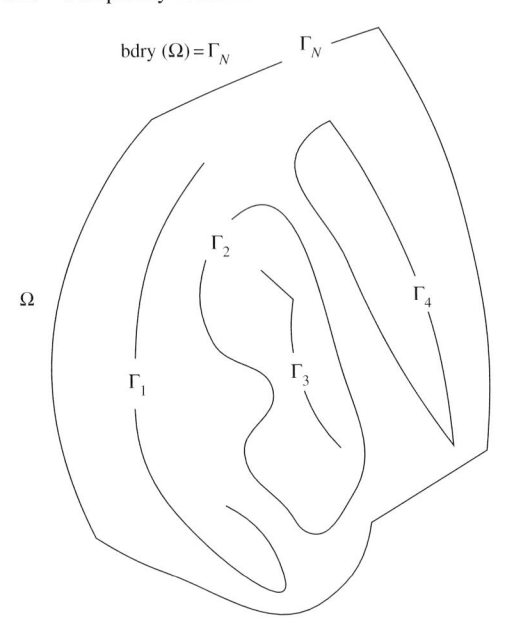

Figure 3.3. The data sets of the polysplines considered in the present book lie on nonintersecting $(n-1)$-dimensional manifolds.

imposed solely on $\cup \Gamma_j$. So we will be very close to the one-dimensional situation of odd-degree splines where the data set and the set of the knot-points coincide, excluding the boundary which only carries data. We will be interested in developing the last data concept, which we call the polyharmonicity data concept.

32 Multivariate polysplines

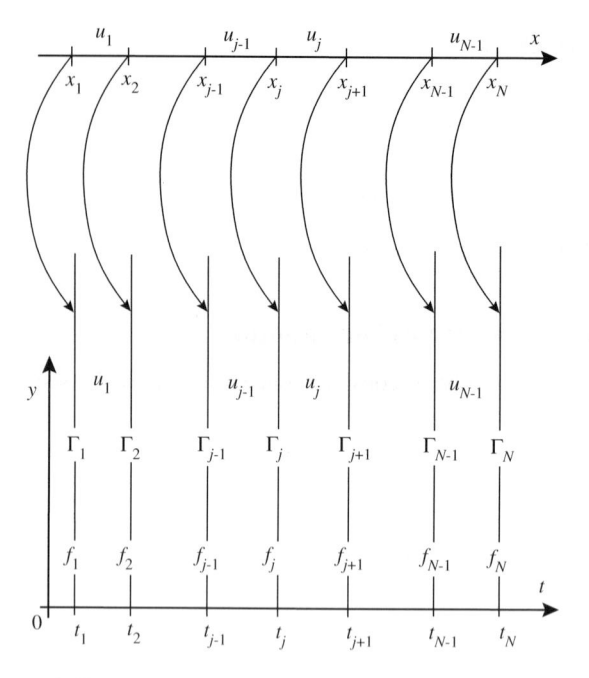

Figure 3.4. The symbolic map showing how the one-dimensional case is generalized to the two-dimensional case.

Which are the *simplest geometric configurations* which might be imagined following this data concept in the plane $\mathbb{R}^2$?

3.1.1 "Parallel lines" or "strips"

The first case is obtained if we replace the points x_j through parallel straight lines (see Figure 3.4).

More precisely, if we put $(t, y) \in \mathbb{R}^2$, and we have the numbers

$$t_1 < \cdots < t_N$$

then the lines will be given by

$$\Gamma_j = \{(t, y) \in \mathbb{R}^2 : t = t_j\}$$

for all $j = 1, \ldots, N$.

The splines will be assembled from pieces $u_j(x)$ defined on the strip Strip_j between every two neighboring lines Γ_j and Γ_{j+1}, the last given by

$$\text{Strip}_j = \{(t, y) \in \mathbb{R}^2 : t_j < t < t_{j+1}\}$$

for $j = 1, 2, \ldots, N - 1$. The data c_j from the one-dimensional case will be replaced by a set of functions $f_j(x)$, each defined on the line Γ_j, for all $j = 1, \ldots, N$. So we have

The two-dimensional case: data and smoothness 33

the generalization scheme

$$x_j \longrightarrow \Gamma_j$$
$$c_j \longrightarrow f_j(x)$$
$$u_j(x) \longrightarrow u_j(x),$$

which is completely depicted in Figure 3.4.

3.1.2 "Concentric circles" or "annuli"

Another possibility would be to consider concentric circles Γ_j (see Figure 3.5). More precisely, if the radii are

$$0 < r_1 < \cdots < r_N$$

we consider the circles

$$\Gamma_j = \{(r, \varphi) \in \mathbb{R}^2 : r = r_j \text{ and } 0 \leq \varphi \leq 2\pi\}$$

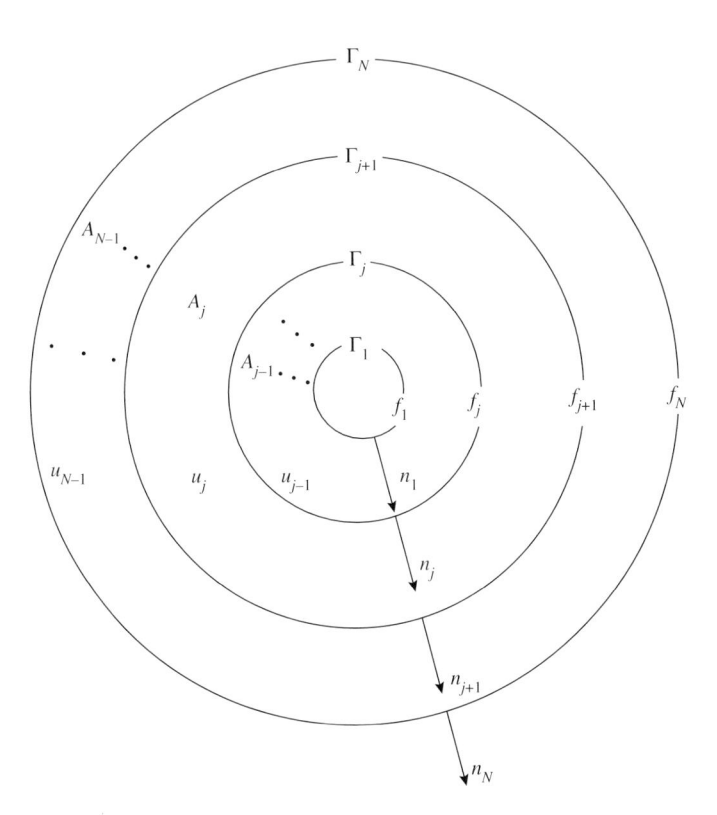

Figure 3.5. All elements of the polyspline with data on concentric circles are shown. The data functions f_j are assigned on the circles Γ_j, the pieces u_j of the polyspline are defined on the annuli A_j, also the unit normal vectors n_j to Γ_j.

for all $j = 1, \ldots, N$. The data will be some functions $f_j(x)$, defined on Γ_j, for $j = 1, \ldots, N$.

In this case the spline $u(x)$ will consist of pieces $u_j(x)$ defined in the annulus A_j between the circles Γ_j and Γ_{j+1} for $j = 1, \ldots, N-1$, the annulus being given by

$$A_j = \{(r, \varphi) \in \mathbb{R}^2 : r_j < r < r_{j+1} \text{ and } 0 \le \varphi \le 2\pi\}.$$

We have the same generalization scheme as in the "parallel lines" case:

$$x_j \longrightarrow \Gamma_j$$
$$c_j \longrightarrow f_j(x)$$
$$u_j(x) \longrightarrow u_j(x).$$

Let us note that there are two possibilities: to consider a function $u_0(x)$ inside the smallest circle $B(0; r_1)$ so that the spline $u(x)$ will be defined in the whole ball

$$B(0; r_N),$$

or to assume that the spline is defined only in the annulus

$$A = \{(r, \varphi) \in \mathbb{R}^2 : r_1 < r < r_N \text{ and } 0 \le \varphi \le 2\pi\}.$$

Let us note that in the first *Ansatz*[1] of "parallel lines" the domains contained between each pair of neighboring lines are noncompact.

In the present book we will be mainly concerned with these two cases and their multivariate generalization. As will become clear in the following chapters, the first *Ansatz* is technically simpler.

3.2 The smoothness concept according to the polyharmonic paradigm

We now pay attention to the data sets Γ_j which are surfaces of dimension $n-1$. We have two cases.

3.2.1 The "strips"

In the "parallel lines" or "strips" case we have the lines

$$\Gamma_j = \{(t, y) \in \mathbb{R}^2 : t = t_j\}$$

for $j = 1, 2, \ldots, N$ and for the numbers $t_1 < t_2 < \cdots < t_N$.

---

[1] The German word Ansatz is often used for "setting" or "framework", or "approach" related to special formula. Carl de Boor has told me that some time ago, the American/British math community decided that there was no good English equivalent to 'Ansatz', as you can tell by the fact that it has become an acceptable word (often written 'ansatz', as in 'We make the following 'ansatz") in English/American math papers.

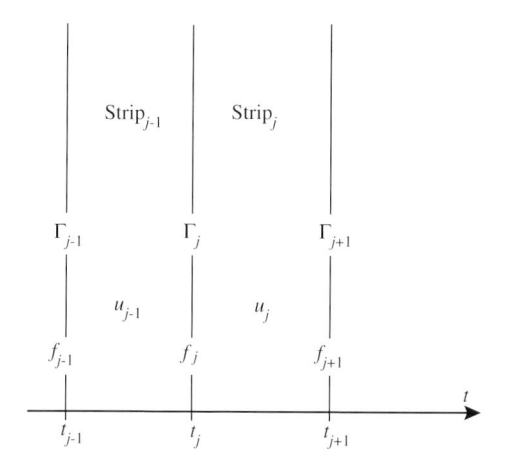

Figure 3.6. The mutual position of strips and parallel lines Γ_j.

As said above, the splines $u(x)$ which we will consider are functions which are assembled from different pieces $u_j(x)$ defined on the strips

$$\text{Strip}_j = \{x = (t, y) \in \mathbb{R}^2 : t_j < t < t_{j+1}\},$$

(see Figure 3.4, p. 32).

We would like to introduce the smoothness conditions which have to be satisfied on these data curves Γ_j satisfying

$$\Gamma_j = \overline{\text{Strip}_j} \cap \overline{\text{Strip}_{j-1}}, \quad \text{for } j = 2, 3, \ldots, N-1,$$

which are also interfaces between every pair of neighboring pieces $u_j(x)$ (see Figure 3.6).

First, it is clear that the continuity requirement on $u(x)$ means that

$$u_j(t_{j+1}, y) = u_{j+1}(t_{j+1}, y) \quad \text{for all } y \text{ in } \mathbb{R},$$

and for all $j = 1, 2, \ldots, N-2$. It should be noted that this is a direct generalization of the continuity condition of the one-dimensional splines that we had in the first row of equalities (2.2), p. 20.

Now the question is what should be the analog to the C^1-smoothness? The answer is almost evident. We will define the C^1-smoothness by imposing smoothness with respect to the normal derivative $\partial/\partial n$ on the straight line Γ_j, where we have used $\vec{n} = \vec{n_x}$ to denote a unit normal vector to Γ_j at the point $x \in \Gamma_j$. Since there are only two possible directions (the axis t and its opposite), for simplicity, let us choose for all x and all Γ_j the same vector $\vec{n}$ having the direction of the axis t, i.e.

$$\vec{n} = (1, 0).$$

So the smoothness (matching) condition reads as

$$\frac{\partial u_j(x)}{\partial n} = \frac{\partial u_{j+1}(x)}{\partial n} \tag{3.1}$$

for every point $x \in \Gamma_j$ and for all $j = 2, 3, \ldots, N - 1$. By the choice of the normal $\vec{n}$ in the coordinates $x = (t, y)$ condition (3.1) becomes

$$\frac{\partial u_j(t_{j+1}, y)}{\partial t} = \frac{\partial u_{j+1}(t_{j+1}, y)}{\partial t} \quad \text{for all } y \text{ in } \mathbb{R}, \text{ and for } j = 2, 3, \ldots, N - 1.$$

The interpolation conditions will be now

$$u(t_j, y) = u_j(t_j, y) = f_j(y) \quad \text{for all } y \text{ in } \mathbb{R}, \text{ for all } j = 1, \ldots, N - 1;$$
$$u(t_N, y) = u_{N-1}(t_N, y) = f_N(y) \quad \text{for all } y \text{ in } \mathbb{R},$$

and they correspond to the interpolation conditions in the one-dimensional case in (2.3), p. 21.

Going further in this direction, it is clear that the one-dimensional concept of spline with smoothness C^k will be generalized through the conditions

$$\frac{\partial^l u_j(x)}{\partial n^l} = \frac{\partial^l u_{j+1}(x)}{\partial n^l} \quad \text{for} \quad l = 0, \ldots, k$$

for every point $x \in \Gamma_j$, and all $j = 2, 3, \ldots, N - 1$.

Written in the coordinates $x = (t, y)$, these conditions become

$$\frac{\partial^l u_j(t_{j+1}, y)}{\partial t^l} = \frac{\partial^l u_{j+1}(t_{j+1}, y)}{\partial t^l} \quad \text{for} \quad l = 0, \ldots, k.$$

3.2.2 The "annuli"

Now let Γ_j be concentric circles, centered at 0, i.e.

$$\Gamma_j = S(0; r_j),$$

where as above we take the radii satisfying $0 < r_1 < \cdots < r_N$. The splines $u(x)$ we wish to define here are assembled from different pieces $u_j(x)$ defined in the annuli A_j lying between every pair of neighboring circles, namely on

$$A_j = \{(r, \varphi) : r_j < r < r_{j+1}\},$$

(see Figure 3.7).

The smoothness conditions will be satisfied on the data surfaces Γ_j, also given by

$$\Gamma_j = \overline{A_j} \cap \overline{A_{j-1}},$$

whence it is clear that the continuity requirement on $u(x)$ means that

$$u_j(x) = u_{j-1}(x) \quad \text{for all } x \text{ in } \Gamma_j,$$

and for all $j = 1, \ldots, N - 2$. Written in polar coordinates $x = (r, \varphi)$, it becomes

$$u_j(r_j, \varphi) = u_{j-1}(r_j, \varphi) \quad \text{for all } 0 \leq \varphi \leq 2\pi.$$

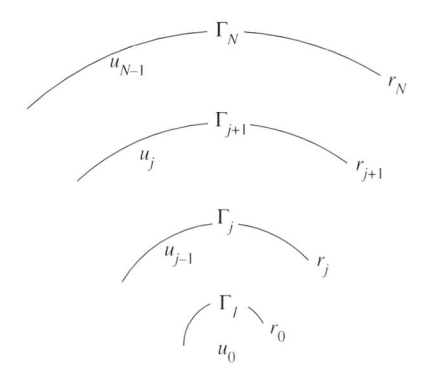

Figure 3.7. The neighborhood of the pieces of the polyspline when the data are on concentric circles.

Obviously, this is a direct generalization of the continuity condition of the one-dimensional splines that we had in the first row of equalities (2.2) p. 20.

The analog to the C^1-smoothness is the same as in the "parallel lines" case. Let $\partial/\partial n$ denote the normal derivative on the circle Γ_j, where $\vec{n} = \vec{n_x}$ denotes the exterior unit normal vector to Γ_j at the point $x \in \Gamma_j$. Namely, we require

$$\frac{\partial u_j(x)}{\partial n} = \frac{\partial u_{j+1}(x)}{\partial n} \qquad (3.2)$$

for every point $x \in \Gamma_j$ and for all $j = 1, 2, \ldots, N-2$. Since this normal derivative coincides with the radial direction, i.e.

$$x = r \cdot \vec{n},$$

in polar coordinates $x = (r, \varphi)$ conditions (3.2) become

$$\frac{\partial u_j(r_j, \varphi)}{\partial r} = \frac{\partial u_{j-1}(r_j, \varphi)}{\partial r} \qquad \text{for all } 0 \leq \varphi \leq 2\pi,$$

and all $j = 1, 2, \ldots, N-1$.

The interpolation conditions will be

$$u(r_j, \varphi) = u_j(r_j, \varphi) = f_j(r_j, \varphi) \quad \text{for all } 0 \leq \varphi \leq 2\pi,$$

and all $j = 1, \ldots, N-1$, and

$$u(r_N, \varphi) = u_{N-1}(r_N, \varphi) = f_N(r_N, \varphi) \quad \text{for all } 0 \leq \varphi \leq 2\pi.$$

Obviously they correspond to the interpolation conditions in the one-dimensional case in formula (2.3), p. 21.

The **smoothness** of order k will be generalized through conditions

$$\frac{\partial^l u_j(x)}{\partial n^l} = \frac{\partial^l u_{j-1}(x)}{\partial n^l} \quad \text{for all } l = 0, \ldots, k,$$

for every point $x \in \Gamma_j$, and all $j = 1, 2, \ldots, N-1$. In polar coordinates $x = (r, \varphi)$ these conditions read as:

$$\frac{\partial^l u_j(r_j, \varphi)}{\partial r^l} = \frac{\partial^l u_{j-1}(r_j, \varphi)}{\partial r^l} \quad \text{for all } 0 \leq \varphi \leq 2\pi, \text{ all } l = 0, \ldots, k,$$

and all $j = 1, 2, \ldots, N-1$ (see Figure 3.7, p. 37).

We will return to the present point since we now have to concentrate on the choice of the proper class of functions $u_j(x)$ from which the multivariate splines will be assembled.

Chapter 4

The objects concept: harmonic and polyharmonic functions in rectangular domains in $\mathbb{R}^2$

Now we come to the most sophisticated part of our story – the choice of the class of functions to which the pieces $u_j(x)$ of the splines we are dreaming about belong. The main point of this choice is that this class of functions has to be rich and flexible enough to satisfy the above smoothness and interpolation conditions.

Recall that the **Laplace operator** Δ in $\mathbb{R}^2$ is given by

$$\Delta = \frac{\partial^2}{\partial x_1^2} + \frac{\partial^2}{\partial x_2^2},$$

where $x = (x_1, x_2)$ is a point in $\mathbb{R}^2$. The function $h(x)$ is called **harmonic** in the domain D if it satisfies the Laplace equation

$$\Delta h(x) = 0 \quad \text{for } x \text{ in } D.$$

For $p = 2$, we have the operator $\Delta^2 = \Delta\Delta$ which is called **biharmonic**. It is given by

$$\Delta^2 h(x) = \left(\frac{\partial^2}{\partial x_1^2} + \frac{\partial^2}{\partial x_2^2}\right)\left(\frac{\partial^2}{\partial x_1^2} + \frac{\partial^2}{\partial x_2^2}\right) h(x)$$

$$= \frac{\partial^4 h(x)}{\partial x_1^4} + 2\frac{\partial^4 h(x)}{\partial x_1^2 \partial x_2^2} + \frac{\partial^4 h(x)}{\partial x_2^4}.$$

A function $h(x)$ is **biharmonic** in the domain D if it satisfies the fourth-order equation

$$\Delta^2 h(x) = 0 \quad \text{for} \quad x \text{ in } D.$$

The **polyharmonic** operator of order $p \geq 2$ is equal to the p times iterated operator Δ, i.e. is defined by

$$\Delta^p h(x) = \left(\frac{\partial^2}{\partial x_1^2} + \frac{\partial^2}{\partial x_2^2} \right)^p h(x),$$

and the function $h(x)$ is a **polyharmonic of order** $p \geq 1$ in the domain D if it satisfies the equation

$$\Delta^p h(x) = 0 \quad \text{for } x \text{ in } D.$$

Let us proclaim right at the start that our **Ansatz** will be the following: the *polysplines* of order p are assembled from pieces of *polyharmonic functions* of order p which join smoothly up to order $2p - 2$ on the surfaces Γ_j.[1]

Postponing the reasoning behind our **Ansatz**, we will spend some time explaining the main qualitative properties of the polyharmonic functions which will play an essential role in our study. In the present chapter we will consider the *strips case* in more detail; we will consider the *annuli case* in Chapter 7. In the *strips case* we will simplify matters by considering *periodic data*. We will also consider the special case of constant boundary data since they reduce the polysplines to one-dimensional polynomial splines and thus throw a bridge to the one-dimensional case. In order to lighten the exposition we have left several details for Chapter 10.

4.1 Harmonic functions in strips or rectangles

Since we will be interested in the *variables* $x = (t, y)$ which we have adopted in the parallel lines case we have

$$\Delta = \frac{\partial^2}{\partial t^2} + \frac{\partial^2}{\partial y^2}.$$

The general approach to finding the solutions of the Laplace equation is by expanding it into "elementary solutions with separated variables"

$$h(x) = \sum_k T_k(t) Y_k(y).$$

Now let us try to find the general form of the elementary harmonic function with separated variables, i.e. having the form

$$h(t, y) = T(t) Y(y).$$

Applying the Laplace operator to both sides immediately gives

$$0 = \Delta h(t, y) = T''(t) Y(y) + T(t) Y''(y).$$

---

[1] As a matter of fact, if we dislike using a *deus ex machina* argument, it is not so easy to explain our choice of objects. So far this choice is strongly motivated by the fact that there is a lot of evidence showing that the polyharmonic functions of order p form a Chebyshev system in the multivariate case. That is the essence of the polyharmonic paradigm briefly outlined in Section 1.5, p. 10.

The last implies
$$\frac{T''(t)}{T(t)} = -\frac{Y''(y)}{Y(y)} = \text{Const} = C.$$

Exercise 4.1 *Prove that the following three cases are possible:*

- $C > 0$. *Then the solution $T(t)Y(y)$ has the form*
$$e^{\pm\sqrt{C}t}\cos\sqrt{C}y, \quad \text{or} \quad e^{\pm\sqrt{C}t}\sin\sqrt{C}y.$$

- $C = 0$. *Then the solution $T(t)Y(y)$ is one of the functions*
$$1, t, y, ty.$$

- $C < 0$. *Then the solution $T(t)Y(y)$ has the form*
$$e^{\pm\sqrt{-C}y}\cos\sqrt{-C}t, \quad \text{or} \quad e^{\pm\sqrt{-C}y}\sin\sqrt{-C}t.$$

Hint: *The above equation $\Delta h(t,y) = 0$ is equivalent to the two ordinary differential equations $T''(t) = CT(t)$ and $Y''(y) = -CY(y)$.*

Let us note that using the definitions for the hyperbolic sine and cosine functions
$$\sinh u = \frac{e^u - e^{-u}}{2},$$
$$\cosh u = \frac{e^u + e^{-u}}{2},$$
one may write another combination of linearly independent solutions, e.g. in the case $C > 0$, they will be
$$\sinh\sqrt{C}t \cdot \cos\sqrt{C}y, \quad \cosh\sqrt{C}t \cdot \cos\sqrt{C}y, \quad \text{or}$$
$$\sinh\sqrt{C}t \cdot \sin\sqrt{C}y, \quad \cosh\sqrt{C}t \cdot \sin\sqrt{C}y.$$

Let us note that $\sinh(0) = 0$.

Perhaps the simplest to describe is the space of harmonic functions in the strip which are periodic with respect to the variable y. For that reason we confine ourselves to this class of harmonic functions.

Our main task is to analyze the functions $h(t, y)$ which are harmonic in the strip
$$\text{Strip}(a,b) := \{(t, y) : a < t < b\},$$
and are 2π-periodic there with respect to the variable y, i.e. to satisfy
$$h(t, y + 2\pi\ell) = h(t, y) \quad \text{for } a < t < b \text{ and } y \text{ in } \mathbb{R},$$
and for every integer ℓ (see Figure 4.1).

42   Multivariate polysplines

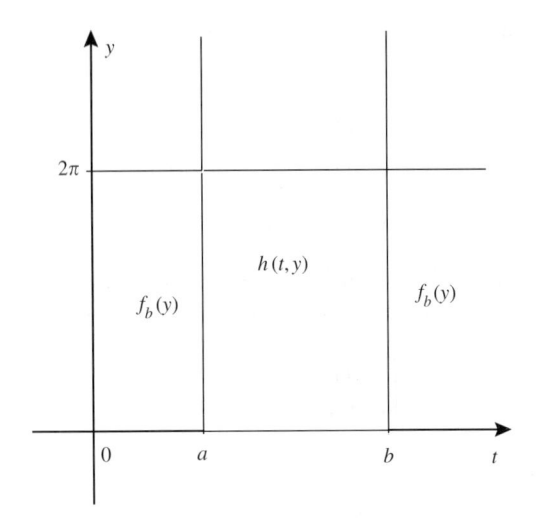

Figure 4.1.   The Dirichlet data.

Due to the periodicity in y we may expand the function $h(t, y)$ with respect to the variable y in a trigonometric Fourier series. For every fixed t satisfying $a < t < b$, we have the Fourier series expansion

$$h(t, y) = \frac{u_0(t)}{2} + \sum_{k=1}^{\infty} (u_k(t) \cos ky + v_k(t) \sin ky), \quad (4.1)$$

where the Fourier coefficients $u_k(t)$ and $v_k(t)$ are given by

$$u_k(t) = \frac{1}{\pi} \int_0^{2\pi} h(t, y) \cos ky \, dy \quad \text{for } k = 0, 1, 2, \ldots,$$

$$v_k(t) = \frac{1}{\pi} \int_0^{2\pi} h(t, y) \sin ky \, dy \quad \text{for } k = 1, 2, \ldots.$$

The reader may consult Chapter 12, in particular Theorem 12.3, p. 210, where the classical results on Fourier series are available.

Assuming that we may differentiate the Fourier series, let us apply the Laplace operator to both sides of equality (4.1). We obtain

$$0 = \Delta h(t, y)$$
$$= \frac{u_0''(t)}{2} + \sum_{k=1}^{\infty} [(u_k''(t) - k^2 u_k(t)) \cos ky + (v_k''(t) - k^2 v_k(t)) \sin ky].$$

After equating all terms to zero we obtain the following system of equations for all $a < t < b$,

$$u_0''(t) = 0,$$
$$u_k''(t) - k^2 u_k(t) = 0 \quad \text{for } k = 1, 2, \ldots,$$
$$v_k''(t) - k^2 v_k(t) = 0 \quad \text{for } k = 1, 2, \ldots.$$

Exercise 4.2 *Prove that the general solution of the ordinary differential equation $w''(t) = k^2 w(t)$ in the interval (a, b) is a linear combination of the two functions*

$$\begin{cases} \{1, t\} & \text{for } k = 0, \\ \{e^{kt}, e^{-kt}\} & \text{for } k \neq 0. \end{cases}$$

Thus for some constants C_1 and C_2 we have

$$w(t) = \begin{cases} C_1 + C_2 t & \text{for } k = 0, \\ C_1 e^{kt} + C_2 e^{-kt} & \text{for } k \neq 0, \end{cases} \quad (4.2)$$

for all $t \in (a, b)$.

Note that this result is independent of the interval (a, b).

Let us note that the case $k = 0$ coincides with the one-dimensional Laplace equation

$$\Delta h(t) = h''(t) = 0.$$

Now we are ready to understand the structure of the space of periodic harmonic functions in the strip.

4.2 "Parametrization" of the space of periodic harmonic functions in the strip: the Dirichlet problem

Obviously, the boundary values of the periodic harmonic function $h(t, y)$ on the edges of the strip, the sets $\{(t, y) \in \mathbb{R}^2 : t = a\}$ and $\{(t, y) \in \mathbb{R}^2 : t = b\}$, also have to be 2π-periodic.

Now let us assume that a function, which is in fact a couple of functions, $f(y) = (f_a(y), f_b(y))$, be given such that the two functions $f_a(y)$ and $f_b(y)$ are 2π-periodic, and both are in $L_2(0, 2\pi)$. We consider the so-called *periodic Dirichlet problem* in the strip: find a function $h(t, y)$, 2π-periodic in the variable y and such that, for all y in $\mathbb{R}$, the following holds:

$$\Delta h(t, y) = 0 \quad \text{for } a < t < b, \quad (4.3)$$
$$h(a, y) = f_a(y), \quad (4.4)$$
$$h(b, y) = f_b(y), \quad (4.5)$$

where $h(t, y)$ is 2π-periodic in y (see Figure 4.1).

44   Multivariate polysplines

Let us expand the functions $f_a(y)$, $f_b(y)$ in trigonometric Fourier series

$$f_a(y) = \frac{f_{a,0}}{2} + \sum_{k=1}^{\infty} [f_{a,k} \cos ky + \tilde{f}_{a,k} \sin ky],$$

$$f_b(y) = \frac{f_{b,0}}{2} + \sum_{k=1}^{\infty} [f_{b,k} \cos ky + \tilde{f}_{b,k} \sin ky].$$

Just as above, we are looking for a solution in the form

$$h(t, y) = \frac{u_0(t)}{2} + \sum_{k=1}^{\infty} [u_k(t) \cos ky + v_k(t) \sin ky].$$

We have seen that it satisfies the following equations for all $a < t < b$:

$$u_0''(t) = 0,$$
$$u_k''(t) - k^2 u_k(t) = 0 \quad \text{for } k = 1, 2, \ldots,$$
$$v_k''(t) - k^2 v_k(t) = 0 \quad \text{for } k = 1, 2, \ldots,$$

and the boundary conditions (4.4)–(4.5), p. 43, imply for all y in $\mathbb{R}$ the following equalities,

$$h(a, y) = f_a(y) = \frac{f_{a,0}}{2} + \sum_{k=1}^{\infty} \left[f_{a,k} \cos ky + \tilde{f}_{a,k} \sin ky \right],$$

$$h(b, y) = f_b(y) = \frac{f_{b,0}}{2} + \sum_{k=1}^{\infty} \left[f_{b,k} \cos ky + \tilde{f}_{b,k} \sin ky \right].$$

Note that we assume that the Fourier series expansion of $h(t, y)$ holds at the endpoints a and b. Hence, by comparing the coefficients of the expansion of $h(a, y)$ with those of $f_a(y)$ we obtain the following boundary conditions for the functions $u_k(t)$:

$$u_k(a) = f_{a,k}, \quad u_k(b) = f_{b,k}, \quad \text{for } k = 0, 1, 2, \ldots.$$

For the functions $v_k(t)$ we likewise obtain

$$v_k(a) = \tilde{f}_{a,k}, \quad v_k(b) = \tilde{f}_{b,k}, \quad \text{for } k = 1, 2, 3, \ldots.$$

By Exercise 4.2, p. 43, we see that in order to find the functions $u_k(t)$ we have to solve a linear system for the coefficients A_k, B_k, where

$$u_0(t) = A_0 + B_0 t,$$
$$u_k(t) = A_k e^{kt} + B_k e^{-kt} \quad \text{for } k = 1, 2, 3, \ldots,$$

and likewise for the coefficients C_k, D_k of functions $v_k(t)$, since by Exercise 4.2 we have

$$v_k(t) = C_k e^{kt} + D_k e^{-kt} \quad \text{for } k = 1, 2, 3, \ldots.$$

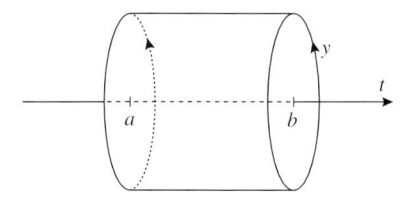

Figure 4.2. The variable $y \in \mathbb{S}^1$ is here "periodized".

Thus, at least formally, we see how to solve completely constructively the periodic Dirichlet problem in the strip. This approach will be further applied in Chapter 5, p. 57, to compute effectively the *polysplines*.

Exercise 4.3 *Let the boundary data be constant, i.e. for some constant M_1 and M_2*

$$f_a(y) = M_1 \quad \text{for } y \text{ in } \mathbb{R},$$
$$f_b(y) = M_2 \quad \text{for } y \text{ in } \mathbb{R}.$$

Prove that the solution to the Dirichlet problem (4.3)–(4.5), is

$$h(t, y) = C_1 + C_2 t,$$

where C_1 and C_2 are constants satisfying

$$\frac{C_1 + C_2 a}{2} = M_1, \quad \frac{C_1 + C_2 b}{2} = M_2.$$

Remark 4.4 1. The above Dirichlet problem (4.3)–(4.5), p. 43, is a Dirichlet problem on a cylinder with a boundary equal to the union of two circles (see Figure 4.2).

2. If the boundary functions f_a and f_b are not only 2π-periodic but also continuous then the solution $h(t, y)$ obtained above by expansion in Fourier series is also continuous. An elementary treatment of this question has been given by Weinberger [64, p. 95, 105].

3. The important question is: for which class of functions $f_a(y)$ and $f_b(y)$ is it possible to solve this boundary value problem and to which space does the solution belong in terms of Sobolev space or Hölder space? This question is studied by the theory of elliptic boundary value problems and results are given in Chapter 23, p. 461.

4. The above solution is good in the approximation sense since we may compute the above Fourier coefficients, $u_k(t)$ and $v_k(t)$, only for $k \leq N$, for a sufficiently large N, and obtain a good approximation to the solution. Weinberger [64, p. 105] considers the convergence for $N \to \infty$.

5. We have considered above the rectangle $[a, b] \times [0, 2\pi]$ but one may easily turn all arguments to the rectangle $[a, b] \times [0, L]$ for an arbitrary number $L > 0$, where we will have the expansion of the L-periodic in t function $h(t, y)$ in trigonometric Fourier series of the following form:

$$h(t, y) = \frac{u_0(t)}{2} + \sum_{k=1}^{\infty} \left[u_k(t) \cos \frac{2\pi}{L} k y + v_k(t) \sin \frac{2\pi}{L} k y \right].$$

46   *Multivariate polysplines*

Conclusion 4.5 *For every periodic Dirichlet data (f_a, f_b) there exists a periodic harmonic function which solves the Dirichlet problem (4.3)–(4.5), p. 43. Thus the set of periodic Dirichlet data in a certain sense "parametrizes" the space of the periodic harmonic functions on the strip. The basic objects which we have discussed are the harmonic functions in the strip which are periodic with respect to the variable y. The space of all such functions is infinite-dimensional but it has the remarkable property that it may be "parametrized" by the Dirichlet data, i.e. the boundary values of the functions $h(t, y)$ for $\{t = a\}$ and $\{t = b\}$. The boundary values of such functions $h(t, y)$ exhaust the space $C(\mathbb{S}) \times C(\mathbb{S})$.*

4.3 "Parametrization" of the space of periodic polyharmonic functions in the strip: the Dirichlet problem

In order to gradually increase the complexity of our study let us first consider the biharmonic case which is very important from a practical point of view as will become clear later in the applications of polysplines to smoothing *magnetic data* and to *computer-aided geometric design* in Chapter 6, p. 67. We recommend that the reader interested in the algorithmic aspects check the details below carefully, since they will play an essential role in the computation of polysplines.

4.3.1   The biharmonic case

So we consider the periodic biharmonic functions in the strip. These are the functions satisfying

$$\Delta^2 h(t, y) = \left(\frac{\partial^2}{\partial t^2} + \frac{\partial^2}{\partial y^2} \right)^2 h(t, y) = 0 \quad \text{for all } a < t < b \text{ and } y \text{ in } \mathbb{R},$$

and

$$h(t, y + 2\pi \ell) = h(t, y) \quad \text{for all } a < t < b \text{ and } y \text{ in } \mathbb{R},$$

and for all integers ℓ.

Due to the periodicity in the variable y we again expect a solution in the form

$$h(t, y) = \frac{u_0(t)}{2} + \sum_{k=1}^{\infty} [u_k(t) \cos ky + v_k(t) \sin ky]. \tag{4.6}$$

We apply the biharmonic operator Δ^2 to both sides of the above equality and obtain

$$0 = \Delta^2 h(t, y)$$

$$= \frac{u_0^{(4)}(t)}{2} + \sum_{k=1}^{\infty} \left[\left(\frac{\partial^2}{\partial t^2} - k^2 \right)^2 u_k(t) \cdot \cos ky + \left(\frac{\partial^2}{\partial t^2} - k^2 \right)^2 v_k(t) \cdot \sin ky \right].$$

Equating all terms to zero we obtain the following equalities for all $a < t < b$:

$$u_0^{(4)}(t) = 0,$$

$$\left(\frac{d^2}{dt^2} - k^2\right)^2 u_k(t) = 0 \quad \text{for } k = 1, 2, 3, \ldots,$$

$$\left(\frac{d^2}{dt^2} - k^2\right)^2 v_k(t) = 0 \quad \text{for } k = 1, 2, 3, \ldots.$$

Note that the first equation is obtained from the second for $k = 0$, and it represents the one-dimensional case of the biharmonic operator, since $\Delta^2 = d^4/dt^4$ in $\mathbb{R}$. These are ordinary differential equations and enjoy the following properties.

Exercise 4.6 *For every fixed real number k the general solution of the ordinary differential equation*

$$\left(\frac{\partial^2}{\partial t^2} - k^2\right)^2 w(t) = w^{(4)}(t) - 2k^2 w''(t) + k^4 w(t) = 0, \quad \text{for } a < t < b,$$

is a linear combination of the four functions which, depending on $k = 0$ or $k \neq 0$, are

$$\begin{cases} \{1, t, t^2, t^3\} & \text{for } k = 0, \\ \{e^{kt}, te^{kt}, e^{-kt}, te^{-kt}\} & \text{for } k \neq 0. \end{cases}$$

Thus for every solution $w(t)$ there exist constants C_j and D_j for which

$$w(t) = \begin{cases} C_0 + C_1 t + C_2 t^2 + C_3 t^3 & \text{for } k = 0, \\ C_0 e^{kt} + C_1 t e^{kt} + D_0 e^{-kt} + D_1 t e^{-kt} & \text{for } k \neq 0. \end{cases}$$

We note that this result is independent of the interval (a, b).

Now we assume that the functions $f_a^0(y)$, $f_a^1(y)$ and $f_b^0(y)$, $f_b^1(y)$ be given for y in $\mathbb{R}$. We assume that they are 2π-periodic in y, i.e.

$$f_a^i(y + 2\pi\ell) = f_a^i(y) \quad \text{for all } y \text{ in } \mathbb{R} \text{ and integer } \ell,$$

$$f_b^i(y + 2\pi\ell) = f_b^i(y) \quad \text{for all } y \text{ in } \mathbb{R} \text{ and integer } \ell,$$

for $i = 0, 1$ (see Figure 4.3).

We can now formulate the *Dirichlet problem*. Find a 2π-periodic in y function $h(t, y)$ such that it satisfies

- first, the biharmonic equation

$$\Delta^2 h(t, y) = 0 \quad \text{for all } a < t < b \text{ and } y \text{ in } \mathbb{R}, \tag{4.7}$$

and

48 Multivariate polysplines

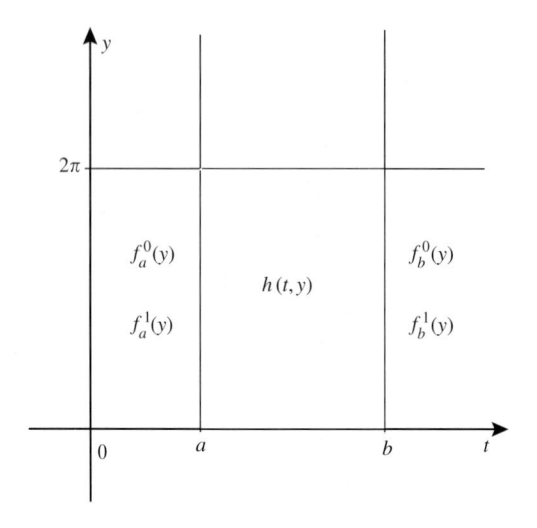

Figure 4.3. The data for the biharmonic Dirichlet problem.

- secondly, the boundary conditions, for all y in $\mathbb{R}$,

$$h(a, y) = f_a^0(y), \quad \frac{\partial h(a, y)}{\partial t} = f_a^1(y), \qquad (4.8)$$

$$h(b, y) = f_b^0(y), \quad \frac{\partial h(b, y)}{\partial t} = f_b^1(y). \qquad (4.9)$$

We now write the expansion in trigonometric Fourier series of the functions $f_a^\ell(y)$, $f_b^\ell(y)$, for $\ell = 0, 1$, namely, for all y in $\mathbb{R}$,

$$f_a^\ell(y) = \frac{f_{a,0}^\ell}{2} + \sum_{k=1}^{\infty} [f_{a,k}^\ell \cos ky + \widetilde{f}_{a,k}^\ell \sin ky], \qquad (4.10)$$

$$f_b^\ell(y) = \frac{f_{b,0}^\ell}{2} + \sum_{k=1}^{\infty} [f_{b,k}^\ell \cos ky + \widetilde{f}_{b,k}^\ell \sin ky]. \qquad (4.11)$$

We are looking for the solution in the form (4.6), p. 46, where by Exercise 4.6, for every t in the interval $a < t < b$ the Fourier coefficients $u_k(t)$, $v_k(t)$ are given by

$$u_0(t) = C_{0,0} + C_{0,1}t + C_{0,2}t^2 + C_{0,3}t^3,$$

$$u_k(t) = C_{k,0}e^{kt} + C_{k,1}te^{kt} + D_{k,0}e^{-kt} + D_{k,1}te^{-kt} \quad \text{for } k = 1, 2, \ldots,$$

$$v_k(t) = \widetilde{C}_{k,0}e^{kt} + \widetilde{C}_{k,1}te^{kt} + \widetilde{D}_{k,0}e^{-kt} + \widetilde{D}_{k,1}te^{-kt} \quad \text{for } k = 1, 2, \ldots.$$

Note that we assume that the expansion of $h(t, y)$ holds at the endpoints a and b, and may be differentiated with respect to the variable t, i.e. we have, for all y in $\mathbb{R}$, the following equalities:

$$\frac{\partial h(a, y)}{\partial t} = \frac{u'_0(a)}{2} + \sum_{k=1}^{\infty} [u'_k(a) \cos ky + v'_k(a) \sin ky],$$

$$\frac{\partial h(b, y)}{\partial t} = \frac{u'_0(b)}{2} + \sum_{k=1}^{\infty} [u'_k(b) \cos ky + v'_k(b) \sin ky].$$

From the boundary conditions (4.8) and (4.9), p. 48, by comparing the terms in the Fourier series of $h(a, y)$, $\partial h(a, y)/\partial t$, $h(b, y)$, $\partial h(b, y)/\partial t$ with those of the Fourier expansions of the functions $f_a^0(y)$, $f_a^1(y)$ and $f_b^0(y)$, $f_b^1(y)$, respectively, we obtain the following boundary conditions for the functions $u_k(t)$ and $v_k(t)$:

$$\left.\begin{array}{ll} u_k(a) = f^0_{a,k}, & u'_k(a) = f^1_{a,k}, \\ u_k(b) = f^0_{b,k}, & u'_k(b) = f^1_{b,k}, \end{array}\right\} \text{ for } k = 0, 1, 2, \ldots, \text{ and}$$

$$\left.\begin{array}{ll} v_k(a) = \tilde{f}^0_{a,k}, & v'_k(a) = \tilde{f}^1_{a,k}, \\ v_k(b) = \tilde{f}^0_{b,k}, & v'_k(b) = \tilde{f}^1_{b,k}, \end{array}\right\} \text{ for } k = 1, 2, 3, \ldots.$$

Remark 4.7 We have come to the basic observation of the present chapter. For every $k = 0, 1, 2, \ldots$, we see that the function $u_k(t)$ (and the function $v_k(t)$, for $k = 1, 2, 3, \ldots$) may be found by the above equations since the number of unknown constants $C_{k,j}$, $D_{k,j}$, which determine the function $u_k(t)$ (respectively, $\tilde{C}_{k,j}$, $\tilde{D}_{k,j}$ for $v_k(t)$) is exactly equal to four and also the number of the preceding equations is equal to four. This linear system is nondegenerate. We may justify this by using a readymade result. The main point is that the system of four functions $w(t)$ in Exercise 4.6, p. 47, forms a Chebyshev system. Next we apply Theorem 11.20, p. 196, from the Appendix on Chebyshev systems. According to this theorem, the functions $u_k(t)$ and $v_k(t)$ are uniquely determined.

The main conclusion of the above is that for every *two* boundary data $f^0 = (f_a^0(y), f_b^0(y))$ and $f^1 = (f_a^1(y), f_b^1(y))$ we have a unique biharmonic function $h(t, y)$ which satisfies the Dirichlet problem (4.7)–(4.9), p. 48. Without going too deep into the theory of function spaces, we also have, roughly speaking, the reverse result: the boundary values provide us with a couple

$$f^0 = (h(a, y), h(b, y)),$$

$$f^1 = \left(\frac{\partial h(a, y)}{\partial t}, \frac{\partial h(b, y)}{\partial t}\right).$$

Thus the couples f^0, f^1 "parametrize" the space of periodic biharmonic functions in the strip.

4.3.2 The polyharmonic case

Having considered the most important special case, which is that of the biharmonic polysplines, we proceed to the general one – for arbitrary $p \geq 1$.

50 Multivariate polysplines

In the same manner we may treat the periodic polyharmonic functions in the strip, i.e. the functions $h(t, y)$ satisfying

$$\Delta^p h(t, y) = \left(\frac{\partial^2}{\partial t^2} + \frac{\partial^2}{\partial y^2}\right)^p h(t, y) = 0 \quad \text{for all } a < t < b \text{ and } y \text{ in } \mathbb{R}.$$

Due to the periodicity in the variable y we again expect a solution in the form

$$h(t, y) = \frac{u_0(t)}{2} + \sum_{k=1}^{\infty} [u_k(t) \cos ky + v_k(t) \sin ky].$$

After applying the Laplace operator to both sides we obtain the equality

$$0 = \Delta^p h(t, y)$$
$$= \frac{u_0^{(2p)}(t)}{2} + \sum_{k=1}^{\infty}\left[\left(\frac{\partial^2}{\partial t^2} - k^2\right)^p u_k(t) \cdot \cos ky + \left(\frac{\partial^2}{\partial t^2} - k^2\right)^p v_k(t) \cdot \sin ky\right].$$

Equating all terms to zero we obtain for all $a < t < b$ the following equalities:

$$u_0^{(2p)}(t) = 0,$$

$$\left(\frac{\partial^2}{\partial t^2} - k^2\right)^p u_k(t) = 0 \quad \text{for } k = 1, 2, \ldots,$$

$$\left(\frac{\partial^2}{\partial t^2} - k^2\right)^p v_k(t) = 0 \quad \text{for } k = 1, 2, \ldots.$$

These are ordinary differential equations and enjoy the following properties.

Exercise 4.8 *For every fixed k the general solution of the ordinary differential equation*

$$\left(\frac{\partial^2}{\partial t^2} - k^2\right)^p w(t) = 0$$

is a linear combination of the $2p$ functions

$$\begin{cases} \{1, t, \ldots, t^{2p-1}\} & \text{for } k = 0, \\ \{e^{kt}, te^{kt}, \ldots, t^{p-1}e^{kt}, e^{-kt}, te^{-kt}, \ldots, t^{p-1}e^{-kt}\} & \text{for } k \neq 0. \end{cases}$$

Thus, for some constants C_j and D_j, we have

$$w(t) = \begin{cases} \sum_{j=0}^{2p-1} C_j t^j & \text{for } k = 0, \\ \sum_{j=0}^{p-1} C_j t^j e^{kt} + \sum_{j=0}^{p-1} D_j t^j e^{-kt} & \text{for } k \neq 0. \end{cases}$$

As in the previous cases we note that the result is independent of the interval (a, b).

Now we assume that the $2p$ functions $f_a^0(y), f_a^1(y), \ldots, f_a^{p-1}(y)$ and $f_b^0(y)$, $f_b^1(y), \ldots, f_b^{p-1}(y)$ be given for y in $\mathbb{R}$. Here the index a means that the function is defined on the edge $\{t = a\}$ of the strip, and the index b indicates a function defined on the edge $\{t = b\}$ of the strip. It is natural to group these $2p$ functions into p couples $f^0 = (f_a^0(y), f_b^0(y))$, $f^1 = (f_a^1(y), f_b^1(y)), \ldots, f^{p-1} = (f_a^{p-1}(y), f_b^{p-1}(y))$ and to consider every couple as boundary data on the boundary of the strip.

We may formulate the *polyharmonic Dirichlet problem*: find a function $h(t, y)$ on the strip satisfying $h(t, y) \in C^{2p}(\text{Strip}) \cap C^{p-1}(\overline{\text{Strip}})$, and the following:

- The function $h(t, y)$ is polyharmonic,

$$\Delta^p h(t, y) = 0 \quad \text{for } a < t < b \text{ and } y \text{ in } \mathbb{R}.$$

- It is subject to the boundary conditions, holding for all y in $\mathbb{R}$, and $\ell = 0, 1, \ldots, p-1$,

$$\frac{\partial^\ell h(a, y)}{\partial t^\ell} = f_a^\ell(y), \tag{4.12}$$

$$\frac{\partial^\ell h(b, y)}{\partial t^\ell} = f_b^\ell(y). \tag{4.13}$$

By Exercise 4.8, p. 50, the Fourier coefficients $u_k(t)$ and $v_k(t)$ are given, for all $a < t < b$, by the equalities

$$u_0(t) = \sum_{j=0}^{2p-1} C_{0,j} t^j,$$

$$u_k(t) = \sum_{j=0}^{p-1} C_{k,j} t^j e^{kt} + \sum_{j=0}^{p-1} D_{k,j} t^j e^{-kt} \quad \text{for } k = 1, 2, \ldots,$$

$$v_k(t) = \sum_{j=0}^{p-1} \widetilde{C}_{k,j} t^j e^{kt} + \sum_{j=0}^{p-1} \widetilde{D}_{k,j} t^j e^{-kt} \quad \text{for } k = 1, 2, \ldots.$$

We proceed exactly as in the *biharmonic* case. Let us denote for every $\ell = 0, 1, \ldots, p-1$, just as in equalities (4.10–4.11), p. 48, the trigonometric Fourier coefficients of the functions $f_a^\ell(y)$ by $\{f_{a,k}^\ell, \widetilde{f}_{a,k}^\ell\}$, and those of the functions $f_b^\ell(y)$ by $\{f_{b,k}^\ell, \widetilde{f}_{b,k}^\ell\}$. Then the unknown constants $C_{k,j}, D_{k,j}, \widetilde{C}_{k,j}, \widetilde{D}_{k,j}$ are found from the boundary conditions

$$u_k^{(\ell)}(a) = f_{a,k}^\ell, \quad u_k^{(\ell)}(b) = f_{b,k}^\ell, \quad \text{for } \ell = 0, 1, \ldots, p-1, \quad \text{and } k = 0, 1, 2, \ldots,$$

$$v_k^{(\ell)}(a) = \widetilde{f}_{a,k}^\ell, \quad v_k^{(\ell)}(b) = \widetilde{f}_{b,k}^\ell, \quad \text{for } \ell = 0, 1, \ldots, p-1, \quad \text{and } k = 1, 2, 3, \ldots.$$

52 Multivariate polysplines

Using the biharmonic functions we can conclude that the space of the periodic polyharmonic functions in the strip is "parametrized" by the p boundary data

$$f^0 = (f_a^0(y), f_b^0(y)),$$
$$f^1 = (f_a^1(y), f_b^1(y)),$$
$$\dots$$
$$f^{p-1} = (f_a^{p-1}(y), f_b^{p-1}(y)).$$

So we have, roughly speaking, p boundary "function-parameters".

Conclusion 4.9 *The polyharmonic functions in the strip which are* **periodic** *with respect to the variable y are of infinite-dimensional space. Similar to the case of harmonic functions they may be "parametrized" using the boundary data $\{(\partial^\ell h(a, y)/\partial t^\ell), (\partial^\ell h(b, y)/\partial t^\ell), \text{ for } \ell = 0, 1, \dots, p-1\}$, where the harmonic functions are obtained for $p = 1$. It follows that in a certain sense this space has a "dimension" which is p times larger than that of the harmonic functions.*

4.4 Nonperiodicity in y

In the previous sections we have discussed only harmonic and polyharmonic functions in the strip which are periodic in the variable y. It is also quite natural to have polyharmonic functions which satisfy some boundary conditions on the upper and lower edges of a rectangle. Let us finish with the following remark which has important applications.

Remark 4.10 *1. First, for an arbitrary number $L > 0$ assume that the Dirichlet data $f_a(y), f_b(y)$ be given for $y \in [0, L]$. We consider the following* **harmonic Dirichlet problem**: *find a function $h(t, y)$ such that*

- *the Laplace equation is satisfied,*

$$\Delta h(t, y) = 0 \quad \text{for all } a < t < b \text{ and } y \text{ in } (0, L),$$

- *the boundary conditions on the side edges hold,*

$$\left.\begin{array}{l} h(a, y) = f_a(y), \\ h(b, y) = f_b(y), \end{array}\right\} \quad \text{for all } 0 \leq y \leq L,$$

- *the boundary conditions on the top and bottom edges hold,*

$$\left.\begin{array}{l} h(t, 0) = 0, \\ h(t, L) = 0, \end{array}\right\} \quad \text{for all } a < t < b.$$

We assume that the Dirichlet boundary conditions are continuous, i.e.

$$f_a(0) = f_a(L) = f_b(0) = f_b(L).$$

This problem is solved in the same way as problem (4.3)–(4.5), p. 43, but the function $h(t, y)$ is expanded in a sine series,

$$h(t, y) = \sum_{k=1}^{\infty} v_k(t) \sin \frac{k\pi y}{L}.$$

See Chapter 12, p. 209, for the possibility of expanding the boundary data in a uniformly convergent sine Fourier series.

2. Now assume that we have the functions $g_0(t)$ and $g_1(t)$ given for all t satisfying $a \leq t \leq b$. The **harmonic Dirichlet problem** is understood as finding a function $h(t, y)$ which satisfies

- the Laplace equation

$$\Delta h(t, y) = 0 \quad \text{for all } a < t < b \text{ and } y \text{ in } (0, L), \tag{4.14}$$

- the boundary conditions

$$\left. \begin{array}{l} h(a, y) = f_a(y), \\ h(b, y) = f_b(y), \end{array} \right\} \quad \text{for all } 0 \leq y \leq L, \tag{4.15}$$

and
- the boundary conditions

$$\left. \begin{array}{l} h(t, 0) = g_0(t), \\ h(t, L) = g_1(t), \end{array} \right\} \quad \text{for all } a \leq t \leq b. \tag{4.16}$$

We assume that the Dirichlet data are continuous on the whole boundary of the rectangle, i.e. the following conditions hold:

$$f_a(0) = g_0(a), \quad f_a(L) = g_1(a),$$

$$f_b(0) = g_0(b), \quad f_b(L) = g_1(b)$$

(see Figure 4.4).

The standard method of solution is as follows. We first find a function in the form

$$h_0(t, y) = \alpha + \beta t + \gamma y + \delta t y,$$

which is obviously harmonic, and which satisfies

$$h_0(a, 0) = f_a(0), \quad h_0(b, 0) = f_b(0),$$
$$h_0(a, L) = f_a(L), \quad h_0(b, L) = f_b(L).$$

These four conditions are sufficient to find the constants α, β, γ, δ. We subtract the function h_0 from the data f_a, f_b, g_0, g_1, and obtain the new data $\tilde{f}_a$, $\tilde{f}_b$, $\tilde{g}_0$, $\tilde{g}_1$. Further, for these data, due to the symmetry of the variables t and y we may split the problem into two problems. We have $h(t, y) - h_0(t, y) = h_1(t, y) + h_2(t, y)$, where $h_1(t, y)$ solves

54 Multivariate polysplines

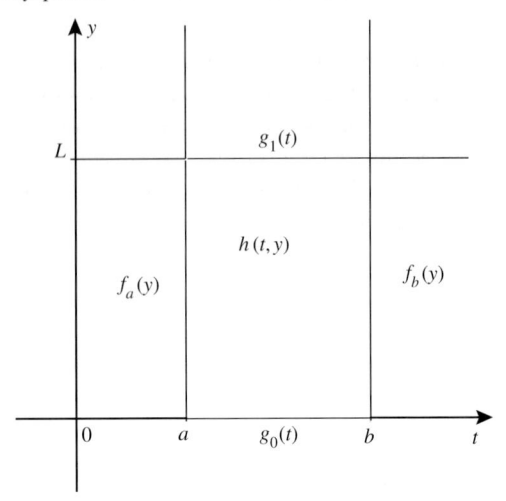

Figure 4.4. The Dirichlet data in the nonperiodic case.

the Dirichlet problem (4.14)–(4.16), p. 53, for the data $\widetilde{f}_a$, $\widetilde{f}_b$, and we then apply the same method to finding the function $h_2(t, y)$ for the data $\widetilde{g}_0$, $\widetilde{g}_1$ but we exchange the role of the variables t and y, and the data $(\widetilde{f}_a, \widetilde{f}_b)$ is replaced by $(\widetilde{g}_0, \widetilde{g}_1)$.

For sufficiently smooth data functions $g_0(t)$, $g_1(t)$ there exists a unique harmonic function $h(t, y)$ which satisfies the Dirichlet boundary conditions on all sides of the rectangle and the solution is expanded in a trigonometric Fourier series as in (4.6), p. 46, which is differentiable up to the boundary. Let us note that, in view of the above, every solution of the problem is given through a sum of two Fourier series. For the rectangle $[0, L_1] \times [0, L_2]$ we have

$$h(t, y) = h_0(t, y) + \sum_{k=1}^{\infty} v_k(t) \sin \frac{k\pi y}{L_2} + \sum_{k=1}^{\infty} \widetilde{v}_k(y) \sin \frac{k\pi t}{L_1},$$

where the functions $v_k(t)$ and $\widetilde{v}_k(y)$ are linear combinations of the type (4.2), p. 43.

3. The **polyharmonic Dirichlet problem** without periodicity is popular. So instead of assuming that $h(t, y)$ is 2π-periodic in y we may impose p boundary conditions for $y = 0$, and other p conditions for $y = 2\pi$. An example would be the following Dirichlet boundary conditions: let the functions $g_0^\ell(t)$ and $g_1^\ell(t)$ be given for $\ell = 0, 1, \ldots, p-1$ and for all t satisfying $a \le t \le b$. Then the Dirichlet boundary conditions are:

- conditions on the side edges (4.12), p. 51, and
- conditions on the lower and the upper edges, for all $a \le t \le b$ and $\ell = 0, 1, \ldots, p-1$,

$$\frac{\partial^\ell h(t, 0)}{\partial y^\ell} = g_0^\ell(t), \tag{4.17}$$

$$\frac{\partial^\ell h(t, 2\pi)}{\partial y^\ell} = g_1^\ell(t). \tag{4.18}$$

For sufficiently smooth data functions $g_0^\ell(t)$, $g_1^\ell(t)$ there exists a unique polyharmonic function $h(t, y)$ which satisfies the Dirichlet boundary conditions on all sides of the rectangle. No such easy decomposition into two Fourier series as in the harmonic case is available. The biharmonic case is of basic importance for elasticity theory. See Timoshenko and Goodier [59, Chapter 3.24] and Muskhelishvili [38] for solutions to the boundary value problems for the biharmonic equation.

Chapter 5

Polysplines on strips in $\mathbb{R}^2$

We now come to the main point of part I, a definition of polysplines in the simplest data set, which is a *finite* union of parallel straight lines.

Recall the geometrical situation outlined in Figure 3.4, p. 32, where we have the parallel straight lines $\Gamma_1, \Gamma_2, \ldots, \Gamma_N$ given by

$$\Gamma_j = \{(t, y) \in \mathbb{R}^2 : t = t_j \text{ and } y \in \mathbb{R}\},$$

where the numbers t_j satisfy $t_1 < t_2 < \cdots < t_N$.

The functions $h_j(t, y)$ are defined between the lines Γ_j and Γ_{j+1} for $j = 1, 2, \ldots, N-1$.

It is clear that every reasonable concept of multivariate splines would need some set of smoothness conditions holding across the interface lines Γ_j.

Up to the present point we have provided sufficient background justification and we can now define the main notion of *polysplines* in the present book.

Definition 5.1 *Let the parallel straight lines Γ_j be given as above for $j = 1, \ldots, N$, and let the integer $p \geq 1$. Let the functions $h_j(t, y)$ be given in the strip*

$$\text{Strip}_j = \{x = (t, y) : t_j < t < t_{j+1}\}$$

for $j = 1, \ldots, N-1$, and satisfy the following three conditions:

1. *The function $h_j(t, y)$ is polyharmonic of order p, i.e.*

$$\Delta^p h_j(t, y) = 0 \quad \text{for all } t_j < t < t_{j+1} \text{ and } y \text{ in } \mathbb{R},$$

   for $j = 1, \ldots, N-1$.
2. *The functions $h_j(t, y)$ belong to the class $C^{2p-2}(\overline{\text{Strip}_j})$. The bar over Strip_j means smoothness, including the boundary of Strip_j.*
3. *The following equalities hold:*

$$\frac{\partial^k}{\partial t^k} h_j(t_{j+1}, y) = \frac{\partial^k}{\partial t^k} h_{j+1}(t_{j+1}, y) \quad \text{for all } y \text{ in } \mathbb{R}, \tag{5.1}$$

and for all $k = 0, 1, \ldots, 2p - 2$, and $j = 1, \ldots, N - 2$. The boundary values are limits of $(\partial^k / \partial t^k) h_j(t, y)$ at points $t \in (t_j, t_{j+1})$ and $t \longrightarrow t_{j+1}$.

We will say that the function $h(t, y)$, which is equal to $h_j(t, y)$ on the strip Strip$_j$ for $j = 1, \ldots, N - 1$, is a **polyspline** of order p in the strip

$$\text{Strip} = \text{Strip}(t_1, t_N) = \{(t, y) \in \mathbb{R}^2 : t_1 < t < t_N \text{ and } y \in \mathbb{R}\}$$

and has **knot-surfaces** (or, more precisely, **knot-lines**) Γ_j.

For $j = 1, \ldots, N - 1$, the smoothness up to the boundary of Strip$_j$ means that the derivatives of the function $h_j(t, y)$ exist and are continuous on both lines

$$\Gamma_j \cup \Gamma_{j+1} = \partial \,\text{Strip}_j$$

which compose the boundary of Strip$_j$.

Remark 5.2 *1. Conditions (2) and (3) of Definition 5.1 mean that $h(x) \in C^{2p-2}(\overline{\text{Strip}})$. Indeed, we have $h(x) \in C^{2p-2}(\Gamma_j)$ for all $j = 1, 2, \ldots, N$, since $h(x) \in C^{2p-2}(\text{Strip}_j)$. This means that all derivatives up to order $2p - 2$ in the direction tangential to the lines Γ_j are continuous. The continuity up to order $2p - 2$ in the normal direction $\partial/\partial t$ is precisely condition (5.1).*

2. Continuing the first remark, let us say that the smoothness requirement $h_j \in C^{2p-2}(\overline{\text{Strip}_j})$ is too stringent but it simplifies the situation to use it. We will consider the case when the polysplines belong to Sobolev or Hölder space on the strips Strip$_j$ and other matching conditions hold, in Chapter 20, p. 409.

Now let us assume that we are given some data functions $f_j(y)$ defined on the real axis $\mathbb{R}$ for $j = 1, \ldots, N$, and that the functions $c_\ell(y), d_\ell(y), \ell = 0, 1, \ldots, p - 1$, are also defined on $\mathbb{R}$.

Definition 5.3 *The polyspline $h(t, y)$ of order p defined in the Strip(t_1, t_N) will be called an* **interpolation polyspline** *if the following conditions hold:*

1. *The* **interpolation** *equalities hold on all straight lines Γ_j*

$$h(t_j, y) = f_j(y) \quad \text{for all } y \text{ in } \mathbb{R}, \tag{5.2}$$

for $j = 1, \ldots, N$.

2. *The following* **boundary conditions** *hold on the straight lines Γ_1 and Γ_N, for all y in $\mathbb{R}$ and $\ell = 1, \ldots, p - 1$:*

$$\frac{\partial^\ell}{\partial t^\ell} h(t_1, y) = \frac{\partial^\ell}{\partial t^\ell} h_1(t_1, y) = c_\ell(y), \tag{5.3}$$

$$\frac{\partial^\ell}{\partial t^\ell} h(t_N, y) = \frac{\partial^\ell}{\partial t^\ell} h_{N-1}(t_N, y) = d_\ell(y). \tag{5.4}$$

If the functions $f_j(y), c_\ell(y), d_\ell(y)$ of the variable y are 2π-periodic we call the corresponding polyspline **periodic with respect to** y.

Let us remark that the *boundary conditions* (5.3) and (5.4) on the straight lines Γ_1, Γ_N are optional in the above definition since they have nothing to do with the smoothness of the polysplines inside Strip(t_1, t_N). So far they provide the uniqueness of the interpolation polyspline. They are always imposed if we need a unique solution to the interpolation problem. We might also choose another type of boundary conditions.

We will consider special cases with gradually increasing complexity in order to allow the reader to realize the relation between the above definition of polysplines and one-dimensional spline theory.

5.1 Periodic harmonic polysplines on strips, $p = 1$

For $p = 1$, by Definition 5.1, p. 57, we obtain polysplines $h(x) = h(t, y)$ which are naturally termed *harmonic* since all pieces $h_j(x)$ satisfy $\Delta u_j(x) = 0$.

Now let us check Definition 5.1. The function $h(t, y)$ given in the strip

$$\text{Strip}(t_1, t_N) = \{(t, y) \in \mathbb{R}^2 : t_1 < t < t_N, \text{ and } y \in \mathbb{R}\}$$

is polyspline of order $p = 1$ if and only if

$$\Delta u(t, y) = 0 \quad \text{for all } t_1 < t < t_N \text{ and } t \neq t_j \text{ for } j = 2, \ldots, N-1,$$

and it satisfies the *global smoothness* condition

$$h(t, y) \in C(\overline{\text{Strip}(t_1, t_N)}).$$

The last implies continuity, in particular on the straight lines Γ_j, i.e.

$$h_j(t_{j+1}, y) = h_{j+1}(t_{j+1}, y) \quad \text{for all } y \text{ in } \mathbb{R},$$

and $j = 1, 2, \ldots, N-2$. Now these are obviously the analog to the *linear spline* continuity conditions (2.5), p. 22. Consequently, the "smoothness condition" for harmonic polysplines simply means their continuity in the strip Strip(t_1, t_N).

The boundary conditions (5.3) and (5.4), p. 58, are missing here which again corresponds to the one-dimensional linear splines (see (2.5), p. 22).

According to Definition 5.3 the *interpolation harmonic polyspline* $h(t, y)$ satisfies

$$h(t_j, y) = f_j(y) \quad \text{for } y \text{ in } \mathbb{R} \text{ and } j = 1, 2, \ldots, N,$$

which is the analog to the one-dimensional interpolation condition (2.3), p. 21.

We see that we have $N - 1$ harmonic functions $h_j(t, y)$ in the strips Strip$_j$ which satisfy Dirichlet boundary conditions. This is quite similar to the case of one-dimensional linear splines in Section 2.2, p. 21.

Remark 5.4 *Finally, it is now clear that for periodic Dirichlet data* $f_j(y)$, $j = 1, 2, \ldots, N$, *we may compute the whole spline* $h(t, y)$ *by computing all pieces* $h_j(t, y)$ *completely separately, thus profiting from the results of Section 4.2, p. 43, on the Dirichlet problem for the Laplace equation.*

5.2 Periodic biharmonic polysplines on strips, $p = 2$

For $p = 2$ by Definition 5.1, p. 57, we obtain polysplines $h(x) = h(t, y)$ which are naturally called *biharmonic* since all pieces $h_j(x)$ satisfy $\Delta^2 h_j(x) = 0$.

According to Definition 5.1 the function $h(t, y)$ given in the strip

$$\text{Strip}(t_1, t_N) = \{(t, y) \in \mathbb{R}^2 : t_1 < t < t_N, \text{ and } y \in \mathbb{R}\}$$

is a polyspline of order $p = 2$ if and only if

$$\Delta^2 h(t, y) = 0 \quad \text{for } t_1 < t < t_N \text{ but } t \neq t_j \text{ for all } j = 2, \ldots, N-1,$$

and if it satisfies the *global smoothness* condition

$$h(x) \in C^2(\overline{\text{Strip}(t_1, t_N)}).$$

In particular, the last condition implies smoothness across the straight lines Γ_j, i.e. for all y in $\mathbb{R}$, and for $j = 1, 2, \ldots, N-2$, the following equalities hold:

$$\left. \begin{aligned} h_j(t_{j+1}, y) &= h_{j+1}(t_{j+1}, y), \\ \frac{\partial}{\partial t} h_j(t_{j+1}, y) &= \frac{\partial}{\partial t} h_{j+1}(t_{j+1}, y), \\ \frac{\partial^2}{\partial t^2} h_j(t_{j+1}, y) &= \frac{\partial^2}{\partial t^2} h_{j+1}(t_{j+1}, y). \end{aligned} \right\} \qquad (5.5)$$

Here we note that these are obviously the analog to the *cubic spline* smoothness conditions (2.2), p. 20.

The boundary conditions (5.3) and (5.4), p. 58, become

$$\left. \begin{aligned} \frac{\partial}{\partial t} h(t_1, y) &= \frac{\partial}{\partial t} h_1(t_1, y) = c(y), \\ \frac{\partial}{\partial t} h(t_N, y) &= \frac{\partial}{\partial t} h_{N-1}(t_N, y) = d(y), \end{aligned} \right\} \quad \text{for all } y \text{ in } \mathbb{R}, \qquad (5.6)$$

and these are obviously the analog to the one-dimensional boundary conditions (2.4), p. 21, for the cubic splines.

According to Definition 5.3, p. 58, the *interpolation biharmonic polyspline* $h(t, y)$ satisfies

$$h(t_j, y) = f_j(y) \quad \text{for all } y \text{ in } \mathbb{R} \text{ and } j = 1, 2, \ldots, N.$$

This is the analog to the one-dimensional interpolation condition (2.3), p. 21.

5.2.1 The smoothness scale of the polysplines

As we saw in Section 5.1, p. 59, we only have the condition that the polyspline $h(t, y)$ be continuous, i.e. belong to C^0. The biharmonic polysplines of the present section belong to C^2. The point is that

> there are no polysplines whose smoothness is exactly C^1

and, more generally, there are *no polysplines of exactly odd-order smoothness*. Further, the polysplines of order p will have smoothness C^{2p-2}. This means that the polysplines provide only even-order smoothness. This is one of the specific features of the polysplines. On one hand it is related to the ellipticity of the polyharmonic operator and on the other hand, as we have already mentioned, it is related to the even-order L-splines which have coinciding data and knot sets.

5.3 Computing the biharmonic polysplines on strips

We will restrict ourselves to the case of 2π-periodic interpolation data $f_j(y)$ and boundary data $c_1(y), d_1(y)$ which is important for the practice of *data smoothing* and *CAGD* (see Chapter 6). We present below a full set of formulas which represent all one needs for the construction of the algorithm for the *practical computation of the periodic biharmonic polysplines on strips*.

The 2π-periodicity means that for all integers ℓ and all y in $\mathbb{R}$ we have the equalities

$$f_j(y + 2\pi\ell) = f_j(y) \quad \text{for all } j = 1, 2, \ldots, N,$$
$$c(y + 2\pi\ell) = c(y),$$
$$d(y + 2\pi\ell) = d(y).$$

For the computation of the biharmonic polysplines on strips we will take advantage of the formulas of Section 4.3, p. 46, where we considered the solution of the Dirichlet problem for the biharmonic equation.

Remark 5.5 *First, we assume that the biharmonic polyspline $h(t, y)$ which satisfies the interpolation and boundary conditions exists. The conditions on the data f_j, c, d which provide the existence of a "reasonably smooth" solution $h(t, y)$ will be thoroughly studied in Chapter 20, p. 409, in the terms of Sobolev and Hölder spaces.*

Following the arguments of Section 4.3, and especially applying formula (4.6), p. 46, we also assume that the pieces $h_j(t, y)$ of the polyspline $h(t, y)$ for all $j = 1, 2, \ldots, N - -1$, will be representable by a trigonometric Fourier series

$$h_j(t, y) = \frac{u_{j,0}(t)}{2} + \sum_{k=1}^{\infty} (u_{j,k}(t) \cos ky + v_{j,k}(t) \sin ky),$$

for $t_j < t < t_{j+1}$, and y in $\mathbb{R}$, and that this series is differentiable up to the boundary of the strip Strip_j. From the results of Exercise 4.6, p. 47, we know that all functions $u_{j,k}(t)$ and $v_{j,k}(t)$ are solutions of the ordinary differential equation

$$\left(\frac{\partial^2}{\partial t^2} - k^2\right)^2 w(t) = 0 \quad \text{for } t_j < t < t_{j+1},$$

hence, are linear combinations of the form

$$w(t) = \begin{cases} C_0 + C_1 t + C_2 t^2 + C_3 t^3 & \text{for } k = 0, \\ C_0 e^{kt} + C_1 t e^{kt} + D_0 e^{-kt} + D_1 t e^{-kt} & \text{for } k \neq 0. \end{cases}$$

For every $j = 1, 2, \ldots, N$, let us write the trigonometric Fourier series of the data

$$f_j(y) = \frac{f_{j,0}}{2} + \sum_{k=1}^{\infty} [f_{j,k} \cos ky + \tilde{f}_{j,k} \sin ky] \quad \text{for all } y \text{ in } \mathbb{R},$$

and the trigonometric Fourier series of the boundary data, for all y in $\mathbb{R}$,

$$c(y) = \frac{c_0}{2} + \sum_{k=1}^{\infty} [c_k \cos ky + \tilde{c}_k \sin ky],$$

$$d(y) = \frac{d_0}{2} + \sum_{k=1}^{\infty} [d_k \cos ky + \tilde{d}_k \sin ky].$$

First, let us introduce some convenient notation. Let us define for every $k = 0, 1, 2, \ldots$, the function

$$U_k(t) := (u_{1,k}(t), u_{2,k}(t), \ldots, u_{N-1,k}(t))$$

to coincide with the function $u_{j,k}(t)$ on the interval (t_j, t_{j+1}) for all $j = 1, 2, \ldots, N-1$ (see Figure 5.1). For all $k = 1, 2, \ldots$ we likewise define the function

$$V_k(t) := (v_{1,k}(t), v_{2,k}(t), \ldots, v_{N-1,k}(t))$$

to coincide with the function $v_{j,k}(t)$ on the interval (t_j, t_{j+1}) for all $j = 1, 2, \ldots, N-1$.

We now substitute the above Fourier series for $h_j(t, y)$ into the smoothness conditions (5.5), p. 60, and into the boundary conditions (5.6), p. 60, and compare the coefficients. We see that the newly defined functions $U_k(t)$ and $V_k(t)$ satisfy the following properties:

1. The following *smoothness* conditions are satisfied:

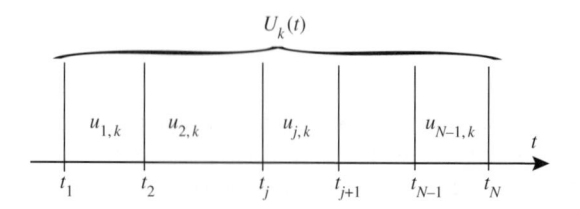

$$U_k(t) \text{ is in } C^2(t_1, t_N), \quad \text{for } k = 0, 1, 2, \ldots,$$

$$V_k(t) \text{ is in } C^2(t_1, t_N), \quad \text{for } k = 1, 2, 3, \ldots.$$

Figure 5.1. The function $U_k(t)$ is assembled from the pieces $u_{j,k}(t)$.

2. The following *interpolation* conditions are satisfied, for $j = 1, 2, \ldots, N$:

$$U_k(t_j) = f_{j,k} \quad \text{for } k = 0, 1, 2, \ldots,$$
$$V_k(t_j) = \widetilde{f}_{j,k} \quad \text{for } k = 1, 2, 3, \ldots.$$

3. The following *boundary* conditions hold:

$$\frac{\partial U_k(t_1)}{\partial t} = c_k \quad \text{for } k = 0, 1, 2, \ldots,$$

$$\frac{\partial U_k(t_N)}{\partial t} = d_k \quad \text{for } k = 0, 1, 2, \ldots,$$

$$\frac{\partial V_k(t_1)}{\partial t} = \widetilde{c}_k \quad \text{for } k = 1, 2, 3, \ldots,$$

$$\frac{\partial V_k(t_N)}{\partial t} = \widetilde{d}_k \quad \text{for } k = 1, 2, 3, \ldots.$$

4. For every $t \in [t_j, t_{j+1}]$ and for $j = 1, 2, \ldots, N-1$, we have the representation

$$U_k(t) = \begin{cases} C_0^{0,j} + C_1^{0,j} t + D_0^{0,j} t^2 + D_1^{0,j} t^3 & \text{for } k = 0, \\ C_0^{k,j} e^{kt} + C_1^{k,j} t e^{kt} + D_0^{k,j} e^{-kt} + D_1^{k,j} t e^{-kt} & \text{for } k \neq 0, \end{cases}$$

and

$$V_k(t) = \widetilde{C}_0^{k,j} e^{kt} + \widetilde{C}_1^{k,j} t e^{kt} + \widetilde{D}_0^{k,j} e^{-kt} + \widetilde{D}_1^{k,j} t e^{-kt} \quad \text{for } k \neq 0.$$

All these conditions represent a linear system from which we can find the coefficients $C_0^{k,j}, C_1^{k,j}, D_0^{k,j}, D_1^{k,j}, \widetilde{C}_0^{k,j}, \widetilde{C}_1^{k,j}, \widetilde{D}_0^{k,j}, \widetilde{D}_1^{k,j}$.

Remark 5.6 1. *There is a ready-made framework for the functions $U_k(t), V_k(t)$. Since all functions $U_k(t), V_k(t)$ satisfy the equation*

$$\left(\frac{\partial^2}{\partial t^2} - k^2 \right)^2 w(t) = 0 \quad \text{for } t \neq t_j, \text{ with } j = 1, 2, \ldots, N,$$

conditions 1. and 4. above mean that the functions $U_k(t), V_k(t)$ belong to the class of L-splines, and even to a more regular class – Chebyshev splines, with respect to the operator

$$L = \left(\frac{\partial^2}{\partial t^2} - k^2 \right)^2.$$

The theory of L-splines and Chebyshev splines is discussed in Chapter 11, p. 187. If the reader is coming across the notion of Chebyshev splines for the first time, we have to say that these splines have almost all of the advantageous properties of the classical polynomial splines. For instance, there are compactly supported splines, called here T B-splines, and they play the same important role in computing the Chebyshev splines as the

B-splines do in the polynomial case. There are recurrence relations for computing the TB-splines completely analogous to the polynomial case etc. Schumaker [50] contains all the fundamental results in this area and is the most complete source to date. It does not contain the recurrency relations, which are available in Dyn and Ron [14], see also Lyche [32].

2. We should comment on the regularity of the interpolation polyspline $h(t, y)$. The main question will be to specify the spaces of the boundary data f_a^i, f_b^i for which the solution $h(t, y)$ has derivatives up to a certain order which are in L_2 or in Hölder spaces. These questions are beyond the scope of the present discussion and cannot be answered in elementary terms. We will answer this question in Chapter 20, p. 409, in terms of Sobolev and Hölder spaces of the boundary data.

3. It should be noted that it is possible to remove the 2π-periodicity in y and replace it with p boundary conditions for $y = 0$ and other p boundary conditions for $y = 2\pi$.

4. Finally, it is clear that the polysplines with 2π-periodic data may be naturally considered as objects on a cylinder as in Section 4.2 (see Figure 4.2), p. 45.

Exercise 5.7 Write down the conditions satisfied by the functions $U_k(t)$, $V_k(t)$ in the three-harmonic case when $p = 3$, and in the tetra-harmonic case – when $p = 4$. Use the TB-splines of Chapter 11, p. 187, for the numerical solution of the interpolation problem.

5.4 Uniqueness of the interpolation polysplines

As we have already noted, the boundary conditions satisfied by the interpolation polysplines provide uniqueness. Let us prove this.

Proposition 5.8 *In Definition 5.3, p. 58, for the interpolation polysplines, the data functions involved f_j, c_ℓ, d_ℓ are identically zero, i.e. for all y in $\mathbb{R}$ we have*

$$f_j(y) = 0 \quad \text{for } j = 1, 2, \ldots, N,$$
$$c_\ell(y) = 0 \quad \text{for } \ell = 1, 2, \ldots, p-1,$$
$$d_\ell(y) = 0 \quad \text{for } \ell = 1, 2, \ldots, p-1.$$

Let us assume that a 2π-periodic in y polyspline $h(t, y)$ of order p with such data has a trigonometric Fourier series which is differentiable up to the boundary $p - 1$ times. Then

$$h(t, y) = 0 \quad \text{for all } t_1 \leq t \leq t_N \text{ and } y \text{ in } \mathbb{R}.$$

Proof Following the chain of arguments of Section 5.3 we obtain the representation

$$h(t, y) = \frac{U_0(t)}{2} + \sum_{k=1}^{\infty} (U_k(t) \cos ky + V_k(t) \sin ky),$$

where the functions $U_k(t)$ and $V_k(t)$ are L-splines for the operator

$$L = \left(\frac{d^2}{dt^2} - k^2 \right)^p$$

with knots $t_1, t_2, \ldots, t_N$ and satisfy zero interpolation and boundary conditions. Indeed, let us denote by $f_{j,k}$, $\widetilde{f}_{j,k}$, $c_{\ell,k}$, $\widetilde{c}_{\ell,k}$, $d_{\ell,k}$, $\widetilde{d}_{\ell,k}$ the coefficients of the trigonometric Fourier series of the functions $f_j(y)$, $c_\ell(y)$, $d_\ell(y)$. Obviously, they are all zero and hence the corresponding data for the L-splines $U_k(t)$, $V_k(t)$ are all zero. Here we apply the uniqueness of the interpolation L-splines (Chebyshev splines) provided in Theorem 11.28, p. 200, which implies that for all t with $t_1 \leq t \leq t_N$, the following holds:

$$U_k(t) = 0 \quad \text{for } k = 0, 1, 2, \ldots,$$
$$V_k(t) = 0 \quad \text{for } k = 1, 2, 3, \ldots.$$

Consequently, $h(t, y) \equiv 0$. ∎

Remark 5.9 *In the case of "even-order polysplines", i.e for $p = 2s$, where s is an integer, we can successfully mimic the one-dimensional identity of Lemma 2.5, p. 26, and we prove a similar general identity, in Theorem 20.7, p. 416. As a byproduct we can then provide an elegant proof of the uniqueness of the interpolation polysplines that are of order $p = 2s$.*

Exercise 5.10 Prove a Holladay-type theorem in the biharmonic case.

Chapter 6

Application of polysplines to magnetism and CAGD

One may say that all arguments provided so far in the present book, which prove the advantageous properties of the polysplines (and thus of the *polyharmonic paradigm*) by comparing them with other smoothing methods, are speculative in character since they appeal mainly to purely mathematical criteria for beauty and naturalness.

This would be the normal reaction for example of the people working with practical data. For them we have prepared two case studies where the superiority of the polysplines over *Kriging*, *Minimum curvature*, and *Radial Basis Functions* (RBFs) (not to talk about polynomial splines) is irrefutable.

6.1 Smoothing airborne magnetic field data

First we consider an interesting application to magnetic data. The case concerns the airborne data (collected through airplanes) over the Cobb Offset (in the ocean near California) where the cooled magma creates a natural magnetic anomaly. Due to the reversals of the magnetic field the neighboring layers of the magma (going somewhat North–South) have opposite signs and thus the data *oscillate wildly*.[1] The 13 tracks of the airplanes are approximately horizontal (East–West), i.e. transversal to the magma layers with nearly 200 data points on each. At these points the magnetic field (the so-called total value) has been measured and they are seen in Figure 6.1. According to the usual terminology in approximation theory these data are "scattered".

Figure 6.2 provides a sample of data on a vertical straight line (going North–South), where we see how strong the oscillation in this direction is. (So far this is not the worst oscillation in the North–South direction!)

---

[1] I have been provided with this data set by Dr. Richard O. Hansen, Pearson&de Ridder&Johnson at Denver, Colorado.

68 *Multivariate polysplines*

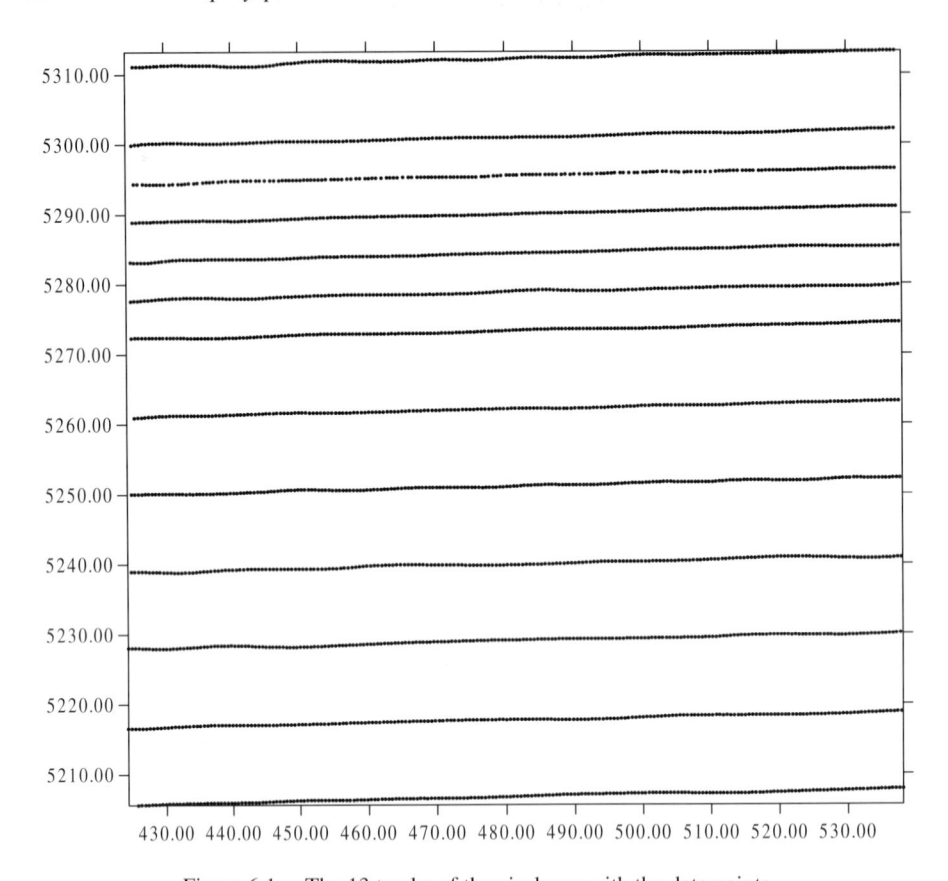

[figure: plot with y-axis from 5210.00 to 5310.00 and x-axis from 430.00 to 530.00 showing 13 tracks]

Figure 6.1. The 13 tracks of the airplanes with the data points.

Let us note that due to the mentioned scattered character of the data and their high oscillation *the test with magnetic data is one of the most difficult tests for every smoothing method!*

Next we provide the figures which show that the polysplines perform distinctly better than well established methods in the area of *magnetic explorations* such as *kriging, minimum curvature, thin plate splines,* and last but not least *RBFs* (the polynomial splines fail completely in such tests!).[2]

Figure 6.3 shows the result of the application of the *polysplines* to the Cobb Offset data. We provide the "level curves" of the graph of the interpolation polyspline. We have "posted" also the 13 data tracks.

When we apply *kriging, minimum curvature*, or *RBFs* to the same data, the result does not differ essentially for the three methods (see Figure 6.4).

On both pictures we have "posted" the locations of the original data by small points. Comparing both pictures we see that on the second one, Figure 6.4, one may easily recognize the thirteen parallel lines where the data points lie since the data points are

---

[2] These methods are incorporated in professional software for geodesists, geophysicists, geographers, etc., e.g. the program "SURFER" [19].

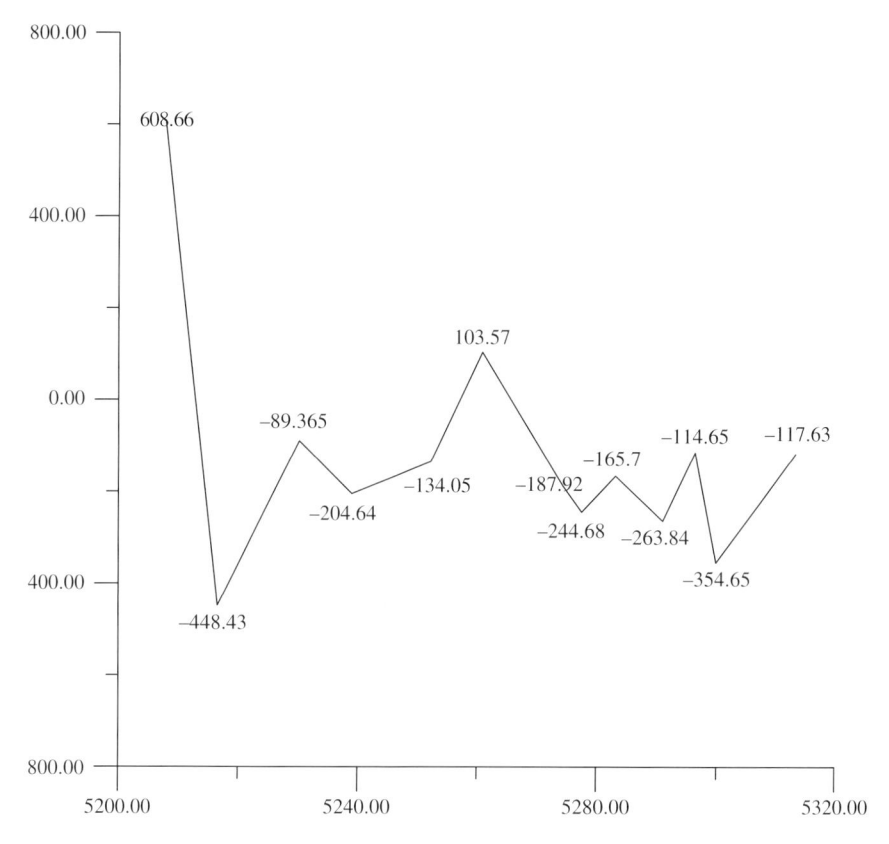

Figure 6.2. Easternmost vertical line of the data, connected by a linear function. This corresponds to $x \approx 425$ on Figure 6.1.

now local extrema (maxima or minima) of the smoothing function. Also the typical "pock marks" located at the extremal points are seen, which is a typical vermin effect of the smoothing with *kriging, minimum curvature* and *RBFs*. We have encircled with two ellipses two such locations with pock marks. However, we see that in Figure 6.4, created with polysplines, these effects are almost invisible! The polysplines minimize the artifacts while the kriging, minimum curvature and RBFs cannot get rid of them.

An important advantage of the polysplines in the above applications is that the result is an **interpolation** polyspline! All the above methods give only approximations at the data points.[3] This *interpolation property* of the polysplines is much more important for computer-aided geometric design (CAGD) which will be considered in Section 6.2.

The conclusion may be drawn that, at least for magnetic data, the polysplines show definite superiority.

---

[3] At least only these implementations are available. There are implementations using (polynomial) splines which interpolate the data but they fail to produce a nice result in the magnetic data case due to the high oscillation of the data. The reader may check e.g. that the polynomial spline algorithms available in the NAG-algorithms in Matlab in www.nag.com/advisory/gamswww/k1a1b.html, are approximation but not interpolation.

70 *Multivariate polysplines*

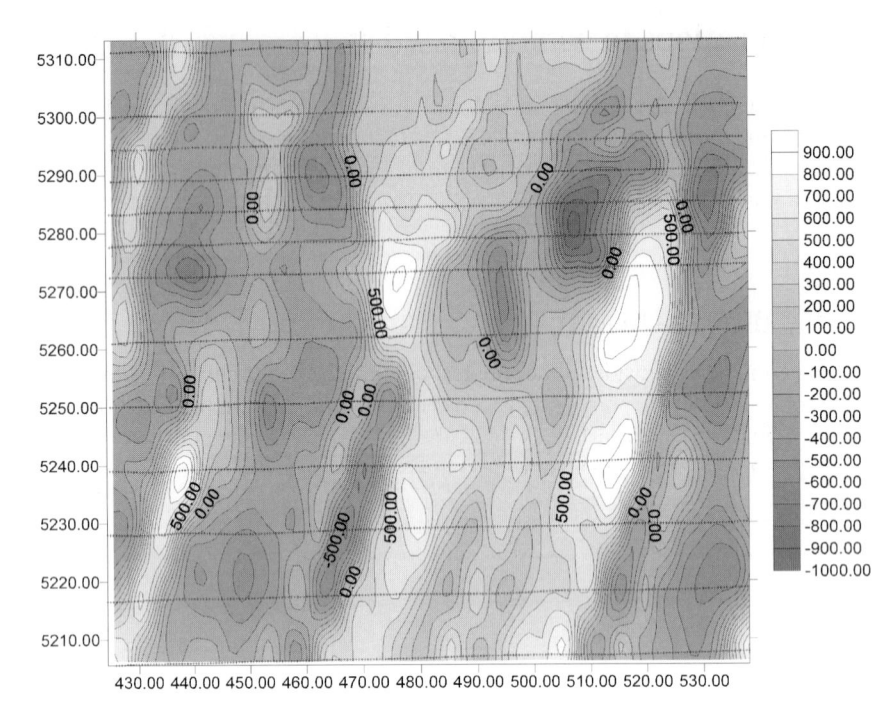

Figure 6.3. Result of applying the polysplines to the Cobb Offset data-level curves of the interpolation polyspline.

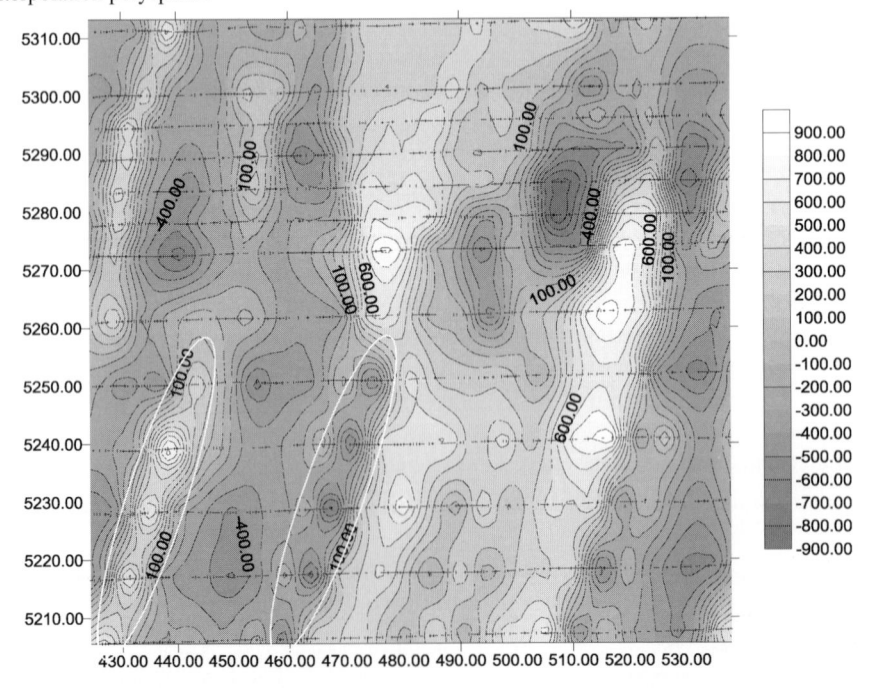

Figure 6.4. Result of applying radial basis functions (level curves). The two ellipses surround areas with strong "pock marks".

6.2 Applications to computer-aided geometric design

One might speculate that the polysplines are extremely successful in the applications to magnetic field data owing to the similarity in the physical nature of the functions. Indeed, the magnetic field is related to harmonic functions and the polysplines which we have used are composed of biharmonic functions. The experiments below will put an end to such doubts.

6.2.1 Parallel data lines Γ_j

The first experiment we consider is the simplest one where all data curves Γ_j are parallel lines. There are 200 sample points (x_i, y_i) on every line Γ_j. So we have 7×200 sample points (see Figure 6.5). The surface is prescribed at the points of these curves, i.e. the value z_i is given at (x_i, y_i). They are uniform. This is a situation typical for the CAGD

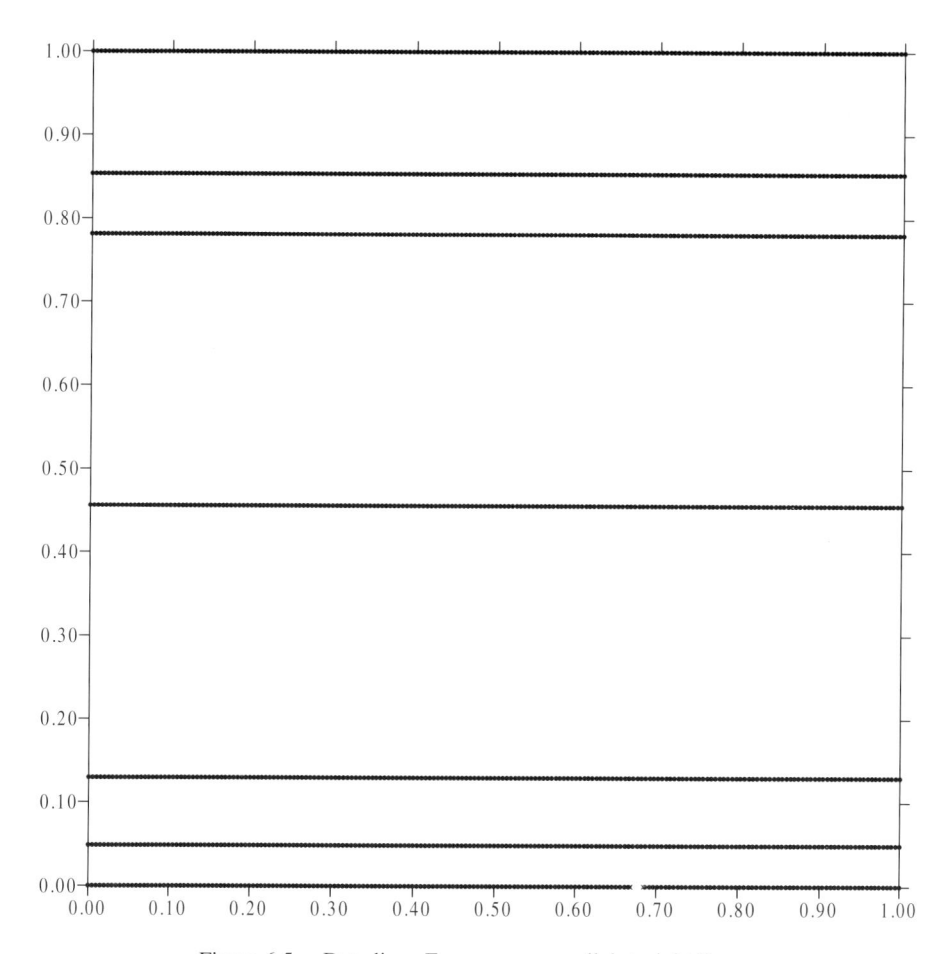

Figure 6.5. Data lines Γ_j are seven parallel straight lines.

72   Multivariate polysplines

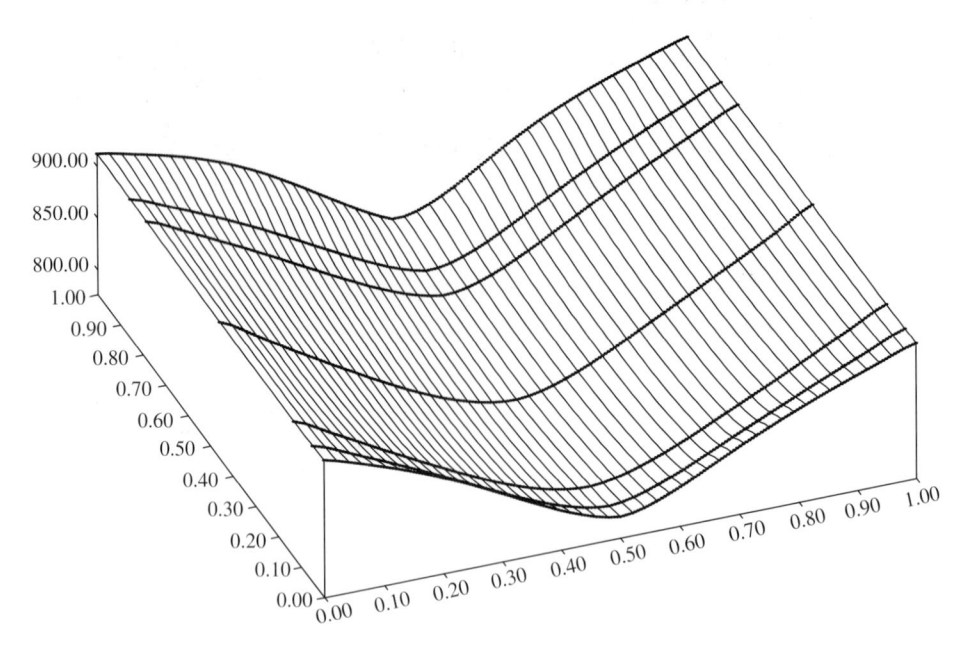

Figure 6.6.   Biharmonic polyspline (surface) defined on the rectangle $[0, 1] \times [0, 1]$. Sample data points are shown on the surface u.

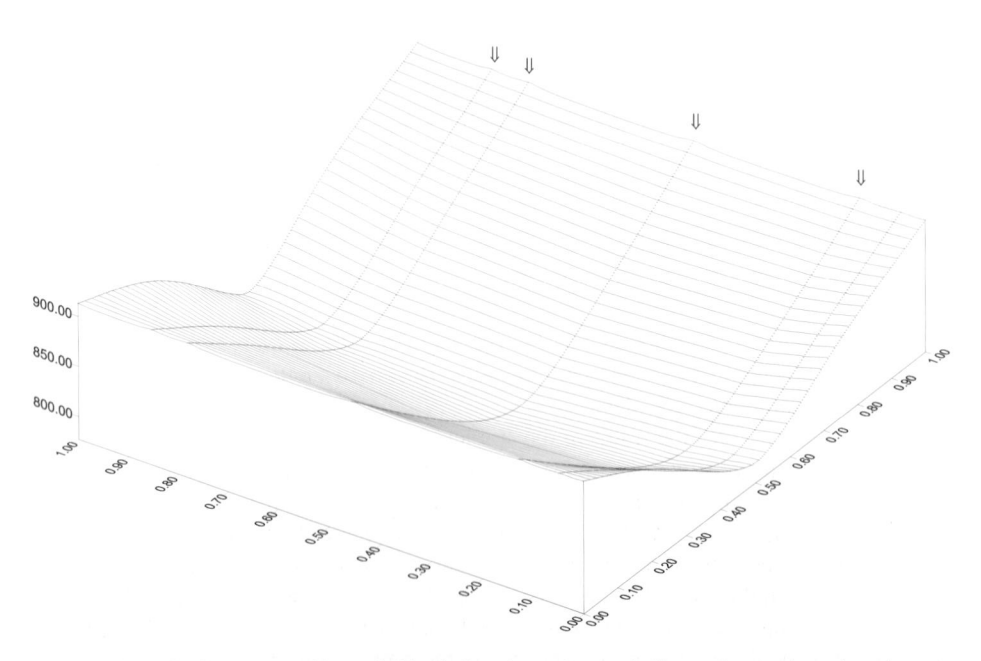

Figure 6.7.   Surfaces created by radial basis functions. Arrows indicate places where there is non smoothness locations (non C^2-smoothness). Obviously these places are near the data lines Γ_j. They are very well visible on the border of Γ_j.

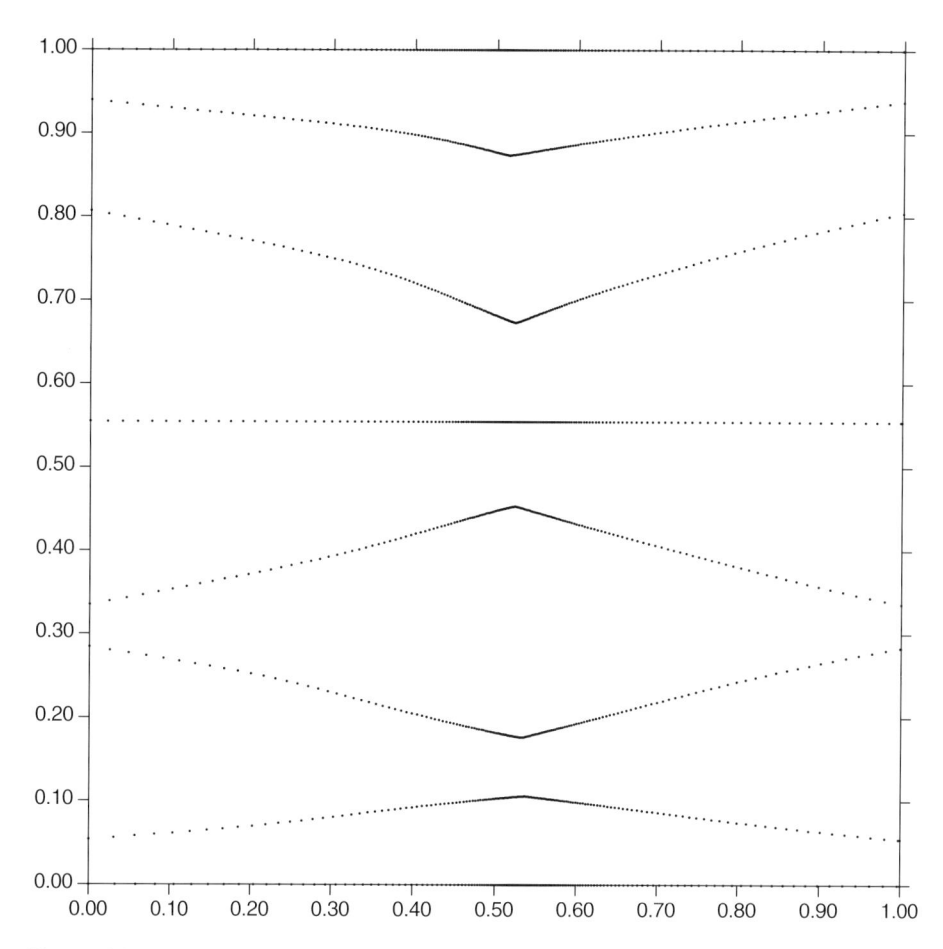

Figure 6.8. The data lines Γ_j are 8 curves with 200 sample points on each which are non–uniformly distributed on the curves.

where the "control curves" are prescribed (perhaps through control points) and we may supply the above points in a sufficient quantity. Thus the situation differs from the one we had in airborne magnetic data.

Figure 6.6 provides the result of the application of the polysplines, and Figure 6.7 provides the result of smoothing the same data set with RBFs. (The result is almost the same if we use kriging or minimum curvature.) We see that the result of the smoothing is reasonable and does not differ essentially from the result obtained using polysplines if one does not deepen. But a closer scrutiny reveals some non-smoothness locations which are essential for CAGD and make the result unsuitable for computerized design. They are indicated by arrows. Since the data do not oscillate as in the case of magnetic data, we provide the graphs of the functions in the usual "surface map" form.

Let us note that such a result would meet the standards in magnetism research but CAGD is much more sensitive to the small details of the form.

74 *Multivariate polysplines*

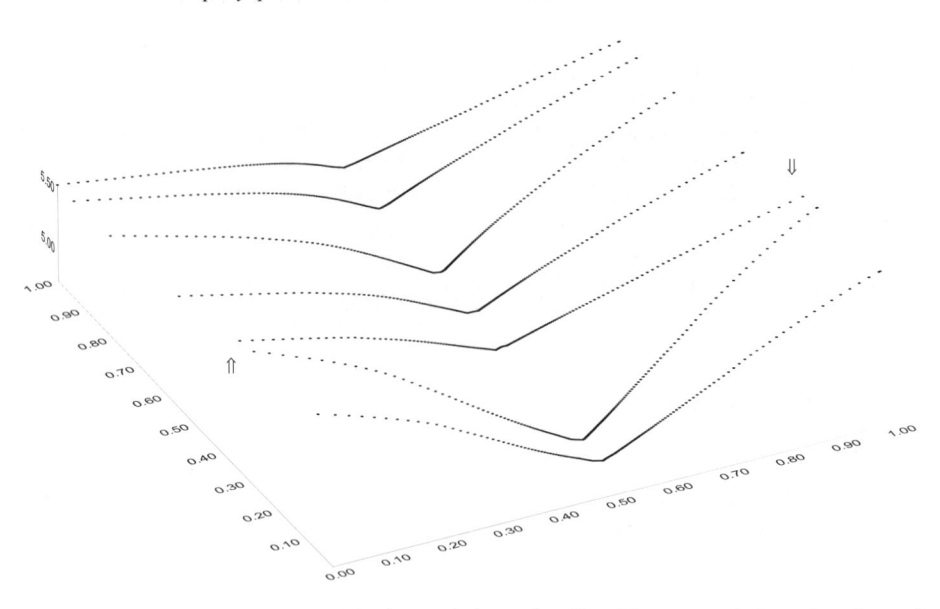

Figure 6.9. This is the surface of the interpolation polyspline. The arrows indicate locations of big tension – hence of big curvature.

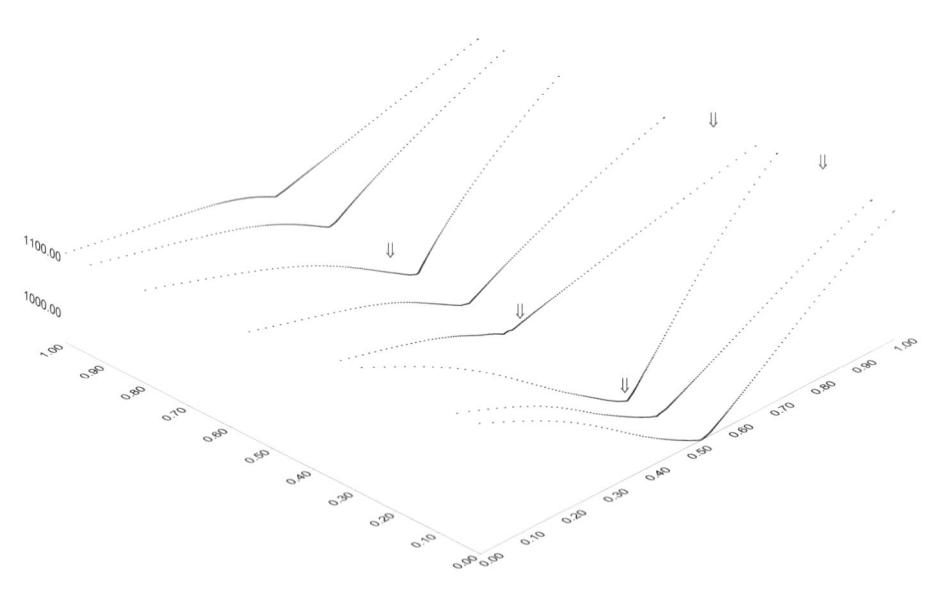

Figure 6.10. This is the surface created by kriging. The arrows show locations with obvious non-smoothness.

6.2.2 Nonparallel data curves Γ_j

The major strength of the theory of polysplines as presented in Part IV is that it allows extending data which are measured on curvilinear boundaries.

The next experiment is with data samples on eight curves Γ_j. Figure 6.8 provides the location of the sample points (x_i, y_i).

The result of the application of the interpolation biharmonic *polysplines* to the data is provided by Figure 6.9. We have indicated with arrows two locations where, due to the data, the curvature is greater. The result of applying kriging to the same data (neither RBFs nor minimum curvature gives a better result), shown in Figure 6.10, indicates that this problem is really difficult.

We have indicated with arrows the rather unpleasant for the eye roughnesses.

6.3 Conclusions

The above experiments and also other experiments carried out for similar data bring us to the following conclusions (which of course do not have the meaning of rigorous mathematical statements and extend only to the classes of data which have been considered):

- The polysplines definitely perform *better* than kriging, RBFs, minimum curvature etc., in magnetic field data problems. Unlike these methods they do not expose any "pock marks" or "line effects".
- The polysplines perform *better* than Kriging, RBFs and minimum curvature in CAGD problems where the controlling curves have parallel projections on the coordinate plane (x, y).[4]
- The polysplines perform *much better* than the above methods if the controlling curves do not have parallel projections on the coordinate plane (x, y).
- The above experiments and many others show that the biharmonic polysplines apparently have some shape-preserving properties. They do not oscillate more than the data in the magnetic field data case. The same is also true in the CAGD case where shape-preserving property is extremely important.

Remark 6.1 *The theoretical comparison between the polysplines and RBFs will be considered in the next volume to this book. Let us note only that the polysplines and the polyharmonic splines (of Madych) of the same order are close relatives due to the second Green formula (20.9), p. 422, by which the polysplines may be expressed in every subdomain where they are polyharmonic functions.*

---

[4] The comparison with the polynomial spline methods also favours the polysplines. The results are not given here.

Chapter 7

The objects concept: harmonic and polyharmonic functions in annuli in $\mathbb{R}^2$

We now carry out an analysis similar to that of Chapter 6 but for the spherical domains – annulus and ball – in $\mathbb{R}^2$. This is somewhat more sophisticated and has a generalization in $\mathbb{R}^n$ by means of the spherical harmonics. We will pay some attention to the *radially symmetric* case since its computation reduces to a one-dimensional L-spline and thus links the one-dimensional case and the multivariate generalization, which is particularly convenient for newcomers.

7.1 Harmonic functions in spherical (circular) domains

We now consider harmonic functions in spherically symmetric domains – the circle and the annulus in $\mathbb{R}^2$. Since we consider the circular case it will be more convenient to consider the *Laplace operator* Δ in terms of polar coordinates, namely $x = x(r, \varphi) = (x_1(r, \varphi), x_2(r, \varphi))$, where

$$x_1 = r \cos \varphi,$$
$$x_2 = r \sin \varphi,$$

with $r \geq 0$ and $0 \leq \varphi \leq 2\pi$. We have the inverse transform

$$r = r(x) = \sqrt{x_1^2 + x_2^2},$$
$$\varphi = \varphi(x) = \arctan \frac{x_2}{x_1}.$$

78   Multivariate polysplines

So let a function $u(x_1, x_2)$ be given. Then we put
$$v(r, \varphi) = u(r \cos \varphi, r \sin \varphi),$$
and
$$v(r(x), \varphi(x)) = u(x),$$
where we assume that the function $u(x)$ is twice differentiable in its domain of definition. As we will prove in Chapter 10, the **Laplace operator** has the following form in polar coordinates:
$$\Delta u(x) = \left(\frac{\partial^2}{\partial r^2} + \frac{1}{r} \frac{\partial}{\partial r} + \frac{1}{r^2} \frac{\partial^2}{\partial \varphi^2} \right) v(r, \varphi) \qquad (7.1)$$
$$= \frac{1}{r} \frac{\partial}{\partial r} \left(r \frac{\partial}{\partial r} v(r, \varphi) \right) + \frac{1}{r^2} \frac{\partial^2}{\partial \varphi^2} v(r, \varphi).$$

In the last expression the part depending only on the variable r is called the radial part of the Laplace operator and is denoted by Δ_r
$$\Delta_r = \frac{\partial^2}{\partial r^2} + \frac{1}{r} \frac{\partial}{\partial r} = \frac{1}{r} \frac{\partial}{\partial r} \left(r \frac{\partial}{\partial r} \right).$$

As usual, we put
$$\Delta_\theta = \frac{\partial^2}{\partial \varphi^2},$$
and obtain the representation
$$\Delta = \Delta_r + \frac{1}{r^2} \Delta_\theta.$$
Now let the function $f(x)$ be radially symmetric[1], i.e.
$$f(x) = F(|x|),$$
then it is obvious that
$$\Delta f(x) = \Delta_r F(|x|).$$

Exercise 7.1 *1. The function $f(x) = 1$ is radially symmetric and harmonic for all $r \geq 0$. The function $f(x) = \log r$ is harmonic for all $r > 0$ but not for $r = 0$! Prove that every radially symmetric harmonic function $u(x)$ in the annulus $A_{a,b} = \{(r, \varphi) : a < r < b\}$ is a linear combination of these two, i.e. has the representation*
$$u(x) = C_1 + C_2 \log r$$
for some constants C_1, C_2.

---
[1] Note that the two terms "radially symmetric" and "spherically symmetric" are sometimes used as synonyms especially when the dimension is $n \geq 3$. The last term often refers to domains.

Prove that in the ball $B(0; R)$ every radially symmetric harmonic function $u(x)$ is a constant
$$u(x) = C_1.$$
Hint: Use the fact that Δ_r is an ordinary differential operator of second order, and has two linearly independent solutions.

2. The following functions are harmonic:
$$r^k \cos k\varphi, \qquad r^k \sin k\varphi \quad (k = 0, 1, 2, \ldots). \tag{7.2}$$
Hint: Use the representation of the operator Δ in polar coordinates (7.1), p. 78.

3. The functions
$$r^{-k} \cos k\varphi, \qquad r^{-k} \sin k\varphi \quad (k = 1, 2, \ldots), \tag{7.3}$$
are harmonic for every $r > 0$. (Here we have to include the function $\log r$ which is not harmonic in the whole plane.) Hint: Check directly using (7.1), p. 78.

Remark 7.2 We see that the only radially symmetric harmonic functions are the constants (obtained from above for $k = 0$) and $\log r$.

The functions in (7.2) are harmonic homogeneous polynomials of degree k and they are a basis for the space of all harmonic homogeneous polynomials of degree k, a fact thoroughly studied in Chapter 10, especially Section 10.5, p. 136.

Moreover, as we will prove in Corollary 10.31, p. 167, every function $h(x)$ which is harmonic in the circle $B(0; R)$ may be expanded in a unique way in a series
$$h(x) = \sum_{k=0}^{\infty} a_k \left(\frac{r}{R}\right)^k \cos k\varphi + \sum_{k=1}^{\infty} b_k \left(\frac{r}{R}\right)^k \sin k\varphi, \tag{7.4}$$
where the convergence is uniform on every compact set strictly contained in the ball $B(0; R)$.

In general terms, we will also prove the representation of harmonic functions in the annulus, see Proposition 10.29, p. 166.

Let us provide here a direct proof of the above representation (7.4) and even of a more general result – the representation of a harmonic function in the annulus. The proof is somewhat heuristic in nature. This kind of heuristics will be necessary for further developments.

7.1.1 Harmonic functions in the annulus

Let us take a function $h(r, \theta)$ which is harmonic in the annulus
$$A_{a,b} = \{(r, \varphi) : a < r < b\},$$
i.e.
$$\Delta h(r, \varphi) = \left(\frac{\partial^2}{\partial r^2} + \frac{1}{r}\frac{\partial}{\partial r} + \frac{1}{r^2}\frac{\partial^2}{\partial \varphi^2}\right) h(r, \varphi) = 0 \quad \text{for} \quad a < r < b.$$

80 *Multivariate polysplines*

We assume throughout that $h \in C^2(A_{a,b})$. Let us expand the function $h(r, \varphi)$ in a Fourier series for every fixed r satisfying $a < r < b$. This is possible since the function $v(\varphi) = h(r, \varphi)$ is C^2 for every r with $a < r < b$, hence $v(\varphi)$ is in $L_2(\mathbb{S}^1)$.[2] We have

$$h(r, \varphi) = \frac{u_0(r)}{2} + \sum_{k=1}^{\infty} [u_k(r) \cos k\varphi + v_k(r) \sin k\varphi]. \tag{7.5}$$

We apply the Laplace operator on both sides and obtain

$$0 = \left(\frac{\partial^2}{\partial r^2} + \frac{1}{r} \frac{\partial}{\partial r} + \frac{1}{r^2} \frac{\partial^2}{\partial \varphi^2} \right) h(r, \varphi)$$

$$= \frac{\Delta_r u_0(r)}{2} + \sum_{k=1}^{\infty} \left[\left(\Delta_r - \frac{k^2}{r^2} \right) u_k(r) \cos k\varphi + \left(\Delta_r - \frac{k^2}{r^2} \right) v_k(r) \sin k\varphi \right].$$

We may equate every term to zero and we obtain the equations

$$\left(\Delta_r - \frac{k^2}{r^2} \right) u_k(r) = 0 \quad \text{for } a < r < b \tag{7.6}$$

and $k = 0, 1, 2, \ldots$.

The operator $(\Delta_r - (k^2/r^2))$ will play an important role in our investigation and for that reason we will give it a name, where for simplicity we drop the dimension n,

$$\boxed{L_{(k)} = \Delta_r - \frac{k^2}{r^2}.} \tag{7.7}$$

We will call it a spherical operator.

Since this is an ordinary differential operator of second order it has two linearly independent solutions. It is easy to check that the following functions are its solutions

$$r^k, \quad r^{-k} \quad \text{for } k \neq 0, \text{ and} \tag{7.8}$$

$$1, \quad \log r \quad \text{for } k = 0. \tag{7.9}$$

Hence, for some constants $c_k, d_k, \tilde{c}_k, \tilde{d}_k$ we obtain the equalities

$$\begin{cases} u_0(r) = c_0 + d_0 \log r, \\ u_k(r) = c_k r^k + d_k r^{-k} & \text{for } k = 1, 2, 3, \ldots, \\ v_k(r) = \tilde{c}_k r^k + \tilde{d}_k r^{-k} & \text{for } k = 1, 2, 3, \ldots. \end{cases} \tag{7.10}$$

---

[2] We will not provide argumentation on the convergence of the Fourier series that follows since this will be discussed exhaustively in Chapter 10.

Finally, we obtain the representation (7.5) above in the form,

$$h(r, \varphi) = \frac{c_0 + d_0 \log r}{2} \qquad (7.11)$$

$$+ \sum_{k=1}^{\infty} \left[(c_k r^k + d_k r^{-k}) \cos k\varphi + (\tilde{c}_k r^k + \tilde{d}_k r^{-k}) \sin k\varphi \right]$$

$$= \frac{d_0}{2} \log r + h_{\text{int}}(r, \varphi) + h_{\text{ext}}(r, \varphi).$$

Here we have denoted by

$$h_{\text{int}}(r, \varphi) = \frac{c_0}{2} + \sum_{k=1}^{\infty} [c_k \cos k\varphi + \tilde{c}_k \sin k\varphi] \cdot r^k$$

the part which is harmonic inside the circle $B(0; b)$ and by

$$h_{\text{ext}}(r, \varphi) = \sum_{k=1}^{\infty} [d_k \cos k\varphi + \tilde{d}_k \sin k\varphi] \cdot r^{-k}$$

the function which is harmonic in the exterior of the circle $B(0; a)$.

Since only the functions r^k for $k = 0, 1, 2, \ldots$, are continuous in the circle $B(0; R)$ it follows that

$$u_k(r) = \alpha_k r^k.$$

This proves the representation (7.4).

In addition, we have seen above, that every function $h(x)$ which is harmonic in the exterior of $B(0; R)$ and satisfies the condition $h(x) - c \log |x| \xrightarrow{|x| \to \infty} 0$ for some constant c, may be represented by the series

$$h(x) = \sum_{k=0}^{\infty} a_k \left(\frac{R}{r}\right)^k \cos k\varphi + \sum_{k=1}^{\infty} b_k \left(\frac{R}{r}\right)^k \sin k\varphi.$$

Based on the above results one may prove the following.

Exercise 7.3 *Every function $h(x)$ which is harmonic in the annulus $A_{a,b}$ may be represented as a sum of two functions and a logarithmic term*

$$h(x) = h_i(x) + h_e(x) + C \log r, \qquad (7.12)$$

where $h_i(x)$ is harmonic in the interior of the ball $B(0; b)$ and $h_e(x)$ is harmonic in the exterior of the ball $B(0; a)$ and approaches zero for $|x| \to 0$.

Similar representation holds for an arbitrary "compact domain with holes" in $\mathbb{R}^n$, see Axler et al., [3, Theorem 9.7, p. 173].

We finish this section by providing an important representation of the operator $L_{(k)}$ defined by equality (7.7), p. 80, which shows that it is a product of two first-order operators.

Proposition 7.4 *The following representation holds for every $k \geq 0$:*

$$L_{(k)}f(r) = \frac{1}{r^{k+1}}\frac{d}{dr}r^{2k+1}\frac{d}{dr}\frac{1}{r^k}f(r) \tag{7.13}$$

$$= r^{k-1}\frac{d}{dr}r^{-2k+1}\frac{d}{dr}r^k f(r).$$

The proof follows by a direct computation. We leave it as an exercise for the reader.

7.1.2 "Parametrization" of the space of harmonic functions in the annulus and the ball: the Dirichlet problem

The most important fact about the functions harmonic in domains and in particular the annulus is the solubility of the so-called *Dirichlet problem*. It plays a major role in understanding the *structure of the space of harmonic functions* and, as we will see later, of the *space of polyharmonic functions*.

Let us start with the one-dimensional case where things are simple to explain. In the one-dimensional case, $n = 1$, the linear functions $h(t) = C_1 + C_2 t$, depending on two real parameters C_1 and C_2, exhaust all solutions of the Laplace equation which in this case is simply the second derivative equal to zero, namely,

$$\Delta h(t) = \frac{d^2 h(t)}{dt^2} = 0.$$

This is a two-dimensional linear subspace of $C([a, b])$. The function $h(t)$ is determined uniquely by its values $h(a)$ and $h(b)$, which are "boundary values" on the interval (a, b). They may be considered as a "parameter", $(h(a), h(b)) \in \mathbb{R}^2$ for the set of all linear functions considered on the interval (a, b).

Let us consider the case $n \geq 2$. The set of all functions that are harmonic on the annulus

$$A_{a,b} = \{(r, \varphi) : a < r < b\},$$

i.e. the functions satisfying

$$\Delta h(x) = 0 \quad \text{for all } x \text{ in } A_{a,b},$$

is an infinite-dimensional subspace of all continuous functions in the closed annulus, $C(\overline{A_{a,b}})$. Let $f_a(\varphi)$ and $f_b(\varphi)$ be two functions that are continuous on the circles $S(0; a)$ and $S(0; b)$, respectively. Thus we have a function $f(x) = (f_a(x), f_b(x))$ continuous on the boundary of the annulus $A_{a,b}$, the last being the union of the two circles

$$\partial A_{a,b} = S(0; a) \cup S(0; b).$$

For simplicity we will use the notations $f_a(x) = f_a(\varphi)$ and $f_b(x) = f_b(\varphi)$ for $x = (r\cos\varphi, r\sin\varphi)$ and $r = a$ or $r = b$, respectively (see Figure 7.1).

The so-called **Dirichlet problem** for the **Laplace operator** in the *annulus* $A_{a,b}$ (or *harmonic Dirichlet problem*) consists in the following: *find* a harmonic function

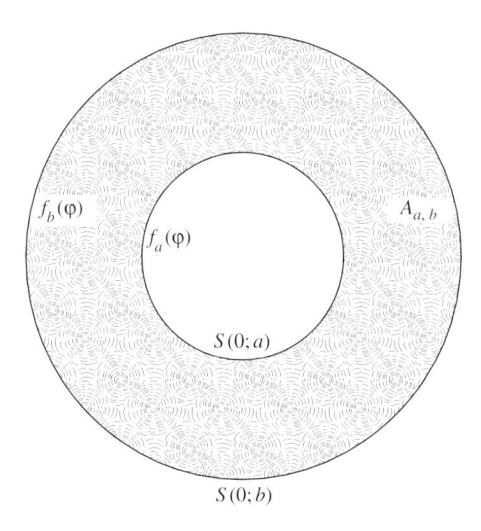

Harmonic and polyharmonic functions in annuli in $\mathbb{R}^2$

Figure 7.1. Dirichlet data on the boundary of the annulus.

$h(x) = h_f(x) = h_f(r, \varphi)$ in the annulus which takes on the prescribed boundary values f_a, f_b. In other words, the following conditions hold:

1. The *harmonic* equation

$$\Delta h_f(x) = 0 \quad \text{for all } x \text{ in } A_{a,b}. \tag{7.14}$$

2. The *Dirichlet boundary conditions* written in polar coordinates

$$\begin{cases} h_f(a, \varphi) = f_a(\varphi) & \text{for all } 0 \leq \varphi \leq 2\pi, \\ h_f(b, \varphi) = f_b(x) & \text{for all } 0 \leq \varphi \leq 2\pi. \end{cases} \tag{7.15}$$

The function

$$f(x) = (f_a(x), f_b(x))$$

which we normally assume to be continuous, i.e. to be in $C(S(0; a)) \times C(S(0; b))$, is called *Dirichlet data* and may be used as a "function-parameter" for the space of all continuous harmonic functions in the annulus $A_{a,b}$ which is analogous to the "parameter" $(h(a), h(b))$ in the one-dimensional case.

Proposition 7.5 is a classical result [3, 64].

Proposition 7.5 *For every two continuous functions $f_a(\varphi)$, $f_b(\varphi)$ the Dirichlet problem (7.14, 7.15) has a unique continuous solution $h(r, \varphi)$.*

Now let us show how to compute the solution $h_f(x)$ by using Fourier's method. We will see that there is a full analogy with the solution of the Dirichlet problem in the rectangle which we carried out in Section 4.2, p. 43.

84 *Multivariate polysplines*

We expand the data functions $f_a(\varphi)$ and $f_b(\varphi)$ in trigonometric Fourier series

$$f_a(\varphi) = \frac{f_{a,0}}{2} + \sum_{k=1}^{\infty} [f_{a,k} \cos k\varphi + \widetilde{f}_{a,k} \sin k\varphi],$$

$$f_b(\varphi) = \frac{f_{b,0}}{2} + \sum_{k=1}^{\infty} [f_{b,k} \cos k\varphi + \widetilde{f}_{b,k} \sin k\varphi].$$

We assume, quite naturally, that the expansion in trigonometric Fourier series of the solution $h_f(x)$ provided by formula (7.5), p. 80, is continuous up to the boundary of the annulus $A_{a,b}$, i.e. the following equalities hold:

$$h_f(a, \varphi) = \frac{u_0(a)}{2} + \sum_{k=1}^{\infty} [u_k(a) \cos k\varphi + v_k(a) \sin k\varphi] \quad \text{for } 0 \leq \varphi \leq 2\pi,$$

$$h_b(b, \varphi) = \frac{u_0(b)}{2} + \sum_{k=1}^{\infty} [u_k(b) \cos k\varphi + v_k(b) \sin k\varphi] \quad \text{for } 0 \leq \varphi \leq 2\pi.$$

We recall that the functions $u_k(r)$, $v_k(r)$ satisfy ordinary differential equations (7.6), p. 80, and are linear combinations given by formulas (7.10), p. 80 containing coefficients $c_k, d_k, \widetilde{c}_k, \widetilde{d}_k$.

Now we substitute all these trigonometric Fourier series into the Dirichlet boundary conditions (7.15) – $h_f(a, \varphi) = f_a(\varphi)$ and $h_f(b, \varphi) = f_a(b)$ for $0 \leq \varphi \leq 2\pi$. We compare the coefficients in front of $\cos k\varphi$ and $\sin k\varphi$ and obtain the following equalities:

$$u_k(a) = f_{a,k}, \quad u_k(b) = f_{b,k}, \quad \text{for } k = 0, 1, 2, \ldots$$

$$v_k(a) = \widetilde{f}_{a,k}, \quad v_k(b) = \widetilde{f}_{b,k}, \quad \text{for } k = 1, 2, 3, \ldots$$

Thus for every function $u_k(r)$ (or $v_k(r)$) we obtain a linear system of second order with respect to the two unknown coefficients c_k, d_k (or $\widetilde{c}_k, \widetilde{d}_k$). The last is easily seen to be solvable for all $a \neq b$, see Exercise 7.6.

Exercise 7.6 1. *Let us define the functions $w_1(r) = 1$, $w_2(r) = \log r$ or $w_1(r) = r^k$, $w_2(r) = r^{-k}$ for $k = 1, 2, 3, \ldots$. Then the determinant*

$$\det \begin{bmatrix} w_1(a) & w_2(a) \\ w_1(b) & w_2(b) \end{bmatrix} \neq 0$$

for all numbers a, b with $0 < a < b$.

2. *Let the Dirichlet data f_a and f_b be constant functions, i.e.*

$$f_a(\varphi) = M_1, \quad f_b(\varphi) = M_2,$$

for some constants M_1 and M_2. Then the solution $h_f(x)$ of the Dirichlet problem in the annulus (7.14, 7.15), p. 83, is given by

$$h_f(x) = C_1 + C_2 \log r$$

where
$$C_1 + C_2 \log a = M_1,$$
$$C_1 + C_2 \log b = M_2.$$

Hint: Use the result of Exercise 7.1, (1), p. 78, that all radially symmetric harmonic functions in the annulus $A_{a,b}$ are of the type
$$C_1 + C_2 \log r.$$

Remark 7.7 *Compare Exercise 7.6, (2) with the one-dimensional case, where $h_f(x) = C_1 + C_2 x$.*

One has to be very careful in carrying the analogy between the cases $n = 1$ and $n \geq 2$ through!

Let us indicate some differences between $n = 1$ and $n \geq 2$, which mainly arise from the fact that the space of harmonic functions in every domain $D \subset \mathbb{R}^n$, is infinite-dimensional for $n \geq 2$.

For $n \geq 2$ the ball in $\mathbb{R}^n$ has a boundary with only one connected component, namely the sphere,
$$\partial B(0; R) = S(0; R),$$
while in $\mathbb{R}$ we have $B(0; R) = (-R, R)$, and the boundary consists of two points – two components,
$$\partial B(0; R) = \{-R, R\}.$$

7.1.3 The Dirichlet problem in the ball

For the functions harmonic in the ball $B(0; R)$ in $\mathbb{R}^n$ we can also solve the so-called *Dirichlet problem*.

Let the function $f(x) = f(\varphi)$ be continuous on the boundary $S(0; R) = \partial B(0; R)$ of $B(0; R)$. Find a harmonic function $h(x)$ in the ball $B(0; R)$ which is usually denoted by $h_f(x) = h_f(r, \varphi)$ and such that
$$h_f(r, \varphi) = f(\varphi) \quad \text{for all } 0 \leq \varphi \leq 2\pi.$$

The following is a classical result [3].

Proposition 7.8 *For every continuous function $f(\varphi)$ there exists a unique continuous solution $h_f(x)$ to the Dirichlet problem in the ball.*

The function $f(\varphi)$ is termed the *Dirichlet data* and may be used as a "function-parameter" for parametrizing the space of continuous harmonic functions in the ball. This makes a difference from the case of the annulus where there were two "function-parameters". This phenomenon *does not have an analog* when $n = 1$ since the linear functions on every compact domain form a two-dimensional space.

A consequence of this observation is that the radially symmetric harmonic functions in the ball are only the constants, i.e. are of the form $h(x) = C_1$, while, in the annulus, they are of the form $h(x) = C_1 + C_2 \log r$.

7.1.4 An important change of the variable, $v = \log r$

The special form of the solutions to equation $(\Delta_r - (k^2/r^2))u_k(r) = 0$ given by (7.8) and (7.9), p. 80, suggests the variable change $v = \log r$. The last transforms these solutions into the functions

$$e^{kv}, \quad e^{-kv}, \quad \text{for } k \geq 1, \text{ and to}$$
$$1, \quad v, \quad \text{for } k = 0.$$

These functions are obviously solutions $w(v)$ to the equation with constant coefficients

$$M_{k,1}\left(\frac{d}{dv}\right) w(v) = \left(\frac{d}{dv} - \lambda_1\right)\left(\frac{d}{dv} - \lambda_2\right),$$

where

$$\lambda_1 = -k,$$
$$\lambda_2 = k,$$

for all $k \geq 0$, and where we have defined the polynomial

$$M_{k,1}(z) = (z - \lambda_1)(z - \lambda_2);$$

here the subindex 1 in $M_{k,1}$ stands for the degree of the operator $\Delta = \Delta^1$. We see that this is a *unified way* to write the two cases, $k \geq 1$ and $k = 0$ and $k = 0$ is the only exception since then $\lambda_1 = \lambda_2$.

We will later see that this is also an important transform in the case of the polyharmonic operator Δ^p where we have operators with constant coefficients $M_{k,p}(d/dv)$.

7.2 Biharmonic and polyharmonic functions

The annulus will play an important role in the space of polyharmonic functions, and especially the space of biharmonic functions.

As we have already said in Chapter 4, p. 39, a function $u(x)$ is called *biharmonic* in the domain D if and only if it satisfies the equation of fourth order

$$\Delta^2 u(x) = 0 \quad \text{for } x \text{ in } D.$$

The function $u(x)$ is called *polyharmonic of order* p if and only if

$$\Delta^p u(x) = 0 \quad \text{for } x \text{ in } D.$$

Evidently, every harmonic function is also biharmonic. But, on the other hand, we have functions which are biharmonic but not harmonic.

Exercise 7.9 1. Check directly that the function r^2 is biharmonic for all $r \geq 0$ but the function $r^2 \log r$ is biharmonic only for $r > 0$, and both are not harmonic. Hint: Use the fact that $\Delta = \Delta_r = d^2/dr^2 + (1/r)(d/dr)$.

2. Prove that the following functions and their linear combinations are the only radially symmetric biharmonic functions:

$$1, \quad \log r,$$
$$r^2, \quad r^2 \log r.$$

Hint: *Recall that the first two are simply the harmonic radially symmetric functions. The operator Δ_r^2 is an ordinary differential operator of fourth order and has four linearly independent solutions which will form the basis. Hence, every function $u(x)$ biharmonic in the annulus $A_{a,b}$ which is radially symmetric there permits the representation*

$$u(x) = C_1 + C_2 \log r + C_3 r^2 + C_4 r^2 \log r$$

for some constants C_1, C_2, C_3, C_4.

We see that this corresponds to the cubic polynomials in the one-dimensional case which satisfy

$$u^{(4)}(t) = 0$$

and have the form

$$u(t) = C_1 + C_2 t + C_3 t^2 + C_4 t^3,$$

i.e. the basis functions are $\{1, t, t^2, t^3\}$.

3. Prove that every radially symmetric biharmonic function in the circle $B(0; R)$ is represented as

$$u(x) = C_1 + C_3 r^2.$$

4. Check the more general case: the functions

$$r^{k+2} \cos k\varphi, \quad r^{k+2} \sin k\varphi \quad (k = 0, 1, 2, \ldots),$$

are biharmonic for all $r \geq 0$ but the functions

$$r^{-k+2} \cos k\varphi, \quad r^{-k+2} \sin k\varphi \quad (k = 1, 2, 3 \ldots),$$

are biharmonic only for $r > 0$, and all these functions are not harmonic.

We advise the reader to consult Sections 10.14 and 10.11 for the general case of spherical harmonics and representation of polyharmonic functions in the annulus, where we study all the solutions with separated variables.

Below we will count all linearly independent solutions of $\Delta^2 h(x) = 0$ that are of the above "separated variables form".

Now let us turn to the biharmonic and polyharmonic functions in the annulus.

7.2.1 Polyharmonic functions in annulus and circle

Recall operator $L_{(k)} = \Delta_r - (k^2/r^2)$ which we have introduced in formula (7.7). Although the superindex p sometimes appears to be rather obscure, for simplicity we will introduce the iterated operator through the notation

$$\boxed{L_{(k)}^p := [L_{(k)}]^p.}$$

88　Multivariate polysplines

We have the following.

Proposition 7.10 *Let $h(x) = h(r, \varphi)$ be polyharmonic of order p in the annulus $A_{a,b}$, i.e satisfies the equation*

$$\Delta^p h(x) = 0 \quad \text{for } x \text{ in } A_{a,b}.$$

Then it permits the representation in trigonometric Fourier series,

$$h(r, \varphi) = \frac{u_0(r)}{2} + \sum_{k=1}^{\infty} [u_k(r) \cos k\varphi + v_k(r) \sin k\varphi], \quad (7.16)$$

where the functions $u_k(r)$, $v_k(r)$ satisfy the equation

$$L_{(k)}^p u_k(r) = 0, \quad \text{for } k = 0, 1, 2, \ldots,$$

$$L_{(k)}^p v_k(r) = 0, \quad \text{for } k = 1, 2, 3, \ldots,$$

and all r with $a < r < b$.

Proof　Since the polyharmonic functions are real-analytic in the domain of definition, just as in formula (7.5), p. 80, for every r satisfying $a < r < b$ we may expand the function $h(r, \varphi)$ in a trigonometric Fourier series,

$$h(r, \varphi) = \frac{u_0(r)}{2} + \sum_{k=1}^{\infty} [u_k(r) \cos k\varphi + v_k(r) \sin k\varphi],$$

and this representation is infinitely differentiable. After applying the polyharmonic operator Δ^p to both sides we obtain

$$0 = \Delta^p h(r, \varphi)$$

$$= \sum_{\ell=-\infty}^{\infty} \left[\cos k\varphi \cdot \left(\Delta_r - \frac{k^2}{r^2} \right)^p u_k(r) + \sin k\varphi \cdot \left(\Delta_r - \frac{k^2}{r^2} \right)^p v_k(r) \right]$$

$$= \sum_{\ell=-\infty}^{\infty} \left[\cos k\varphi \cdot [L_{(k)}]^p u_k(r) + \sin k\varphi \cdot [L_{(k)}]^p v_k(r) \right].$$

Equating as before all terms of the series to zero we obtain the equations

$$L_{(k)}^p u_k(r) = 0, \quad \text{for } k = 0, 1, 2, \ldots, \quad (7.17)$$

$$L_{(k)}^p v_k(r) = 0, \quad \text{for } k = 1, 2, 3, \ldots, \quad (7.18)$$

both for $a < r < b$, which proves the proposition.　■

7.2.2 The set of solutions of $L_{(k)}^P u(r) = 0$

It should be noted that this ordinary differential equation with variable coefficients can be solved explicitly, and that the set of solutions has a relatively simple description. Let us denote by $U_{k,p}$ the set of solutions of the equation

$$L_{(k)}^P u(r) = 0 \tag{7.19}$$

which are C^∞ for $r > 0$, i.e. we put

$$U_{k,p} := \left\{ u \in C^\infty(\mathbb{R}_+) \text{ such that } L_{(k)}^P u(r) = 0 \text{ for all } r > 0 \right\}.$$

Since the *ordinary differential operator* $L_{(k)}^P$ is of order $2p$ the number of linearly independent solutions to equation (7.19) is precisely $2p$, i.e

$$\dim(U_{k,p}) = 2p.$$

In order to give the reader an idea of the simplest solutions of the equation $L_{(k)}^P u(r) = 0$, we provide some special elements of the space $U_{k,p}$ in the following exercise.

Exercise 7.11 1. Check that the function $u_0(r) = r^{2p-2}$ is a solution to (7.19) for $k = 0$ and for all $r \geq 0$, but the function $u_0(r) = r^{2p-2} \log r$ is a solution for $k = 0$ only for $r > 0$. Both functions are not solutions to $L_{(0)}^j f(r) = 0$ for $j = 1, \ldots, p-1$. Hint: Here and below use representation (7.13), p. 82, of the operator $L_{(k)}$.
2. Check that the functions

$$v_k(r) = u_k(r) = \begin{cases} r^{-k+2p-2}, \\ r^{k+2p-2}, \end{cases}$$

are solutions to equation (7.19) but not to $L_{(k)}^j f(r) = 0$ for $j = 1, \ldots, p-1$.
3. Check that the two groups of functions

$$u_k(r) = \begin{cases} r^{-k}, & r^{-k+2}, & \ldots, & r^{-k+2p-2}; \\ r^k, & r^{k+2}, & \ldots, & r^{k+2p-2}; \end{cases} \tag{7.20}$$

are solutions to equation (7.19). Hint: The functions $r^{-k+2j-2}$ and r^{k+2j-2} are solutions to $L_{(k)}^j f(r) = 0$.
4. From the above conclude that the functions

$$u_k(r) \cos k\varphi, \qquad u_k(r) \sin k\varphi,$$

are polyharmonic of order p.

The solutions in the first and second rows of Exercise 7.11, (3), may overlap! (see Figure 7.2). Hence, this is not the full number of linearly independent solutions to (7.19). Recall the simplest case of $p = 1$ (the harmonic functions) for $k = 0$, implying that

90   Multivariate polysplines

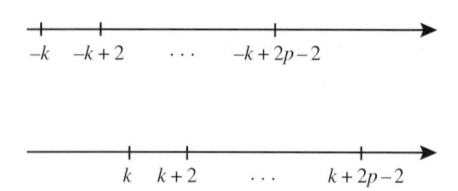

```
+----+----+----------+---------->
-k  -k+2   ...    -k+2p-2
```

```
+----+----+----------+---------->
k   k+2    ...     k+2p-2
```

Figure 7.2. The two groups of exponents in (7.20) are shown. They will overlap if and only if $k \leq -k + 2p - 2$.

$r^k = 1$. In addition to this, we also have the solution $\log r$. Approximately the same situation holds for arbitrary $p \geq 1$.

We now provide a complete description of the simplest basis of the set $U_{k,p}$.

Proposition 7.12 *For all integers $p \geq 1$ and $k \geq 0$ the following equalities hold:*

- *Let $0 \leq k \leq p - 1$. Then*

$$U_{k,p} = \text{lin} \left\{ \begin{array}{l} r^{-k}, r^{-k+2}, \ldots, r^{k+2p-2}; \\ r^k \log r, r^{k+2} \log r, \ldots, r^{-k+2p-2} \log r \end{array} \right\}.$$

- *Let $k \geq p$. Then*

$$U_{k,p} = \text{lin} \left\{ r^{-k}, r^{-k+2}, \ldots, r^{-k+2p-2}; r^k, r^{k+2}, \ldots, r^{k+2p-2} \right\}.$$

We leave the *direct proof* as an exercise for the reader; another approach to the proof based on *Chebyshev systems* is provided in the following subsection. For the direct proof one has to check that all functions listed are solutions to equation $L_{(k)}^p u(r) = 0$, and that they are linearly independent.

We now need to comment on the case $k \leq p - 1$. The logarithmic terms $r^{k+2\ell} \log r$ appear precisely for the overlapping parts of the two series of solutions in (7.20), p. 89. Indeed, such an overlapping means that for some other integer ℓ_1 with $0 \leq \ell_1 \leq p - 1$ we have

$$-k + 2\ell_1 = k + 2\ell,$$

hence

$$r^{-k+2\ell_1} = r^{k+2\ell}.$$

Thus the solutions of the form $r^{k+2\ell} \log r$ compensate this overlapping. In the case of $k \geq p$ no such overlap is possible in the two series of solutions in (7.20) since then $-k + 2p - 2 < k$.

7.2.3   The operators $L_{(k)}^p (d/dr)$ generate an Extended Complete Chebyshev system

A strategic observation for the whole book, especially concerning the so-called spherical polysplines, is that the operators $L_{(k)}^p (d/dr)$ generate an *Extended Complete Chebyshev system (ECT)* on every semi-infinite interval $[\varepsilon, \infty)$ with $\varepsilon > 0$. Indeed, due to equality (7.13), p. 82, we obtain the following representation of the operator $L^P_{(k)}(d/dr)$:

$$L^P_{(k)}f(r) = \underbrace{\frac{1}{r^{k+1}}\frac{d}{dr}r^{2k+1}\frac{d}{dr}\frac{1}{r^k}\cdots\frac{1}{r^{k+1}}\frac{d}{dr}r^{2k+1}\frac{d}{dr}\frac{1}{r^k}}_{p}f(r).$$

This shows that the ordinary differential operator $L = r^{k+1}L^P_{(k)}$ belongs to the class of differential operators which generate an *ECT-system* on every interval $[\varepsilon, \infty)$ with $\varepsilon > 0$, see Chapter 11, p. 187, especially formulas (11.1) and (11.2), p. 187. The C^∞ solutions on the interval $[\varepsilon, \infty)$ to equation

$$r^{k+1}L^P_{(k)}f(r) = 0$$

coincide with those of

$$L^P_{(k)}f(r) = 0,$$

and hence with the set $U_{k,p}$. It follows that the set $U_{k,p}$ represents an *ECT-system* on the interval $[\varepsilon, \infty)$.

The functions $w_i(r)$ in the defining formula (11.1), p. 187, are given by

$$w_1(r) = r^k,$$

$$w_3(r) = w_5(r) = \cdots = w_{2p-1}(r) = r^{2k+1},$$

$$w_2(r) = w_4(r) = \cdots = w_{2p}(r) = \frac{1}{r^{2k+1}}$$

and the operators in formula (11.1) are given by

$$D_1 = \frac{\partial}{\partial r}\frac{1}{r^k},$$

$$D_3 = D_5 = \cdots = D_{2p-1} = \frac{\partial}{\partial r}\frac{1}{r^{2k+1}},$$

$$D_2 = D_4 = \cdots = D_{2p} = \frac{\partial}{\partial r}r^{2k+1}.$$

Thus

$$Lf(r) = r^{k+1}L^P_{(k)}f(r) = D_{2p}\cdots D_1 f(r).$$

Hence, we may apply all the results on the *Extended Complete Chebyshev systems* available in Chapter 11, p. 187.

Exercise 7.13 *Prove Proposition 7.12, p. 90. Hint: Use the fact that the operator $L = r^{k+1}L^P_{(k)}$ generates an ECT-system.*

7.3 "Parametrization" of the space of polyharmonic functions in the annulus and ball: the Dirichlet problem

In Section 7.1.2, p. 82, we saw that the space of all continuous harmonic functions in the annulus and the ball is "parametrized" through the boundary value function thanks to the solubility of the Dirichlet problem. We now provide a similar parametrization for the *polyharmonic* functions of order p in the *annulus* and in the *ball*. Certainly, things are not so simple here if one wants to describe the set of all continuous polyharmonic functions.

7.3.1 The one-dimensional case

Let us first consider the one-dimensional case, $n = 1$. Then the polyharmonic equation becomes
$$\Delta^p h(t) = \frac{d^{2p} h(t)}{dt^{2p}} = 0,$$
and its general solution is given by
$$h(t) = \sum_{j=0}^{2p-1} C_j t^j. \tag{7.21}$$

The space of all such functions is $2p$-dimensional. It is possible to find all coefficients C_j if the following information is available:

$$h(a), \ h'(a), \ \ldots, \ h^{(p-1)}(a),$$
$$h(b), \ h'(b), \ \ldots, \ h^{(p-1)}(b).$$

7.3.2 The biharmonic case

Let us consider the space of *biharmonic* functions in the *annulus* $A_{a,b}$. Let us assume that the following four functions are given:

$f_a^0(\varphi), f_a^1(\varphi)$ defined on $S(0; a)$, i.e. for all $0 \le \varphi \le 2\pi$,

$f_b^0(\varphi), f_b^1(\varphi)$ defined on $S(0; b)$, i.e. for all $0 \le \varphi \le 2\pi$.

For simplicity we will simultaneously use the variable x on $S(0; a)$ (or on $S(0; b)$) which is equivalent to the "local" variable φ and we will write $f_a^i(x) = f_a^i(\varphi)$ for x on $S(0; a)$ and $f_b^i(x) = f_b^i(\varphi)$ for x on $S(0; b)$.

The so-called *biharmonic Dirichlet problem* in the *annulus* $A_{a,b}$ consists of finding a function $h(x)$ in $A_{a,b}$ which satisfies the following conditions:

1. the *biharmonic equation*
$$\Delta^2 h(r, \varphi) = 0 \quad \text{for all } a < r < b \text{ and } 0 \le \varphi \le 2\pi,$$

2. the Dirichlet *boundary conditions*

$$h(a, \varphi) = f_a^0(\varphi) \\ \frac{\partial h(a,\varphi)}{\partial r} = f_a^1(\varphi) \Bigg\} \quad \text{for } 0 \leq \varphi \leq 2\pi, \tag{7.22}$$

$$h(b, \varphi) = f_b^0(\varphi) \\ \frac{\partial h(b,\varphi)}{\partial r} = f_b^1(\varphi) \Bigg\} \quad \text{for } 0 \leq \varphi \leq 2\pi. \tag{7.23}$$

(It should be noted that for all $x = (a, \varphi) \in S(0; a)$ the derivative $\partial/\partial r$ coincides with $-\partial/\partial n$ and for $x = (b, \varphi) \in S(0; b)$ the derivative $\partial/\partial r$ coincides with $\partial/\partial n$, where $\vec{n}$ is the unit exterior vector at the point $x \in \partial A_{a,b}$, see Figure 7.3).

Similarly, one formulates the *biharmonic Dirichlet problem* in the *ball* $B(0; b)$. It consists in finding a function $h(x)$ in $B(0; b)$ which satisfies the following conditions:

1. the *biharmonic equation*

$$\Delta^2 h(r, \varphi) = 0 \quad \text{for all } 0 \leq r < b \text{ and } 0 \leq \varphi \leq 2\pi,$$

2. the Dirichlet *boundary conditions*

$$h(b, \varphi) = f_b^0(\varphi) \\ \frac{\partial h(b,\varphi)}{\partial r} = f_b^1(\varphi) \Bigg\} \quad \text{for } 0 \leq \varphi \leq 2\pi.$$

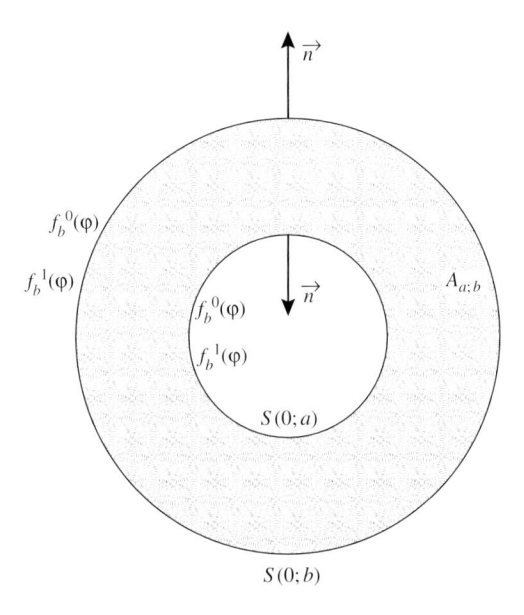

Figure 7.3. The Dirichlet data on the annulus and direction of the normal derivatives on both parts of the boundary.

94   Multivariate polysplines

Let us mention a sufficient condition for the solvability of the above problem:[3] let $\psi(x)$ be a function in $C^4(\mathbb{R}^2)$. If the functions f_a^i, f_b^i are defined by the equalities

$$f_a^0(\varphi) = \psi(a,\varphi), \qquad f_a^1(\varphi) = \frac{\partial \psi(a,\varphi)}{r} \quad \text{for } 0 \le \varphi \le 2\pi,$$

$$f_b^0(\varphi) = \psi(b,\varphi), \qquad f_b^1(\varphi) = \frac{\partial \psi(b,\varphi)}{r} \quad \text{for } 0 \le \varphi \le 2\pi,$$

then the Dirichlet problem is solvable and has a trigonometric Fourier series differentiable up to the boundary.

We have gained enough experience in solving the *biharmonic Dirichlet problem* in the *rectangle* in Section 4.3, p. 46. And in Section 7.1.2, p. 80, we have explicitly solved the *harmonic Dirichlet* problem in the *annulus*. Thus, we leave the analogous constructive solution of the *biharmonic Dirichlet problem* in the *annulus* as an exercise for the reader.

Exercise 7.14 *1. Let the function $h(r,\varphi)$ be a solution to the biharmonic Dirichlet problem (7.22), (7.23) in the annulus $A_{a,b}$. Assuming that the trigonometric Fourier series of the biharmonic function $h(r,\varphi)$ is given by formula (7.5), p. 80, i.e.*

$$h(r,\varphi) = \frac{u_0(r)}{2} + \sum_{k=1}^{\infty}[u_k(r)\cos k\varphi + v_k(r)\sin k\varphi], \quad \text{for all } a < r < b,$$

which is differentiable up to the boundary of the annulus $A_{a,b}$, write down the boundary conditions for the functions $u_k(r)$ and $v_k(r)$ at the points a and b. As we have established above, they satisfy equation (7.19), p. 89, i.e. $L_{(k)}^p u(r) = 0$. Prove that these conditions are sufficient to find the functions $u_k(r)$ for $k = 0, 1, 2, \ldots,$ and the functions $v_k(r)$ for $k = 1, 2, 3, \ldots$.

2. Let the function $h(r,\varphi)$ be a solution to the biharmonic Dirichlet problem in the ball $B(0;b)$. Write down the conditions satisfied by the functions $u_k(r)$ and $v_k(r)$. What is the difference for the annulus?

The main conclusion of the solution of the above biharmonic Dirichlet problem in the annulus is that we may "parametrize" the space of *biharmonic* functions (almost all in some sense) through the set of *four* boundary value functions

$$f_a^0(x), \; f_a^1(x), \; f_b^0(x), \; f_b^1(x),$$

which are two boundary functions, $f^0(x)$ and $f^1(x)$, given by

$$f^0(x) = (f_a^0(x), f_b^0(x)),$$

$$f^1(x) = (f_a^1(x), f_b^1(x)).$$

This is very similar to the one-dimensional case, where we have four constants C_j in equality (7.21), p. 92, if $p = 2$. In the case of the ball $B(0; R)$ we only have *two* "function-parameters" since the boundary has only one component.

---

[3] We will not discuss the most general conditions on the data functions f_a^i and f_b^i providing solutions to the above Dirichlet problem. This is done in Chapter 22 in the context of Sobolev and Hölder spaces.

7.3.3 The polyharmonic case

For *polyharmonic* functions of order p in the annulus $A_{a,b}$, we need the "Dirichlet boundary data" which consists of p functions on the sphere $S(0; a)$ and p functions on the sphere $S(0; b)$:

$$f_a^0(x), f_a^1(x), \ldots, f_a^{p-1}(x) \quad \text{on } S(0; a),$$
$$f_b^0(x), f_b^1(x), \ldots, f_b^{p-1}(x) \quad \text{on } S(0; b).$$

Now we may formulate the *polyharmonic Dirichlet problem* in the *annulus*. Find a function $h(x)$ in the annulus $A_{a,b}$ satisfying

1. the polyharmonic equation

$$\Delta^p h(x) = 0 \quad \text{for } x \text{ in } A_{a,b},$$

2. the Dirichlet boundary conditions. The *Dirichlet problem* consists of finding a polyharmonic function $h(x)$ which satisfies

$$\frac{\partial^j h(x)}{\partial r^j} = f_a^j(x) \quad \text{for all } x \text{ in } S(0; a),$$

where $j = 0, 1, \ldots, p-1$, and

$$\frac{\partial^j h(x)}{\partial r^j} = f_b^j(x) \quad \text{for all } x \text{ in } S(0; b),$$

where $j = 0, 1, \ldots, p-1$.

We have a sufficient condition for solubility of this problem, see the footnote on p. 94.

Proposition 7.15 *The polyharmonic Dirichlet problem in the annulus $A_{a,b}$ can be solved for boundary data f_a^j, f_b^j obtained in the following way:*

$$f_a^j(x) = \frac{\partial^j \psi(x)}{\partial r^j} \quad \text{for all } x \text{ in } S(0; a),$$

$$f_b^j(x) = \frac{\partial^j \psi(x)}{\partial r^j} \quad \text{for all } x \text{ in } S(0; b),$$

for every $j = 0, 1, \ldots, p-1$, where the function $\psi(x)$ is in $C^{2p}(\mathbb{R}^2)$.

We conclude that the functions f_a^j, f_b^j are $2p$ "function-parameters" which describe the set of (almost) all functions polyharmonic of order p in the annulus $A_{a,b}$. They may be grouped into p "boundary functions":

$$f^0(x) = (f_a^0(\varphi), f_b^0(\varphi)),$$
$$f^1(x) = (f_a^1(\varphi), f_b^1(\varphi)),$$
$$\ldots$$
$$f^{p-1}(x) = (f_a^{p-1}(\varphi), f_b^{p-1}(\varphi)).$$

This is very similar to the one-dimensional case where we have $2p$ constants C_j.

Similarly, if instead of the annulus $A_{a,b}$ we consider the ball $B(0; R)$, we only have p "function-parameters".

The following exercise may be solved in a manner almost identical with the case of the periodic polyharmonic functions in strips in Section 5.3, p. 61.

Exercise 7.16 *1. Assuming, as in Exercise 7.14, p. 94, that the solution of the polyharmonic Dirichlet problem has a trigonometric Fourier series*

$$h(r, \varphi) = \frac{u_0(r)}{2} + \sum_{k=1}^{\infty} [u_k(r) \cos k\varphi + v_k(r) \sin k\varphi], \quad \textit{for all } a < r < b,$$

which is $p - 1$ times differentiable up to the boundary of the annulus $A_{a,b}$ write down the conditions satisfied by the functions u_k, v_k at the endpoints a, b. Prove that these conditions are sufficient for the unique determination of $u_k(r)$, for $k = 0, 1, 2, \ldots$, and $v_k(r)$ for $k = 1, 2, 3, \ldots$.

7.3.4 Another approach to "parametrization": the Almansi representation

We have seen that the Dirichlet problem for the polyharmonic equation provides a "natural parametrization" of the space of polyharmonic functions of order p. This will be of great importance in our further study. There is another beautiful, and also very natural, way to parametrize the space of polyharmonic functions of order p. It is provided by the so-called Almansi formula, which we will prove in Chapter 10, Theorem 10.51, p. 184.

The *Almansi theorem for the ball* states that every function $h(x)$ which is *biharmonic* in the ball $B(0; R)$ may be represented as follows:

$$h(x) = h_0(x) + |x|^2 h_1(x),$$

where $h_0(x)$, $h_1(x)$ are two functions harmonic in the ball $B(0; R)$, and $|x|^2 = x_1^2 + x_2^2$.

More generally, for every integer $p \geq 1$ Almansi's theorem states that if the function $h(x)$ is *polyharmonic* of order p in the ball $B(0; R)$ then it admits the following representation:

$$h(x) = h_0(x) + |x|^2 h_1(x) + \cdots + |x|^{2(p-1)} h_{p-1}(x), \qquad (7.24)$$

where the functions $h_j(x)$ are harmonic in the ball $B(0; R)$.

Roughly speaking, the Almansi representation shows that the function $h(x)$ polyharmonic of order p has p "function-parameters" for which one may choose the boundary values of the harmonic functions $h_j(x)$. This information is essentially equivalent to that of the Dirichlet problem, where we also have p boundary data.

There is also a version of the Almansi formula in the annulus which shows a "parametrization" similar to that of the Dirichlet problem. Let us mention this result, at least in the two-dimensional case: The *Almansi representation* formula for the polyharmonic functions of order p in the *annulus* $A_{a,b}$ with $a > 0$, has been proved by Vekua [62].

Proposition 7.17 Let the function $h(x, y)$ be polyharmonic of order $p \geq 1$ in $A_{a,b}$. Then it permits the following representation:

$$u(x, y) = \sum_{j=0}^{p-1} \omega_j(x, y) r^{2j} + \left\{ \sum_{k=0}^{p-1} (P_k(r) \cos k\varphi + Q_k(r) \sin k\varphi) \right\} \log r, \quad (7.25)$$

where the functions $\omega_j(x, y)$ are harmonic in the annulus $A_{a,b}$ and the polynomials P_k and Q_k are of the type

$$a_0 r^k + a_1 r^{k+2} + \cdots + a_{p-k-1} r^{2p-2-k}.$$

Here we have the polar coordinates defined, as usual, by the complex number representation

$$(x, y) = r \cdot e^{i\varphi} \in \mathbb{R}^2.$$

(See Chapter 10, Theorem 10.46, p. 179, for the general case.) Note that the terms with $k = 0$ may be dropped since $\log r$ is a harmonic function and the sum

$$(a_0 + a_1 r^2 + \cdots + a_{p-k-1} r^{2p-2}) \log r$$

is of the same type as the first sum in (7.25).

We will use this formula extensively. It shows that every function in the annulus has roughly speaking $2p$ "function-parameters", which are the boundary values of the harmonic functions $\omega_j(x, y)$ on both pieces of the boundary of $A_{a,b}$, but also a number of discrete parameters

$$2 \sum_{k=0}^{p-1} (p - k),$$

which are the coefficients of the polynomials P_k and Q_k, for $k = 0, \ldots, p - 1$. Let us note that this *Almansi formula cannot* be obtained directly from the representation of the harmonic functions in the annulus in (7.12), p. 81, in Chapter 7.

7.3.5 Radially symmetric polyharmonic functions

In order to give the reader a more precise feeling for the parameters which describe the space of polyharmonic functions of order p we will consider the case of radially symmetric functions.

Let us use the Almansi formula to find the form of the radially symmetric polyharmonic functions of order p in a simple way which differs from the argumentation of Exercise 8.13, p. 115.

- Let $h(x) = \tilde{h}(r)$ be a radially symmetric and polyharmonic function of order p in the ball $B(0; R)$. It is clear that in the Almansi formula (7.24), p. 96, the harmonic functions h_j will also be radially symmetric. The result of Exercise 7.1, (1), on p. 78, says that

$$h_j(x) = C_j$$

for some constants C_j. Thus we come to the representation

$$h(x) = \sum_{j=0}^{p-1} C_j r^{2j}$$

which is a direct link to the one-dimensional case formula (7.21), p. 92, but with p parameters fewer.

- Now let $h(x) = \tilde{h}(r)$ be radially symmetric and polyharmonic of order p in the annulus $A_{a,b}$. Then by the Almansi formula of Vekua (7.25), p. 97, we see that all terms have to be radially symmetric, hence,

$$h(x) = \sum_{j=0}^{p-1} C_j r^{2j} + P_0(r) \log r;$$

here

$$P_0(r) = a_0 + a_1 r^2 + \cdots + a_{p-1} r^{2p-2}.$$

The number of parameters is precisely $2p$, which corresponds well with the one-dimensional case formula (7.21), p. 92.

Let us finish this section by noting that the Almansi formula holds for the polyharmonic functions in the ball as well as in the annulus in $\mathbb{R}^n$; see Chapter 10, Theorem 10.51, p. 184, and Theorem 10.46, p. 179.

The analogy of the Almansi formula with the one-dimensional Taylor formula has been discussed in Section 1.5, p. 10, devoted to the *polyharmonic paradigm*.

7.3.6 Another proof of the representation of radially symmetric polyharmonic functions

There is an alternative approach to the solution of the equation

$$\Delta_r^p u(r) = L_{(0)}^p u(r) = 0,$$

which after the change of the variable by $v = \log r$ may be found easily. Indeed, according to formula (8.6), p. 111, applied for $k = 0$ and $n = 2$, we have

$$\Delta_r^p f(\log r) = e^{-2pv} \prod_{j=0}^{p-1} \left(\frac{d}{dv} - 2j \right)^2 f(v),$$

where $v = \log r$. Since every root of the equation

$$M_{0,p}(z) = \prod_{j=0}^{p-1} (z - 2j)^2 = 0$$

is double, it follows that the fundamental set of solutions to

$$\prod_{j=0}^{p-1}\left(\frac{d}{dv}-2j\right)^2 f(v) = 0$$

is given by the set of functions

$$e^{2jv}, \quad \text{for } j = 0, 1, \ldots, p-1;$$
$$ve^{2jv}, \quad \text{for } j = 0, 1, \ldots, p-1.$$

After making the inverse transform we find that the functions

$$r^{2j}, \quad \text{for } j = 0, 1, \ldots, p-1;$$
$$\log r \cdot r^{2j}, \quad \text{for } j = 0, 1, \ldots, p-1,$$

are a fundamental set of solutions for the operator $L_{(k)}^p$. It follows that every solution to

$$\Delta_r^p u(r) = L_{(k)}^p u(r) = 0$$

is a linear combination of these functions. We have obtained the same result relying on the Almansi representation.

Chapter 8

Polysplines on annuli in $\mathbb{R}^2$

We proceed to the definition of polysplines when the *break-lines* are concentric circles.

Recall the overall situation which was outlined in Figure 3.5, p. 31, where the concentric circles $\Gamma_1, \Gamma_2, \ldots, \Gamma_N$ are increasing, and the functions h_j are defined between the circles Γ_j and Γ_{j+1}.

It is clear that every reasonable concept of multivariate splines would need some smoothness conditions across the *interface* circles $\Gamma_j = S(0; r_j)$. As before, we assume that the radii r_j satisfy $r_1 < r_2 < \cdots < r_N$.

Up to the present point we have provided sufficient justification and we can give the following definition of the main notion of *polyspline* in this setting:

Definition 8.1 *Let the circles Γ_j be given for $j = 1, \ldots, N$, and the integer $p \geq 1$. Let the functions $h_j(x)$ be given in the annulus*

$$A_j = \{(r, \varphi) : r_j < r < r_{j+1} \text{ and } 0 \leq \varphi \leq 2\pi\}$$

for $j = 1, \ldots, N-1$, and the function $h_0(x)$ be given in the ball $B(0; r_1)$ and satisfy the following:

1. $h_j(x)$ *is polyharmonic of order p, i.e.*

$$\Delta^p h_j(x) = 0 \quad \text{for } x \text{ in } A_j$$

for $j = 1, \ldots, N-1$, and

$$\Delta^p h_0(x) = 0 \quad \text{for } x \text{ in } B(0; r_1).$$

2. $h_j(x)$ *belongs to the class $C^{2p-2}(\overline{A_j})$, where $\overline{A_j}$ means differentiability up to the boundary of A_j, and $h_0 \in C^{2p-2}(\overline{B(0; r_1)})$.*

3. *The following equalities hold*

$$\frac{\partial^k}{\partial r^k} h_j(x) = \frac{\partial^k}{\partial r^k} h_{j+1}(x) \quad \text{for } x \text{ in } \Gamma_{j+1} \text{ and } k = 0, 1, \ldots, 2p-2, \quad (8.1)$$

where the boundary values are limits of $(\partial^k/\partial r^k)h_j(y)$ at points y in the interior of A_j, and $y \longrightarrow x$.

We will say that the function $h(x)$, which is equal to $h_j(x)$ on the annulus A_j for $j = 1, \ldots, N-1$, and to $h_0(x)$ in the ball $B(0; r_1)$, is a **polyspline** of order p **supported in the ball** $B(0; r_N)$ and has **knot-surfaces** (more precisely, knot-circles) Γ_j, $j = 1, 2, \ldots, N-1$.

Ignoring the ball $B(0; r_1)$ we will say that the function $h(x)$ which is equal to $h_j(x)$ on the annulus A_j for $j = 1, \ldots, N-1$, is a **polyspline** of order p **supported in the annulus**

$$A = A_{r_1, r_N} := \{(r, \varphi) : r_1 < r < r_N \text{ and } 0 \leq \varphi \leq 2\pi\}$$

and has **knot-surfaces** (more precisely, knot-circles) Γ_j, $j = 2, 3, \ldots, N-1$.

For $j = 1, \ldots, N-1$, the smoothness up to the boundary of A_j means that the derivatives of the function $h_j(x)$ exist and are continuous on both spheres

$$\Gamma_j \cup \Gamma_{j+1} = \partial A_j$$

which compose the boundary of A_j, and likewise the derivatives of the function $h_0(x)$ exist and are continuous on the sphere Γ_1 which is the boundary of the ball $B(0; r_1)$.

Remark 8.2 1. Condition (3) of Definition 8.1 means that $h(x)$ belongs to $C^{2p-2}(\overline{B(0; r_N)})$. Indeed, we have $h(x) \in C^{2p-2}(\Gamma_j)$ for all $j = 1, 2, \ldots, N$, since $h(x) \in C^{2p-2}(\overline{A_j})$ and $h(x) \in C^{2p-2}(\overline{B(0; r_1)})$. This means that all derivatives up to order $2p-2$ in directions tangent to the sphere Γ_j are continuous. The continuity up to order $2p-2$ in the normal direction $\partial/\partial r$ is precisely condition (8.1).

2. Here we consider only classical smoothness polysplines, with $h_j \in C^{2p-2}(\overline{A_j})$. Later, in Chapter 20, p. 409, devoted to the existence of polysplines, we will consider the case when the polysplines belong to Sobolev space on the annuli A_j.

Now let us assume that we are given some data functions $f_j(x)$ defined on the circles Γ_j for $j = 1, \ldots, N$.

Definition 8.3 The polyspline $h(x)$ of order p (supported in the ball $B(0; r_N)$ or in the annulus A) will be called an **interpolation polyspline** if the following equalities hold on all circles:

$$h(x) = f_j(x) \qquad \text{for } x \text{ in } \Gamma_j \qquad (8.2)$$

for $j = 1, \ldots, N$.

If the polyspline $h(x)$ is supported in the ball $B(0; r_N)$ then it has to satisfy the following boundary conditions on the greatest circle Γ_N:

$$\frac{\partial^\ell}{\partial r^\ell} h(x) = \frac{\partial^\ell}{\partial r^\ell} h_{N-1}(x) = d_\ell(x) \qquad \text{for } x \text{ in } \Gamma_N, \qquad (8.3)$$

for all $\ell = 1, \ldots, p-1$.

If the polyspline $h(x)$ is supported in the annulus

$$A(r_1, r_N) = \{(r, \varphi) : r_1 < r < r_N \text{ and } 0 \leq \varphi \leq 2\pi\}$$

then it has to satisfy the following boundary conditions on the smallest and on the greatest circles Γ_1, Γ_N:

$$\frac{\partial^\ell}{\partial r^\ell} h(x) = \frac{\partial^\ell}{\partial r^\ell} h_{N-1}(x) = d_\ell(x) \qquad \text{for } x \text{ in } \Gamma_N, \tag{8.4}$$

$$\frac{\partial^\ell}{\partial r^\ell} h(x) = \frac{\partial^\ell}{\partial r^\ell} h_1(x) = b_\ell(x) \qquad \text{for } x \text{ in } \Gamma_1, \tag{8.5}$$

for all $\ell = 1, \ldots, p-1$.

Let us remark that the *boundary conditions* (8.3) on the largest sphere $\Gamma_N = \partial B(0; r_N)$ are optional in this definition (or in the case of polysplines supported in the annulus $A(r_1, r_N)$ – the boundary conditions (8.4)–(8.5) on Γ_1 and Γ_N), since they have to be imposed solely to obtain the unique solution of the interpolation problem. We might simply choose another type of boundary conditions as well.

Here we will appeal to the analogy with the polysplines on strips. The special case $p = 1$, of *harmonic polysplines on annuli*, is treated in completely the same way as that of harmonic polysplines on strips. The harmonic polysplines are *only* continuous functions, not even in C^1 across the interfaces Γ_j! We immediately come to the *biharmonic polysplines* as the simplest non-trivial case.

8.1 The biharmonic polysplines, $p = 2$

For $p = 2$ we have polyspline which it is natural to call *biharmonic*. According to Definition 8.3 the function $h(x) = h(r, \varphi)$ is a polyspline of order $p = 2$ supported in the ball $B(0; r_N)$ if and only if

$$\Delta^2 h(r, \varphi) = 0 \qquad \text{for } 0 \leq r < r_N \text{ but } r \neq r_j \text{ for } j = 1, 2, \ldots, N-1,$$

and the *global smoothness* condition

$$h(x) \in C^2(\overline{B(0; r_N)}).$$

The last encompasses the following three conditions:

$$h_j(r_{j+1}, \varphi) = h_{j+1}(r_{j+1}, \varphi) \qquad \text{for } 0 \leq \varphi \leq 2\pi,$$

$$\frac{\partial}{\partial r} h_j(r_{j+1}, \varphi) = \frac{\partial}{\partial r} h_{j+1}(r_{j+1}, \varphi) \qquad \text{for } 0 \leq \varphi \leq 2\pi,$$

$$\frac{\partial^2}{\partial r^2} h_j(r_{j+1}, \varphi) = \frac{\partial^2}{\partial r^2} h_{j+1}(r_{j+1}, \varphi) \qquad \text{for } 0 \leq \varphi \leq 2\pi.$$

These are obviously the analog to the cubic spline smoothness conditions (2.2), p. 20.

The boundary conditions (8.3) reduce to only one:

$$\frac{\partial}{\partial r} h(r_N, \varphi) = \frac{\partial}{\partial r} h_{N-1}(r_N, \varphi) = d_1(\varphi) \qquad \text{for } 0 \leq \varphi \leq 2\pi,$$

104 *Multivariate polysplines*

and it is obviously the analog to the one-dimensional boundary condition (2.4), p. 21, for the cubic splines.

The *interpolation* biharmonic polyspline satisfies

$$h(r_j, \varphi) = f_j(\varphi) \qquad \text{for all } 0 \leq \varphi \leq 2\pi \text{ and } j = 1, 2, \ldots, N,$$

which is the analog to the one-dimensional interpolation condition (2.3), p. 21.

8.2 Radially symmetric interpolation polysplines

Now we give examples of interpolation polysplines starting with the simplest setting.

Let us make some important remarks about the radial part of the polyharmonic operator. On spherically symmetric functions the Laplace operator reduces to its radial part, i.e. if $f(x) = F(|x|)$, then

$$\Delta f(x) = \Delta_r F(|x|),$$

where

$$\Delta_r = \frac{1}{r} \frac{\partial}{\partial r} r \frac{\partial}{\partial r}.$$

For arbitrary $p \geq 1$ we have likewise

$$\Delta^p f(x) = \Delta_r^p F(|x|).$$

It should be noted that the operator Δ_r^p is equal to $L_{(0)}^p$. Here we refer to our observation in Section 7.2.3, p. 90, that the operator $L_{(k)}^p$ generates an *Extended Complete Chebyshev (ECT-) system*. Here for the operator $\Delta_r^p = L_{(0)}^p$ we have the special case $k = 0$. It has a very suitable form, namely

$$\Delta_r^p f(r) = \underbrace{\Delta_r \cdot \Delta_r \cdots \Delta_r}_{p}$$

$$= \frac{1}{r} \frac{\partial}{\partial r} r \frac{\partial}{\partial r} \cdot \frac{1}{r} \frac{\partial}{\partial r} r \frac{\partial}{\partial r} \cdots \frac{1}{r} \frac{\partial}{\partial r} r \frac{\partial}{\partial r} \cdot \frac{1}{r} \frac{\partial}{\partial r} r \frac{\partial}{\partial r} f(r)$$

and generates an ECT-system for $r > 0$. This shows that the ordinary differential operator $L = \Delta_r^p$ belongs to the class of differential operators which generate an *ECT-system*. It has representing differential operators, see Chapter 11, formulas (11.1) and (11.2), p. 187, given by

$$D_1 = \frac{\partial}{\partial r}$$

and

$$D_{2\nu+1} = \frac{\partial}{\partial r} \frac{1}{r}, \qquad \text{for } \nu = 1, \ldots, p-1,$$

$$D_{2\nu} = \frac{\partial}{\partial r} r, \qquad \text{for } \nu = 1, \ldots, p-1.$$

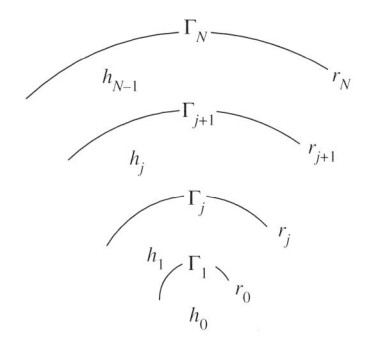

Figure 8.1. The configuration of data sets for a polyspline on annuli.

Thus
$$L = \Delta_r^p = \frac{1}{r} D_{2p} \cdots D_1.$$

1. We consider the data functions $f_j(x)$ for $j = 1, 2, \ldots, N$, which are constants on the circles $\Gamma_j = \{|x| = r_j\}$, i.e.

$$f_j(x) = C_j \quad \text{for } x \text{ in } \Gamma_j,$$

holds for all $j = 1, \ldots, N$. Let us find the *interpolation polyspline* $h(x)$, i.e. a function satisfying

$$h(x) = C_j \quad \text{for } x \text{ in } \Gamma_j$$

for all $j = 1, \ldots, N$.

2. Let us consider the *polyspline* $h(x)$ *supported* in the ball $B(0; r_N)$, i.e. including the piece h_0 inside the smallest ball $B(0; r_1)$. It is plausible and intuitively clear that the function $h(x)$ has to be *radially symmetric*, i.e. we may write

$$h_j(x) = \widetilde{h}_j(r) = \widetilde{h}_j(|x|)$$

for every piece $j = 0, 1, \ldots, N - 1$ (see Figure 8.1).

Proposition 8.4 *Let* $h(r, \varphi)$ *be a polyspline supported in the annulus* $A(r_1, r_N)$ *with all data being constant, i.e.* $f_j(x) = C_j$, *and* $d_\ell(\varphi) = D_\ell$, $b_\ell(\varphi) = B_\ell$ *for all* $0 \le \varphi \le 2\pi$, *and* $\ell = 1, 2, \ldots, p - 1$; *here* C_j, D_ℓ, B_ℓ *are constants. Then the polyspline* $h(r, \varphi)$ *is radially symmetric.*

2. *A similar result holds for a polyspline* $h(r, \varphi)$ *supported in the ball* $B(0; r_N)$.

We will construct a solution which is radially symmetric and the uniqueness will imply the proof of this proposition.

Let us consider the case of the *polysplines* supported in the *ball* $B(0; r_N)$. Here we refer to the results of Section 7.3.5, p. 97, where we proved that for every $j = 1, \ldots, N - 1$, in the annulus A_j the function $\widetilde{h}_j(r)$ permits the representation

$$\widetilde{h}_j(r) = \sum_{\nu=0}^{p-1} \omega_{j,\nu} r^{2\nu} + P_{j,0}(r) \log r,$$

for some constants $\omega_{j,\nu}$, where the polynomials

$$P_{j,0}(r) = a_{j,0} + a_{j,1}r^2 + \cdots + a_{j,p-1}r^{2p-2}$$

for other constants $a_{j,s}$. The function $\tilde{h}_0(r)$ in the ball $B(0; r_1)$ permits the representation

$$\tilde{h}_0(r) = \sum_{\nu=0}^{p-1} \omega_{0\nu} r^{2\nu},$$

for some constants $\omega_{0\nu}$.

We see that all functions $\tilde{h}_j(r)$ belong to the set $U_{0,p}$ which is given by

$$U_{0,p} = \lin\{1, r^2, \ldots, r^{2p-2}, \log r, r^2 \log r, \ldots, r^{2p-2} \log r\}.$$

Recall that the last is an *Extended Complete Chebyshev system* generated by the operator $L = \Delta_r^p$.

3. On the other hand the matching conditions (8.1), in Definition 8.1 of polysplines imply that

$$\frac{d^\ell}{dr^\ell} \tilde{h}_j(r_{j+1}) = \frac{d^\ell}{dr^\ell} \tilde{h}_{j+1}(r_{j+1}) \quad \text{for} \quad \ell = 0, \ldots, 2p-2,$$

for $j = 0, 1, \ldots, N-2$.

The interpolation conditions imply

$$\tilde{h}(r_{j+1}) = \tilde{h}_j(r_{j+1}) = C_j \quad \text{for} \quad j = 0, \ldots, N-1.$$

4. We recognize immediately that we are within the *Chebyshev splines* framework, where the operator $L = \Delta_r^p$. Indeed, we saw that the function $h_0(r)$ is a linear combination only of the functions

$$\{1, r^2, \ldots, r^{2p-2}\}$$

which are *half of the functions* forming a basis of the space $U_{0,p}$. It follows that we are precisely within the class of *left-sided natural Chebyshev splines* introduced in Chapter 11, Definition 11.35, p. 205, (or see Schumaker [50, Chapter 9.8, p. 396][1]).

Thus we have a natural Chebyshev spline in the interval $[0, r_N]$ to which we can apply the whole theory available in the above reference, in particular there are compactly supported Chebyshev splines (the so-called TB-splines) and thus the function $\tilde{h}(r)$ is a computable object.

In the case of a polyspline $h(r, \varphi)$ defined in the annulus $A(r_1, r_N)$, reasoning as the above shows that we have a standard Chebyshev spline in the interval $[r_1, r_N]$.

---

[1] In this reference "left-sided" or "right-sided" natural splines are not available but are an almost evident variation of the notion of natural spline.

8.2.1 Applying the change of variable $v = \log r$

Things may be simplified after we introduce the fundamental change of the variables. We put
$$v = \log r, \qquad r = e^v.$$
As we saw above, the space of functions $\mathcal{U}_{0,p}$ becomes
$$\widetilde{\mathcal{U}}_{0,p} := \{1, e^{2v}, \ldots, e^{(2p-2)v}, v, ve^{2v}, \ldots, ve^{(2p-2)v}\},$$
where the system of functions $\widetilde{\mathcal{U}}_{0,p}$ is the set of solutions for the ordinary differential operator
$$M_{0,p} = \prod_{v=0}^{p-1} \left(\frac{d}{dv} - 2v\right)^2.$$

Again we are within the framework of *left-sided natural Chebyshev splines* but with the operator $L = M_{0,p}$. The knots will be
$$v_j = \log r_j \quad \text{for} \quad j = 1, \ldots, N,$$
and the unknown spline will be
$$f(v) = h(e^v)$$
defined in the interval $(-\infty, \log r_N]$.

Finally, the boundary conditions in Definition 8.3, p. 102, of the interpolation polyspline imply the boundary condition on the right end $r_N = \log v_N$, namely
$$\frac{d^\ell}{dr^\ell} h_{N-1}(r_N) = D_\ell \qquad (\ell = 1, \ldots, p-1),$$
where D_ℓ are the constants. But we have
$$\frac{d}{dr} f(\log r) = \frac{1}{r} \frac{d}{dv} f(v),$$
i.e.
$$\frac{d}{dv} = r \frac{d}{dr}.$$
Hence, we obtain
$$f(\log r) = h(r),$$
$$f'(\log r) = r h'(r),$$
$$f''(\log r) = r \frac{d}{dr} r \frac{d}{dr} h(r) = r h'(r) + r^2 h''(r),$$
$$\ldots$$
$$f^{(s)}(\log r) = \left(r \frac{d}{dr}\right)^s h(r) = \sum_{j=0}^{s} \alpha_{sj} h^{(j)}(r),$$

for appropriate coefficients α_{sj}. This shows that all boundary conditions for $h(r)$, in particular at the point r_N, will be translated to boundary conditions for $f^{(s)}(\log r_N)$. Now by the theory of *natural Chebyshev splines* it follows that the function $f(\log r) = h(r)$ is uniquely determined.

Thus we determine the functions $h_j(e^v)$ for all $j = 0, \ldots, N-1$. For the original function $h(r)$ we have the interpolation conditions

$$h(e^{v_j}) = C_j \quad \text{for} \quad j = 1, \ldots, N.$$

A minor difference will arise if we consider the case of the interpolation spherical polyspline $h(x)$ supported in the annulus $A = \{(r, \varphi) : r_1 < r < r_N \text{ and } 0 \leq \varphi \leq 2\pi\}$. Then we do not have a piece $h_0(x)$ in the smallest ball $B(0; r_1)$ but we have to pay the price of having the boundary conditions (8.4) and (8.5), p. 103, on the smallest circle $\Gamma_1 = \partial(B(0; r_1))$. Hence, the above reduction holds completely in the present case but now we will obtain an L-spline on the interval $[r_1, r_N]$ with boundary conditions on both ends r_1 and r_N.

8.2.2 The radially symmetric biharmonic polysplines

In the case $p = 2$ we obtain the pieces of the biharmonic polysplines

$$h_j(r) = \omega_{j,0} + \omega_{j,1} r^2 + (a_{j,0} + a_{j,1} r^2) \log r$$

for $j = 1, 2, \ldots, N-1$, and

$$h_0(r) = \omega_{0,0} + \omega_{0,1} r^2.$$

The matching conditions become

$$h_j(r_{j+1}) = h_{j+1}(r_{j+1}),$$
$$h'_j(r_{j+1}) = h'_{j+1}(r_{j+1}),$$
$$h''_j(r_{j+1}) = h''_{j+1}(r_{j+1}),$$

for $j = 0, 1, \ldots, N-2$.

The interpolation conditions are

$$h_j(r_j) = C_j \quad \text{for } j = 1, 2, \ldots, N-1,$$
$$h_{N-1}(r_N) = C_N.$$

The boundary condition becomes

$$\frac{\partial}{\partial r} h_{N-1}(r_N) = D_1.$$

Thus we see that the above completely resembles the one-dimensional cubic spline setting formulated in (2.1)–(2.4), pp. 19–21.

Finally, the above case is a basic example. The main features of the polysplines an annuli are reflected in the above consideration.

8.3 Computing the polysplines for general (nonconstant) data

It is interesting that the consideration of the nonsymmetric data requires a study of *Chebyshev splines* for the ordinary differential operators $L_{(k)}^p$ for arbitrary $k \geq 0$, and this study is the same as for the case $k = 0$. We proceed in a way very similar to the one we followed for the polysplines on strips in Section 5.3, p. 61.

For simplicity we consider biharmonic ($p = 2$) polysplines $h(x)$ supported in the ball $B(0; r_N)$. Assume that the data functions have expansions in trigonometric Fourier series

$$f_j(\varphi) = \frac{f_{j,0}}{2} + \sum_{k=1}^{\infty} [f_{j,k} \cos k\varphi + \widetilde{f}_{j,k} \sin k\varphi].$$

As we saw above, on every interval $[r_j, r_{j+1}]$ the polyspline $h(x)$ which we are looking for has the Fourier expansion

$$h_j(r, \varphi) = \frac{U_{j,0}(r)}{2} + \sum_{k=1}^{\infty} \left[U_{j,k}(r) \cos k\varphi + V_{j,k}(r) \sin k\varphi \right].$$

The functions U_k and V_k are assembled from the pieces $U_{j,k}$ and $V_{j,k}$ on every interval $[r_j, r_{j+1}]$.

We now provide the construction of the polysplines on annuli which is useful for the practical computation.

Exercise 8.5 *1. Prove that the interpolation polyspline $h(x)$ supported in the ball $B(0; r_N)$ with data $f_j(\varphi)$ has an expansion in a trigonometric Fourier series given by*

$$h(r, \varphi) = \frac{U_0(r)}{2} + \sum_{k=1}^{\infty} [U_k(r) \cos k\varphi + V_k(r) \sin k\varphi], \quad \text{for } r \neq r_j, \ j = 1, 2, \ldots, N$$

where the functions $U_k(r)$, $V_k(r)$ are left-sided natural L-splines for the operator $L = L_{(k)}^2$ in the interval $(0, r_N]$ (as such they are C^2 functions) with interpolation data

$$U_k(r_j) = f_{j,k} \quad \text{for } j = 0, 1, \ldots, N \text{ and } k = 0, 1, 2, \ldots,$$
$$V_k(r_j) = \widetilde{f}_{j,k} \quad \text{for } j = 1, 2, \ldots, N \text{ and } k = 1, 2, 3, \ldots.$$

The boundary data are

$$\frac{\partial}{\partial r} U_k(r_N) = d_{1,k} \quad \text{for } k = 0, 1, 2, \ldots,$$
$$\frac{\partial}{\partial r} V_k(r_N) = \widetilde{d}_{1,k} \quad \text{for } k = 1, 2, 3, \ldots,$$

which are the Fourier coefficients in the expansions of the boundary data:

$$d_1(\varphi) = \frac{d_{1,0}}{2} + \sum_{k=1}^{\infty} \left[d_{1,k} \cos k\varphi + \widetilde{d}_{1,k} \sin k\varphi \right].$$

2. If the polyspline $h(x)$ is supported *in the* annulus $A(r_1, r_N)$ *then the functions* $U_k(r)$, $V_k(r)$ *are Chebyshev splines for the operator* $L_{(k)}^p$, *and the boundary conditions satisfied by the functions* $U_k(r)$, $V_k(r)$ *are, in addition to the above, also those on the smallest circle* $S(0; r_1)$, i.e.

$$\frac{\partial}{\partial r} U_k(r_N) = b_{1,k} \quad \text{for } k = 0, 1, 2, \ldots,$$

$$\frac{\partial}{\partial r} V_k(r_N) = \widetilde{b_{1,k}} \quad \text{for } k = 1, 2, 3, \ldots,$$

where

$$b_1(\varphi) = \frac{b_{1,0}}{2} + \sum_{k=1}^{\infty} \left[b_{1,k} \cos k\varphi + \widetilde{b_{1,k}} \sin k\varphi \right].$$

8.4 The uniqueness of interpolation polysplines on annuli

In Proposition 5.8, p. 64, we proved the uniqueness of the interpolation polysplines on strips. Following the same idea we may prove the uniqueness of polysplines supported in a ball or supported in an annulus.

Proposition 8.6 *1. In Definition 8.3, p. 102, of the interpolation polyspline supported in the ball let the data functions involved* f_j, d_ℓ *be identically zero, i.e. we have*

$$f_j = 0 \quad \text{on } \Gamma_j \quad \text{for } j = 1, 2, \ldots, N,$$
$$d_\ell = 0 \quad \text{on } \Gamma_N \quad \text{for } \ell = 1, 2, \ldots, p - 1.$$

Then the interpolation polyspline h of Definition 8.1 is identically zero.

2. If we consider the interpolation polyspline h supported in the annulus and the data functions of Definition 8.3 f_j, d_ℓ *and* $\widetilde{d_\ell}$ *are zero, then we have the same conclusion:* $h \equiv 0$.

The proof is very similar to that of Proposition 5.8, p. 64, about polysplines on strips.

Exercise 8.7 *Prove Proposition 8.6. Hint: If we have a polyspline supported in the* annulus *then its computation is reduced to the computation of Chebyshev splines, and its uniqueness will follow from their uniqueness. If we have polysplines supported in* the ball, *then the uniqueness is reduced to uniqueness of left-sided natural Chebyshev splines.*

Exercise 8.8 *Prove a Holladay-type theorem: Assume that a biharmonic polyspline h exists which interpolates the given data. Then it is a solution to the following minimization problem*

$$\inf \int_D (\Delta u)^2 \, dx,$$

where the infimum is taken over all functions u which satisfy the same interpolation conditions as h. Hint: *Have a look at the most general Holladay-type result in Part IV, Theorem 20.14, p. 425.*

8.5 The change $v = \log r$ and the operators $M_{k,p}$

Here we will concentrate on an essential simplification of the representation of a polyharmonic function in the annulus. The *main result* below is that after the change $v = \log r$ the coefficients $u_k(r)$, $v_k(r)$ of the trigonometric Fourier series (7.16), p. 88, become solutions to an equation with constant coefficients. (The special case $k = 0$ has been considered in subsection 8.2.1.) This also simplifies the computation of the interpolation polysplines. Although this topic will be studied in generality and in detail in Chapter 10, due to its importance for Parts II and III we will spend some time explaining the two-dimensional case.

In order to make things easier to comprehend we will again change the variable as follows:
$$v = \log r;$$
obviously, the inverse transform is
$$r = e^v.$$
The point is that the operator $L_{(k)}^p$ is transformed (up to a multiple) into another operator with constant coefficients. More precisely, for an arbitrary function $f \in C^{2p}$ we have the following equality (see Chapter 10, Theorem 10.34, p. 168):
$$L_{(k)}^p f(\log r) = e^{-2pv} M_{k,p}\left(\frac{d}{dv}\right) f(v); \tag{8.6}$$
here the polynomial $M_{k,p}(z)$ is defined as follows:
$$M_{k,p}(z) = \prod_{j=1}^{2p}(z - \lambda_j) \tag{8.7}$$
and the constants λ_j are given by
$$\begin{aligned}&\lambda_1 = -k, \quad \lambda_2 = -k+2, \quad \ldots, \quad \lambda_p = -k+2p-2,\\ &\lambda_{p+1} = k, \quad \lambda_{p+2} = k+2, \quad \ldots, \quad \lambda_{2p} = k+2p-2,\end{aligned} \tag{8.8}$$
(see Figure 8.2).

Clearly we have to write
$$\lambda_j = \lambda_{j,k}$$
but we will omit the index k in order to avoid overcomplex notation.

Obviously, if for some function $u(r)$ the equation
$$L_{(k)}^p u(r) = 0$$

112 Multivariate polysplines

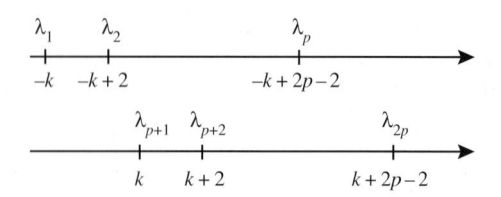

Figure 8.2. There may be overlapping between the two groups of the numbers λ_j.

is satisfied, then according to the above equality (8.6) after making the inverse change of variable, $r = e^v$, we see that for the function

$$f(v) = u(e^v)$$

the equation

$$e^{-2pv} M_{k,p}\left(\frac{d}{dv}\right) f(v) = 0$$

is satisfied. Hence, since $e^{-2pv} \neq 0$, it follows

$$M_{k,p}\left(\frac{d}{dv}\right) f(v) = 0. \tag{8.9}$$

The conclusion is that the solutions to the last equation provide the solutions to equation (7.19), p. 89, by the inverse change.

Let us denote by $\widetilde{\mathcal{U}}_{k,p}$ the set of C^∞ solutions to equation (8.9), i.e.

$$\widetilde{\mathcal{U}}_{k,p} := \left\{ f \in C^\infty(\mathbb{R}) \text{ such that } M_{k,p}\left(\frac{d}{dv}\right) f(v) = 0 \text{ for all } v \text{ in } \mathbb{R} \right\}.$$

We provide a basis for the set $\widetilde{\mathcal{U}}_{k,p}$ in the simplest cases $p = 1, 2, 3$, in the following example.

Example 8.9 *1. For $p = 1$ we have the case of harmonic functions and we have provided the solutions of the equation $M_{k,1}(d/dv)w(v) = 0$ in Section 7.1.4, p. 86.*

2. For the biharmonic case, $p = 2$, by (8.8) we have the following table with a basis of $\widetilde{\mathcal{U}}_{k,p}$:

$k =$	polynomial $M_{k,2}(z) =$	basis of $\widetilde{\mathcal{U}}_{k,p}$
0	$z^2(z-2)^2$	$1, v, e^{2v}, ve^{2v}$
1	$(z+1)(z-1)^2(z-3)$	$e^{-v}, e^v, ve^v, e^{3v}$
2	$(z+2)z(z-2)(z-4)$	$e^{-2v}, 1, e^{2v}, e^{4v}$
3	$(z+3)(z+1)(z-3)(z-5)$	$e^{-3v}, e^{-v}, e^{3v}, e^{5v}$

Every solution of $M_{k,2}(d/dv)w(v) = 0$ is represented in the form

$$w(v) = C_1 w_1(v) + C_2 w_2(v) + C_3 w_3(v) + C_4 w_4(v),$$

where the functions w_j denote the four functions belonging to the set $\widetilde{\mathcal{U}}_{k,p}$.

3. For the case $p = 3$, by (8.8) we have the following table with a basis of $\widetilde{\mathcal{U}}_{k,p}$:

$k =$	polynomial $M_{k,3}(z) =$	basis of $\widetilde{\mathcal{U}}_{k,p}$
0	$z^2(z-2)^2(z-4)^2$	$1, v, e^{2v}, ve^{2v}, e^{4v}, ve^{4v}$
1	$(z+1)(z-1)^2(z-3)^2(z-5)$	$e^{-v}, e^v, ve^v, e^{3v}, ve^{3v}, e^{5v}$
2	$(z+2)z(z-2)^2(z-4)(z-6)$	$e^{-2v}, 1, e^{2v}, ve^{2v}, e^{4v}, e^{6v}$
3	$(z+3)(z+1)(z-3)(z-5)(z-1)(z-7)$	$e^{-3v}, e^{-v}, e^{3v}, e^{5v}, e^v, e^{7v}$

Following from the above examples we can observe that the polynomial $M_{k,p}(z)$ has multiple roots for several ks.

Now for an arbitrary function $h(r, \varphi)$ polyharmonic of order p in the annulus $A_{a,b}$ let us put

$$H(v, \varphi) = h(e^v, \varphi) \quad \text{for } v \text{ in } \mathbb{R} \text{ and } 0 \leq \varphi \leq 2\pi.$$

Then the expansion in trigonometric Fourier series (7.5), p. 80, of the function $h(r, \varphi)$ will be transformed into that of the function $H(v, \varphi)$, i.e.

$$H(v, \varphi) = h(e^v, \varphi) = \frac{u_0(e^v)}{2} + \sum_{k=1}^{\infty} [u_k(e^v) \cos k\varphi + v_k(e^v) \sin k\varphi]$$

$$= \frac{U_0(v)}{2} + \sum_{k=1}^{\infty} [U_k(v) \cos k\varphi + V_k(v) \sin k\varphi],$$

where we have evidently put

$$U_k(v) = u_k(e^v), \qquad V_k(v) = v_k(e^v).$$

Exercise 8.10 Prove that the functions $U_k(v)$, for $k = 0, 1, 2, \ldots$, and the functions $V_k(v)$, for $k = 1, 2, 3, \ldots$, satisfy the equation

$$M_{k,p}\left(\frac{d}{dv}\right) f(v) = 0.$$

Hint: *Use the fact proved above that the functions $u_k(r)$, $v_k(r)$ satisfy the equation*

$$L_{(k)}^p \left(\frac{d}{dr}\right) w(r) = 0.$$

Conclusion 8.11 From Exercise 8.10 we see that all functions $U_k(v)$, $V_k(v)$ belong to the set $\widetilde{\mathcal{U}}_{k,p}$. They are linear combinations of the basis of $\widetilde{\mathcal{U}}_{k,p}$ which we have described in the cases $p = 2, 3$ in Example 8.9, p. 112.

8.6 The fundamental set of solutions for the operator $M_{k,p}(d/dv)$

The main point here is that the operator $M_{k,p}(d/dv)$ has constant coefficients and its set of fundamental solutions is simpler to describe. Indeed, for every λ_j there are only two possibilities.

- The first is that λ_j be a simple root of the equation

$$M_{k,p}(z) = 0.$$

This is equivalent to the fact that $(z - \lambda_j)$ divides $M_{k,p}(z)$ but $(z - \lambda_j)^2$ does not divide $M_{k,p}(z)$. Then the function $e^{\lambda_j v}$ belongs to the set of fundamental solutions of equation (8.9).

- The second possibility is that the root λ_j is a double root, i.e.

$$M_{k,p}(\lambda_j) = M'_{k,p}(\lambda_j) = 0.$$

This is equivalent to the fact that $(z - \lambda_j)^2$ divides $M_{k,p}(z)$. Then the function $e^{\lambda_j v}$ belongs to the set of fundamental solutions of the equation (8.9) as well as the function $v e^{\lambda_j v}$.

Evidently, in order to obtain, for some $k \geq 0$, the multiple root λ_j of the polynomial $M_{k,p}(z)$, the two sets of λ_j in (8.8) have to overlap, which happens precisely if we have the inequality

$$-k + 2p - 2 \geq k,$$

i.e. if

$$k \leq p - 1.$$

By definition every fundamental set of solutions of equation

$$M_{k,p}(d/dv)w(v) = 0$$

has exactly $2p$ linearly independent functions. We will be interested in those having the form $e^{\lambda_j v}$ and $v e^{\lambda_j v}$. The above elementary arguments prove the following proposition.

Proposition 8.12 *The function $e^{\lambda_j v}$ belongs to $\widetilde{\mathcal{U}}_{k,p}$ if $M_{k,p}(\lambda_j) = 0$. The function $v e^{\lambda_j v}$ belongs to $\widetilde{\mathcal{U}}_{k,p}$ if $M_{k,p}(\lambda_j) = M'_{k,p}(\lambda_j) = 0$. So we have the following rigorous description of a basis of the set $\widetilde{\mathcal{U}}_{k,p}$ of solutions of the equation $M_{k,p}(d/dv)w(v) = 0$, namely*

$$\widetilde{\mathcal{U}}_{k,p} = \text{lin}\left\{ \bigcup_{j=1}^{2p} \{e^{\lambda_j v} : \text{for } M_{k,p}(\lambda_j) = 0, \ M'_{k,p}(\lambda_j) \neq 0\} \cup \bigcup_{j=1}^{2p} \{e^{\lambda_j v}, v e^{\lambda_j v} : \text{for } M_{k,p}(\lambda_j) = 0, \ M'_{k,p}(\lambda_j) = 0\} \right\}.$$

(see Figure 8.2, p. 112.)

Now let us return to the space of the solutions of the operator $L_{(k)}^p$. By the discussion in the previous section based on formula (8.6), p. 111, we saw that by the inverse change the function $e^{\lambda_j v}$ is mapped into r^{λ_j} and the function $v e^{\lambda_j v}$ is mapped into $\log r \cdot r^{\lambda_j}$. Now Proposition 8.12 provides a full description of the space of solutions of equation $L_{(k)}^p u(r) = 0$, as a linear hull of such terms.

In particular, for $k = 0$ we have $L_{(0)}^p u(r) = \Delta_r^p u(r)$, where Δ_r is the radial part of the Laplace operator defined in (7.1), p. 78. We can easily find the basis for the set $U_{k,p}$.

Exercise 8.13 *1. Prove that all radially symmetric polyharmonic functions in the annulus in $\mathbb{R}^2$ are linear combinations of the system of functions*

$$U_{0,p} := \operatorname{lin}\{1, r^2, \ldots, r^{2p-2}, \log r, r^2 \log r, \ldots, r^{2p-2} \log r\}.$$

Hint: *By the definition of the λ_js in (8.8), p. 111, see that $M_{0,p}(z) = z^2(z-2)^2 \cdots (z - 2p + 2)^2$, find a basis of the set $\widetilde{U}_{0,p}$ and apply the variable change $r = e^v$ to obtain the basis of $U_{0,p}$.*

2. Prove that every radially symmetric polyharmonic function $h(x)$ of order p in the ball $B(0; R)$ admits the representation

$$h(x) = \sum_{j=0}^{p-1} C_j r^{2j}$$

for some constants C_j. Hint: *The only continuous solutions of the ordinary differential equation $\Delta_r^p f(r) = 0$ are r^{2j}, for $j = 0, 2, \ldots, p-1$. Those having $\log$ terms are not continuous.*

3. Prove that every radially symmetric polyharmonic function $h(x)$ of order p in the annulus $A_{a,b}$ admits the representation

$$h(x) = \sum_{j=0}^{p-1} (C_j + D_j \log r) r^{2j}$$

for some constants C_j, D_j.

Chapter 9

Polysplines on strips and annuli in $\mathbb{R}^n$

In the previous chapters we have demonstrated that polysplines are a natural type of multivariate spline. Relying upon that background we will define polysplines on strips and annuli in $\mathbb{R}^n$ which differ only slightly from the two-dimensional case. We will also rely on the existence results in Part IV. Respectively, it is assumed that the reader has some basic familiarity with the notions of Sobolev and Hölder spaces (the definition of which are provided in an appendix, Chapter 23). Let us note that one may prove the existence directly by establishing appropriate estimates for the "coordinate" Chebyshev splines which appear by the reduction of the polysplines, but this is considerably more technical.

We will be again concerned with the two special classes of polysplines but for arbitrary dimension $n \geq 2$. We will provide a way to compute them. In the case of *polysplines on strips* we will show that by applying the Fourier transform their computation is reduced to one-dimensional *Chebyshev splines*. We assume that the reader is familiar with the properties of the Fourier transform which we provide in Chapter 12, p. 209. On the other hand, for *polysplines on annuli* we use essentially the results proved in the compendium on spherical harmonics and polyharmonic functions in Chapter 10, p. 129, in particular the representation of polyharmonic functions in the annulus using spherical harmonics. We show that the computation of the polysplines on annuli is reduced to the computation of infinitely many one-dimensional *Chebyshev splines*. These results are basically the same as those of the previous chapters. The difference is that here we will be very precise in the formulation of the results and their proofs.

In order to make a smooth transition from $\mathbb{R}^2$ to $\mathbb{R}^n$, we wanted to avoid interrupting the exposition with information on spherical harmonics and polyharmonic functions, and so we have placed it in Chapter 10. We would recommend that readers acquaint themselves with the main results there while studying the polysplines on annuli.

In this chapter the point of importance is that the **knot-surfaces** (or *break-surfaces*) are of finite number and are generally speaking "non-equidistant"; the meaning of the last will become completely clear when we begin to study the so-called *cardinal polysplines* in Part II.

9.1 Polysplines on strips in $\mathbb{R}^n$

We assume that N hyperplanes $\Gamma_1, \Gamma_2, \ldots, \Gamma_N$ in $\mathbb{R}^n$ are given, namely we assume that the numbers t_j satisfying

$$t_1 < t_2 < \cdots < t_N$$

are given, and the parallel hyperplanes Γ_j in $\mathbb{R}^n$ are defined for $j = 1, \ldots, N$, by putting

$$\Gamma_j := \{(t, y) \in \mathbb{R}^n : t = t_j \text{ and } y \in \mathbb{R}^{n-1}\}.$$

This is visualized as in the two-dimensional case by Figure 3.4, on p. 32.

The pieces of the polyspline – the functions h_j – are defined between the hyperplanes Γ_j and Γ_{j+1} for $j = 1, 2, \ldots, N - 1$.

Similar to the two-dimensional polysplines of Part I, introduced by Definition 5.1, p. 57, we have the following definition of *polysplines* on strips.

Definition 9.1 *Let the parallel hyperplanes Γ_j be given as above for $j = 1, \ldots, N$, and let the integer $p \geq 1$. Let the functions h_j be given in the strip*

$$\text{Strip}_j = \{x = (t, y) \in \mathbb{R}^n : t_j < t < t_{j+1}\}$$

for $j = 1, \ldots, N - 1$, and satisfy the following conditions:

1. *The function h_j is polyharmonic of order p, i.e.*

$$\Delta^p h_j = 0 \quad \text{in Strip}_j$$

for $j = 1, \ldots, N - 1$.

2. *The functions h_j belong to the class $C^{2p-2}(\overline{\text{Strip}_j})$. (The bar over Strip_j means smoothness including the boundary of Strip_j).*

3. *The following equalities hold*

$$\frac{\partial^k}{\partial t^k} h_j(t_{j+1}, y) = \frac{\partial^k}{\partial t^k} h_{j+1}(t_{j+1}, y) \quad \text{for all } y \text{ in } \mathbb{R}^{n-1}, \tag{9.1}$$

and for all $k = 0, 1, \ldots, 2p - 2$, and $j = 1, \ldots, N - 2$. The derivatives exist on the boundary by the previous point.

We will say that the function h, which is equal to h_j on the set Strip_j for $j = 1, \ldots, N - 1$, is a **class C^{2p-2} polyspline on strips** of order p in the set

$$\text{Strip} = \text{Strip}(t_1, t_N) = \{(t, y) \in \mathbb{R}^n : t_1 < t < t_N \text{ and } y \in \mathbb{R}^{n-1}\}$$

and has **knot-hyperplanes** Γ_j.

We will say that h is a **Sobolev class polyspline** if every function h_j belongs to the Sobolev space $H^{2p}(\text{Strip}_j)$ for $j = 1, \ldots, N - 1$, and the matching conditions (9.1) are understood as an equality between traces of functions in Sobolev space, see Theorem 23.4, p. 468.

Now let us assume that we are given some data functions $f_j(y)$ defined on the space $\mathbb{R}^{n-1}$ for $j = 1, \ldots, N$, and also the functions $c_\ell(y), d_\ell(y), \ell = 0, 1, \ldots, p-1$, defined on $\mathbb{R}^{n-1}$.

Definition 9.2 *The polyspline h of order p defined in the* Strip(t_1, t_N) *will be called an* **interpolation polyspline** *if the following conditions hold:*

1. *The* **interpolation** *equalities hold on all hyperplanes* Γ_j:

$$h(t_j, y) = f_j(y) \quad \text{for all } y \text{ in } \mathbb{R}^{n-1}, \tag{9.2}$$

for $j = 1, \ldots, N$.

2. *The following* **boundary conditions** *hold on the hyperplanes* Γ_1 *and* Γ_N, *for all y in* $\mathbb{R}$ *and* $\ell = 1, \ldots, p - 1$:

$$\frac{\partial^\ell}{\partial t^\ell} h(t_1, y) = \frac{\partial^\ell}{\partial t^\ell} h_1(t_1, y) = c_\ell(y), \tag{9.3}$$

$$\frac{\partial^\ell}{\partial t^\ell} h(t_N, y) = \frac{\partial^\ell}{\partial t^\ell} h_{N-1}(t_N, y) = d_\ell(y). \tag{9.4}$$

If the functions $f_j(y), c_\ell(y), d_\ell(y)$ *of the variable* $y = (y_1, y_2, \ldots, y_{n-1})$ *are* 2π-*periodic in every variable* y_i *for* $i = 1, 2, \ldots, n - 1$, *we will call the corresponding polyspline* **periodic with respect to** y.

In the case of a polyspline of Sobolev smoothness the equalities (9.2)–(9.4) are understood as equalities between traces.

It should be noted that as in the two-dimensional case the *boundary conditions* (9.3) and (9.4) on the straight lines Γ_1, Γ_N are optional.

9.1.1 Polysplines on strips with data periodic in y

As in the two-dimensional case we may completely study the case of the data which are periodic in y. The treatment does not differ essentially from the two-dimensional case. We generalize the uniqueness of interpolation polysplines which we proved in Proposition 5.8, p. 64. So far, to prove the existence we refer to some rather general existence results which will be proved in Part IV.

Let us introduce for every multi-index $\xi = (\xi_1, \xi_2, \ldots, \xi_{n-1}) \in \mathbb{R}^{n-1}$ the Fourier transform with respect to the variables y, by equality

$$\widehat{h}(t, \xi) := \frac{1}{(2\pi)^{n-1}} \int_0^{2\pi} \cdots \int_0^{2\pi} e^{-i(y_1 \xi_1 + \cdots + y_{n-1}\xi_{n-1})} h(t, y) dy_1 \cdots dy_{n-1}. \tag{9.5}$$

As a direct application of the existence results of Part IV we obtain Theorem 9.3.

Theorem 9.3 *1. Let the functions* $f_j(y), c_\ell(y), d_\ell(y)$ *be* 2π*-periodic in every variable* y_i *for* $i = 1, 2, \ldots, n - 1$. *Let* $r \geq 0$ *be a real number and* $f_j \in H^{2p+r-\frac{1}{2}}(\mathbb{T}^{n-1})$, *for* $j = 1, 2, \ldots, N$, *and* $c_\ell \in H^{2p+r-\ell-\frac{1}{2}}(\mathbb{T}^{n-1})$, $d_\ell \in H^{2p+r-\ell-\frac{1}{2}}(\mathbb{T}^{n-1})$ *for* $\ell = 1, 2, \ldots, p - 1$, *where* $\mathbb{T}^{n-1}$ *denotes the* $(n-1)$-*dimensional torus (see subsection 23.1.2, p. 463). Then there exists a unique interpolation Polyspline* $h = h(t, y)$ *which is* 2π-*periodic in y and such that* $h \in H^{2p+r}(\widetilde{\text{Strip}_j})$ *for* $j = 1, 2, \ldots, N - 1$, *i.e.* h

120 *Multivariate polysplines*

is a Sobolev class periodic polyspline in $\widetilde{\text{Strip}}(t_1, t_N)$. (We use the notation $\widetilde{\text{Strip}}(a, b)$ for the set $(a, b) \times \mathbb{T}^{n-1}$.)

If for an $r > 0$ which is not an integer the data belong to Hölder space, namely $f_j \in C^{2p+r}(\Gamma_j)$, $c_\ell \in C^{2p+r-\ell}(\Gamma_1)$ and $d_\ell \in C^{2p+r-\ell}(\Gamma_N)$, then there exists a unique periodic class C^{2p-2} polyspline h such that $h \in C^{2p+r}\left(\widetilde{\text{Strip}}_j\right)$ for $j = 1, 2, \ldots, N$.

2. For every fixed multi-index $\kappa = (k_1, k_2, \ldots, k_{n-1}) \in \mathbb{Z}^{n-1}$ the Fourier coefficient $\widehat{h}(t, \kappa)$ is a Chebyshev spline *for the ordinary differential operator*

$$L_{|\kappa|}\left(\frac{d}{dt}\right) = \left(\frac{d^2}{dt^2} - k_1^2 - \cdots - k_{n-1}^2\right)^p, \qquad (9.6)$$

i.e. $L_{|\kappa|}(d/dt)\widehat{h}(t, \kappa) = 0$ for $t \neq t_j$ with $j = 1, 2, \ldots, N$, and $\widehat{h}(t, \kappa) \in C^{2p-2}(t_1, t_N)$. In addition $\widehat{h}(t, \kappa)$ satisfies for every $\kappa \in \mathbb{Z}^{n-1}$ the interpolation *and* boundary *conditions*

$$\widehat{h}(t_j, \kappa) = \widehat{f}_j(\kappa) \qquad \text{for } j = 1, 2, \ldots, N,$$

$$\frac{\partial^\ell}{\partial t^\ell}\widehat{h}(t_1, \kappa) = \widehat{c}_\ell(\kappa) \qquad \text{for } \ell = 1, 2, \ldots, p-1,$$

$$\frac{\partial^\ell}{\partial t^\ell}\widehat{h}(t_N, \kappa) = \widehat{d}_\ell(\kappa) \qquad \text{for } \ell = 1, 2, \ldots, p-1.$$

Proof We work in the Sobolev space setting; the Hölder space setting is almost identical.

(1) First, let us prove the uniqueness. It will be proved in a way very similar to the proof of Proposition 5.8, p. 64, in the two-dimensional case, but we will take care of all details.

Since the polyharmonic functions are real-analytic functions inside the domain of their definition we have $h \in C^\infty(\text{Strip}_j)$ for every $j = 1, 2, \ldots, N-1$. Hence, by the Sobolev embedding theorem in the torus $\mathbb{T}^{n-1}$, see Proposition 23.2, p. 464, we know that the Fourier series representation

$$h(t, y) = \sum_{k_1=-\infty}^{\infty} \cdots \sum_{k_{n-1}=-\infty}^{\infty} \widehat{h}(t, \kappa) e^{i(y_1 k_1 + \cdots + y_{n-1} k_{n-1})} \qquad (9.7)$$

may be infinitely often differentiated with respect to the variables y for all $t \neq t_j$ with $j = 1, 2, \ldots, N$. For the differentiability in t, we obtain $(\partial/\partial t)\widehat{h}(t, \kappa)$ directly from the definition of the Fourier–Laplace coefficients in formula (9.5), p. 119. On the other hand the function $(\partial/\partial t)h(t, y)$ has a unique Fourier series representation, hence $(\partial/\partial t)\widehat{h}(t, \kappa) = \widehat{(\partial/\partial t)h}(t, \kappa)$.

Thus for $t \neq t_j$ with $j = 1, 2, \ldots, N$, we obtain

$$0 = \Delta^p h(t, y) = \sum_{\kappa \in \mathbb{Z}^{n-1}} \left(\frac{\partial^2}{\partial t^2} - k_1^2 - \cdots - k_{n-1}^2\right)^p \widehat{h}(t, \kappa) e^{iy \cdot \kappa}.$$

Hence, for $t \neq t_j$ with $j = 1, 2, \ldots, N$, we have the equation

$$\left(\frac{\partial^2}{\partial t^2} - k_1^2 - \cdots - k_{n-1}^2\right)^p \widehat{h}(t, \kappa) = 0 \quad \text{for all } \kappa \in \mathbb{Z}^{n-1},$$

which proves that, as a function of t, $\widehat{h}(t, \kappa)$ is a piecewise solution of the equation $L_{|\kappa|}g = 0$. Since the function $h(t, y)$ takes the data in the sense of traces in Sobolev spaces, we refer to Proposition 23.6, p. 469, which implies that its discrete Fourier transform $\widehat{h}(t, \kappa)$ takes on precisely the Fourier transforms of the data – $\widehat{f}_j(\kappa)$, $\widehat{c}_\ell(\kappa)$ and $\widehat{d}_\ell(\kappa)$.

Now the uniqueness of the polyspline $h(t, y)$ follows through the uniqueness of the interpolation Chebyshev splines $\widehat{h}(t, \kappa)$ for every $\kappa \in \mathbb{Z}^{n-1}$, provided in Theorem 11.28, p. 200.

(2) The existence of the polyspline $h(t, y)$ will now follow by Theorem 22.6, p. 447, see also the remarks in Section 22.2, p. 448. It is a consequence of a more general framework on elliptic boundary value problems. ∎

The above theorem provides us with an algorithm for the computation of the interpolation polyspline. We first compute for every $\kappa \in \mathbb{Z}^{n-1}$ the interpolation Chebyshev splines $\widehat{h}(t, \kappa)$ which exist by Theorem 11.28, p. 200. After that we find the function h through the series

$$h(t, y) = \sum_{\kappa \in \mathbb{Z}^{n-1}} \widehat{h}(t, \kappa) e^{i(y_1 k_1 + \cdots + y_{n-1} k_{n-1})}.$$

We also refer to Section 23.1.2, p. 463, where the coefficients of the Fourier series expansion $\widehat{h}(t, \kappa)$ provide an exact measure of the Sobolev space H^s to which a function h on the torus belongs.

9.1.2 Polysplines on strips with compact data

By reasoning in the above way one may treat the case when the data are compactly supported in y or have some fast decay in y. We will confine ourselves to the case of compactly supported data.

Let us introduce for every multi-index $\xi = (\xi_1, \xi_2, \ldots, \xi_{n-1}) \in \mathbb{R}^{n-1}$ the Fourier transform $\widehat{h}(t, \xi)$ with respect to the variables y, by the same equality (9.5), p. 119, but the integral is taken over $\mathbb{R}^{n-1}$.".

As a direct application of the existence results of Part IV we obtain the following.

Theorem 9.4 *Let the data functions $f_j(y)$, $c_\ell(y)$, $d_\ell(y)$ have compact supports in $\mathbb{R}^{n-1}$. Let all the assumptions of Theorem 9.3 hold. Then there exists a unique interpolation polyspline h such that $h \in H^{2p+r}(\text{Strip}_j)$ for $j = 1, 2, \ldots, N-1$, i.e. h is a Sobolev smoothness polyspline in the domain $\text{Strip}(t_1, t_N)$. The conclusions of point (2) of Theorem 9.3 hold where the integer multi-index $\kappa \in \mathbb{Z}^{n-1}$ involved has to be replaced by the multi-index $\xi \in \mathbb{R}^{n-1}$.*

Adopting the assumptions of Theorem 9.3 if $r > 0$ is not an integer there exists a class C^{2p-2} polyspline h such that $h \in C^{2p+r}(\overline{\text{Strip}_j})$ for $j = 1, 2, \ldots, N$.

The proof is similar to that of Theorem 9.3 in the periodic case with more emphasis on the uniqueness part. We refer also to Section 22.3, p. 450, where the existence part of the proof is justified as a consequence of a more general setting, and the *a priori* estimates of the solution are provided.

9.1.3 The case $p = 2$

Let $p = 2$. The data satisfy $f_j \in H^{3(1/2)}(\Gamma_j)$, $c_1 \in H^{2(1/2)}(\mathbb{R}^{n-1})$ and $d_1 \in H^{2(1/2)}(\mathbb{R}^{n-1})$ (We may take in particular $c_1 = d_1 = 0$.). We apply the Fourier transform. Then for every $\xi \in \mathbb{R}^{n-1}$ the conditions in Definition 9.1, p. 118, become:

$$\left(\frac{\partial^2}{\partial t^2} - |\xi|^2\right)^2 \widehat{h}_j(t, \xi) = 0 \quad \text{for } t_{j-1} < t < t_j \text{ and } j = 2, \ldots, N;$$

$$\widehat{h}_j(t_j, \xi) = \widehat{h}_{j+1}(t_j, \xi) = \widehat{f}_j(\xi) \quad \text{for } j = 2, \ldots, N-1;$$

$$\frac{\partial}{\partial t}\widehat{h}_j(t_j, \xi) = \frac{\partial}{\partial t}\widehat{h}_{j+1}(t_j, \xi) \quad \text{for } j = 2, \ldots, N-1;$$

$$\frac{\partial^2}{\partial t^2}\widehat{h}_j(t_j, \xi) = \frac{\partial^2}{\partial t^2}\widehat{h}_{j+1}(t_j, \xi) \quad \text{for } j = 2, \ldots, N-1.$$

For every $\xi \in \mathbb{R}^{n-1}$ the function $\widehat{h}(t, \xi)$ is a one-dimensional interpolation L-spline defined by the operator $L = L_\xi = ((\partial^2/\partial t^2) - |\xi|^2)^2$, which can be computed by reference to Theorem 11.28, p. 200, quoted in the previous section. Thus the function $\widehat{h}(t, \xi)$ is of the form

$$C_1 e^{|\xi|t} + C_2 t e^{|\xi|t} + C_3 e^{-|\xi|t} + C_4 t e^{-|\xi|t}$$

on every interval (t_{j-1}, t_j) for $j = 2, 3, \ldots, N$. The polyspline components $h_j(t, x)$ themselves are obtained by applying the inverse Fourier transform to $\widehat{h}_j(t, \xi)$.

It should be noted that the existence and uniqueness of polysplines in the case of data without any restrictions on their growth at infinity has other features which will not be discussed here.

9.2 Polysplines on annuli in $\mathbb{R}^n$

We proceed to the polysplines on annuli. As in the two-dimensional case there are two possibilities – when the support for the polyspline is a ball and when it is an annulus.

In principle things are very similar to the case of periodic polysplines on strips. The difference is that here the polysplines $h(r\theta)$ will be expanded in the *Laplace series* through spherical harmonics $Y_{k,\ell}(\theta)$ in the variable θ which will replace the Fourier series for $h(t, y)$ in formula (9.7), p. 120.

The polysplines on annuli will have as *knot-surfaces* (or *interfaces*) the spheres $\Gamma_j = S(0; r_j)$, where the radii satisfy $r_1 < r_2 < \cdots < r_N$. We simply reproduce with minor changes Definition 8.1, p. 101, of the two-dimensional case. (See Figure 3.5, p. 33, from the two-dimensional case which also applies to the present situation.)

Definition 9.5 *Let $\Gamma_j = S(0; r_j)$ for $j = 1, \ldots, N$, and let the integer $p \geq 1$. Let the functions $h_j(x)$ be given in the annulus*

$$A_j := \{x = r\theta \in \mathbb{R}^n : r_j < r < r_{j+1} \text{ and } \theta \in \mathbb{S}^{n-1}\}$$

for $j = 1, \ldots, N-1$, and $h_0(x)$ be given in the ball $B(0; r_1)$ and satisfy the following:

1. *h_j is polyharmonic of order p, i.e.*

$$\Delta^p h_j = 0 \quad \text{in } A_j$$

for $j = 1, \ldots, N-1$, and

$$\Delta^p h_0 = 0 \quad \text{in } B(0; r_1).$$

2. *h_j belongs to the class $C^{2p-2}(\overline{A_j})$, which means differentiability up to the boundary of A_j.*
3. *The following equalities hold*

$$\frac{\partial^k}{\partial r^k} h_j = \frac{\partial^k}{\partial r^k} h_{j+1} \quad \text{on } \Gamma_{j+1} \text{ and } k = 0, 1, \ldots, 2p-2, \qquad (9.8)$$

where the boundary values on Γ_{j+1} exist by the previous point.

We will say that the function h which is equal to h_j on the annulus A_j for $j = 1, \ldots, N-1$, and to h_0 in the ball $B(0; r_1)$ is a **class C^{2p-2} polyspline** of order p **supported in the ball** $B(0; r_N)$ and has **knot-surfaces** Γ_j, $j = 1, 2, \ldots, N-1$.

Ignoring the ball $B(0; r_1)$ we will say that the function h which is equal to h_j on the annulus A_j for $j = 1, \ldots, N-1$, is a **class C^{2p-2} polyspline** of order p **supported in the annulus**

$$A(r_1, r_N) := \{x = r\theta \in \mathbb{R}^n : r_1 < r < r_N \text{ and } \theta \in \mathbb{S}^{n-1}\}$$

and has **knot-surfaces** Γ_j, $j = 2, 3, \ldots, N-1$.

We will say that the polyspline h is of **Sobolev class** H^{2p} if $h_j \in H^{2p}(A_j)$ for all $j = 1, 2, \ldots, N-1$, and the above equalities (9.8) are understood as equalities between traces in Sobolev space.

Recall all remarks which we have made after the introduction of the polysplines on annuli in the two-dimensional case in Definition 8.1, p. 101.

Assuming that some data functions f_j are given on the spheres Γ_j, as in Definition 8.3, p. 102, of the two-dimensional case, we now define the **interpolation polysplines**.

Definition 9.6 *The polyspline h of order p (in the ball $B(0; r_N)$ or in the annulus $A(r_1, r_N)$) will be called an **interpolation polyspline** if the following equalities hold:*

$$h = f_j \quad \text{on } \Gamma_j \text{ for } j = 1, \ldots, N. \qquad (9.9)$$

If the polyspline h is supported **in the ball** $B(0; r_N)$ then it has to satisfy the following boundary *conditions on the greatest sphere* Γ_N:

$$\frac{\partial^\ell}{\partial r^\ell} h = \frac{\partial^\ell}{\partial r^\ell} h_{N-1} = d_\ell \quad \text{on } \Gamma_N, \tag{9.10}$$

for all $\ell = 1, \ldots, p-1$.

If the polyspline h is supported **in the annulus**

$$A(r_1, r_N) = \{ x = r\theta \in \mathbb{R}^n : r_1 < r < r_N \text{ and } \theta \in \mathbb{S}^{n-1} \}$$

then it has to satisfy the following boundary *conditions on the smallest and on the greatest spheres* Γ_1, Γ_N:

$$\frac{\partial^\ell}{\partial r^\ell} h = \frac{\partial^\ell}{\partial r^\ell} h_{N-1} = d_\ell \quad \text{on } \Gamma_N, \tag{9.11}$$

$$\frac{\partial^\ell}{\partial r^\ell} h = \frac{\partial^\ell}{\partial r^\ell} h_1 = \tilde{d}_\ell \quad \text{on } \Gamma_1, \tag{9.12}$$

for all $\ell = 1, \ldots, p-1$.

As in the two-dimensional case the *boundary conditions* (9.10) on the largest sphere $\Gamma_N = \partial B(0; r_N)$ for a polyspline supported in the ball (or (9.11)–(9.12) if the polyspline is supported in the annulus) are optional in the above definition.

Now the time has come to apply the whole science of spherical harmonics which we have developed in Chapter 10.

The treatment of the polysplines on annuli in $\mathbb{R}^n$ is rather similar to the two-dimensional case. The only difference is that we have to employ the spherical harmonics on $\mathbb{S}^{n-1}$ which generalize the trigonometric polynomials on $\mathbb{S}$. This fact is widely discussed in Chapter 10, p. 129.

We assume that the set

$$\{ Y_{k,\ell} : k = 0, 1, 2, \ldots; \; \ell = 1, 2, \ldots, d_k \}$$

of spherical harmonics is an orthonormal basis of the whole space of spherical harmonics as in formula (10.13), p. 149. Let the function h be given in the ball or in the annulus. Let us introduce for every pair of integers $k \geq 0$ and $\ell = 1, 2, \ldots, d_k$ the coefficient

$$\widehat{h}(r, k, \ell) := \int_{\mathbb{S}^{n-1}} h(r\theta) Y_{k,\ell}(\theta) \, d\theta \tag{9.13}$$

in the *Laplace* series of h defined in formula (10.14), p. 150.

For the existence we will need the general results of Part IV.

Theorem 9.7 1. *For the real number* $r \geq 0$ *let the functions* $f_j(\theta)$, $d_\ell(\theta)$ *satisfy* $f_j \in H^{2p+r-(1/2)}(\Gamma_j)$, *for* $j = 1, 2, \ldots, N$, *and* $d_\ell \in H^{2p+r-\ell-(1/2)}(\Gamma_N)$, *for*

$\ell = 1, 2, \ldots, p - 1$ (for polysplines with support in the annulus we also require $\widetilde{d}_\ell \in H^{2p+r-\ell-(1/2)}(\Gamma_1)$). Then:

- There exists a unique interpolation polyspline $h(x)$ supported **in the ball** $B(0; r_N)$ satisfying the data f_j, d_ℓ, and such that $h \in H^{2p+r}(A_j)$ for $j = 1, 2, \ldots, N - 1$, and $h \in B(0; r_1)$, i.e. $h(x)$ is a Sobolev smoothness polyspline.
- There exists a unique interpolation polyspline $h(x)$ supported **in the annulus** $A_{a,b}$ satisfying the data f_j, d_ℓ, $\widetilde{d}_\ell$, and such that $h \in H^{2p+r}(A_j)$ for $j = 1, 2, \ldots, N - 1$, i.e. $h(x)$ is a Sobolev smoothness polyspline.

For a number $r > 0$ which is not an integer let the data satisfy $f_j \in C^{2p+r}(\Gamma_j)$ and $d_\ell \in C^{2p+r-\ell}(\Gamma_j)$ (or $\widetilde{d}_\ell \in C^{2p+r-\ell}(\Gamma_j)$). Then there exists a class C^{2p-2} polyspline h supported in the ball $B(0; r_N)$ such that $h \in C^{2p+r}(\overline{A_j})$ for $j = 1, 2, \ldots, N$, and $h \in C^{2p+r}(\overline{B(0; r_1)})$, and satisfying the interpolation and boundary data. Thus there exists a class C^{2p-2} polyspline h supported in the annulus A_{r_1, r_N} with $h \in C^{2p+r}(\overline{A_j})$ for $j = 1, 2, \ldots, N$, and which satisfies the interpolation and boundary data.

2. If h is a polyspline supported in the ball, then for every pair of integers $k \geq 0$ and $\ell = 1, 2, \ldots, d_k$, the Laplace coefficient $\widehat{h}(e^v, k, \ell)$ considered as a function of the variable $v = \log r$ is a left-sided Chebyshev spline (in the sense of Definition 11.35, p. 205) for the ordinary differential operator defined by formula (10.26), p. 169,

$$M_{k,p}\left(\frac{d}{dv}\right) = \prod_{l=0}^{p-1}\left(\frac{d}{dv} - (k + 2l)\right) \prod_{l=1}^{p}\left(\frac{d}{dv} - (-n - k + 2l)\right),$$

i.e.

- $M_{k,p}(d/dv)\widehat{h}(e^v, k, \ell) = 0$ on the interval $(\log r_j, \log r_{j+1})$ for $j = 1, 2, \ldots, N-1$,
- $\prod_{l=0}^{p-1}((d/dv) - (k + 2l))\widehat{h}(e^v, k, \ell) = 0$ on the interval $(-\infty, \log r_1)$,
- $\widehat{h}(e^v, k, \ell) \in C^{2p-2}(-\infty, \log r_N)$.
- In addition $\widehat{h}(e^v, k, \ell)$ satisfies the interpolation and boundary conditions (reformulated in the variable v)

$$\widehat{h}(r_j, k, \ell) = \widehat{f}_j(k, \ell) \qquad \text{for } j = 1, 2, \ldots, N,$$

$$\frac{\partial^l}{\partial r^l}\widehat{h}(r_N, k, \ell) = \widehat{d}_l(k, \ell) \qquad \text{for } l = 1, 2, \ldots, p - 1.$$

Here we have denoted by $\widehat{f}_j(k, \ell)$, $\widehat{d}_l(k, \ell)$, respectively, the coefficients of the Laplace series of the functions f_j and d_l.

3. If the polyspline h is supported **in the annulus** $A_{a,b}$ then we have similar conclusions to above: the function $\widehat{h}(e^v, k, \ell)$ considered as a function of the variable $v = \log r$ is an interpolation Chebyshev spline with respect to the operator $M_{k,p}(d/dv)$ in the

interval $(\log r_1, \log r_N)$ and the interpolation *and* boundary *conditions are*

$$\widehat{h}(r_j, k, \ell) = \widehat{f_j}(k, \ell) \quad \text{for } j = 1, 2, \ldots, N,$$

$$\frac{\partial^l}{\partial r^l}\widehat{h}(r_1, k, \ell) = \widehat{d_l}(k, \ell) \quad \text{for } l = 1, 2, \ldots, p-1,$$

$$\frac{\partial^l}{\partial r^l}\widehat{h}(r_N, k, \ell) = \widehat{d_l}(k, \ell) \quad \text{for } l = 1, 2, \ldots, p-1.$$

Proof (1) Let us consider the case of polysplines supported in a ball in the Sobolev space setting. We will first prove the uniqueness. It will be proved in a way very similar to the proof of Proposition 8.6, p. 110, in the two-dimensional case, but we will take care of all details.

By Theorem 10.39, p. 173, the function $h(x)$ has the following representation in the annulus A_j for $j = 1, 2, \ldots, N-1$, and in the ball $B(0; r_1)$:

$$h(x) = \sum_{k=0}^{\infty} \sum_{\ell=1}^{d_k} \widehat{h}(r, k, \ell) Y_{k,\ell}(\theta) \quad \text{in } A_j \text{ and } B(0; r_1),$$

where the coefficient $\widehat{h}(r, k, \ell)$ is given by (9.13). For every $k = 0, 1, 2, \ldots,$ and $\ell = 1, 2, \ldots, d_k$, the functions $\widehat{h}(r, k, \ell)$ are solutions to equation

$$L_{(k)}^p g = 0 \quad \text{in } [r_j, r_{j+1}] \text{ for } j = 1, 2, \ldots, N-1, \text{ and in } [0, r_1].$$

We have to prove that $\widehat{h}(r, k, \ell) \in C^{2p-2}([0, r_N])$ and that the interpolation and boundary conditions are satisfied.

By Proposition 23.6, p. 469, it follows that the function $h(c\theta)$ of $\theta \in \mathbb{S}^{n-1}$ is a continuous function in $c \in [r_j, r_{j+1}]$ (respectively, in $c \in (0, r_1]$) in the space $H^{2p-(1/2)}(\mathbb{S}^{n-1})$. This implies the following representation of the radial derivatives for $i = 0, 1, \ldots, 2p-2$,

$$\frac{\partial^i}{\partial r^i} h(x) = \sum_{k=0}^{\infty} \sum_{\ell=1}^{d_k} \frac{\partial^i}{\partial r^i} \widehat{h}(r, k, \ell) Y_{k,\ell}(\theta) \quad \text{in } A_j \text{ and } B(0; r_1).$$

The smoothness of the function h implies that

$$\frac{\partial^i}{\partial r^i} \widehat{h}(r, k, \ell) = \frac{\widehat{\partial^i h}}{\partial r^i}(r, k, \ell) \quad \text{for } i = 0, 1, \ldots, 2p-2.$$

Hence, $\widehat{h}(r, k, \ell)$ belongs to $C^{2p-2}(0, r_N)$, and the interpolation and boundary conditions hold.

The fact that $\widehat{h}(e^v, k, \ell)$ satisfies the equation $M_{k,p}\widehat{h}(e^v, k, \ell)g(v) = 0$ for $v = \log r \neq \log r_j$ with $j = 1, 2, \ldots, N$, follows from the fundamental Theorem 10.39, p. 173, and Theorem 10.42, p. 177, which establish the representation of polyharmonic functions through spherical harmonics in the annulus and in the ball. On the other hand in the ball $B(0; r_1)$ the functions $\widehat{h}(r, k, \ell)$ satisfy equation (10.29), p. 173, i.e.

$$M_{k,p}^+ g(v) = 0 \qquad \text{for } r = e^v < r_1.$$

We refer to Chapter 11 on Chebyshev splines, Definition 11.35, p. 205, of the left-sided natural Chebyshev spline and to Remark 11.36, p. 205. From Remark 11.35 one sees that the operator $M_{k,p}$ generates a Chebyshev system $\mathcal{U}_{2p}$ and we have a subsystem $\mathcal{U}_p$ generated by the operator $M_{k,p}^+$. Thus we see that $\widehat{h}(e^v, k, \ell)$ as a function of v is a left-sided natural Chebyshev spline.

Now the uniqueness of the polyspline $h(x)$ follows from the uniqueness of the interpolation left-sided natural Chebyshev splines $\widehat{h}(e^v, k, \ell)$ for all pairs $k \geq 0$, $\ell = 1, 2, \ldots, d_k$, provided in Theorem 11.40, p. 207.

(2) The existence of the polyspline $h(x)$ now will follow by Theorem 22.6, p. 447; see also the remarks thereafter in Section 22.2, p. 448. It is a consequence of a more general framework on elliptic boundary value problems.

In a similar way we proceed for the case of polysplines supported in an annulus. The only difference is that we have to refer to the uniqueness of the usual Chebyshev splines generated by the operator $M_{k,p}$ for variable v satisfying $r_1 \leq r = e^v \leq r_N$.

(3) The uniqueness of the polysplines in Hölder space is easier to prove since the traces on the concentric spheres are easier to estimate. ∎

The above theorem provides us with an algorithm for the computation of the interpolation polyspline. We first compute for every $\kappa \in \mathbb{Z}^{n-1}$ the interpolation Chebyshev splines $\widehat{h}(t, \kappa)$ and after that find the function $h(t, y)$ through the series:

$$h(t, y) = \sum_{\kappa \in \mathbb{Z}^{n-1}} \widehat{h}(t, \kappa) e^{i(y_1 k_1 + \cdots + y_{n-1} k_{n-1})}.$$

We refer also to Section 23.1.3, p. 464, where the coefficients in the expansion in spherical harmonics provide an exact measure of the Sobolev space to which a function on the sphere belongs.

9.2.1 Biharmonic polysplines in $\mathbb{R}^3$ and $\mathbb{R}^4$

Let us consider the case $n = 3$ and $p = 2$. According to Theorem 10.39, p. 173, we have the expansion in a Fourier–Laplace series containing the basis functions for $k = 0$, $k = 1$, $k = 2$, $k = 3$, listed in the following table:

k	r^{-k-1}	r^{-k+1}	r^k	r^{k+2}
0	r^{-1}	r	1	r^2
1	r^{-2}	1	r	r^3
2	r^{-3}	r^{-1}	r^2	r^4
3	r^{-4}	r^{-2}	r^3	r^5

They represent the basis for the Chebyshev splines for every k.

Let $n = 4$ and $p = 2$. The basis functions are:

k	r^{-k-2}	r^{-k}	r^k	r^{k+2}
0	r^{-2}	$\log r$	1	r^2
1	r^{-3}	r^{-1}	r	r^3
2	r^{-4}	r^{-2}	r^2	r^4
3	r^{-5}	r^{-3}	r^3	r^5

9.2.2 An "elementary" proof of the existence of interpolation polysplines

It will be useful to use "elementary" means to prove the existence of interpolation polysplines in Theorem 9.3, p. 119, and Theorem 9.7, p. 124. Let us explain what we mean in the case of the periodic polysplines on strips treated in Theorem 9.3. Following the reduction there, first, we have to compute the L-splines $\widehat{h}(t, \kappa)$ for every $\kappa \in \mathbb{Z}^{n-1}$, where the operator $L = L_\kappa$ is given by formula (9.6), p. 120, and then prove the convergence of the series $h(t, y) = \sum_\kappa \widehat{h}(t, \kappa) e^{iy \cdot \kappa}$ in formula (9.7), p. 120. It is clear that the series will be convergent if the coefficient $\widehat{h}(t, \kappa)$ has a fast decay for $\kappa \to \infty$. It is a classical fact from Fourier analysis that the function $h(t, y)$ will have more derivatives if this decay is faster. This decay depends on the Fourier coefficients $\widehat{f}_j(\kappa)$, $\widehat{c}_\ell(\kappa)$ and $\widehat{d}_\ell(\kappa)$ used for the computation of $\widehat{h}(t, \kappa)$. Thus $\widehat{h}(t, \kappa)$ will decay faster if these coefficients decay faster with $\kappa \to \infty$. As said already the Fourier coefficients $\widehat{f}_j(\kappa)$, $\widehat{c}_\ell(\kappa)$ and $\widehat{d}_\ell(\kappa)$ decay faster if the functions f_j, c_ℓ and d_ℓ are smoother. Thus we have the chance to obtain, in an "elementary way", the relation between the smoothness of the data functions and the solution $h(t, y)$ which is proved (using abstract arguments from Part IV) in Theorem 9.3 for the cases of Hölder space data and Sobolev space data. As is clear from above, such an elementary proof has to make a subtle estimate of the one-dimensional L-splines $\widehat{h}(t, \kappa)$ depending on the parameter κ. The first step will be to estimate the corresponding compactly supported splines, the so-called TB-splines for every $L = L_\kappa$. One also has to use the decay of the Fourier coefficients of a function in a Sobolev space on the torus $\mathbb{T}^n$ and on the sphere $\mathbb{S}^n$ provided in Subsection 23.1.2, p. 463, and Section 23.1.3, p. 464, respectively.

The same idea will also work to make an elementary proof of Theorem 9.7, about the existence of interpolation polysplines on annuli.

In Part II we will provide such estimates of the TB-splines for the case of *cardinal polysplines*, i.e. when we have *equidistant* t_js for the polysplines on strips, and when $\log r_j$ are *equidistant* for the polysplines on annuli. In the case of finitely many non-equidistant t_j's these estimates are rather subtle.

For standard material on separation of variables in elliptic equations we recommend the book of B. Budak, A. Samarskii, A. Tikhonov [7], Chapter IV, Section 4, problem 80.

Chapter 10

Compendium on spherical harmonics and polyharmonic functions

10.1 Introduction

The main purpose of this compendium is to present basic results about the representation of polyharmonic functions in the spherically symmetric domains – ball and annulus. As a necessary prelude we will provide an exposition of the basics of the theory of *spherical harmonics*. Although it takes more space, the theory of spherical harmonics is very important for the theory of polysplines (when the break-surfaces are concentric spheres); it seemed appropriate to have an independent exposition of the main results which would be within the capacity of a wider audience.

The Compendium contains many well known results but also some which are far from popular, like the *Almansi representation of polyharmonic functions* in the *annulus* (a result which apparently belongs to *Vekua* in $\mathbb{R}^2$ and to *Sobolev* in $\mathbb{R}^n$). There is also an important new result to be applied to the polysplines, concerning the representation of the polyharmonic operator Δ^p by spherical harmonics. It says that if a function $u(x)$ is represented through the *Fourier–Laplace* series

$$u(x) = \sum_{k=0}^{\infty} \sum_{\ell=0}^{d_k} \psi_{k,\ell}(\log r) Y_{k,\ell}(\theta) \quad \text{for } a < |x| < b,$$

then for all $a < r < b$ holds equality

$$\Delta^p u(x) = e^{-2pv} \sum_{k=0}^{\infty} \sum_{\ell=0}^{d_k} M_{k,p} \psi(v) Y_{k,\ell}(\theta) \quad \text{for } v = \log r,$$

where the operator $M_{k,p}(d/dv)$ is an ordinary differential operator with constant coefficients, see Theorem 10.34, p. 169.

10.2 Notations

We will work in the Euclidean space $\mathbb{R}^n$ where $x = (x_1, \ldots, x_n) \in \mathbb{R}^n$ and the usual metric $|x| = \sqrt{\sum_{j=0}^n x_j^2}$. We will consider the *multipower* and the *real differential operators* given by

$$x^\alpha = x_1^{\alpha_1} x_2^{\alpha_2} \ldots x_n^{\alpha_n}$$

$$D^\alpha = \frac{\partial^{|\alpha|}}{\partial x_1^{\alpha_1} \partial x_2^{\alpha_2} \ldots \partial x_n^{\alpha_n}}$$

where the integer multi-indexes $\alpha = (\alpha_1, \alpha_2, \ldots, \alpha_n) \in \mathbb{Z}^n$ have nonnegative entries $\alpha_i \geq 0$, and $|\alpha| = \alpha_1 + \alpha_2 + \cdots + \alpha_n$. We put $\alpha! = \alpha_1! \alpha_2! \ldots \alpha_n!$, where as usual $0! = 1$. The Laplace operator (sometimes called Laplacian) is given by

$$\Delta = \sum_{j=0}^\infty \frac{\partial^2}{\partial x_j^2}$$

and the polyharmonic operator of order $p \geq 1$ is defined inductively as $\Delta^p = \Delta \Delta^{p-1}$. It will be further convenient to use for $p = 0$, $\Delta^0 = id$, where id denotes the identity operator.

We will often use open balls with center x_0 and radius $r > 0$ defined by

$$B(x_0; r) = B_n(x_0; r) := \{x \in \mathbb{R}^n : |x - x_0| < r\}$$

where we use the subindex n if we want to emphasize the dimension of the space. Its volume is

$$v_n = \frac{\pi^{n/2}}{\Gamma((n/2) + 1)} = \frac{2\pi^{n/2}}{n \Gamma(n/2)}. \tag{10.1}$$

The sphere with center x_0 and radius $r > 0$ is defined by

$$S(x_0; r) = S_n(x_0; r) := \partial B(x_0; r) = \{x \in \mathbb{R}^n : |x - x_0| = r\}.$$

It should be noted that we write $S_{n-1}(x_0; r)$ in order to stress that it is a $(n-1)$-dimensional manifold. The unit sphere in $\mathbb{R}^n$ will often be used and is given the notation

$$\mathbb{S}^{n-1} = S_{n-1}(0; 1)$$

where the dimension n is distinguished. Its area, which is $(n-1)$-dimensional, is

$$\sigma_{n-1} = |\mathbb{S}^{n-1}| = n v_n = \frac{2\pi^{(n/2)}}{\Gamma(n/2)}. \tag{10.2}$$

For later use we put

$$\Omega_n = \begin{cases} \dfrac{1}{(n-2)\sigma_n} = \dfrac{\Gamma(n/2)}{(n-2)2\pi^{(n/2)}} = \dfrac{\Gamma((n/2) - 1)}{4\pi^{(n/2)}} & \text{for } n \geq 3 \\ \dfrac{1}{\sigma_2} = \dfrac{1}{2\pi} & \text{for } n = 2. \end{cases} \tag{10.3}$$

(see Helms [22].)

In $\mathbb{R}^n$ we have the relation between the volume of the ball and the area of the sphere expressed through the following formula

$$\frac{1}{R^{n-1}\sigma_n} \int_{S(x_0;R)} 1\, d\sigma(z) = \frac{1}{\sigma_n} \int_{\mathbb{S}^{n-1}} 1\, d\sigma(\theta).$$

Here we have used the fact that the area element $\sigma(\cdot)$ on the sphere $S(x_0; R)$ is related to the area element on the unit sphere $\mathbb{S}^{n-1} = S(0; 1)$ through the equality

$$d\sigma(z) = R^{n-1} d\sigma_1(\theta) \quad \text{for } z \in S(x_0; R).$$

It will be useful to recall that the surface area of $S(x_0; R)$ is $R^{n-1}\sigma_{n-1}$ so that we have the norming

$$\int_{S(x_0;R)} 1\, d\sigma(x) = R^{n-1}\sigma_n.$$

10.3 Spherical coordinates and the Laplace operator

Let us introduce the spherical coordinates in $\mathbb{R}^n$ by putting $r = |x|$, $\theta = x/|x|$. The unit vector θ may be expressed through the "Euler angles" ϑ_i, $i = 1, 2, \ldots, n-2$, and φ by the formulas:

$$\theta_1 = \cos \vartheta_1$$
$$\theta_2 = \sin \vartheta_1 \cos \vartheta_2$$
$$\theta_2 = \sin \vartheta_1 \sin \vartheta_2 \cos \vartheta_3$$
$$\vdots$$
$$\theta_{n-2} = \sin \vartheta_1 \sin \vartheta_2 \cdots \sin \vartheta_{n-3} \cos \vartheta_{n-2}$$
$$\theta_{n-1} = \sin \vartheta_1 \sin \vartheta_2 \cdots \sin \vartheta_{n-3} \sin \vartheta_{n-2} \cos \varphi$$
$$\theta_n = \sin \vartheta_1 \sin \vartheta_2 \cdots \sin \vartheta_{n-3} \sin \vartheta_{n-2} \sin \varphi$$

with

$$0 \leq \vartheta_i \leq \pi \quad (i = 1, \ldots, n-2)$$
$$0 \leq \varphi \leq 2\pi$$

Geometrically, the variable ϑ_1 is the angle between the vector x and the axis x_1 and is given by

$$\vartheta_1 = \arccos \frac{x_1}{|x|}.$$

This would simplify the computations if we take a fixed direction to coincide with the axis x_1. For details see Triebel [60, formula (31.20), Lemma 31.2, and around].

132 *Multivariate polysplines*

Now it is easy to compute the volume element [60, Lemma 31.2, and Remark 31.3] (in Triebel's notation this is the function $\tilde{g}(\theta)$)

$$dx = r^{n-1} \, d\theta \, dr$$
$$= r^{n-1} \sin \vartheta_{n-2} \sin^2 \vartheta_{n-3} \cdots \sin^{n-2} \vartheta_1 \, d\varphi \, d\vartheta_1 \, d\vartheta_2 \cdots d\vartheta_{n-2} \, dr.$$

We can compute the volume of the unit ball $B(0; 1)$ which we have already provided in the previous section by formula (10.1), see the computation by Treves [61, p. 74]. The surface of the sphere $\mathbb{S}^{n-1}$ is also provided in formula (10.2), p. 130.

In these coordinates the Laplace operator is given by [60, Lemma 31.2],

$$\Delta f(r, \theta) = \Delta_r f + \frac{1}{r^2} \Delta_\theta f, \qquad (10.4)$$

where Δ_r denotes the so-called radial part of the operator and Δ_θ denotes the spherical Laplacian (*Laplace–Beltrami*) operator. They are given by the formulas

$$\Delta_r f = \frac{1}{r^{n-1}} \frac{\partial}{\partial r} r^{n-1} \frac{\partial}{\partial r} = \frac{\partial^2}{\partial r^2} + \frac{n-1}{r} \frac{\partial}{\partial r} \qquad (10.5)$$

$$\Delta_\theta f = \frac{1}{\sin^{n-2} \vartheta_1} \frac{\partial}{\partial \vartheta_1} \sin^{n-2} \vartheta_1 \frac{\partial}{\partial \vartheta_1} \qquad (10.6)$$

$$+ \frac{1}{\sin^2 \vartheta_1} \frac{1}{\sin^{n-3} \vartheta_2} \frac{\partial}{\partial \vartheta_2} \sin^{n-3} \vartheta_2 \frac{\partial}{\partial \vartheta_2}$$

$$\cdots$$

$$+ \frac{1}{\sin^2 \vartheta_1 \sin^2 \vartheta_2 \cdots \sin^2 \vartheta_{n-2}} \frac{\partial^2}{\partial \varphi^2}.$$

In the case of $\mathbb{R}^2$ we have the representation given by Budak et al. [7, p. 392], or Magnus et al. [35, p. 487],

$$\Delta_r = \frac{\partial^2}{\partial r^2} + \frac{1}{r} \frac{\partial}{\partial r},$$

$$\Delta_\theta = \frac{\partial^2}{\partial \varphi^2},$$

$$\Delta = \Delta_r + \frac{1}{r^2} \Delta_\theta,$$

where

$$\theta_1 = \cos \varphi, \quad \theta_2 = \sin \varphi.$$

For $\mathbb{R}^3$ we have the representation [61, p. 315],

$$\Delta_r = \frac{\partial^2}{\partial r^2} + \frac{2}{r}\frac{\partial}{\partial r},$$

$$\Delta_\theta = \frac{1}{\sin\vartheta}\frac{\partial}{\partial\vartheta}\sin\vartheta + \frac{1}{\sin^2\vartheta}\frac{\partial^2}{\partial\varphi^2},$$

$$\Delta = \Delta_r + \frac{1}{r^2}\Delta_\theta,$$

where

$$\theta_1 = \cos\vartheta_1,$$
$$\theta_2 = \sin\vartheta_1\cos\varphi,$$
$$\theta_3 = \sin\vartheta_1\sin\varphi.$$

The cylindrical coordinates are defined in $\mathbb{R}^n$ by putting $x = (x', z)$ and changing x' to spherical coordinates in $\mathbb{R}^{n-1}$. In cylindrical coordinates in $\mathbb{R}^n$ we have

$$\Delta f = \Delta_{(n-1)} f + \frac{\partial^2}{\partial z^2} f$$

where $\Delta_{(n-1)}$ denotes the Laplace operator in $\mathbb{R}^{n-1}$. In particular in $\mathbb{R}^3$ [7, p. 392],

$$\Delta f(r,\varphi,z) = \frac{1}{r}\frac{\partial}{\partial r}r\frac{\partial}{\partial r} + \frac{1}{r^2}\frac{\partial^2}{\partial\varphi^2} + \frac{\partial^2}{\partial z^2}.$$

The operator of *Laplace–Beltrami* Δ_θ is *self-adjoint* on the sphere $\mathbb{S}^{n-1}$. Indeed, for proving this we may apply directly the theorem of de Rham (generalization of integration by parts formula for manifolds). This result may also be obtained directly by using the usual Green formula. One considers simply functions in a spherical layer (annulus) which contains $\mathbb{S}^{n-1}$, and the functions do not depend on the variable r. Let us provide this proof.

Theorem 10.1 *The operator Δ_θ is self-adjoint on the sphere $\mathbb{S}^{n-1}$, in particular it is symmetric: Let $u, v \in C^2(\mathbb{S}^{n-1})$ be two arbitrary functions. Then*

$$\int_{\mathbb{S}^{n-1}} \Delta_\theta u(\theta) v(\theta)\, d\theta = \int_{\mathbb{S}^{n-1}} u(\theta)\Delta_\theta v(\theta)\, d\theta.$$

Proof We prove the symmetry, the self-adjointness follows easily. Let us consider the annular domain

$$A = \left\{x : \frac{1}{2} < |x| < 2\right\}.$$

We define the extensions of functions u, v on A

$$u(x) = u\left(\frac{x}{|x|}\right), \quad v(x) = v\left(\frac{x}{|x|}\right).$$

Obviously, these two functions satisfy

$$u, v \in C^2(A),$$

and they are independent of the variable r.

We apply the usual symmetric Green formula, and obtain

$$\int_A (\Delta u(x) \cdot v(x) - u(x) \cdot \Delta v(x))\, dx = \int_{\partial A} \left(\frac{\partial}{\partial n} u(x) \cdot v(x) - u(x) \cdot \frac{\partial}{\partial n} v(x) \right) dx,$$

where $\partial/\partial n$ is the exterior normal to the domain A. It should be noted that this means that for $|x| = 2$ we have

$$\frac{\partial}{\partial n} = \frac{\partial}{\partial r},$$

and for $|x| = 1/2$ we have

$$\frac{\partial}{\partial n} = -\frac{\partial}{\partial r}.$$

This implies

$$\frac{\partial}{\partial n} u(x) = \frac{\partial}{\partial n} v(x) = 0 \quad \text{for } x \text{ in } \partial A.$$

Alternatively, we can use the representation

$$\Delta = \Delta_r + \frac{1}{r^2} \Delta_\theta,$$

where $x = r\theta$. Since the functions u, v are independent of r we obtain

$$\Delta u(r\theta) = \frac{1}{r^2} \Delta_\theta u(r\theta),$$

$$\Delta v(r\theta) = \frac{1}{r^2} \Delta_\theta v(r\theta).$$

Hence, from the Green formula follows

$$\int_A (\Delta_\theta u(x) \cdot v(x) - u(x) \cdot \Delta_\theta v(x))\, dx = 0.$$

This proves our statement. ∎

In our further exposition we follow in part Seeley [51], Stein and Weiss [57], and also Axler *et al.* [2].

10.4 Fourier series and basic properties

Let us recall the basic properties of the one-dimensional Fourier series, which will serve as a point of view in order to obtain the spherical harmonics.

As usual, we identify the circle $\mathbb{S}^1 = \{z \in \mathbb{C} : |z| = 1\}$ and the interval $[0, 2\pi]$. The set of complex-valued functions f defined on $\mathbb{S}^1$ is identified with the set of functions defined and 2π-periodic on the interval $[0, 2\pi]$. We supply the space $L_2(\mathbb{S})$ with the scalar product

$$\langle f, g \rangle := \int_0^{2\pi} f(\varphi)\overline{g(\varphi)}\, d\varphi.$$

For every real-valued function $f \in L_2(0, 2\pi)$ we have the trigonometric Fourier expansion (see equality (12.2), p. 210)

$$f(\varphi) = \frac{a_0}{2} + \sum_{k=1}^{\infty}(a_k \cos k\varphi + b_k \sin k\varphi).$$

Clearly, the periodicity for the functions in $L_2(0, 2\pi)$ does not play any essential role in the convergence of the Fourier series.

According to the general theory of Fourier expansions with respect to an arbitrary orthogonal system, we see that the above is exactly the expansion with respect to the orthonormal system of functions

$$1, \quad \frac{1}{\sqrt{\pi}} \cos k\varphi, \quad \frac{1}{\sqrt{\pi}} \sin k\varphi \quad (k \geq 1).$$

The orthonormality of this system follows by the well known relations

$$\int_0^{2\pi} \cos k\varphi \sin l\varphi = 0$$

$$\int_0^{2\pi} \cos k\varphi \cos l\varphi = \begin{cases} 0, & k \neq l \\ \pi, & k = l \end{cases}$$

$$\int_0^{2\pi} \sin k\varphi \sin l\varphi = \begin{cases} 0, & k \neq l \\ \pi, & k = l. \end{cases}$$

The main problem solved by the *spherical harmonics* is the generalization of the Fourier series but on the sphere $\mathbb{S}^n$.

There is another straightforward generalization of the one-dimensional Fourier series on the n-dimensional torus, namely on the set

$$\mathbb{T}^n := \underbrace{\mathbb{S}^1 \times \mathbb{S}^1 \times \cdots \times \mathbb{S}^1}_{n} = [\{z \in \mathbb{C} : |z| = 1\}]^n.$$

These are simply the multiple Fourier series given by

$$f(\varphi_1, \varphi_2, \ldots, \varphi_n) = \sum_{|\alpha|=0}^{\infty} a_\alpha e^{i(\alpha_1\varphi_1 + \cdots + \alpha_n\varphi_n)}$$

A detailed treatment has been given by Stein and Weiss [57, Chapter VII].

136 Multivariate polysplines

10.5 Finding the point of view

We have to find the point of view from which the spherical harmonics are a natural generalization of the Fourier series. For that reason let us consider the functions

$$1, \quad r^k \cos k\varphi, \quad r^k \sin k\varphi \quad \text{for } k \geq 1.$$

10.5.1 The functions $r^k \cos k\varphi$ and $r^k \sin k\varphi$ are harmonic for $k \geq 0$

We claim that these are harmonic functions in the plane $\mathbb{R}^2$. Indeed, we may check it directly thanks to the separation of the variables, and since the operator Δ_r depends only on $\partial/\partial r$, and the operator Δ_θ depends only on the variable φ:

$$\Delta(r^k \cos k\varphi) = \left(\Delta_r + \frac{1}{r^2} \Delta_\theta \right)(r^k \cos k\varphi)$$

$$= \cos k\varphi \cdot \Delta_r(r^k) + \frac{r^k}{r^2} \Delta_\theta(\cos k\varphi)$$

$$= \cos k\varphi \cdot \left(\frac{\partial^2}{\partial r^2} + \frac{1}{r} \frac{\partial}{\partial r} \right)(r^k) + \frac{r^k}{r^2} \frac{\partial^2}{\partial \varphi^2}(\cos k\varphi)$$

$$= \cos k\varphi \cdot (k(k-1) + k) r^{k-2} - k^2 r^{k-2} \cos k\varphi$$

$$= 0,$$

and similarly for $r^k \sin k\varphi$ we see that

$$\Delta(r^k \sin k\varphi) = 0.$$

10.5.2 The functions $r^k \cos k\varphi$ and $r^k \sin k\varphi$ are polynomials

The functions $r^k \cos k\varphi$ and $r^k \sin k\varphi$ are polynomials. Indeed, we have the representation

$$(x_1 + ix_2)^k = (re^{i\varphi})^k = r^k e^{ik\varphi} = r^k \cos k\varphi + i \cdot r^k \sin k\varphi,$$

and also

$$(x_1 + ix_2)^k = \sum_{\ell=0}^{k} \binom{k}{\ell} x_1^{k-\ell} (ix_2)^\ell = \sum_{\ell=0}^{k} i^\ell \binom{k}{\ell} x_1^{k-\ell} x_2^\ell$$

$$= \sum_{s=0}^{[k/2]} (-1)^s \binom{k}{2s} x_1^{k-2s} x_2^{2s}$$

$$+ i \sum_{s=0}^{[(k-1)/2]} (-1)^s \binom{k}{2s+1} x_1^{k-2s-1} x_2^{2s+1}.$$

Hence

$$r^k \cos k\varphi = \sum_{s=0}^{[k/2]} (-1)^s \binom{k}{2s} x_1^{k-2s} x_2^{2s},$$

$$r^k \sin k\varphi = \sum_{s=0}^{[(k-1)/2]} (-1)^s \binom{k}{2s+1} x_1^{k-2s-1} x_2^{2s+1}.$$

10.5.3 The functions $r^k \cos k\varphi$ and $r^k \sin k\varphi$ are homogeneous of degree $k \geq 0$

Another important property which is evident is that the polynomials $r^k \cos k\varphi$ and $r^k \sin k\varphi$ are homogeneous of degree $k \geq 0$, i.e. satisfy

$$P(tx_1, tx_2) = t^k P(x_1, x_2)$$

for every number $t \neq 0$.

10.5.4 The functions $r^k \cos k\varphi$ and $r^k \sin k\varphi$ are a basis of the homogeneous harmonic polynomials of degree k

One may see directly that every polynomial $P(x_1, x_2)$ which is harmonic, i.e. satisfies $\Delta P(x_1, x_2) = 0$, is a linear combination of these two polynomials. In other words, these two polynomials form a basis for the harmonic polynomials.

Problem

Prove this statement through a direct computation of $\Delta P(x_1, x_2) = 0$. One has to use the representation

$$P(x_1, x_2) = a_k x_1^k + a_{k-1} x_1^{k-1} x_2 + \cdots + a_2 x_1^2 x_2^{k-2} + a_1 x_1^1 x_2^{k-1} + a_0 x_2^k.$$

We will present an indirect proof later by computing the dimension of the space of harmonic homogeneous polynomials of degree k.

Another argument is the following: let $P(x)$ be harmonic polynomial, i.e. $\Delta P(x) = 0$. Since it has the representation in components of homogeneity,

$$P(x) = \sum_{k=0}^{N} P_k(x_1, x_2),$$

it follows that $\Delta P_k(x) = 0$. On the other hand we have $x_1 = r \cos \varphi$, and $x_2 = r \sin \varphi$, which gives

$$P_k(x_1, x_2) = r^k \sum_{\nu=-k}^{k} c_\nu e^{i\nu\varphi}.$$

But all functions $r^k e^{i\nu\varphi}$ are linearly independent, hence every $r^k e^{i\nu\varphi}$ has to be harmonic. We apply the Laplace operator, which in $\mathbb{R}^2$ takes the form

$$\Delta = \Delta_r + \frac{1}{r^2}\Delta_\theta = \frac{\partial^2}{\partial r^2} + \frac{1}{r}\frac{\partial}{\partial r} + \frac{1}{r^2}\frac{\partial^2}{\partial \varphi^2},$$

to $r^k e^{i\nu\varphi}$ and obtain $\Delta r^k e^{i\nu\varphi} = 0$ if and only if $\nu^2 = k^2$. Hence, $\nu = k$ or $\nu = -k$. It follows that

$$P_k(x_1, x_2) = c_{-k} z^{-k} + c_k z^k.$$

10.5.5 The multidimensional Ansatz

We now make the fundamental conclusion that the functions

$$\cos k\varphi, \quad \sin k\varphi \quad \text{for } k \geq 1,$$

are restrictions on the circle $r = 1$ of the harmonic polynomials $r^k \cos k\varphi$ and $r^k \sin k\varphi$ which are homogeneous of degree k. And the entire classical Fourier analysis on the unit circle profits from these functions.

The multivariate *Ansatz*[1] in $\mathbb{R}^n$ will be the following: We will consider the *harmonic homogeneous polynomials* of degree k in $\mathbb{R}^n$. We will take some **basis**. For $r = 1$ these functions play a role similar to $\cos k\varphi$, $\sin k\varphi$ on the sphere $\mathbb{S}^{n-1}$ and generate Fourier analysis on the sphere which is essentially different to the one which we have on the torus $\mathbb{T}^n$. These functions are called **solid spherical harmonics**. Their restrictions on the sphere, when $r = 1$ are simply called spherical harmonics or **surface spherical harmonics**.

10.6 Homogeneous polynomials in $\mathbb{R}^n$

Working now in $\mathbb{R}^n$ let us denote by Hom_k the space of all *homogeneous polynomials* of degree $k \geq 0$, i.e.

$$\text{Hom}_k = \{P_k(x) : P_k(tx) = t^k P_k(x) \quad \text{for every } t \neq 0\}.$$

For every $P_k \in \text{Hom}_k$ we have evidently the representation

$$P_k(x) = \sum_{|\alpha|=k} a_\alpha x^\alpha.$$

The dimension of this space is evidently equal to the number of monomials

$$\dim(\text{Hom}_k) = \#\{x^\alpha : |\alpha| = k\}.$$

---

[1] The meaning of Ansatz is explained in the footnote on p. 32.

Spherical harmonics and polyharmonic functions 139

Using combinatorial arguments one can easily compute

$$\#\{x^\alpha : |\alpha| = k\} = \binom{n+k-1}{k} = \binom{n+k-1}{n-1} = \frac{(n+k-1)!}{(n-1)!k!}.$$

Continuing our program we consider the space of harmonic polynomials in Hom_k. Let us denote it by

$$\text{HHom}_k = \{P_k \in \text{Hom}_k : \Delta P_k = 0\}.$$

We would like to know its dimension. Its direct computation is rather difficult. For that reason we will find later an indirect method to compute $\dim(\text{HHom}_k)$ by using the Gauss representation.

10.6.1 Examples of homogeneous polynomials

We have seen that in $\mathbb{R}^2$ the polynomials $r^k \cos k\varphi$ and $r^k \sin k\varphi$ are homogeneous and we have stated that they are a basis for Hom_k.

Let us consider another example which is not trivial: Hom_2 in $\mathbb{R}^3$. Every element $P \in \text{Hom}_2$ has the representation

$$P(x, y, z) = a_{11}x^2 + a_{12}xy + a_{13}xz + a_{22}y^2 + a_{23}yz + a_{33}z^2,$$

and the dimension

$$\dim(\text{Hom}_2) = 6.$$

By definition $P \in \text{HHom}_2$ if and only if

$$\Delta P = 0,$$

or, if written in detail,

$$\Delta P(x, y, z) = 2(a_{11} + a_{22} + a_{33}) = 0.$$

It follows that every $P \in \text{HHom}_2$ has the representation

$$P(x, y, z) = a_{11}x^2 + a_{12}xy + a_{13}xz + a_{22}y^2 + a_{23}yz + (-a_{11} - a_{22})z^2$$
$$= a_{11}(x^2 - z^2) + a_{12}xy + a_{13}xz + a_{22}(y^2 - z^2) + a_{23}yz.$$

This shows that

$$\dim(\text{HHom}_2) = 5.$$

There is a fundamental representation of *Gauss* which is a cornerstone for the theory of spherical harmonics.

10.7 Gauss representation of homogeneous polynomials

The Gauss representation which we present is not trivial even in the two-dimensional case! For that reason let us start with the two-dimensional case.

10.7.1 Gauss representation in $\mathbb{R}^2$

Let us consider only $k \geq 2$, since for $k = 0$ we know that Hom_0 only contains constants, and also in Hom_1
$$P_1(x, y) = ax + by,$$
which evidently satisfies $\Delta P_1 = 0$. Thus we have
$$\mathrm{HHom}_0 = \mathrm{Hom}_0,$$
$$\mathrm{HHom}_1 = \mathrm{Hom}_1.$$

Evidently this applies to an arbitrary dimension $n \geq 2$, and we have
$$\dim(\mathrm{HHom}_0) = 1,$$
$$\dim(\mathrm{HHom}_1) = n.$$

We consider the arbitrary polynomial $P_k \in \mathrm{Hom}_k$,
$$P_k(x, y) = \sum_{\ell=0}^{k} a_\ell x^\ell y^{k-\ell}.$$

As usual, $r = |x|$. Let us see that we have
$$P_k(x, y) = h_k(x) + r^2 h_{k-2}(x) + \cdots + r^{2s} h_{k-2s}(x), \tag{10.7}$$

where $s = [k/2]$, and h_j are harmonic polynomials which are homogeneous of order j. Actually, we will see that the representation holds for every $\ell = 0, \ldots, k$:
$$x^\ell y^{k-\ell} = r^k \cdot \sum_{j=0}^{[k/2]} [a_j \cos(k - 2j)\varphi + b_j \sin(k - 2j)\varphi].$$

Indeed, we put
$$x = r \cos \varphi, \qquad y = r \sin \varphi.$$

Now we have to see that the equivalent statement, i.e. that for some constants a_j, b_j, holds
$$(\cos \varphi)^\ell (\sin \varphi)^{k-\ell} = \sum_{j=0}^{[k/2]} [a_j \cos(k - 2j)\varphi + b_j \sin(k - 2j)\varphi].$$

Problem

Prove the last formula by induction in k. Use the formulas [7, p. 749], $\cos x \cos y = (1/2) \cos(x+y) + (1/2) \cos(x-y)$, $\sin x \sin y = -(1/2) \cos(x+y) + (1/2) \cos(x-y)$, $\sin x \cos y = (1/2) \sin(x + y) + (1/2) \sin(x - y)$.

The main point is that Gauss representation (10.7), p. 140, holds for an arbitrary dimension.

10.7.2 Gauss representation in $\mathbb{R}^n$

Theorem 10.2 *Let $P_k(x)$ be an element of Hom_k, i.e. a homogeneous polynomial of degree $k \geq 0$. Then we have the following Gauss representation*

$$P_k(x) = h_k(x) + r^2 h_{k-2}(x) + \cdots + r^{2s} h_{k-2s}(x),$$

where $s = [k/2]$ and h_j are harmonic polynomials which are homogeneous of order j.

We follow the elegant proof of Stein and Weiss [57, Chapter IV].

Before doing it we make some simple but important remarks concerning the action of the Laplace operator Δ on the homogeneous polynomials. Let $P_k = P_k(x) \in \mathrm{Hom}_k$. Then for $k \geq 2$ it is easy to see that $\Delta P_k \in \mathrm{Hom}_{k-2}$. Indeed, for every $y \in \mathbb{R}^n$ and $t \neq 0$ we may differentiate the equality

$$P_k(ty_1, \ldots, ty_n) = P_k(ty) = t^k P_k(y)$$

with respect to y. For every $j = 1, \ldots, n$ we obtain

$$\frac{\partial}{\partial x_j} P_k(ty) = t \left(\frac{\partial}{\partial x_j} P_k \right)(ty) = t^k \frac{\partial}{\partial x_j} P_k(y),$$

hence,

$$\left(\frac{\partial}{\partial x_j} P_k \right)(ty) = t^{k-1} \frac{\partial}{\partial x_j} P_k(y).$$

The last shows that the polynomial

$$\frac{\partial}{\partial x_j} P_k \in \mathrm{Hom}_{k-1},$$

i.e. every differentiation decreases the degree of homogeneity by one.

However, the multiplication with $|x|^2$ plays the opposite role to the action of the operator Δ. If $P_k \in \mathrm{Hom}_k$ then the polynomial

$$|x|^2 P_k(x) = (x_1^2 + \cdots + x_n^2) P_k(x)$$

satisfies

$$|x|^2 P_k(x) \in \mathrm{Hom}_{k+2}.$$

Indeed,

$$|tx|^2 P_k(tx) = t^{2+k} |x|^2 P_k(x)$$

for every $t \neq 0$.

There is a useful inner product on the finite-dimensional space HHom_k where we will see that the operators Δ and the multiplication by $|x|^2$ are indeed adjoint.[2]

---

[2] On a finite-dimensional space all norms are equivalent, and the choice of a special norm has purely algebraic consequences.

If $P_k, Q_k \in \text{HHom}_k$ and

$$P_k(x) = \sum_{|\alpha|=k} a_\alpha x^\alpha$$

$$Q_k(x) = \sum_{|\alpha|=k} b_\alpha x^\alpha,$$

then their *inner product* is given by

$$\langle P_k, Q_k \rangle = \sum_{|\alpha|=k} \alpha! \cdot a_\alpha \cdot \overline{b_\alpha}.$$

It is easy to check that this is an inner product since $\langle P_k, P_k \rangle = \sum_{|\alpha|=k} \alpha! \cdot |a_\alpha|^2$, and $\langle P_k, P_k \rangle = 0$ implies $P_k = 0$. There is another representation of this inner product, namely

$$\langle P_k, Q_k \rangle = P_k \left(\frac{\partial}{\partial x_1}, \ldots, \frac{\partial}{\partial x_n} \right) \overline{Q_k(x)},$$

where we have denoted the differential operator

$$P_k(D_x) = P_k \left(\frac{\partial}{\partial x_1}, \ldots, \frac{\partial}{\partial x_n} \right) = \sum_{|\alpha|=k} a_\alpha \frac{\partial^{|\alpha|}}{\partial x_1^{\alpha_1} \cdots \partial x_n^{\alpha_n}}.$$

For example if we have the polynomial

$$P_{2k}(x) = |x|^{2k},$$

then

$$P_k \left(\frac{\partial}{\partial x_1}, \ldots, \frac{\partial}{\partial x_n} \right) = \Delta^k,$$

where Δ^k denotes the polyharmonic operator of order k.

A useful property of this inner product is that the operator Δ and the multiplication by $|x|^2$ are indeed *adjoint* with respect to it. If we let the polynomials $P_k \in \text{HHom}_k$ and $Q_{k-2} \in \text{HHom}_{k-2}$, then their inner product is (we write the variables x in order to make the notation clear)

$$\langle \Delta P_k(x), Q_{k-2}(x) \rangle = \overline{\langle Q_{k-2}(x), \Delta P_k(x) \rangle} \qquad (10.8)$$

$$= \overline{Q_{k-2}(D_x) \overline{\Delta P_k(x)}}$$

$$= \overline{\Delta Q_{k-2}(D_x) \overline{P_k(x)}}$$

$$= \overline{\langle |x|^2 Q_{k-2}(x), P_k(x) \rangle}$$

$$= \langle P_k(x), |x|^2 Q_{k-2}(x) \rangle.$$

Essentially, we have used the simple fact that the operator Δ and $Q_{k-2}(D_x)$ commute since they have constant coefficients.

Spherical harmonics and polyharmonic functions 143

Proof of Theorem 10.2 Evidently, every polynomial of degree < 2 is harmonic. So we may assume that $k \geq 2$.

(1) We have seen that the operator of differentiation $\partial/\partial x_j$ maps the set

$$\mathrm{HHom}_k \longrightarrow \mathrm{HHom}_{k-1},$$

hence the operator Δ maps $\mathrm{HHom}_k \longrightarrow \mathrm{HHom}_{k-2}$. The remarkable thing about the operator Δ is that it maps HHom_k on HHom_{k-2}.

(2) Let us prove this statement assuming that the opposite is true, i.e. the set

$$\mathrm{HHom}_{k-2} \setminus \Delta(\mathrm{HHom}_k) \neq \varnothing,$$

and contains a nonzero element. Then by the theorem for orthogonal projection (in the finite-dimensional case it is evident) it follows that there exists an element $R_{k-2} \in \mathrm{HHom}_{k-2} \setminus \Delta(\mathrm{HHom}_k)$ with $R_{k-2} \neq 0$, and

$$R_{k-2} \perp \Delta(\mathrm{HHom}_k).$$

The last means that

$$0 = \langle \Delta P_k, R_{k-2} \rangle,$$

for every $P_k \in \mathrm{HHom}_k$. By the adjointness of the operators Δ and multiplication by $|x|^2$ in (10.8), p. 142, it follows that

$$0 = \langle |x|^2 R_{k-2}, P_k(x) \rangle.$$

We put now

$$P_k(x) = |x|^2 R_{k-2}.$$

It follows that $P_k = 0$, hence $R_{k-2} = 0$. This contradiction implies that

$$\Delta(\mathrm{HHom}_k) = \mathrm{HHom}_{k-2}.$$

(3) In Hom_k we have evidently two subspaces: HHom_k and

$$|x|^2 \mathrm{Hom}_{k-2} = \{|x|^2 P_{k-2} : P_{k-2} \in \mathrm{Hom}_{k-2}\}.$$

In the previous point we have proved that

$$|x|^2 \mathrm{Hom}_{k-2} = |x|^2 \Delta \mathrm{Hom}_k.$$

We see that the direct sum of the two subspaces exhausts Hom_k. Indeed, if we take some $Q_k \in \mathrm{Hom}_k$ such that it is orthogonal to $|x|^2 \mathrm{Hom}_{k-2} = |x|^2 \Delta \mathrm{Hom}_k$ then we have

$$\langle Q_k, |x|^2 \Delta P_k \rangle = 0$$

for every $P_k \in \mathrm{Hom}_k$. Since the operators Δ and the multiplication by $|x|^2$ are mutually adjoint, the last implies

$$0 = \langle Q_k, |x|^2 \Delta P_k \rangle = \langle \Delta Q_k, \Delta P_k \rangle,$$

and taking $P_k = Q_k$ gives $\Delta Q_k = 0$, i.e. $Q_k \in \mathrm{HHom}_k$. ■

144 Multivariate polysplines

Corollary 10.3 *Let $P(x)$ be an arbitrary polynomial which satisfies the polyharmonic equation*

$$\Delta^k P = 0.$$

Then it is possible to expand it as

$$P(x) = \sum_{j=0}^{k-1} |x|^{2j} h_j(x),$$

where $h_j(x)$ are harmonic polynomials.

The immediate consequence of the Gauss representation is that we can compute the dimension d_k of the spaces HHom_k for all integers $k \geq 2$.

Let us note the evident fact that

$$\dim(\mathrm{HHom}_0) = 1,$$
$$\dim(\mathrm{HHom}_1) = n.$$

This is evident since $\mathrm{HHom}_k = \mathrm{Hom}_k$ for $k = 0, 1$. Now let us proceed to the case $k \geq 2$.

Proposition 10.4 *For $k \geq 2$ the dimension $d_k = \dim(\mathrm{HHom}_k)$ is given by a polynomial of degree $n - 2$ in k, namely,*

$$d_k = \frac{1}{(n-2)!}(n + 2k - 2)(n + k - 3) \cdots (k + 1).$$

Proof Since every element $P_k \in \mathrm{Hom}_k$ has the representation

$$P_k(x) = h_k(x) + r^2 h_{k-2}(x) + \cdots + r^{2s} h_{k-2s}(x),$$

we see that all terms are linearly independent and $h_j(x)$ runs the space HHom_j having dimension d_j. It follows that

$$\dim(\mathrm{Hom}_k) = d_k + d_{k-2} + \cdots + d_{k-2s},$$

where $s = [k/2]$. Since $[(k-2)/2] = [k/2] - 1$, the same equality for $k - 2$ is

$$\dim(\mathrm{Hom}_{k-2}) = d_{k-2} + d_{k-4} + \cdots + d_{k-2s}.$$

We subtract and obtain by simple manipulations

$$d_k = \dim(\mathrm{Hom}_k) - \dim(\mathrm{Hom}_{k-2}) \tag{10.9}$$

$$= \binom{n+k-1}{k} - \binom{n+k-3}{k-2} = \frac{(n+k-1)!}{(n-1)!k!} - \frac{(n+k-3)!}{(n-1)!(k-2)!}$$

$$= \frac{(n+k-3)!}{(n-1)!(k-2)!}\left(\frac{(n+k-1)(n+k-2)}{k(k-1)} - 1\right)$$

$$= \frac{(n+k-3)!}{(n-1)!(k-2)!} \cdot \frac{(n+2k-2)(n-1)}{k(k-1)}$$

$$= \frac{(k+n-3)!}{(n-2)!k!} \cdot (n+2k-2) \tag{10.10}$$

$$= \frac{1}{(n-2)!}(n+2k-2)(n+k-3)\cdots(k+1),$$

which shows that $d_k = h_{n-2}(k)$ is a polynomial of degree $n-2$ in k. ∎

Corollary 10.5 *For $k \longrightarrow \infty$, the dimension $d_k = \dim(\mathrm{HHom}_k)$ satisfies*

$$d_k \approx \frac{2}{(n-2)!}k^{n-2}. \tag{10.11}$$

In particular, for $n = 2$ we obtain

$$d_k = \frac{(k+1)!}{k!} - \frac{(k-1)!}{(k-2)!} = 2.$$

For $n = 3$ we obtain

$$d_k = \binom{k+2}{k} - \binom{k}{k-2} = \frac{(k+2)!}{k!2!} - \frac{k!}{(k-2)!2!}$$

$$= \frac{(k+2)(k+1)}{2} - \frac{k(k-1)}{2}$$

$$= 2k+1.$$

Thus we have the following corollary.

Corollary 10.6 *In $\mathbb{R}^2$ and for $k \geq 1$ the polynomials $r^k \cos k\varphi$ and $r^k \sin k\varphi$ form a basis of the space HHom_k. In $\mathbb{R}^3$ for all $k \geq 2$ the dimension satisfies*

$$\dim(\mathrm{HHom}_k) = 2k+1.$$

10.8 Gauss representation: analog to the Taylor series, the polyharmonic paradigm

In Section 1.5, p. 10, we announced the so-called *polyharmonic paradigm*. Since we are now at the heart of the subject it will be instructive to recall, at the risk of some repetition, its manifestation by the Gauss representation.

The Gauss representation may be interpreted as a version of the one-dimensional Taylor expansion. Indeed, for the solutions of the equation

$$\frac{d^{k+1}}{dt^k} f(t) = 0,$$

which are in fact the polynomials of degree k, we have the Taylor representation

$$f(t) = \sum_{j=0}^{k} h_j t^j.$$

It should be noted that the constants α_j satisfy the equation

$$\frac{d}{dt} \alpha_j = 0.$$

Now let us replace formally

$$\frac{d}{dt} \longrightarrow \Delta.$$

Thus for the polynomial solutions of the equation

$$\Delta^{k+1} f(x) = 0$$

we have the Gauss representation

$$f(x) = \sum_{j=0}^{k} h_j(x) |x|^{2j}.$$

Here the "coefficients" h_j satisfy

$$\Delta h_j(x) = 0.$$

This shows an interesting analogy which is one of the leading threads of this book.

10.8.1 The Almansi representation

A *Gauss-type representation* holds for a much wider class than the polynomials. A much more general class was considered by the Italian mathematician *Almansi* who, in 1899, proved the following theorem.

Theorem 10.7 *Let the function $f(x)$ be polyharmonic of order $p \geq 1$ in the ball $B(0; R)$, i.e. it satisfies*

$$\Delta^p f = 0 \quad \text{in } B(0; R).$$

Then there exist uniquely determined functions $f_j(x)$, for $j = 0, \ldots, p-1$, which are harmonic in the domain D and such that the following representation holds:

$$f(x) = f_0(x) + f_1(x)|x|^2 + \cdots + f_{p-1}(x)|x|^{2p-2} \quad \text{for } x \in B(0; R).$$

Spherical harmonics and polyharmonic functions 147

Let us note that the difference between the Gauss and the Almansi representations is that the last holds only in the ball $B(0; R)$ since the function f may have singularities on the boundary of $B(0; R)$. Here we do not also have the distinction of the homogeneity degrees.

We will postpone the proof of this result until the end of this chapter.

We now return to the Gauss representation in order to make the most of what it can tell us about the structure of the space of the spherical harmonics.

10.9 The sets $\mathcal{H}_k$ are eigenspaces for the operator Δ_θ

Very often the restriction of the elements of HHom$_k$ on the sphere $r = 1$ are called as we have already said **(surface) spherical harmonics** and are denoted by

$$\mathcal{H}_k := \{ f : |x|^k f(\theta) \in \text{HHom}_k \}. \tag{10.12}$$

Let us note that although the set

$$\bigcup_{k=0}^{\infty} \mathcal{H}_k$$

consists only of restrictions of the elements of the set

$$\bigcup_{k=0}^{\infty} \text{HHom}_k$$

the correspondence is one-to-one. Indeed, by the maximum principle, if two harmonic functions coincide on the sphere $S(0; R) = \partial B(0; R)$ then they also coincide in the interior of $B(0; R)$. So as polynomials they coincide everywhere.

From the Gauss representation (or from Corollary 10.3, p. 144) we immediately obtain the following corollary.

Corollary 10.8 *For every polynomial $P(x)$ there exists a finite sequence of harmonic homogeneous polynomials $h_k(x) \in$ HHom$_k$ such that*

$$P(x) = \sum_k h_k(x) \quad \text{for } |x| = 1.$$

An important fact is that the harmonic homogeneous polynomials satisfy a special equation related to the Laplace operator.

Lemma 10.9 *Let $P_k(x) \in$ HHom$_k$. Then*

$$\Delta_\theta P_k(\theta) = -k(n + k - 2) P_k(\theta),$$

i.e. the function $Y_k(\theta) = P_k(\theta)$ is an eigenfunction of the operator Δ_θ on the sphere $\mathbb{S}^{n-1}$.

148 Multivariate polysplines

Proof Let us apply the Laplace operator to $P_k(x) = |x|^k P_k(\theta)$. Due to the separation of the variables we have obtained in formulas (10.4) and (10.5), p. 132, the representation

$$\Delta_r = \frac{\partial^2}{\partial r^2} + \frac{n-1}{r}\frac{\partial}{\partial r},$$

$$\Delta = \Delta_r + \frac{1}{r^2}\Delta_\theta.$$

Hence, for all x in $\mathbb{R}^n$ we obtain

$$0 = \Delta P_k(x) = \left(\Delta_r + \frac{1}{r^2}\Delta_\theta\right) r^k P_k(\theta)$$

$$= P_k(\theta) \cdot \Delta_r(r^k) + \frac{r^k}{r^2}\Delta_\theta(P_k(\theta))$$

$$= P_k(\theta) \cdot \left(k(k-1)r^{k-2} + (n-1)kr^{k-2}\right) + r^{k-2}\Delta_\theta(P_k(\theta)).$$

This implies, after putting $r = 1$, that

$$\Delta_\theta P_k(\theta) = -k(n+k-2)P_k(\theta) \quad \text{on } \mathbb{S}^{n-1},$$

and since by the definition $Y_k(\theta) = P_k(\theta)$, we find that the functions Y_k are eigenfunctions of the operator Δ_θ on the sphere $\mathbb{S}^{n-1}$ with eigenvalues $-k(n+k-2)$. ∎

There is one more remarkable feature of the different spaces $\mathcal{H}_k$ and $\mathcal{H}_l$ for $k \neq l$. They are mutually orthogonal in the L_2-scalar product:

Lemma 10.10 *Let $Y_k \in \mathcal{H}_k$ and $Y_l \in \mathcal{H}_l$ be two arbitrary spherical harmonics such that $k \neq l$. Then*

$$\int_{\mathbb{S}^{n-1}} Y_k(\theta) Y_l(\theta) \, d\theta = 0.$$

Proof Using the eigenvalue property of Lemma 10.9, we obtain the equalities

$$I := \int_{\mathbb{S}^{n-1}} Y_k(\theta) Y_l(\theta) \, d\theta = \frac{-1}{k(n+k-2)} \int_{\mathbb{S}^{n-1}} \Delta_\theta Y_k(\theta) Y_l(\theta) \, d\theta.$$

By the self-adjointness of the operator Δ_θ, Theorem 10.1, p. 133, we obtain

$$I = \frac{-1}{k(n+k-2)} \int_{\mathbb{S}^{n-1}} Y_k(\theta) \Delta_\theta Y_l(\theta) \, d\theta$$

$$= \frac{l(n+l-2)}{k(n+k-2)} \int_{\mathbb{S}^{n-1}} Y_k(\theta) \Delta_\theta Y_l(\theta) \, d\theta.$$

This implies that $I = 0$; otherwise we should have

$$k(n+k-2) = l(n+l-2).$$

The last is impossible since the function $\sigma(t) = t(n+t-2)$ has derivative $\sigma'(t) = 2t + n - 2 > 0$ for $n \geq 2$ and $t > 0$, and as such is strictly increasing. ∎

10.10 Completeness of the spherical harmonics in $L_2(\mathbb{S}^{n-1})$

Now with respect to the scalar product

$$\langle f, g \rangle = \int_{\mathbb{S}^{n-1}} f(\theta)\overline{g(\theta)}\, d\theta$$

we may orthonormalize the elements in every set $\mathcal{H}_k$. We will denote an orthonormal basis in $\mathcal{H}_k$ by

$$\{Y_{k,1}(\theta), Y_{k,2}(\theta), \ldots, Y_{k,d_k}(\theta)\}. \tag{10.13}$$

We have also seen that all $\mathcal{H}_k$ are mutually orthogonal. Thus if we let $k \geq 0$, we obtain an orthonormal basis for the whole set $\cup \mathcal{H}_k$.

Let us note that in the two-dimensional case since

$$d_0 = 1, \; d_1 = d_2 = d_3 = \cdots = 2,$$

the above orthonormal basis is given by

$$Y_{0,1}(\theta) = 1,$$

$$Y_{1,1}(\theta) = \frac{1}{\sqrt{\pi}} \cos \varphi, \qquad Y_{1,2}(\theta) = \frac{1}{\sqrt{\pi}} \sin \varphi,$$

$$Y_{2,1}(\theta) = \frac{1}{\sqrt{\pi}} \cos 2\varphi, \qquad Y_{2,2}(\theta) = \frac{1}{\sqrt{\pi}} \sin 2\varphi,$$

$$\vdots$$

$$Y_{k,1}(\theta) = \frac{1}{\sqrt{\pi}} \cos k\varphi, \qquad Y_{k,2}(\theta) = \frac{1}{\sqrt{\pi}} \sin k\varphi,$$

$$\vdots$$

Using this result we can prove the following fundamental theorem which provides the analog to the Fourier series representation.

Theorem 10.11 *1. The set*

$$\left[\bigcup_{k=0}^{\infty} \mathcal{H}_k\right]_{\mathrm{lin}},$$

where $[\cdot]_{\mathrm{lin}}$ denotes the linear hull (finite linear combinations of elements), is dense in $L_2(\mathbb{S}^{n-1})$ and in the space of the continuous functions $C(\mathbb{S}^{n-1})$ in their norms, respectively.

2. *Further, there exists a unique representation in infinite series (called the* Laplace series*)*

$$f(\theta) = \sum_{k=0}^{\infty} Y_k(\theta) \tag{10.14}$$

$$= \sum_{k=0}^{\infty} \sum_{\ell=1}^{d_k} \alpha_{k,\ell} Y_{k,\ell}(\theta) \quad \text{for } \theta \in \mathbb{S}^{n-1},$$

and the convergence is in the L_2 norm on $\mathbb{S}^{n-1}$.

Proof As is well known from integration theory on compact spaces the space $C(\mathbb{S}^{n-1})$ is dense in the space $L_2(\mathbb{S}^{n-1})$.[3]

Due to this, assuming some $\varepsilon > 0$ be given, we may choose a function $g \in C(\mathbb{S}^{n-1})$ such that

$$\int_{\mathbb{S}^{n-1}} (f(\theta) - g(\theta))^2 \, d\theta < \frac{\varepsilon}{2}.$$

It remains to approximate the function g through elements of $\cup_{k=0}^{\infty} \mathcal{H}_k$. For that purpose we apply the Stone–Weierstrass theorem to the sphere $\mathbb{S}^{n-1}$.[4]

Thus after applying the Stone–Weierstrass theorem we obtain a polynomial P whose degree is not important for us, and such that

$$\max_{\theta \in \mathbb{S}^{n-1}} |g(\theta) - P(\theta)| < \frac{\varepsilon}{2}.$$

By Corollary 10.8, p. 147, we have

$$P(\theta) = \sum_k h_k(\theta) \quad \text{for } \theta \in \mathbb{S}^{n-1},$$

where $h_k \in \mathcal{H}_k$, i.e. $P(\theta)$ belongs to $\left[\bigcup_{k=0}^{\infty} \mathcal{H}_k \right]_{\text{lin}}$.

Hence, we obtain

$$\|f - P\|_{L_2(\mathbb{S}^{n-1})} \leq \|f - g\|_{L_2(\mathbb{S}^{n-1})} + \|g - P\|_{L_2(\mathbb{S}^{n-1})}$$

$$\leq \frac{\varepsilon}{2} + \sqrt{\int_{\mathbb{S}^{n-1}} |g(\theta) - P(\theta)|^2 \, d\theta}$$

$$\leq \frac{\varepsilon}{2} + \varepsilon \sqrt{\sigma_{n-1}}.$$

The last tends to 0 when ε tends to 0. This finishes the proof of the first statement.

---

[3] One may also view this as a result on a manifold. In such a case we may act more precisely, by making a partition of unity and prove this density result in every local chart. For a rigorous proof we refer to a standard course on analysis on manifolds, e.g. Narasimhan [38].

[4] See Rudin [48], for the most general Bishop–Stone–Weierstrass theorem, where the classical result follows easily. In order to be able to apply this theorem we have to separate by polynomials every pair of points θ^1 and θ^2 on $\mathbb{S}^{n-1}$. We can obviously do it by a linear function (which is a polynomial). Or we can assume for simplicity that $\theta^1 = (1, 0, \ldots, 0)$ and choose the function in the form $l(x) = a_2 x_2 + \cdots + a_n x_n$. Since θ^2 has to have at least one nonzero component for $j = 2, \ldots, n$, say $j = j_1$, we will choose all other $a_j = 0$ and only $a_{j_1} = 1$. This gives $l(\theta^2) = \theta_{j_1}^2 \neq 0$ but $l(\theta^1) = 0$.

Spherical harmonics and polyharmonic functions 151

In order to prove the second statement, we will need the above results, namely Lemma 10.10. This is a standard argument from the theory of orthogonal expansions in a Hilbert space. In fact, we see that the set of functions

$$B = \bigcup_{k=0}^{\infty} \{Y_{k,1}(\theta), Y_{k,2}(\theta), \ldots, Y_{k,d_k}(\theta)\},$$

is a basis of $L_2(\mathbb{S}^{n-1})$. Let us consider the closure in $L_2(\mathbb{S}^{n-1})$, call it $\mathcal{M}$, of the linear hull of all elements of B. Evidently,

$$\mathcal{M} = clos_{L_2(\mathbb{S}^{n-1})} \left[\bigcup_{k=0}^{\infty} \mathcal{H}_k \right]_{lin}.$$

Assume that $\mathcal{M} \subset L_2(\mathbb{S}^{n-1})$ but $\mathcal{M} \neq L_2(\mathbb{S}^{n-1})$. Then by the theorem about the orthogonal projection in a Hilbert space, there exists a function $f \in L_2(\mathbb{S}^{n-1})$, with $f \neq 0$, which is orthogonal to $\mathcal{M}$, hence to all $Y_{k,\ell}$, i.e.

$$\int_{\mathbb{S}^{n-1}} f(\theta) \overline{Y_{k,\ell}(\theta)} \, d\theta = 0 \quad \text{for } k = 0, 1, 2, \ldots, \text{ and } \ell = 1, \ldots, d_k.$$

Then if we take an arbitrary finite linear combination of elements of B,

$$S(\theta) = \sum \alpha_{k,\ell} Y_{k,\ell}(\theta);$$

evidently $S \in \left[\bigcup_{k=0}^{\infty} \mathcal{H}_k \right]_{lin}$. We obtain

$$\|f - S\|_{L_2} = \|f\|_{L_2} - \langle f, S \rangle - \langle S, f \rangle + \|S\|_{L_2} = \|f\|_{L_2} + \|S\|_{L_2} \geq \|f\|_{L_2} > 0.$$

But by the first part of the theorem this is impossible, since we may approximate every $f \in L_2(\mathbb{S}^{n-1})$ through elements of $\left[\bigcup_{k=0}^{\infty} \mathcal{H}_k \right]_{lin}$. ∎

Now we see that we have a complete analogy with the case of the Fourier series. Indeed, recall that there we have the representation

$$f(\varphi) = \frac{a_0}{2} + \sum_{k=1}^{\infty} (a_k \cos k\varphi + b_k \sin k\varphi) \quad \text{for } 0 \leq \varphi \leq 2\pi,$$

where for $k = 0$ we have only one term, a_0. The meaning of the above theorem is the following expansion in orthonormal basis which generalizes all elements of the Fourier expansion:

$$f(\theta) = \sum_{k=0}^{\infty} \left(\sum_{\ell=1}^{d_k} \alpha_{k,\ell} Y_{k,\ell}(\theta) \right) \quad \text{for } \theta \in \mathbb{S}^{n-1}.$$

10.11 Solutions of $\Delta w(x) = 0$ with separated variables

Let us consider solutions of $\Delta w(x) = 0$ having the form

$$w(x) = u(r)v(\theta).$$

Due to the separation in the Laplace operator we obtain

$$0 = \Delta w(x) = \left(\Delta_r + \frac{1}{r^2}\Delta_\theta\right) w(x) = v(\theta) \cdot \Delta_r u(r) + u(r)\frac{1}{r^2}\Delta_\theta v(\theta).$$

Now let us assume that for some integer $k \geq 0$,

$$v(\theta) = Y_k(\theta).$$

By applying Lemma 10.9, p. 147, we obtain

$$0 = \Delta w(x) = Y_k(\theta) \cdot \Delta_r u(r) - u(r)\frac{1}{r^2}k(n+k-2)Y_k(\theta).$$

It follows that

$$\Delta_r u(r) = \frac{k(n+k-2)}{r^2}u(r),$$

or

$$\left(\frac{\partial^2}{\partial r^2} + \frac{n-1}{r}\frac{\partial}{\partial r}\right)u(r) = \frac{k(n+k-2)}{r^2}u(r). \tag{10.15}$$

We will put

$$L_{(k)}u(r) := \left(\frac{\partial^2}{\partial r^2} + \frac{n-1}{r}\frac{\partial}{\partial r}\right)u(r) - \frac{k(n+k-2)}{r^2}u(r), \tag{10.16}$$

and the operator $L_{(k)}$ will play a central role in our further considerations. Obviously

$$L_{(k)}u(r) = \frac{1}{r^{n-1}}\frac{\partial}{\partial r}r^{n-1}\frac{\partial}{\partial r}u(r) - \frac{k(n+k-2)}{r^2}u(r). \tag{10.17}$$

It is easy to prove that we also have the following representation:

$$L_{(k)}u(r) = \frac{1}{r^{n+k-1}}\frac{d}{dr}\left[r^{n+2k-1}\frac{d}{dr}\left[\frac{1}{r^k}u(r)\right]\right] \tag{10.18}$$

$$= r^{k-1}\frac{d}{dr}\left[\frac{1}{r^{n+2k-3}}\frac{d}{dr}[r^{n+k-2}u(r)]\right].$$

We can directly check the validity of the following proposition.

Spherical harmonics and polyharmonic functions 153

Proposition 10.12 *For $r > 0$, the two linear independent solutions to the ordinary differential equation (10.15), $L_{(k)}u(r) = 0$, are given by $R_1(r) = R_{k,1}(r)$ and $R_2(r) = R_{k,2}(r)$,*

$$\begin{cases} R_1(r) = r^k, & R_2(r) = r^{-n-k+2} & \text{for } k \geq 0 \text{ and } n \geq 3, \\ R_1(r) = r^k, & R_2(r) = r^{-n-k+2} & \text{for } k \geq 1 \text{ and } n = 2, \\ R_1(r) = 1, & R_2(r) = \log r & \text{for } k = 0 \text{ and } n = 2, \end{cases} \quad (10.19)$$

and every solution $u(r)$ has the form

$$u(r) = C_1 R_1(r) + C_2 R_2(r),$$

where C_1, C_2 are arbitrary constants.

Exercise 10.13 *Prove Proposition 10.12.*

Thus we see that the solutions which have separated variables are of the form

$$\begin{cases} r^k Y_k(\theta), & r^{-n-k+2} Y_k(\theta) & \text{for } k \geq 0 \text{ and } n \geq 3, \\ r^k Y_k(\theta), & r^{-n-k+2} Y_k(\theta) & \text{for } k \geq 1 \text{ and } n = 2, \\ 1, & \log r & \text{for } k = 0 \text{ and } n = 2, \end{cases}$$

or, written more briefly, of the form

$$R_i(r) Y_k(\theta) \quad \text{for } i = 1, 2 \text{ and } k \geq 0.$$

Let us recall that for $k = 0$ we have only one spherical harmonic in the set $\mathcal{H}_0$, namely $Y_{k,1}(\theta) = 1$ (or an arbitrary nonzero constant).

10.12 Zonal harmonics $Z_{\theta'}^{(k)}(\theta)$: the functional approach

Here we explain the functional approach to the zonal harmonics. Within this approach one can avoid the explicit definition of the zonal harmonics which is typical for all books on special functions [23,53].

Let us consider the space $\mathcal{H}_k$ endowed with the inner product (Hermitean symmetric)

$$\langle f, g \rangle = \int_{\mathbb{S}^{n-1}} f(\theta) \overline{g(\theta)} \, d\theta.$$

Exercise 10.14 *Prove the statement that $\langle f, g \rangle$ defines an inner product on $\mathcal{H}_k$.*

We consider some fixed point $\theta' \in \mathbb{S}^{n-1}$ and the functional

$$L(f) = f(\theta')$$

on the space $\mathcal{H}_k$. This is obviously a linear functional. As a functional on a finite-dimensional space, it may be represented by an element $Z_{\theta'}^{(k)} \in \mathcal{H}_k$. Namely, we have

$$f(\theta') = \int_{\mathbb{S}^{n-1}} Z_{\theta'}^{(k)}(\theta) \overline{f(\theta)} \, d\theta \tag{10.20}$$

for every $f(\theta) \in \mathcal{H}_k$.

Definition 10.15 *The function $Z_{\theta'}^{(k)}(\theta)$, which for every fixed θ', as an element of $\mathcal{H}_k$, is uniquely defined by equality (10.20), is called a **zonal harmonic**.*

We assume that we are given an orthonormal basis of $\mathcal{H}_k$, denoted by

$$\{Y_{k,1}(\theta), \ldots, Y_{k,d_k}(\theta)\}.$$

Clearly, in the space $\mathcal{H}_k$ with complex coefficients we may choose in this basis only polynomials $|x^k| Y_{k,\ell}(\theta)$ with real coefficients. Here, as usual, $d_k = \dim(\mathcal{H}_k)$.

From Theorem 10.11, p. 149, we immediately obtain Corollary 10.16.

Corollary 10.16 *The components $Y_k(\theta)$ of the Laplace series of the function $h(\theta)$,*

$$h(\theta) = \sum_{k=0}^{\infty} Y_k(\theta) \qquad (\theta \in \mathbb{S}^{n-1})$$

are obtained by the equality

$$Y_k(\theta') = \int_{\mathbb{S}^{n-1}} Z_{\theta'}^{(k)}(\theta) h(\theta) \, d\theta.$$

We have the following seven results.

Proposition 10.17 *1. The zonal harmonic $Z_{\theta'}^{(k)}(\theta)$ permits the representation:*

$$Z_{\theta'}^{(k)}(\theta) = \sum_{\ell=1}^{d_k} Y_{k,\ell}(\theta) \overline{Y_{k,\ell}(\theta')} \quad \text{for } \theta, \theta' \in \mathbb{S}^{n-1}.$$

2. The function $Z_{\theta'}^{(k)}(\theta)$ is real-valued and symmetric, i.e. $Z_{\theta'}^{(k)}(\theta) = Z_{\theta}^{(k)}(\theta')$ for all $\theta, \theta' \in \mathbb{S}^{n-1}$.

3. Let ρ be a rotation of $\mathbb{R}^n$, i.e. it is a linear transform of $\mathbb{R}^n$ which may be represented through a matrix A which is orthogonal, i.e. $AA^T = I$, and $\det(A) = 1$. Then

$$Z_{A\theta'}^{(k)}(A\theta) = Z_{\theta'}^{(k)}(\theta)$$

for every $\theta, \theta' \in \mathbb{S}^{n-1}$.

4. For every $\theta \in \mathbb{S}^{n-1}$ it is the case that

$$Z_{\theta}^{(k)}(\theta) = \frac{d_k}{\sigma_{n-1}}.$$

5. For every $\theta \in \mathbb{S}^{n-1}$ it is the case that
$$Z_\theta^{(k)}(\theta) = \sum_{\ell=1}^{d_k} |Y_{k,\ell}(\theta)|^2 = \frac{d_k}{\sigma_{n-1}}.$$

6. The norm
$$\|Z_{\theta'}^{(k)}\|_{L_2(\mathbb{S}^{n-1})} = \sqrt{\frac{d_k}{\sigma_{n-1}}}.$$

7. For every $\theta, \theta' \in \mathbb{S}^{n-1}$ the inequality
$$|Z_{\theta'}^{(k)}(\theta)| \leq \frac{d_k}{\sigma_{n-1}}$$

holds.

Proof *Proof of 1.* Since $\{Y_{k,\ell}\}$ is a basis of $\mathcal{H}_k$ we may express every element of $\mathcal{H}_k$ as a linear combination. In particular, for every fixed θ' the function $Z_{\theta'}^{(k)}(\theta)$ which belongs to $\mathcal{H}_k$ is expanded as

$$Z_{\theta'}^{(k)}(\theta) = \sum_{\ell=1}^{d_k} \left(Z_{\theta'}^{(k)}, Y_{k,\ell} \right) Y_{k,\ell}(\theta).$$

Now by the definition of $Z_{\theta'}^{(k)}(\theta)$ we have

$$\left(Z_{\theta'}^{(k)}, Y_{k,\ell} \right) = \int_{\mathbb{S}^{n-1}} Z_{\theta'}^{(k)}(\theta) \overline{Y_{k,\ell}(\theta)} \, d\theta = \overline{Y_{k,\ell}(\theta')}.$$

Proof of 2. Since we have chosen a basis $\{Y_{k,\ell}\}$ which is only composed of real-valued functions the proof follows immediately.

Proof of 3. For the proof of this let us note that if $Y_k \in \mathcal{H}_k$ then the function

$$Y_k(A\theta) \in \mathcal{H}_k.$$

Indeed, we have $f(x) = |x|^k Y_k(\theta) \in \text{HHom}_k$ and now since A is orthogonal, it follows that

$$f(Ax) = |Ax|^k Y_k(A\theta)$$
$$= |x|^k Y_k(A\theta),$$

and $Y_k(A\theta), Y_k(A^{-1}\theta) \in \mathcal{H}_k$. Hence, for every $Y_k \in \mathcal{H}_k$ we obtain, after changing the variables $\psi = A\theta$,

$$\int_{\mathbb{S}^{n-1}} Z_{A\theta'}^{(k)}(A\theta) Y_k(\theta) \, d\theta = \int_{\mathbb{S}^{n-1}} Z_{A\theta'}^{(k)}(\theta) Y_k(A^{-1}\psi) \, d\psi$$
$$= Y_k(A^{-1}A\theta') = Y_k(\theta')$$
$$= \int_{\mathbb{S}^{n-1}} Z_{\theta'}^{(k)}(\theta) Y_k(\theta) \, d\theta.$$

The last equality follows by the definition of $Z_{\theta'}^{(k)}(\theta)$. The result follows owing to the uniqueness of $Z_{\theta'}^{(k)}(\theta)$.

Proof of 4. For every two points $\theta, \theta' \in \mathbb{S}^{n-1}$ there obviously exists a rotation ρ represented by an orthogonal matrix A such that

$$A\theta = \theta'.$$

From property (3) it follows that

$$Z_{\theta'}^{(k)}(\theta') = Z_{A\theta}^{(k)}(A\theta) = Z_{\theta}^{(k)}(\theta),$$

hence, $Z_{\theta}^{(k)}(\theta)$ is independent of $\theta \in \mathbb{S}^{n-1}$. From property (1) it follows that

$$Z_{\theta}^{(k)}(\theta) = \sum_{\ell=1}^{d_k} |Y_{k,\ell}(\theta)|^2.$$

Since $\{Y_{k,\ell}\}$ is an orthonormal basis we have $\int_{\mathbb{S}^{n-1}} |Y_{k,\ell}|^2 \, d\theta = 1$, hence

$$d_k = \sum_{\ell=1}^{d_k} \int_{\mathbb{S}^{n-1}} |Y_{k,\ell}(\theta)|^2 \, d\theta = \int_{\mathbb{S}^{n-1}} \sum_{\ell=1}^{d_k} |Y_{k,\ell}(\theta)|^2 \, d\theta$$

$$= Z_{\theta}^{(k)}(\theta) \cdot \sigma_{n-1},$$

or

$$Z_{\theta}^{(k)}(\theta) = \frac{d_k}{\sigma_{n-1}}.$$

We have also proved property (5).

Proof of 6. By the definition of $Z_{\theta'}^{(k)}(\theta)$ we obtain

$$Z_{\theta'}^{(k)}(\theta) = \int_{\mathbb{S}^{n-1}} Z_{\theta'}^{(k)}(\theta'') \cdot Z_{\theta}^{(k)}(\theta'') \, d\theta''. \tag{10.21}$$

For $\theta = \theta'$ we obtain

$$Z_{\theta}^{(k)}(\theta) = \int_{\mathbb{S}^{n-1}} |Z_{\theta}^{(k)}(\theta'')|^2 \, d\theta''$$

$$= \|Z_{\theta}^{(k)}\|_{L_2}^2,$$

hence,

$$\|Z_{\theta}^{(k)}\|_{L_2}^2 = \frac{d_k}{\sigma_{n-1}},$$

which proves our statement.

Proof of 7. By the Cauchy–Schwarz inequality we obtain the following inequality from (10.21):
$$|Z^{(k)}_{\theta'}(\theta)| \leq \|Z^{(k)}_{\theta'}\|_{L_2}\|Z^{(k)}_{\theta}\|_{L_2} = \frac{d_k}{\sigma_{n-1}}.$$

This completes the proof. ∎

Compared with other sources which are oriented towards special functions the last results were obtained in a very elegant way without explicit constants.

We immediately obtain Corollary 10.18.

Corollary 10.18 *For every $Y_k \in \mathcal{H}_k$ we have*
$$|Y_k(\theta)| \leq \sqrt{\frac{d_k}{\sigma_{n-1}}}\|Y_k\|_{L_2} \quad \text{for } \theta \in \mathbb{S}^{n-1}.$$

Proof We have by the definition of $Z^{(k)}_{\theta'}(\theta)$ the equality
$$Y_k(\theta') = \int_{\mathbb{S}^{n-1}} Z^{(k)}_{\theta'}(\theta) Y_k(\theta)\, d\theta,$$

hence, by the Cauchy–Schwarz inequality we obtain
$$|Y_k(\theta')| \leq \|Z^{(k)}_{\theta'}\|_{L_2} \cdot \|Y_k\|_{L_2}$$
$$\leq \sqrt{\frac{d_k}{\sigma_{n-1}}} \cdot \|Y_k\|_{L_2}.$$
∎

As a direct application we obtain a criterion for the uniform convergence of the Laplace series of a function. This improves the very general result of Theorem 10.11, p. 149.

Theorem 10.19 *Let the integer $l > n/2$. Let the function $h \in L_2(\mathbb{S}^{n-1})$ and satisfy*
$$\int_{\mathbb{S}^{n-1}} |\Delta^l h(\theta)|^2\, d\theta < \infty.$$

Then its Laplace series $h = \sum Y_k$ defined in formula (10.14), p. 150, is uniformly convergent.

Proof As we have seen in Corollary 10.16, p. 154, we have the representation
$$Y_k(\theta') = \int_{\mathbb{S}^{n-1}} Z^{(k)}_{\theta'}(\theta) h(\theta)\, d\theta,$$

hence after applying Lemma 10.9, p. 147, to the element $Z^{(k)}_{\theta'}(\theta) \in \mathcal{H}_k$, we obtain
$$Z^{(k)}_{\theta'}(\theta) = \frac{1}{[-k(n+k-2)]^l} \Delta^l_\theta Z^{(k)}_{\theta'}(\theta).$$

158   *Multivariate polysplines*

After replacing this in the integral, we obtain

$$Y_k(\theta') = \frac{1}{[-k(n+k-2)]^l} \int_{\mathbb{S}^{n-1}} \Delta_\theta^l Z_{\theta'}^{(k)}(\theta) h(\theta) \, d\theta.$$

We apply the self-adjointness of the operator Δ_θ, Theorem 10.1, p. 133, and obtain

$$Y_k(\theta') = \frac{1}{[-k(n+k-2)]^l} \int_{\mathbb{S}^{n-1}} Z_{\theta'}^{(k)}(\theta) \Delta_\theta^l h(\theta) \, d\theta.$$

By the Cauchy–Schwarz inequality we obtain

$$|Y_k(\theta')| \leq \frac{1}{[k(n+k-2)]^l} \|Z_{\theta'}^{(k)}\|_{L_2} \|\Delta_\theta^l h\|_{L_2}$$

which by Corollary 10.18, p. 157, gives the estimate

$$|Y_k(\theta')| \leq \frac{1}{[k(n+k-2)]^l} \sqrt{\frac{d_k}{\sigma_{n-1}}} \|\Delta_\theta^l h\|_{L_2}.$$

Now recall formula (10.11), p. 145, by which we have, for $k \to \infty$, the asymptotics

$$d_k \approx \frac{2}{(n-2)!} k^{n-2}.$$

Hence, we obtain the estimate

$$|Y_k(\theta')| \leq C \frac{1}{k^{2l}} k^{(n-2)/2}$$
$$= C k^{(n/2)-1-2l},$$

for some constant $C > 0$. The series $\sum_{k \geq 1} Y_k$ is uniformly convergent if the dominating series

$$\sum_{k=1}^{\infty} C k^{(n/2)-1-2l}$$

is convergent. That is the case for

$$\frac{n}{2} - 1 - 2l < -1 \iff l > \frac{n}{4}.$$

This proves the Theorem. ∎

The best result belongs to Ragozin [45–47].

Theorem 10.20 (Ragozin, [47]) *Let the function* $h \in C^{[(n-1)/2]}(\mathbb{S}^{n-1})$. *Then its Laplace series converges uniformly.*

See the review article on this subject by Kalf [25]. We provide the characterization of the Sobolev spaces on the sphere $\mathbb{S}^{n-1}$ in terms of the coefficients of the Laplace series in Section 23.1.3.

10.12.1 Estimates of the derivatives of $Y_k(\theta)$: Markov–Bernstein-type inequality

The classical inequality of Bernstein is as follows: for every trigonometric polynomial

$$T_N(\varphi) = \frac{a_0}{2} + \sum_{k=1}^{N}(a_k \cos k\varphi + b_k \sin k\varphi)$$

holds

$$|T'_N(\varphi)| \leq N\|T_N\|_{L_\infty(\mathbb{R})} \quad \text{for all } \varphi \text{ in } \mathbb{R}.$$

We have derived the definition of the spherical harmonics by analogy with the trigonometric polynomials. Hence it would be natural to expect a similar inequality to hold.

Theorem 10.21 *Let $Y_k \in \mathcal{H}_k$. Then for every $j = 1, 2, \ldots, n$ the following inequality holds:*

$$\left|\frac{\partial}{\partial \theta_j} Y_k(\theta)\right| \leq k\|Y_k\|_{L_\infty(\mathbb{S}^{n-1})}.$$

This inequality is easy to prove. It suffices to note that the restriction of $Y_k(\theta)$ to a great circle of $\mathbb{S}^{n-1}$ is a univariate trigonometric polynomial.

Bos et al. [5] have proved a more general inequality of that type for algebraic manifolds without boundary in $\mathbb{R}^n$. Sobolev [54, Chapter XI.4, p. 489] proves a weaker estimate directly.

10.13 The classical approach to zonal harmonics

We now follow the classical approach to introducing the zonal harmonics $Z_{\theta'}^{(k)}(\theta)$ by appealing to the expansion of the *Newton potential* in spherical harmonics. This is the historical route by which the spherical harmonics have appeared.

We consider the Newton potential function

$$N(x - x') = \begin{cases} \dfrac{1}{|x - x'|^{n-2}} & (n \geq 3), \\ -\log|x - x'| & (n = 2). \end{cases}$$

It is well known [36, Chapter 11.2, p. 220], that for every two points $x \neq x'$,

$$\Delta_x N(x - x') = \Delta_{x'} N(x - x') = 0.$$

Here Δ_x denotes the Laplace operator with respect to the variable x.

For simplicity, we will work only in the space dimension $n \geq 3$ but the results remain valid in principle for all dimensions $n \geq 2$.

160   Multivariate polysplines

Theorem 10.22 *Let $n \geq 3$. Then for every two points $\theta, \theta' \in \mathbb{S}^{n-1}$, we have the following representation of the zonal harmonics:*

$$Z_{\theta'}^{(k)}(\theta) = c_{n,k} \cdot P_k^\lambda(\theta \cdot \theta'), \tag{10.22}$$

where

$$c_{n,k} = (n + 2k - 2)\Omega_n$$

and the constant Ω_n has been defined in (10.3), p. 130.

Proof   First we prove a useful formula.
(1) We put

$$|x'| = R, \qquad |x| = r.$$

We assume that

$$r < R.$$

Let the angle between x and x' be denoted by γ. By the definition of the scalar product of two vectors in $\mathbb{R}^n$ we have

$$x \cdot x' = |x||x'| \cos \gamma = Rr \cos \gamma.$$

Due to

$$|x - x'|^2 = r^2 - 2rR\cos\gamma + R^2,$$

we obviously have the expansion

$$\frac{1}{|x-x'|^{n-2}} = \frac{1}{R^{n-2}} \frac{1}{\left(1 - 2\frac{r}{R}\cos\gamma + \frac{r^2}{R^2}\right)^{(n/2)-1}} \tag{10.23}$$

$$= \frac{1}{R^{n-2}} \sum_{k=0}^{\infty} \left(\frac{r}{R}\right)^k P_k^\lambda(\cos\gamma)$$

$$= \sum_{k=0}^{\infty} \frac{r^k}{R^{n+k-2}} P_k^\lambda(\cos\gamma),$$

where the polynomial $P_k^\lambda(t)$ is obtained from the Taylor expansion around $t = 0$ of the function

$$\frac{1}{(1 - 2tz + t^2)^\lambda} = \sum_{k=0}^{\infty} t^k P_k^\lambda(z),$$

where we may have an arbitrary number $\lambda > 0$. Obviously, the function $(r/R)^k P_k^\lambda(\cos\gamma)$ of the variable $x = r\theta$ is homogeneous of degree k.

For simplicity we will put

$$\lambda = \frac{n}{2} - 1.$$

(2) We also have the Taylor expansion of the function $1/(|x - x'|^{n-2})$ around $x = 0$, which is convergent, namely

$$\frac{R^{n-2}}{|x - x'|^{n-2}} = R^{n-2} \sum_{k=0}^{\infty} \sum_{|\alpha|=k} \frac{x^{\alpha}}{\alpha!} D_x^{\alpha} \left(\frac{1}{|x - x'|^{n-2}}\right)_{|x=0} \quad (10.24)$$

$$= \sum_{k=0}^{\infty} R_k(x),$$

where $R_k(x)$ are homogeneous polynomials of degree k. Their coefficients depend on x'.

As was said above, the function $N(x - x') = 1/(|x - x'|^{n-2})$ is harmonic in x for $x \neq x'$. This implies that the polynomials $R_k(x)$ are also harmonic, or $R_k(x) \in \text{HHom}_k$. Comparing the two expansions in homogeneous functions provided by (10.23), p. 160, and (10.24), we see that

$$R_k(x) = \frac{1}{R^{n-2}} \left(\frac{r}{R}\right)^k P_k^{\lambda}(\cos \gamma),$$

i.e. the function $P_k^{\lambda}(\cos \gamma)$ as a function of the variable $\theta = x/|x|$ is a restriction of a polynomial of HHom_k, hence it is a spherical harmonic and

$$P_k^{\lambda}(\cos \gamma) \in \mathcal{H}_k.$$

(3) In order to find the relation with the zonal harmonics $Z_{\theta'}^{(k)}(\theta)$ we will apply the so-called second Green formula [36, Chapter 10.6, p. 212] or [3, p. 9] to the ball $B(0; 1)$ having boundary $\partial B(0; 1) = \mathbb{S}^{n-1}$. We have

$$\int_{\mathbb{S}^{n-1}} \left[\frac{\partial u(y)}{\partial n_y} \frac{1}{|x - y|^{n-2}} - u(y) \frac{\partial}{\partial n_y} \left(\frac{1}{|x - y|^{n-2}}\right)\right] d\sigma_y$$

$$= \begin{cases} \frac{1}{\Omega_n} u(x) & \text{for } x \in B(0; 1), \\ 0 & \text{for } x \notin \overline{B(0; 1)}. \end{cases}$$

Here we have used the constant Ω_n introduced in (10.3), p. 130. By $\vec{n_y}$ we have denoted the exterior normal vector to the ball $B(0; 1)$ at the point $y \in \mathbb{S}^{n-1} = \partial B(0; 1)$, hence

$$\frac{\partial}{\partial n_y} = \frac{\partial}{\partial r}.$$

Further,

$$u(x) = r^j Y_j(\theta),$$

for some integer $j \geq 0$, and for clarity we put $r_x = |x|$, $r_y = |y|$, $\theta_y = y/|y|$, and $\theta_x = x/|x|$. Evidently, we have

$$\theta_x \cdot \theta_y = \cos \gamma_{x,y},$$

162 Multivariate polysplines

if $\gamma_{x,y}$ denotes the angle between the vectors x and y. For $y \in \mathbb{S}^{n-1}$ we have

$$\frac{\partial u(y)}{\partial r_y} = j|y|^{j-1} Y_k(\theta) = j Y_j(\theta),$$

and from the expansion (10.23), p. 160, it follows that

$$\frac{1}{|x-y|^{n-2}} = \sum_{k=0}^{\infty} \frac{r_x^k}{r_y^{n+k-2}} P_k^{\lambda}(\cos \gamma_{x,y})$$

$$\frac{\partial}{\partial r_y}\left(\frac{1}{|x-y|^{n-2}}\right) = -\sum_{k=0}^{\infty} (n+k-2) \frac{r_x^k}{r_y^{n+k-1}} P_k^{\lambda}(\cos \gamma_{x,y}).$$

Applying the above mentioned Green formula for every point $x \in B(0; 1)$ we obtain the equality

$$\frac{1}{\Omega_n} r_x^j Y_j(\theta_x) = \int_{\partial B(0;1)} \left[\frac{\partial u(y)}{\partial r_y} \frac{1}{|x-y|^{n-2}} - u(y) \frac{\partial}{\partial r_y}\left(\frac{1}{|x-y|^{n-2}}\right)\right] d\sigma_y$$

$$= \int_{\partial B(0;1)} j Y_j(\theta_y) \left(\sum_{k=0}^{\infty} \frac{r_x^k}{r_y^{n+k-2}} P_k^{\lambda}(\cos \gamma_{x,y})\right)$$

$$+ Y_j(\theta_y) \left(\sum_{k=0}^{\infty} (n+k-2) \frac{r_x^k}{r_y^{n+k-1}} P_k^{\lambda}(\cos \gamma_{x,y})\right) d\sigma_y.$$

Since $r_y = 1$ for $y \in \partial B(0; 1)$, and recalling that $P_k^{\lambda}(\cos \gamma_{x,y})$ is a spherical harmonic with respect to θ_y, we see that for $k \neq j$ we have orthogonal terms, hence we obtain

$$\frac{1}{\Omega_n} r_x^j Y_j(\theta_x) = \int_{\mathbb{S}^{n-1}} Y_j(\theta_y) r_x^j (n+2j-2) P_j^{\lambda}(\cos \gamma_{x,y}) d\sigma_y,$$

which after dividing by r_x implies

$$Y_j(\theta_x) = \Omega_n (n+2j-2) \cdot \int_{\mathbb{S}^{n-1}} P_j^{\lambda}(\cos \gamma_{x,y}) Y_j(\theta_y) d\sigma_y.$$

Due to the uniqueness of the zonal harmonic $Z_{\theta'}^{(j)}(\theta)$ as an element of the space $\mathcal{H}_j$ having this property it follows that for every two points $\theta, \theta' \in \mathbb{S}^{n-1}$, we have

$$Z_{\theta'}^{(j)}(\theta) = c_{n,j} \cdot P_j^{\lambda}(\theta \cdot \theta'),$$

where

$$c_{n,j} = (n+2j-2)\Omega_n.$$

This completes the proof of the theorem. ∎

We now have an alternative way to compute the norm $\|Z_{\theta'}^{(k)}\|_{L_2}$ which we had computed in Proposition 10.17, p. 154. We split the proof into smaller steps as exercises.

Exercise 10.23 *Prove that the following Taylor expansion holds around the point $z = 0$,*

$$\frac{1}{(1-z)^{n-2}} = \sum_{k=0}^{\infty} \frac{(n+k-3)!}{k!(n-3)!} z^k \quad \text{for } |z| < 1.$$

Note that using $\Gamma(p+1) = p!$ for integers $p \geq 0$, one may write the coefficients as

$$\frac{\Gamma(n+k-2)}{\Gamma(k+1)\Gamma(n-2)}.$$

Exercise 10.24 *Prove that for $\lambda = (n/2) - 1$, and for $k \geq 0$,*

$$P_k^\lambda(1) = \frac{(n+k-3)!}{k!(n-3)!}.$$

Hint: *Use the defining formula for the zonal harmonics and use $Y_k(\theta) = P_k^\lambda(\cos\gamma)$, where we consider two points $\theta, \theta' \in \mathbb{S}^{n-1}$, and $\theta \cdot \theta' = \cos\gamma$. As we have seen in Theorem 10.22, p. 158, the function $P_k^\lambda(\cos\gamma)$ as a function of θ is indeed in $\mathcal{H}_k$. One obtains*

$$P_k^\lambda(\theta'' \cdot \theta) = \int_{\mathbb{S}^{n-1}} Z_\theta^{(k)}(\theta') P_k^\lambda(\theta'' \cdot \theta') \, d\theta'.$$

Use the equality (10.22), p. 160, proved above, i.e. $Z_{\theta'}^{(k)}(\theta) = c_{n,k} P_k^\lambda(\theta \cdot \theta')$, and put $\theta = \theta'' = (1, 0, \ldots, 0)$. It follows that

$$P_k^\lambda(1) = \frac{1}{c_{n,k}} \int_{\mathbb{S}^{n-1}} |Z_\theta^{(k)}(\theta')|^2 \, d\theta'.$$

Hence, due to the rotational invariance proved in Proposition 10.17, p. 154, for every $\theta' \in \mathbb{S}^{n-1}$ we have

$$\|Z_{\theta'}^{(k)}\|_{L_2}^2 = c_{n,k} P_k^\lambda(1) = (n+2k-2)\Omega_n \frac{(n+k-3)!}{k!(n-3)!}.$$

For $n \geq 3$ we have $\Omega_n^{-1} = (n-2)\sigma_{n-1}$. On the other hand the value of $\|Z_{\theta'}^{(k)}\|_{L_2}^2$ provided in point (6) of Proposition 10.17, p. 154, is d_k/σ_{n-1}. Now we have to take the expression of d_k in (10.10), p. 145, and we see that both results coincide.

Remark 10.25 We see that working with the functional definition of the zonal harmonics $Z_{\theta'}^{(k)}(\theta)$ provides transparent formulas. The fashion in special functions is always to specify the exact constants and coefficients which makes the formulas very clumsy. One also has to avoid treating the special cases $n = 2$ or $n = 3$ up to the last point where necessary.

164 *Multivariate polysplines*

10.14 The representation of polyharmonic functions using spherical harmonics

First we study the space of solutions of the operator $L_{(k)}$.

Let us recall that for every $h(r\theta) \in L_2(\mathbb{S}^{n-1})$ we have the Laplace series

$$h(x) = h(r\theta) = \sum_{k=0}^{\infty} Y_k(r, \theta), \qquad (10.24a)$$

and by Corollary 10.16, p. 154, we have

$$Y_k(r, \theta) = \int_{\mathbb{S}^{n-1}} Z_{\theta'}^{(k)}(\theta) h(r\theta) d\theta \qquad (10.24b)$$

with

$$Y_k(r, \theta) \in \mathcal{H}_k$$

for every $r > 0$.

We first prove the following property of the polyharmonic operator in the annulus $A_{a,b}$, given by:

$$A_{a,b} = \{x : a < |x| < b\},$$

for two numbers $a, b > 0$.

Lemma 10.26 *Let us assume that for some integer $p \geq 1$, holds $\Delta^p h(x) \in L_2(A_{a,b})$. Then*

$$[L_{(k)}]^p Y_k(r, \theta) = \int_{\mathbb{S}^{n-1}} Z_{\theta'}^{(k)}(\theta) \Delta^p h(r\theta) d\theta.$$

Proof We have

$$\Delta_r Y_k(r, \theta) = \int_{\mathbb{S}^{n-1}} Z_{\theta'}^{(k)}(\theta) \Delta_r h(r\theta) d\theta,$$

which due to:

- the decomposition of Δ,

$$\Delta = \Delta_r + \frac{1}{r^2} \Delta_\theta,$$

- the self-adjointness of the operator Δ_θ in Theorem 10.1, p. 133, and

Spherical harmonics and polyharmonic functions 165

- the eigenvalue property in Lemma 10.9, p. 147, becomes

$$\Delta_r Y_k(r,\theta) = \int_{\mathbb{S}^{n-1}} Z^{(k)}_{\theta'}(\theta)\left(\Delta - \frac{1}{r^2}\Delta_\theta\right) h(r\theta)\, d\theta$$

$$= \int_{\mathbb{S}^{n-1}} Z^{(k)}_{\theta'}(\theta)\Delta h(r\theta)\, d\theta - \frac{1}{r^2}\int_{\mathbb{S}^{n-1}} Z^{(k)}_{\theta'}(\theta)\Delta_\theta h(r\theta)\, d\theta$$

$$= \int_{\mathbb{S}^{n-1}} Z^{(k)}_{\theta'}(\theta)\Delta h(r\theta)\, d\theta - \frac{1}{r^2}\int_{\mathbb{S}^{n-1}} \Delta_\theta Z^{(k)}_{\theta'}(\theta) h(r\theta)\, d\theta$$

$$= \int_{\mathbb{S}^{n-1}} Z^{(k)}_{\theta'}(\theta)\Delta h(r\theta)\, d\theta + \frac{k(n+k-2)}{r^2}\int_{\mathbb{S}^{n-1}} Z^{(k)}_{\theta'}(\theta) h(r\theta)\, d\theta$$

$$= \int_{\mathbb{S}^{n-1}} Z^{(k)}_{\theta'}(\theta)\Delta h(r\theta)\, d\theta + \frac{k(n+k-2)}{r^2} Y_k(r,\theta).$$

Hence,

$$L_{(k)} Y_k(r,\theta) = \int_{\mathbb{S}^{n-1}} Z^{(k)}_{\theta'}(\theta) \Delta h(r\theta)\, d\theta.$$

This proves the statement for $p=1$. We obtain the general case $p \geq 1$ inductively by applying p times the above result. ∎

We have the following representation of the Laplace operator.

Theorem 10.27 *Assume that the function $\Delta h \in L_2(A_{a,b})$. Then the Laplace series for the function $\Delta h(x)$ is, for $a < r < b$ and $\theta \in \mathbb{S}^{n-1}$, given by*

$$\Delta h(x) = \sum_{k=0}^{\infty} L_{(k)} Y_k(r,\theta),$$

and more generally, if $\Delta^p h(x) \in L_2(A_{a,b})$ then

$$\Delta^p h(x) = \sum_{k=0}^{\infty} L^p_{(k)} Y_k(r,\theta) \tag{10.25}$$

for every integer $p \geq 1$. Here $Y_k(r,\theta)$ is defined by formula (10.24b).

Proof We apply the Laplace operator Δ on both sides of (10.24a) and obtain

$$\Delta h(x) = \sum_{k=0}^{\infty} \Delta Y_k(r,\theta),$$

or more generally

$$\Delta^p h(x) = \sum_{k=0}^{\infty} \Delta^p Y_k(r,\theta),$$

for every integer $p \geq 1$. The differentiability of every element $Y_k(r,\theta)$ is justified owing to the representation through zonal harmonics. The convergence of the series follows from Lemma 10.26, p. 164.

Since $\Delta = \Delta_r + (1/r^2)\Delta_\theta$ and $Y_k(r,\theta) \in \mathcal{H}_k$ we obtain by about eigenvalues Lemma 10.9, p. 147, the equality

$$\Delta Y_k(r,\theta) = \left(\Delta_r + \frac{1}{r^2}\Delta_\theta\right) Y_k(r,\theta)$$

$$= \Delta_r Y_k(r,\theta) - \frac{k(n+k-2)}{r^2} Y_k(r,\theta)$$

$$= L_{(k)} Y_k(r,\theta).$$

Obviously the Laplace series for the function $\Delta h(x)$ now becomes

$$\Delta h(x) = \sum_{k=0}^{\infty} L_{(k)} Y_k(r,\theta).$$

This completes the proof. ∎

This is a remarkable formula showing that the Laplace operator and its powers may be split into infinitely many one-dimensional operators $L_{(k)}$ for $k = 0, 1, \ldots$. We immediately obtain the following important corollary.

Corollary 10.28 *If the function h is polyharmonic of order p in the annulus $A_{a,b} = \{x : a < |x| < b\}$, i.e.*

$$\Delta^p h = 0 \quad \text{in } A_{a,b},$$

it follows that the function h has a Laplace series $h(x) = \sum_{k=0}^{\infty} Y_k(r,\theta)$ satisfying

$$L_{(k)}^p Y_k(r,\theta) = 0 \quad \text{for } x = r\theta \in A_{a,b}.$$

Let us analyze the simplest case, $p = 1$.

10.14.1 Representation of harmonic functions using spherical harmonics

Proposition 10.29 *Let the function $h(x)$ be harmonic in the annulus $A_{a,b}$. Then it has the following Laplace series:*

$$h(x) = \sum_{k=0}^{\infty} \left(R_{k,1}(r) Y_k^1(\theta) + R_{k,2}(r) Y_k^2(\theta) \right), \quad \text{for } a < r < b \text{ and } \theta \text{ in } \mathbb{S}^{n-1},$$

where $Y_k^1(\theta), Y_k^2(\theta) \in \mathcal{H}_k$ are two spherical harmonics, and $R_{k,1}(r), R_{k,2}(r)$ are the two linearly independent solutions of the equation $L_{(k)} f(r) = 0$ given by (10.19), p. 153.

Proof As we have seen in Proposition 10.12, p. 152, the general solution of the equation

$$L_{(k)} f(r) = 0$$

in an annulus $A_{a,b}$ has the form

$$f(r) = C_1 R_1(r) + C_2 R_2(r),$$

where C_1, C_2 are two arbitrary constants. Hence, the solution to

$$L_{(k)} Y_k(r, \theta) = 0 \quad \text{for } x = r\theta \in A_{a,b}$$

has the form

$$Y_k(r, \theta) = R_1(r) \cdot g_{k,1}(\theta) + R_2(r) \cdot g_{k,2}(\theta).$$

Now we take two different numbers r_1 and r_2 satisfying $a < r_1 < r_2 < b$. We want to prove that the matrix $[R_i(r_j)]_{i,j}$ is invertible. Here we recall the representation of the operator $L_{(k)}$ in (10.18), p. 152. It shows that $L_{(k)}$ generates a *Chebyshev system*, see Chapter 11, p. 187. Since R_1, R_2 are linearly independent solutions to $L_{(k)} g = 0$ it follows that they belong to the Chebyshev system and by one of the equivalent definitions (see Theorem 11.4, p. 188) of the Chebyshev system the matrix $[R_i(r_j)]_{i,j}$ is invertible. Hence we may express the functions $g_{k,1}(\theta)$, $g_{k,2}(\theta)$ as linear combinations of the functions $Y_k(r_1, \theta)$, $Y_k(r_2, \theta)$. Since $Y_k(r_1, \theta), Y_k(r_2, \theta) \in \mathcal{H}_k$, it also follows that $g_{k,1}(\theta), g_{k,2}(\theta) \in \mathcal{H}_k$. This proves our proposition. ∎

Remark 10.30 *In the above proof we might have referred directly to the basic sources on Chebyshev systems [26, 50]. From [50] we apply Definition (9.1) and formulas (9.4)–(9.5) and the discussion on pages 363–366. From these formulas we see that by (10.18), p. 152, our operator $r^{n+k-1} L_{(k)}^p$ is of the type considered there and its solutions form an Extended Complete Chebyshev system for $r > 0$. Hence, the determinants in question are nonzero.*

Now let us consider the case of the ball $B(0; R) = \{x : |x| < R\}$ instead of the annulus $A_{a,b}$. We have seen that in the ball the only solutions to the equation $L_{(k)} f(r) = 0$ which are C^∞ have the form

$$f(r) = C_1 R_1(r).$$

Thus we can immediately prove Corollary 10.31.

Corollary 10.31 *Let the function $h(x)$ be harmonic in the ball $B(0; R)$. Then it has the following Laplace series:*

$$h(x) = \sum_{k=0}^{\infty} R_1(r) Y_k^1(\theta), \quad \text{for } 0 \le r < R \text{ and } \theta \text{ in } \mathbb{S}^{n-1},$$

where $Y_k^1(\theta) \in \mathcal{H}_k$ is a spherical harmonic, and $R_1(r)$ is the first solution of the equation $L_{(k)} f(r) = 0$ given by (10.19), p. 153, which is continuous for $r = 0$.
In particular, for $n = 2$ we obtain the representation

$$h(x) = h(r, \varphi)$$

$$= c_0 + \sum_{j=1}^{\infty} [c_j \cos j\varphi + d_j \sin j\varphi] \cdot r^j \quad \text{for } a < r < b \text{ and } 0 \le \varphi \le 2\pi.$$

Now it is possible to solve explicitly the Dirichlet problem in the ball and in the annulus by using the above results. See also Mikhlin [36, Chapter 13, Section 5] for this direction. Let us note that the ball differs from the annular domain $A_{0,b}$! In the last

168 Multivariate polysplines

domain the functions $R_2(r)$ are harmonic and continuous, although not on its boundary. In order to obtain a similar characterization of the polyharmonic functions in the annulus we need to study the solutions of the equation $L_{(k)}^p f = 0$.

10.14.2 Solutions of the spherical operator $L_{(k)}^p f(r) = 0$

The main purpose of our investigation is now to prove the following proposition.

Proposition 10.32 *All solutions to equation $L_{(k)}^p f(r) = 0$ which are C^∞ for $r > 0$ are given by*

$$r^{-n-k+2}, r^{-n-k+4}, \ldots, r^{-n-k+2p},$$

$$r^k, r^{k+2}, \ldots, r^{k+2p-2},$$

where there are no coincidences in the two rows. If there is a coincidence of two terms, i.e. if for some indexes j_1 and j_2 satisfying $1 \leq j_1 \leq p$, and $0 \leq j_2 \leq p-1$, the following equality holds:

$$l = -n - k + 2j_1 = k + 2j_2$$

then except for the function r^l we also find that the function

$$r^l \log r$$

is a solution.

It is possible to check this directly owing to the representation (10.18), p. 152, and this is approximately the manner in which it has been treated by Sobolev [54]. There is a more elegant method based on an appropriate change of the variables, $v = \log r$.

Exercise 10.33 *Apply the formulas*

$$\Delta_r^p(r^k) = A_{p,k} r^{k-2p},$$

$$\Delta_r^p(r^k \log r) = A_{p,k} r^{k-2p} \log r + r^{k-2p},$$

see Aronszajn et al. [3, p. 2], to prove the above proposition.

10.14.3 Operator with constant coefficients equivalent to the spherical operator $L_{(k)}^p$

So far the operator $L_{(k)}^p$ has variable coefficients. Here we will give an operator with constant coefficients which is obtained after a suitable change of the variable r.

We make an important simplification of the operator $L_{(k)}$ by putting

$$v = \log r.$$

The inverse transform is

$$r = e^v.$$

Spherical harmonics and polyharmonic functions   169

Theorem 10.34 *Under the change $v = \log r$ the operator $[L_{(k)}^p](d/dr)$ is transformed for $r > 0$ into the operator*
$$e^{-2pv} M_{k,p}\left(\frac{d}{dv}\right),$$
on the whole line $v \in \mathbb{R}$, where
$$M_{k,p}\left(\frac{d}{dv}\right) = \prod_{l=0}^{p-1}\left(\frac{d}{dv} - (k+2l)\right) \prod_{l=1}^{p}\left(\frac{d}{dv} - (-n-k+2l)\right). \quad (10.26)$$

Written more explicitly, for every function $w \in C^{2p}(\mathbb{R})$ we have

$$\boxed{[L_{(k)}^p]\left(\frac{d}{dr}\right) w(\log r) = e^{-2pv} M_{k,p}\left(\frac{d}{dv}\right) w(v).}$$

Proof We will prove this theorem by induction. For every twice-differentiable function $w(v) = w(\log r)$ we evidently obtain
$$\frac{d}{dr} w(\log r) = \frac{1}{r} w'(\log r) = e^{-v} \frac{d}{dv} w(v).$$

This shows that the operator d/dr is transformed into the operator $e^{-v}(d/dv)$.

(1) Let us prove the statement for $p = 1$.

Using representation (10.18), p. 152, we obtain
$$L_{(k)} w(\log r) = \frac{1}{r^{n+k-1}} \frac{d}{dr}\left[r^{n+2k-1} \frac{d}{dr}\left[\frac{1}{r^k} w(\log r)\right]\right]$$
$$= e^{-(n+k-1)v} \cdot e^{-v} \frac{d}{dv}\left[e^{(n+2k-1)v} \cdot e^{-v} \frac{d}{dv}\left[e^{-kv} w(\log r)\right]\right]$$
$$= e^{-(n+k)v} \cdot \frac{d}{dv}\left[e^{(n+2k-2)v} \cdot \frac{d}{dv}\left[e^{-kv} w(\log r)\right]\right].$$

Since for every number λ we have
$$\left(\frac{d}{dt} - \lambda\right) f(t) = e^{\lambda t} \frac{d}{dt}\left[e^{-\lambda t} f(t)\right], \quad (10.27)$$
we obtain
$$L_{(k)} w(\log r) = e^{-2v} \cdot e^{-(n+k-2)v} \frac{d}{dv}\left[e^{(n+k-2)v} \cdot e^{kv} \frac{d}{dv}\left[e^{-kv} w(v)\right]\right]$$
$$= e^{-2v} \cdot \left(\frac{d}{dv} + n + k - 2\right) \cdot \left(\frac{d}{dv} - k\right) w(v),$$

which is the statement of the theorem for $p = 1$.

(2) Assuming that
$$\left[L_{(k)}\left(\frac{d}{dr}\right)\right]^p w(\log r) = e^{-2pv} M_{k,p}\left(\frac{d}{dv}\right) w(v),$$

170　Multivariate polysplines

let us prove that

$$\left[L_{(k)}\left(\frac{d}{dr}\right)\right]^{p+1} w(\log r) = e^{-2(p+1)v} M_{k,p+1}\left(\frac{d}{dv}\right) w(v).$$

Writing the expressions for $M_{k,p}(d/dv)$ and $M_{k,p+1}(d/dv)$ we see that this is equivalent to proving that

$$L_{(k)}\left(\frac{d}{dr}\right) e^{-2pv} = e^{-2(p+1)v}\left(\left(\frac{d}{dv}+n+k-2p-2\right)\cdot\left(\frac{d}{dv}-k-2p\right)\right),$$

or

$$e^{-(n+k)v} \cdot \frac{d}{dv}\left[e^{(n+2k-2)v} \cdot \frac{d}{dv}\left[e^{-kv} \cdot e^{-2pv} w(v)\right]\right]$$
$$= e^{-2(p+1)v}\left(\left(\frac{d}{dv}+n+k-2p-2\right)\cdot\left(\frac{d}{dv}-k-2p\right)\right) w(v).$$

The last is evident since after regrouping we obtain

$$e^{-(n+k)v} \cdot \frac{d}{dv}\left[e^{(n+2k-2)v} \cdot \frac{d}{dv}\left[e^{-kv} \cdot e^{-2pv} w(v)\right]\right] = e^{-(2p+2)v}$$
$$\cdot e^{-(n+k-2p-2)v} \cdot \frac{d}{dv}\left[e^{(n+k-2p-2)v} \cdot e^{(2p+k)v} \cdot \frac{d}{dv}\left[e^{-(2p+k)v} w(v)\right]\right],$$

and the result is obtained again by applying (10.27), p. 169.
This completes the proof. ∎

For simplicity we will write the operator in the following way:

$$M_{k,p}\left(\frac{d}{dv}\right) = \prod_{j=1}^{2p}\left(\frac{d}{dv}-\lambda_j\right),$$

where

$$\begin{cases} \lambda_1 = -n-k+2, \\ \lambda_2 = -n-k+4, \\ \quad\vdots \\ \lambda_p = -n-k+2p, \\ \lambda_{p+1} = k, \\ \lambda_{p+2} = k+2, \\ \quad\vdots \\ \lambda_{2p} = k+2p-2. \end{cases} \quad (10.28)$$

Remark 10.35 *Let us note that the polynomial $M_{k,p}(z)$ has multiple roots only in one case, namely when the dimension is even, i.e. $n = 2m$ for integer m and satisfies*

Spherical harmonics and polyharmonic functions

$n \leq -2k + 2p$. In this case the two groups of values $\{\lambda_1, \ldots, \lambda_p\}$ and $\{\lambda_{p+1}, \ldots, \lambda_{2p}\}$ overlap and $M_{k,p}(z)$ admits the following representation:

$$M_{k,p}(z) = \prod_{l=0}^{p-1}(z-(k+2l)) \prod_{l=1}^{k+n/2-1}(z-(-n-k+2l))$$

$$\times \prod_{l=k+n/2}^{p}(z-(-n-k+2l))$$

$$= \prod_{l=-n-2k+2p+1}^{p-1}(z-(k+2l)) \prod_{l=1}^{k+n/2-1}(z-(-n-k+2l))$$

$$\times \prod_{l=k+n/2}^{p}(z-(-n-k+2l))^2.$$

The overlapping of the two groups of values was illustrated in the two-dimensional case (Figure 8.2, p. 112).

After the change $v = \log r$, let us put

$$\widetilde{Y}_k(\log r, \theta) = Y_k(r, \theta).$$

We immediately obtain Corollary 10.36.

Corollary 10.36 *The polyharmonic operator Δ^p is decomposed in the following way:*

$$\Delta^p h(x) = \sum_{k=0}^{\infty} \Delta^p \widetilde{Y}_k(\log r, \theta) = \sum_{k=0}^{\infty} L_{(k)}^p \widetilde{Y}_k(\log r, \theta)$$

$$\boxed{= e^{-2pv} \sum_{k=0}^{\infty} M_{k,p}\left(\frac{d}{dv}\right) \widetilde{Y}_k(v, \theta)}$$

on functions h having the Laplace series

$$h = \sum_{k=0}^{\infty} Y_k(r, \theta) = \sum_{k=0}^{\infty} \widetilde{Y}_k(\log r, \theta),$$

where $Y_k(r, \theta)$ are defined by equality (10.24b) on p. 162.

This beautiful decomposition of the action of the polyharmonic operator Δ^p will play a basic role in our study of cardinal polysplines and in the wavelet analysis using polysplines.

Let us denote by $N_{M_{k,p}}$ the set of solutions to the ordinary differential equation $M_{k,p} f(v) = 0$, i.e.

$$N_{M_{k,p}} := \{f \in C^{\infty} : M_{k,p} f(v) = 0, \quad -\infty < v < \infty\}.$$

From the above representation it is obvious that the polynomial $M_{k,p}(z)$ contains only factors of the form $(z - \lambda_j)$ or $(z - \lambda_j)^2$. By the usual properties of ordinary differential operators, see Pontryagin [41], the set $N_{M_{k,p}}$ will contain the exponential functions,

$$e^{\lambda_j v} \in N_{M_{k,p}}$$

if and only if $(z - \lambda_j)$ divides the polynomial $M_{k,p}(z)$, and both

$$e^{\lambda_j v}, \; v e^{\lambda_j v} \in N_{M_{k,p}}$$

if $(z - \lambda_j)^2$ also divides the polynomial $M_{k,p}(z)$. In this way we obtain precisely $2p$ functions in the set $N_{M_{k,p}}$ which are linearly independent.

Let us remark that such λ_js with *multiplicity two* occur for some ks if we have the inequality

$$k \leq -n - k + 2p,$$

which is equivalent to

$$2k \leq 2p - n.$$

The last is possible for some $k \geq 0$ if and only if

$$2p - n \geq 0.$$

After changing the variable by putting $v = \log r$ we obtain, in a reverse order, all solutions of the equation $L_{(k)}^p u(r) = 0$. If we denote its set of solutions by $N_{L_{(k)}^p}$, we have Theorem 10.37.

Theorem 10.37 *The set of solutions $N_{L_{(k)}^p}$ of the equation $L_{(k)}^p u(r) = 0$ which are C^∞ for $r > 0$ has a basis*

$$R_{k,j}(r) = R_{k,j}^{(p)}(r), \quad \text{for } j = 1, 2, \ldots, 2p,$$

such that the function r^{λ_j} belongs to $N_{L_{(k)}^p}$ if λ_j occurs only once in the vector $(\lambda_1, \ldots, \lambda_{2p})$, and the two functions r^{λ_j}, $r^{\lambda_j} \log r$ belong to $N_{L_{(k)}^p}$ if λ_j occurs twice in the vector $(\lambda_1, \ldots, \lambda_{2p})$. Hence, every solution of that equation may be written as

$$u(r) = \sum_{j=1}^{2p} C_j R_{k,j}(r),$$

where C_j are constants. Then the functions $R_{k,j}(e^v) = R_{k,j}^{(p)}(e^v)$ are a basis of the space of solutions of the equation $M_{k,p}(d/dv) f(r) = 0$.

We will put $R_{k,j} = R_{k,j}^{(p)}$, where we will often drop the dependence on the parameter p, and define

$$R_{k,1}(r) = r^k, \qquad R_{k,2}(r) = r^{k+2}, \quad \ldots \quad R_{k,p}(r) = r^{k+2p-2},$$

$$R_{k,p+1}(r) = r^{-n-k+2}, \quad R_{k,p+2}(r) = r^{-n-k+4}, \quad \ldots \quad R_{k,2p} = r^{-n-k+2p}.$$

Spherical harmonics and polyharmonic functions 173

The set of solutions of $L_{(k)}^p u(r) = 0$ which are C^∞ for all $r \geq 0$ are only the first p, namely $R_{k,1}(r), R_{k,2}(r), \ldots, R_{k,p}(r)$. If we put

$$M_{k,p}^+ := \prod_{j=p+1}^{2p} \left(\frac{d}{dv} - \lambda_j\right)$$

then the functions $R_{k,j}(e^v)$, $j = 1, 2, \ldots, p$, have the property that they are solutions to

$$M_{k,p}^+\left(\frac{d}{dv}\right) f(v) = 0. \tag{10.29}$$

Taking into account all the above results we immediately obtain Corollary 10.38.

Corollary 10.38 *Let the function $h(x)$ be polyharmonic of order p in the annulus or in the ball. Then it has the representation*

$$h(x) = \sum_{k=0}^{\infty} \sum_{\ell=1}^{d_k} f_{k,\ell}(\log r) Y_{k,\ell}(\theta),$$

where $f_{k,\ell}(\log r) = \widehat{h}_{k,\ell}(r)$ is a solution to $M_{k,p} f_{k,\ell}(v) = 0$. In the case of the ball $f_{k,\ell}(\log r)$ is a solution to $M_{k,p}^+ f_{k,\ell}(v) = 0$.

In order to assist the reader's comprehension we refer to some very simple examples of solutions to ordinary differential equations (ODEs) with constant coefficients, provided in Example 13.4–13.8, p. 225, where the operator L is defined by

$$L = \prod_{j=1}^{z+1} \left(\frac{d}{dv} - \lambda_j\right).$$

10.14.4 Representation of polyharmonic functions in annulus and ball

Now we are ready to characterize a polyharmonic function by means of its Laplace series.

Theorem 10.39 *1. Let the function $h(x)$ be polyharmonic in the annular domain $A_{a,b}$. Then it has the following Laplace series:*

$$h(x) = \sum_{k=0}^{\infty} \left(R_{k,1}(r) Y_k^1(\theta) + \cdots + R_{k,2p}(r) Y_k^{2p}(\theta)\right)$$

$$= \sum_{k=0}^{\infty} \sum_{\ell=1}^{d_k} \left(\alpha_{k,\ell,1} R_{k,1}(r) + \alpha_{k,\ell,2} R_{k,2}(r) + \cdots + \alpha_{k,\ell,2p} R_{k,2p}(r)\right) Y_{k,\ell}(\theta), \tag{10.30}$$

where $Y_k^1(\theta), \ldots, Y_k^{2p}(\theta) \in \mathcal{H}_k$ are $2p$ spherical harmonics, and $R_{k,1}(r), \ldots, R_{k,2p}(r)$ are the $2p$ linearly independent solutions of the equation $L_{(k)} f(r) = 0$ in the interval (a, b) given by Theorem 10.37, p. 172.

2. The series is absolutely and uniformly convergent in every subannulus A_{a_1,b_1} with $a < a_1 < b_1 < b$. A stronger statement holds: let the numbers be given with $a < a_2 < a_1 < b_1 < b_2 < b$. Then recalling the definition of the functions $R_{k,j}$ in Theorem 10.37, p. 172, the separate terms of the series satisfy the estimate

$$|\alpha_{k,\ell,j} r^{k+2j-2}| \leq K \left(\frac{r}{b_2}\right)^k \quad \text{for } a_1 < r < b_1, \text{ and } j = 1, 2, \ldots, p, \quad (10.31)$$

$$|\alpha_{k,\ell,j} r^{-n-k+2j}| \leq K \left(\frac{a_2}{r}\right)^k \quad \text{for } a_1 < r < b_1, \text{ and } j = p+1, p+2, \ldots, 2p, \quad (10.32)$$

where the constant K is independent of the parameters k, j.

Proof By Theorem 10.27, p. 165, we saw that if $\Delta^p h(x) = 0$ then

$$L_{(k)}^p Y_k(r, \theta) = 0.$$

In Theorem 10.37, p. 172, we have seen that such a solution is a linear combination of the form

$$Y_k(r, \theta) = R_{k,1}(r) \cdot g_{k,1}(\theta) + \cdots + R_{k,2p}(r) \cdot g_{k,2p}(\theta),$$

where $R_{k,j}(r)$ are the $2p$ linearly independent solutions of the equation $L_{(k)}^p f(r) = 0$. We will prove that the functions $g_{k,i}$ are spherical harmonics.

Now we take different numbers $r_1, \ldots, r_{2p}$ satisfying

$$a < r_1 < \cdots < r_{2p} < b,$$

and we obtain the matrix

$$[R_{k,j}(r_i)]_{i,j=1}^{2p}$$

which is invertible. The invertibility follows, as we have already indicated in Remark 10.30, p. 167, by referring to the fact that the functions $R_{k,j}(r)$ are elements of a *Chebyshev system* for $r > 0$.

Since the functions $Y_k(r_i, \theta)$ satisfy

$$Y_k(r_i, \theta) \in \mathcal{H}_k \quad \text{for } i = 1, \ldots, 2p,$$

due to the invertibility of the matrix $[R_{k,j}(r_i)]_{i,j=1}^{2p}$ we find that the functions $g_{k,j}(\theta)$ are their linear combinations; hence they satisfy

$$g_{k,j}(\theta) \in \mathcal{H}_k \quad \text{for } j = 1, \ldots, 2p.$$

We put $Y_k^j(\theta) = g_{k,j}(\theta)$ which ends the proof of statement (1) of the theorem.

The proof of statement (2) is split into exercises below and is left to the reader. ■

Spherical harmonics and polyharmonic functions 175

Exercise 10.40 1. First note that as in Theorem 10.19, p. 157, it follows immediately that

$$g_{k,\ell}(r) = g_{k,\ell}^{(p)}(r) := \sum_{j=1}^{2p} \alpha_{k,\ell,j} R_{k,j}(r) = \int_{\mathbb{S}^{n-1}} h(r\theta) Y_{k,\ell}(\theta) \, d\theta. \tag{10.33}$$

As in the same theorem for every N there exists a constant $C_N > 0$ such that the following estimate,

$$|g_{k,\ell}(r)| \leq \frac{C_N}{k^{2N}} \quad \text{for } k \geq 1 \text{ and } a_1 < r < b_1,$$

holds.

2. Then observe that we may neglect the terms with small k and for that reason we may avoid terms having logarithms. Prove the estimates for the case $p = 1$. For every large N we have the expansion

$$h(x) = h_1(x) + \sum_{k=N}^{\infty} \left(\alpha_{k,1,1} r^k + \alpha_{k,1,2} r^{-n-k+2} \right) \cos k\theta$$

$$+ \left(\alpha_{k,2,1} r^k + \alpha_{k,2,2} r^{-n-k+2} \right) \sin k\theta.$$

Using formula (10.33) express the coefficients α through the values $g_{k,\ell}(a_2)$ and $g_{k,\ell}(b_2)$. The estimates (10.31) and (10.32) to be proved now follow directly.

3. We proceed inductively on p. We consider the function $\Delta h(x)$. It has the representation in the form (10.30) given by

$$\Delta h(x) = \sum_{k=0}^{\infty} \sum_{\ell=1}^{d_k} L_{(k)} \left(\frac{d}{dr} \right) g_{k,\ell}^{(p)}(r) Y_{k,\ell}(\theta).$$

Note that for $j = 0, 1, \ldots, p - 1$ we have the equality

$$L_{(k)} \left(\frac{d}{dr} \right) R_{k,j+1}(r) = 2j(n + 2k + 2j - 2) R_{k,j}(r).$$

Similarly, for the last p functions we obtain

$$L_{(k)} \left(\frac{d}{dr} \right) r^{-n-k+2j} = 2(j-1)(-n - 2k + 2j) r^{-n-k+2j-2}$$

for $j = 1, 2, \ldots, p$. To prove this use the representation of the operator $L_{(k)}$ in (10.18), p. 152, and the fact that $R_{k,j}(r) = r^{k+2j-2}$ for $j = 1, 2, \ldots, p$, and $R_{k,p+j}(r) = r^{-n-k+2j}$ for $j = 1, 2, \ldots, p$, as defined in Theorem 10.37, p. 172.

4. Hence

$$g_{k,\ell}^{(p-1)}(r) = L_{(k)}\left(\frac{d}{dr}\right)g_{k,\ell}^{(p)}(r)$$

$$= \sum_{j=2}^{p} 2(j-1)(n+2k+2j-4)\alpha_{k,\ell,j}r^{k+2j-2}$$

$$+ \sum_{j=2}^{p} 2(j-1)(-n-2k+2j)\alpha_{k,\ell,p+j}r^{-n-k+2j-2}.$$

5. By the inductive argument we know that

$$|2(j-1)(n+2k+2j-4)\alpha_{k,\ell,j}r^{k+2j-2}| \leq K\left(\frac{r}{b_2}\right)^k$$

for $a_1 < r < b_1$, and $j = 2, 3, \ldots, p$,

$$|2(j-1)(-n-2k+2j)\alpha_{k,\ell,p+j}r^{-n-k+2j-2}| \leq K\left(\frac{a_2}{r}\right)^k$$

for $a_1 < r < b_1$, and $j = 2, 3, \ldots, p$.

From these inequalities the estimates (10.31)–(10.32) of the theorem for $j = 2, 3, \ldots, p$ follow immediately. By these estimates it follows that the series

$$h_1(x) = \sum_{k=0}^{\infty}\sum_{\ell=1}^{d_k}\left\{\sum_{j=2}^{p}\left[\alpha_{k,\ell,j}R_{k,j}^{(p)}(r) + \alpha_{k,\ell,p+j}R_{k,p+j}^{(p)}(r)\right]\right\} \cdot Y_{k,\ell}(\theta)$$

is absolutely and uniformly convergent in every annulus A_{a_1,b_1} satisfying $\overline{A_{a_1,b_1}} \subset A_{a,b}$.

6. It is clear that the function $h - h_1$ is harmonic and is represented by the series

$$h(x) - h_1(x) = \sum_{k=0}^{\infty}\sum_{\ell=1}^{d_k}\left\{\alpha_{k,\ell,1}R_{k,1}^{(1)}(r) + \alpha_{k,\ell,p+j}R_{k,2}^{(1)}(r)\right\} \cdot Y_{k,\ell}(\theta).$$

We have seen in point (2) that the coefficients satisfy

$$|\alpha_{k,\ell,1}r^k| \leq K\left(\frac{r}{b_2}\right)^k,$$

$$|\alpha_{k,\ell,p+j}r^{-n-k+2}| \leq K\left(\frac{a_2}{r}\right)^k.$$

Thus we have the exponential decay estimates of all coefficients $\alpha_{k,\ell,j}$. This completes the proof of the second part of Theorem 10.39.

Remark 10.41 The second part of Theorem 10.39 enables rearrangements of the terms of the Laplace series.

Spherical harmonics and polyharmonic functions 177

In a similar way we may prove the analogous representation for the ball but the difference is that only half of the functions $R_{k,j}$ are necessary in that case.

Theorem 10.42 *Let the function $h(x)$ be polyharmonic in the ball $B(0; b)$. Then it has the following Laplace series:*

$$h(x) = \sum_{k=0}^{\infty} (R_{k,1}(r) Y_k^1(\theta) + \cdots + R_{k,p}(r) Y_k^p(\theta)) \quad \text{for } 0 \le r < b, \; \theta \in \mathbb{S}^{n-1},$$

where $Y_k^1(\theta), \ldots, Y_k^p(\theta) \in \mathcal{H}_k$ are p spherical harmonics. The functions $R_{k,1}(r)$, $R_{k,2}(r), \ldots, R_p(r)$ are the first p linearly independent solutions of the equation $L_{(k)}^p f(r) = 0$ given by Theorem 10.37, p. 172, which satisfy for $r = e^v$ the equation

$$M_{k,p}^+ \left(\frac{d}{dv} \right) f(e^v) = 0 \quad \text{for } r = e^v < b,$$

where the operator $M_{k,p}^+$ is given by (10.29), p. 173.

The main argument of the proof is that all terms of the series have to be C^∞ at $r = 0$, and these were specified in Theorem 10.37, p. 172.

Remark 10.43 *Due to the above results we see that all solutions of the equation*

$$\Delta^p u = 0$$

with separated variables may be written in the form

$$u(x) = R_{k,j}(r) \cdot Y_{k,\ell}(\theta) \quad \text{for } x = r\theta,$$

where for $j = 1, \ldots, 2p$ the functions $R_{k,j}(r)$ form the basis of the set $N_{L_{(k)}^p}$ provided in Theorem 10.37, p. 172, and for every fixed $k = 0, 1, 2, \ldots$, the spherical harmonics $\{Y_{k,\ell}(\theta)\}_{\ell=1}^\infty \subset \mathcal{H}_k$ are a basis of $\mathcal{H}_k$.

10.15 The operator $r^{n-1} L_{(k)}^p$ is formally self-adjoint

Lemma 10.44 *For every integer $p \ge 1$ the operator $r^{n-1} L_{(k)}^p$ is formally self-adjoint on the interval $(0, \infty)$, or, which is the same, the operator $L_{(k)}^p$ is formally self-adjoint with respect to the weight function r^{n-1}.*

Proof The proof will be by induction in p. We have to prove that if $f, g \in C^\infty(-\infty, \infty)$ and $\operatorname{supp}(f)$ and $\operatorname{supp}(g)$ are contained in $(0, \infty)$ then

$$\int_0^\infty f(r) \cdot r^{n-1} \left\{ L_{(k)} \left(\frac{d}{dr} \right) \right\}^p g(r) \, dr = \int_0^\infty r^{n-1} \left\{ L_{(k)} \left(\frac{d}{dr} \right) \right\}^p [f(r)] \cdot g(r) \, dr.$$

178   Multivariate polysplines

For $p = 1$ by applying formula (10.17), p. 152, for the operator $L_{(k)}$, and by integrating by parts, we obtain the equalities

$$\int_0^\infty f(r) \cdot r^{n-1} L_{(k)}\left(\frac{d}{dr}\right) g(r)\, dr$$

$$= \int_0^\infty f(r) \cdot r^{n-1} \left(\frac{1}{r^{n-1}} \frac{d}{dr} r^{n-1} \frac{d}{dr} - \frac{k(k+n-2)}{r^2}\right) g(r)\, dr$$

$$= \int_0^\infty f(r) \cdot \frac{d}{dr} r^{n-1} \frac{d}{dr} g(r)\, dr - \int_0^\infty f(r) \cdot r^{n-1} \frac{k(k+n-2)}{r^2} g(r)\, dr$$

$$= -\int_0^\infty \frac{d}{dr} f(r) \cdot r^{n-1} \frac{d}{dr} g(r)\, dr - \int_0^\infty f(r) \cdot r^{n-1} \frac{k(k+n-2)}{r^2} g(r)\, dr$$

$$= \int_0^\infty \frac{d}{dr} r^{n-1} \frac{d}{dr} f(r) \cdot g(r)\, dr - \int_0^\infty f(r) \cdot r^{n-1} \frac{k(k+n-2)}{r^2} g(r)\, dr$$

$$= \int_0^\infty r^{n-1} L_{(k)}\left(\frac{d}{dr}\right) f(r) \cdot g(r)\, dr.$$

Now for arbitrary $p \geq 1$ we have

$$\int_0^\infty f(r) \cdot r^{n-1} \left\{L_{(k)}\left(\frac{d}{dr}\right)\right\}^p g(r)\, dr$$

$$= \int_0^\infty r^{n-1} L_{(k)}\left(\frac{d}{dr}\right) f(r) \cdot \left\{L_{(k)}\left(\frac{d}{dr}\right)\right\}^{p-1} g(r)\, dr$$

$$= \int_0^\infty L_{(k)}\left(\frac{d}{dr}\right) f(r) \cdot r^{n-1} \left\{L_{(k)}\left(\frac{d}{dr}\right)\right\}^{p-1} g(r)\, dr$$

which by a simple inductive procedure proves the Lemma. ∎

Now let us make the following remark about the relation between the properties of the operator Δ^p and the operators $L_{(k)}^p$. Since the operator Δ^p is symmetric, for every two functions f and g having compact support and belonging to C^2, we obtain the equality

$$\int \Delta^p f(x) g(x)\, dx = \int f(x) \Delta^p g(x)\, dx.$$

Let us assume that supp f and supp g are contained in the ball $B(0; R)$ for some radius $R > 0$. Then we may expand f and g in spherical harmonics

$$f(x) = \sum_{k=0}^\infty \sum_{\ell=1}^{d_k} f^{k,\ell}(r) Y_{k,\ell}(\theta),$$

$$g(x) = \sum_{k=0}^\infty \sum_{\ell=1}^{d_k} g^{k,\ell}(r) Y_{k,\ell}(\theta).$$

By formula (10.25), p. 165, about the splitting of the operator Δ^p into the actions of the operators $L_{(k)}^p$ we obtain the equality

$$\int \Delta^p f(x) g(x) \, dx = \int_0^\infty \int_{\mathbb{S}^{n-1}} \Delta^p f(x) g(x) r^{n-1} \, dr \, d\theta$$

$$= \int_0^\infty \int_{\mathbb{S}^{n-1}} r^{n-1} \sum_{k=0}^\infty \sum_{\ell=1}^{d_k} L_{(k)}^p f^{k,\ell}(r) Y_{k,\ell}(\theta)$$

$$\cdot \sum_{k=0}^\infty \sum_{\ell=1}^{d_k} g^{k,\ell}(r) Y_{k,\ell}(\theta) \, dr \, d\theta$$

$$= \sum_{k=0}^\infty \sum_{\ell=1}^{d_k} \int_0^\infty r^{n-1} L_{(k)}^p f^{k,\ell}(r) \cdot g^{k,\ell}(r) \, dr.$$

We see that this is an alternative proof of the above lemma.

Remark 10.45 *Running very much ahead, let us note that by the above arguments we may use results already obtained for the operator Δ^p, in particular those in Part IV, as Theorem 20.7, p. 416, about the basic identity of polysplines. Further examples are Theorem 20.9, p. 421, about the uniqueness of polysplines, and the Holladay property for polysplines in Theorem 20.14, p. 425. In the case of a ball or an annulus they provide interesting relations for the L-splines with $L = r^{n-1} L_{(k)}^{2p}$, or with $L = L_{(k)}^{2p}$ but for the weight function r^{n-1}. This will be used later in Chapter 11, p. 187, to prove the existence of interpolation L-splines for $L = L_{(k)}^{2p}$.*

10.16 The Almansi theorem

We will prove the Almansi representation which generalizes the Gauss representation which we have obtained for polynomials. Although we will need the Almansi representation only in the annulus and in the ball, we provide a proof for star-shaped domains owing to its importance for the overall philosophy of the subject of the present book, and since several useful formulas including the polyharmonic operator appear in the process of the proof.

The *order of polyharmonicity* of a function $f(x)$ in a domain D is an integer p such that

$$\Delta^p f(x) = 0 \quad \text{for } x \in D,$$

but for some $y \in D$ we have $\Delta^{p-1} f(y) \neq 0$.

An immediate application of Theorem 10.42, p. 177, and Theorem 10.39, p. 173, is the Almansi formulas for the ball and the annulus. We provide only the formula for the annulus since the one for the ball is provided in Theorem 10.51, p. 184, in the more general case of a star-shaped domain.

Theorem 10.46 *Let the function $f(x)$ be polyharmonic of order $p \geq 1$ in the annulus $A_{a,b} = \{a < |x| < b\}$, i.e. $f(x)$ satisfies*

$$\Delta^p f = 0 \quad \text{in } A_{a,b}.$$

Then there exist functions $f_j(x)$, $j = 0, \ldots, p-1$, which are harmonic in the domain D, such that the following representation holds:

$$f(x) = \sum_{j=0}^{p-1} f_j(x)|x|^{2j} + g(x) \quad \text{for } x \in A_{a,b}, \tag{10.34}$$

with

$$\Delta f_j = 0 \quad \text{in } A_{a,b}.$$

The function $g(x)$ is a finite sum which appears only for even dimensions $n = 2m$ and is given by

$$g(x) = \log r \cdot \sum_{k=0}^{p-m} \sum_{j=0}^{p-m-k} r^{k+2j} Y_k^{(j)}(\theta),$$

where $Y_k^{(j)}(\theta)$ are spherical harmonics of degree k.

For the case $n = 2$, we have provided the above representation in formula (7.25), p. 97, and so the terms with $k = 0$ may be dropped.

Proof The proof is the immediate consequence of Theorem 10.39, p. 173, since we may rearrange the terms in the absolutely convergent series. Indeed, from (10.33), p. 175, we have

$$g_{k,\ell}(r) = \sum_{j=1}^{2p} \alpha_{k,\ell,j} R_{k,j}(r),$$

$$h(x) = \sum_{k=0}^{\infty} \sum_{\ell=1}^{d_k} g_{k,\ell}(r) Y_{k,\ell}(\theta).$$

Since for large enough k we do not have overlapping between the two groups in the solutions $R_{k,j}^{(p)}$ we have

$$R_{k,j}^{(p)}(r) = r^{2j-2} R_{k,1}^{(p)}(r) \quad \text{for } j = 1, 2, \ldots, p,$$

$$R_{k,p+j}^{(p)}(r) = r^{2j-2} R_{k,p+1}^{(p)}(r) \quad \text{for } j = 1, 2, \ldots, p,$$

and the functions $R_{k,1}^{(p)}(r)$ and $R_{k,p+1}^{(p)}(r)$; are harmonic, we obtain the Almansi representation, where

$$f_j(x) = \sum_{k=0}^{\infty} \sum_{\ell=1}^{d_k} \left(\alpha_{k,\ell,j} R_{k,j}^{(p)}(r) + \alpha_{k,\ell,p+j} R_{k,p+j}^{(p)}(r) \right) Y_{k,\ell}(\theta),$$

and we have removed the terms containing $\log r$. This completes the proof. ■

Spherical harmonics and polyharmonic functions

Let us introduce the class of star-shaped domains which are natural for the Almansi-type representations.

Definition 10.47 *The domain D in $\mathbb{R}^n$ is called* star-shaped *with respect to the point $z \in D$ if and only if for every $y \in D$ the whole line segment $[z, y]$ is also contained in D, i.e.*

$$\alpha z + (1 - \alpha) y \in D$$

for every number α with $0 \le \alpha \le 1$.

In order to prove the uniqueness we will need the following subtle result.

Lemma 10.48 *Let the order of polyharmonicity of $f(x)$ in a ball $B(0; R)$ be p. Then the order of polyharmonicity of the function $|x|^2 f(x)$ in the same ball $B(0; R)$ is $p + 1$.*

Proof The proof consists of several technical results.
(1) We apply the following obvious one-dimensional formula

$$[u(t)v(t)]'' = u''(t)v(t) + 2u'(t)v'(t) + u(t)v''(t),$$

and its multivariate consequence,

$$\Delta(u(x)v(x)) = \Delta u(x) \cdot v(x) + 2\nabla u \cdot \nabla v(x) + u \cdot \Delta v(x).$$

As usual, let $r = |x|$. We also apply the notation

$$x \cdot \nabla u(x) = r \frac{\partial}{\partial r} u(x).$$

(2) We compute directly

$$\Delta(r^2 h(x)) = r^2 \Delta h(x) + 4x \cdot \nabla h(x) + 2nh(x)$$

$$= r^2 \Delta h(x) + 4r \frac{\partial}{\partial r} h(x) + 2nh(x).$$

(3) Here we prove that the operator Δ and $r(\partial/\partial r)$ commute up to a certain term, namely,

$$\Delta \left(r \frac{\partial}{\partial r} g(x) \right) = \left(2 + r \frac{\partial}{\partial r} \right) \Delta g(x).$$

Since

$$\frac{\partial^2}{\partial x_k^2} \left(x_j \frac{\partial h}{\partial x_j} \right) = 2 \frac{\partial}{\partial x_k}(x_j) \cdot \frac{\partial^2 h}{\partial x_j \partial x_k} + x_j \frac{\partial^3 h}{\partial x_j \partial^2 x_k}$$

$$= \begin{cases} 2 \cdot \dfrac{\partial^2 h}{\partial x_k^2} + x_k \dfrac{\partial^3 h}{\partial^3 x_k} & \text{for } k = j, \\[2mm] x_j \dfrac{\partial^3 h}{\partial x_j \partial^2 x_k} & \text{for } k \ne j, \end{cases}$$

we see that

$$\Delta(x \cdot \nabla h(x)) = \Delta \left(r \cdot \frac{\partial}{\partial r} h(x) \right) = \sum_{k=1}^{n} \frac{\partial^2}{\partial x_k^2} \left(\sum_{j=1}^{n} x_j \frac{\partial h}{\partial x_j} \right)$$

$$= \sum_{k=1}^{n} \left(2 \cdot \frac{\partial^2 h}{\partial x_k^2} + \sum_{j=1}^{n} x_j \frac{\partial^3 h}{\partial x_j \partial^2 x_k} \right)$$

$$= 2\Delta h(x) + \sum_{j=1}^{n} x_j \frac{\partial}{\partial x_j} \left(\sum_{k=1}^{n} \frac{\partial^2 h}{\partial^2 x_k} \right)$$

$$= 2\Delta h(x) + x \cdot \nabla(\Delta h(x)) = \left(2 + r \frac{\partial}{\partial r} \right) \Delta h(x).$$

(4) Then we prove inductively the formula

$$\Delta^s(r^2 u(x)) = r^2 \Delta^s u(x) + 4sr \frac{\partial}{\partial r} \Delta^{s-1} u(x) \tag{10.35}$$
$$+ 2s(2s + n - 2) \Delta^{s-1} u(x).$$

Indeed, assuming that the above formula holds, let us again apply the Laplace operator, Δ, to both sides of it. We obtain

$$\Delta^{s+1}(r^2 u(x)) = \Delta(r^2 \Delta^s u(x)) + \Delta \left(4sr \frac{\partial}{\partial r} \Delta^{s-1} u(x) \right) + 2s(2s + n - 2) \Delta^s u(x).$$

By point (2) with $h(x) = \Delta^s u(x)$, we obtain

$$\Delta(r^2 \Delta^s u(x)) = r^2 \Delta^{s+1} u(x) + 4r \frac{\partial}{\partial r} \Delta^s u(x) + 2n \Delta^s u(x).$$

From point (3) for $h(x) = \Delta^{s-1} u(x)$ it follows that:

$$\Delta \left(4sr \frac{\partial}{\partial r} \Delta^{s-1} u(x) \right) = 4s \left(2 + r \frac{\partial}{\partial r} \right) \Delta^s u(x).$$

Hence, we obtain

$$\Delta^{s+1}(r^2 u(x)) = r^2 \Delta^{s+1} u(x) + 4r \frac{\partial}{\partial r} \Delta^s u(x) + 2n \Delta^s u(x)$$
$$+ 4s \left(2 + r \frac{\partial}{\partial r} \right) \Delta^s u(x) + 2s(2s + n - 2) \Delta^s u(x)$$
$$= r^2 \Delta^{s+1} u(x) + (4s + 4)r \frac{\partial}{\partial r} \Delta^s u(x) + 2(s + 1)(2s + n) \Delta^s u(x),$$

which is precisely formula (10.35) for $s + 1$.

(5) Now putting $s = p+1$ in the above formulas, we see that from $\Delta^p f(x) = 0$ and formula (10.35) it immediately follows that

$$\Delta^{p+1}(|x|^2 f(x)) = 0.$$

On the other hand, if we assume that also $\Delta^p(|x|^2 f(x)) = 0$ in the ball $B(0; R)$ then it follows from the same formula (10.35) now with $s = p$, that

$$0 = \Delta^p(r^2 f(x)) = r^2 \Delta^p f(x) + 4pr\frac{\partial}{\partial r}\Delta^{p-1} f(x) + 2p(2p+n-2)\Delta^{p-1} f(x),$$

or

$$\left[r\frac{\partial}{\partial r} + \left(p + \frac{n}{2} - 1\right)\right]\Delta^{p-1} f(x) = 0.$$

Let us consider now the *ODE*

$$\left[r\frac{d}{dr} + \left(p + \frac{n}{2} - 1\right)\right]g(r) = 0.$$

It is a first-order *ODE* and its only solution is given by

$$g(r) = Cr^{-(p+(n/2)-1)}.$$

Hence $g(r) = \Delta^{p-1} f(x)$. But that is impossible since $g(r)$ has a singularity at the origin 0 and $\Delta^{p-1} f(x)$ is a smooth function. This contradiction completes the proof. ∎

By repeated application of the above lemma we have the following corollary.

Corollary 10.49 *Let the order of polyharmonicity of $f(x)$ in a ball $B(0; R)$ be p. Then the order of polyharmonicity of the function $|x|^{2s} f(x)$ in the same ball $B(0; R)$ is $p+s$.*

We will need one more independent lemma which is of general interest.

Lemma 10.50 *Let the function $h(x)$ be harmonic in the star-shaped domain D. Then for every number $\alpha > 0$ the function*

$$\psi(x) = \frac{1}{r^\alpha}\int_0^r t^{\alpha-1} h(t\theta)\,dt$$

is again harmonic in the domain D.

Proof We change the variable by putting $\tau = t/r$. We obtain

$$\psi(x) = \int_0^1 \tau^{\alpha-1} h(\tau x)\,d\tau.$$

It is easy to see that we may differentiate under the sign of the integral, hence

$$\Delta\psi(x) = \int_0^1 \tau^{\alpha-1} \Delta h(\tau x)\,d\tau.$$

184 Multivariate polysplines

The last implies
$$\Delta \psi(x) = 0$$
since $h(\tau x)$ is a harmonic function for every real number τ. ∎

We now proceed to the proof of the Almansi representation for star-shaped domains.

Theorem 10.51 *Let D be a domain which is star-shaped with respect to the origin 0. Let the function $f(x)$ be polyharmonic of order $p \geq 1$ in D, i.e. there $f(x)$ satisfies*
$$\Delta^p f(x) = 0.$$
Then there exist the functions $f_j(x)$, $j = 0, \ldots, p - 1$, which are harmonic in the domain D, such that the following representation holds:
$$f(x) = f_0(x) + f_1(x)|x|^2 + \cdots + f_{p-1}(x)|x|^{2p-2} \text{ for } x \in D, \tag{10.36}$$
with
$$\Delta f_j = 0 \quad \text{in } D.$$

Proof We provide an inductive proof. For $p = 1$ the theorem is evident. Assume that the theorem holds for $p - 1$. We will prove it for p. Consider the function $h(x) = \Delta f(x)$. It satisfies the equation
$$\Delta^{p-1} h(x) = 0$$
in D, hence by the inductive assumption it is represented as
$$h(x) = \varphi_0(x) + \varphi_1(x)|x|^2 + \cdots + \varphi_{p-2}(x)|x|^{2p-4},$$
where φ_j are harmonic functions.

Let us provide some useful formulas.

- For $j \geq 1$ we have evidently
$$\frac{\partial^2}{\partial x_k^2} |x|^{2j} = \frac{\partial^2}{\partial x_k^2} \left(\sum_{s=1}^n x_s^2\right)^j = \frac{\partial}{\partial x_k} \left[2x_k j \left(\sum_{s=1}^n x_s^2\right)^{j-1}\right]$$
$$= 2j \left(\sum_{s=1}^n x_s^2\right)^{j-1} + 4x_k^2 j(j-1) \left(\sum_{s=1}^n x_s^2\right)^{j-2}$$
$$= 2j r^{2j-2} + 4x_k^2 j(j-1) r^{2j-4}.$$

- We sum the above in k and obtain
$$\Delta(|x|^{2j}) = \left(\sum_{k=1}^n \frac{\partial^2}{\partial x_k^2}\right)\left(\sum_{s=1}^n x_s^2\right)^j = \sum_{k=1}^n \left[2jr^{2j-2} + 4x_k^2 j(j-1)r^{2j-4}\right]$$
$$= 2jnr^{2j-2} + 4j(j-1)r^{2j-2} = 2j(2j+n-2)r^{2j-2}.$$

- We have
$$\nabla(|x|^{2j}) = 2j|x|^{2j-2}x,$$
where x is now considered as a vector.

Now we assume that we have the representation (10.36). Due to the general formula $\Delta(fg) = \Delta f \cdot g + 2\nabla f \cdot \nabla g + f \cdot \nabla g$, where we put $f = u(x)$ and $g = |x|^{2j}$ and using the above formulas we obtain

$$\Delta(u(x)|x|^{2j}) = |x|^{2j-2}[4jx \cdot \nabla u(x) + 2j(2j + n - 2)u(x)].$$

Hence,

$$\Delta f(x) = \Delta f_0(x) + \Delta(f_1(x)|x|^2) + \cdots + \Delta(f_{p-1}(x)|x|^{2p-2}) \quad (10.37)$$

$$= \sum_{j=1}^{p-1}[4jx \cdot \nabla f_j(x) + 2j(2j + n - 2)f_j(x)]|x|^{2j-2}.$$

This implies that

$$4jx \cdot \nabla f_j(x) + 2j(2j + n - 2)f_j(x) = \varphi_{j-1}(x) \quad \text{for } x \in D.$$

On the left-hand side we have an operator which is essentially one-dimensional, namely we see that the harmonic functions have to satisfy

$$4jx \cdot \nabla f_j(x) + 2j(2j + n - 2)f_j(x) = 4j\frac{1}{r^{j+(n/2)-2}}\frac{\partial}{\partial r}\left(r^{j+(n/2)-1}f_j(x)\right)$$
$$= \varphi_{j-1}(x) \quad \text{for } x \in D.$$

At this point we see that we may apply Lemma 10.50, p. 183, with $\alpha = j + (n/2) - 2$, which provides functions $f_j(x)$ which are harmonic and satisfy the above equalities. It remains to determine the function $f_0(x)$. We simply put

$$f_0(x) = f(x) - f_1(x)|x|^2 - \cdots - f_{p-1}(x)|x|^{2p-2} \quad \text{for } x \in D.$$

That the function $f_0(x)$ is harmonic follows from equality (10.37). ∎

We see that when D is a ball the above proof reduces to the representation of the polyharmonic function through spherical harmonics.

10.17 Bibliographical Notes

A brief exposition of the properties of spherical harmonics may be found in Seeley [51]. We note that in the case of the sphere $\mathbb{S}^2$ there is an exhaustive study of spherical harmonics in Freeden et al. [18]. It is within the framework of Sobolev spaces.

Chapter 11

Appendix on Chebyshev splines

In the present appendix we provide all the results for one-dimensional generalized splines which are necessary in this book. In the main we follow Schumaker [50].

11.1 Differential operators and Extended Complete Chebyshev systems

We will have many applications of the so-called **Chebyshev**[1] splines which are a special class of the L-splines. The last are defined by means of a differential operator L.

Let $I = [a, b]$ be an interval on the real line $\mathbb{R}$. It may be finite or infinite. Let p be an integer and $p \geq 1$. Assume that the functions $w_i \in C^{p-i}(I)$ be positive on I for all $i = 1, 2, \ldots, p$. We define the following generalized *differentiation* operators, [50, p. 365]:

$$D_0 f(t) = f(t) \tag{11.1}$$

$$D_i f(t) = \frac{d}{dt}\left(\frac{f(t)}{w_i(t)}\right) \quad (i = 1, \ldots, p).$$

and the following *differential* operators:

$$L_i = D_i D_{i-1} \ldots D_0 \quad (i = 0, \ldots, p). \tag{11.2}$$

Definition 11.1 *A set of functions* $U_p = \{u_i\}_{i=1}^{p} \in C^{p-1}(I)$, $i = 1, 2, \ldots, p$, *is said to be an **Extended Complete Chebyshev** system, abbreviated as **ECT**-system, if and only if their Wronskian determinants are strictly positive on I, i.e. for every k with $1 \leq k \leq p$*

---

[1] One has to make the necessary linguistic remark that "Chebyshev" = "Tchebycheff". The first is the modern (English) way to write the name and the second is the traditional French way. Hence we have the accepted notation [26, 51] **ECT** = **Extended Complete Tchebycheff**, but **not ECC**.

188 Multivariate polysplines

we have

$$W(u_1, ..., u_k)(t) = \det\left[\frac{d^{j-1}}{dt^{j-1}} u_i(t)\right]_{i,j=1}^{k} > 0 \quad \text{for } t \in I. \tag{11.2a}$$

The linear space $\mathcal{U}_p$ spanned by the elements of U_p is called ECT-space.

We have the following basic result, see Karlin and Studden [28, p. 364, Theorem 9.2, p. 379],

Theorem 11.2 *The space of all solutions of the ordinary differential equation $L_p u(t) = 0$ in the interval I is spanned by the following set of p functions:*

$$\left.\begin{aligned} u_1(t) &= w_1(t) \\ u_2(t) &= w_1(t) \int_a^t w_2(t_2)\, dt_2 \\ &\vdots \\ u_p(t) &= w_1(t) \int_a^t w_2(t_2) \int_a^{t_2} \cdots \int_a^{t_{p-1}} w_p(t_p)\, dt_p \ldots dt_2. \end{aligned}\right\} \tag{11.3}$$

It is an ECT-Space. In particular, for every k with $1 \le k \le p$ the Wronskian determinants satisfy

$$W(u_1, \ldots, u_k)(t) = \det\left[\frac{d^{j-1}}{dt^{j-1}} u_i(t)\right]_{i,j=1}^{k} > 0 \quad (t \in I). \tag{11.4}$$

We immediately obtain Corollary 11.3.

Corollary 11.3 *Let the operator $L = L_1 L_2$ where operators L_1 and L_2 are of the form (11.2) and L_1 is of order p_1. Then the first p_1 elements of the Chebyshev system generated by L using formulas (11.3) belong to the system generated by the operator L_1 by the same formulas (11.3).*

There is a condition which is equivalent to (11.4) and which does not need differentiability of the functions u_i [50, p. 363], and which is usually taken as an definition of a ECT-system.

Theorem 11.4 *Condition (11.4) is equivalent to*

$$D\begin{pmatrix} t_1, \ldots, t_k \\ u_1, \ldots, u_k \end{pmatrix} > 0 \tag{11.5}$$

for all $t_1 \le t_2 \le \cdots \le t_k$ in I and all k with $1 \le k \le p$.
It means that the set of all solutions of $L_p u(t) = 0$ is an ECT-system.

The determinants D are defined by

$$D\begin{pmatrix} t_1, \ldots, t_k \\ u_1, \ldots, u_k \end{pmatrix} := \det\begin{pmatrix} u_1(t_1) & u_2(t_1) & \cdots & u_k(t_1) \\ u_1(t_2) & u_2(t_2) & \cdots & u_k(t_2) \\ \vdots & \vdots & \vdots & \vdots \\ u_1(t_k) & u_2(t_k) & \cdots & u_k(t_k) \end{pmatrix} \tag{11.6}$$

Chebyshev splines 189

if $t_1 < t_2 < \cdots < t_k$ as in [50, Section 2.3]. If there are repeated ts then there will be derivatives, see [50, p. 366] and the lemma below. If we write the determinant in (11.5) in terms of the differential operators L_i, then we have the following result [50, p. 366].

Lemma 11.5 *The determinant D above satisfies the equality*

$$D\begin{pmatrix} t_1, \ldots, t_p \\ u_1, \ldots, u_p \end{pmatrix} = \det\,[L_{d_i} u_j(t_i)]_{i,j=1}^p,$$

where we have taken into account the repetition of some t_is, namely

$$d_i = \max\,\{j : t_i = \cdots = t_{i-j}\}$$

for $i = 1, \ldots, p$.

We will denote by $\mathcal{U}_p$ the space spanned by the functions $\{u_i\}_{i=1}^p$, i.e.

$$\mathcal{U}_p := \mathrm{lin}\,\{u_i, i = 1, 2, \ldots, p\}. \tag{11.7}$$

Theorem 11.6 *If $U_p = \{u_i\}_{i=1}^p$ is an ECT-system then there is a basis of $\mathcal{U}_p$, possibly different from U_p, which is in the canonical form (11.3).*

In the last case one says that the system U_p is a **canonical ECT-system**.

Remark 11.7 *Let us note that the theory of extended Chebyshev systems and especially splines is generalized to the so-called L-splines, where L is a differential operator of order p which is of a more general form than the operator L_p above. So far for the present book, the operators of interest fall into the above category. As we have seen in Section 7.2.3, p. 90, where we considered polysplines on concentric spheres, some of the functions w_j may have zeros at the end of the interval and the operator L_p will have singular coefficients.*

The explicit form of u_i permits us to prove easily that

$$L_j u_i(a) = w_i(a) \delta_{j,i-1} \quad (0 \le j \le i-1,\ 1 \le i \le p)$$

where δ_{st} is the Kronecker symbol.

The *Chebyshev splines* which will be of interest to us are piecewise solutions to operators $L(D)$ which are of the form

$$L(\tau) = \prod_{j=1}^p (\tau - \tau_j)$$

where τ_j are real constants. They are the so-called exponential splines [50, p. 405]. We will often have the case of $\tau_i \ne \tau_j$ for $i \ne j$. In that case we have the following representation:

$$\left(\frac{d}{dt} - \tau_j\right) f(t) = e^{\tau_j t}\left(-\tau_j e^{-\tau_j t} f(t) + e^{-\tau_j t}\frac{d}{dt} f(t)\right)$$

$$= e^{\tau_j t}\frac{d}{dt}\left(e^{-\tau_j t} f(t)\right)$$

190 Multivariate polysplines

hence,

$$L\left(\frac{d}{dt}\right) f(t) = \prod_{j=1}^{p} \left(\frac{d}{dt} - \tau_j\right) f(t) \tag{11.8}$$

$$= \prod_{j=1}^{p} e^{\tau_j t} \frac{d}{dt} e^{-\tau_j t} f(t)$$

$$= e^{\tau_1 t} \frac{d}{dt} e^{(\tau_2 - \tau_1)t} \cdot \frac{d}{dt} e^{(\tau_3 - \tau_2)t} \cdots e^{(\tau_p - \tau_{p-1})t} \cdot \frac{d}{dt} e^{-\tau_p t} f(t)$$

$$= e^{\tau_1 t} \frac{d}{dt} \frac{1}{e^{-(\tau_2 - \tau_1)t}} \cdot \frac{d}{dt} \frac{1}{e^{-(\tau_3 - \tau_2)t}} \cdots \frac{1}{e^{-(\tau_p - \tau_{p-1})t}} \cdot \frac{d}{dt} \frac{1}{e^{\tau_p t}} f(t).$$

So we will comply with the definition of the operator L_p in (11.1) and (11.2), p. 187, if we put

$$w_1(t) = e^{\tau_p t},$$
$$w_2(t) = e^{-(\tau_p - \tau_{p-1})t},$$
$$\vdots$$
$$w_{p-1}(t) = e^{-(\tau_3 - \tau_2)t},$$
$$w_p(t) = e^{-(\tau_2 - \tau_1)t}.$$

Vice versa, if

$$w_j(t) = e^{\gamma_j t} \quad \text{for } j = 1, \ldots, p,$$

then we easily obtain

$$\tau_p = \gamma_1,$$
$$\tau_{p-1} = \gamma_1 + \gamma_2,$$
$$\tau_{p-2} = \gamma_1 + \gamma_2 + \gamma_3,$$
$$\vdots$$
$$\tau_1 = \gamma_1 + \gamma_2 + \cdots + \gamma_p.$$

See [50, p. 405] on exponential splines. Now the solutions to the equation

$$L\left(\frac{d}{dt}\right) f(t) = 0$$

coincide with those of the equation

$$e^{-\tau_1 t} L\left(\frac{d}{dt}\right) f(t) = 0$$

in every interval I.

We have the following precise estimate of the determinants (11.5), p. 188, see [50, Lemma 9.6, p. 367].

Proposition 11.8 *For* $i = 1, \ldots, p$ *we put*

$$\begin{cases} \underline{M}_i = \min_{a \le t \le b} w_i(t), \\ \overline{M}_i = \max_{a \le t \le b} w_i(t). \end{cases} \tag{11.9}$$

Then for every choice of the points t_i satisfying $a \le t_1 \le t_2 \le \cdots \le t_p \le b$ the inequality

$$C_1 V(t_1, t_2, \ldots, t_p) \le D\begin{pmatrix} t_1, \ldots, t_p \\ u_1, \ldots, u_p \end{pmatrix} \le C_2 V(t_1, t_2, \ldots, t_p)$$

holds where V is the Vandermonde determinant and the constants C_1, C_2 depend only on p and the values $\underline{M}_i, \overline{M}_i$. A similar inequality holds for the derivatives of the determinants

$$C_1 \left| \frac{d^k}{dt^k} V(t_1, t_2, \ldots, t_{p-1}, t) \right| \le \left| L_k D\begin{pmatrix} t_1, \ldots, t_p \\ u_1, \ldots, u_p \end{pmatrix} \right|$$

$$\le C_2 \left| \frac{d^k}{dt^k} V(t_1, t_2, \ldots, t_p) \right|.$$

11.2 Divided differences for Extended Complete Chebyshev systems

11.2.1 The classical polynomial case

The notion of divided differences for the usual algebraic polynomials has a good generalization for *ECT*-systems [51, p. 368].

Let us take the one-dimensional divided difference operator [50, p. 46, Theorem 2.50]. Let a system of points $t_1 < \cdots < t_{p+1}$ be given, then the *divided difference operator* of order p is defined by

$$f[t_1, \ldots, t_{p+1}] := \sum_{i=1}^{p+1} \left\{ f(t_i) / \prod_{j=1, j \ne i}^{p+1} (t_i - t_j) \right\} = \sum_{i=1}^{p+1} \frac{f(t_i)}{\omega'(t_i)}, \tag{11.10}$$

where

$$\omega(t) := (t - t_1)(t - t_2) \cdots (t - t_{p+1}).$$

Sometimes it is written as

$$f[t_1, \ldots, t_{p+1}] = [t_1, \ldots, t_{p+1}] f$$

in order to emphasize that it is an operator.

192 Multivariate polysplines

In the case of repeating points

$$\{t_1 \leq t_2 \leq \cdots \leq t_{p+1}\}$$
$$= \left\{ \underbrace{\tau_1 = \cdots = \tau_1}_{l_1} < \cdots < \underbrace{\tau_d = \cdots = \tau_d}_{l_d} \right\}$$

the divided difference operator is defined for functions which have sufficient smoothness, namely for f which has l_i derivatives at the point τ_i, for $i = 1, \ldots, d$.

The following definition is sometimes taken as a theorem and uniquely characterizes the coefficients of the divided difference operator [50, Section 2.7, formula (2.86), p. 45].

Definition 11.9 *The divided difference operator is the (unique) linear functional*

$$f[t_1, \ldots, t_{p+1}] := L(f) = \sum_{i=1}^{d} \sum_{j=1}^{l_i} \alpha_{ij} D^{j-1} f(\tau_i) \tag{11.11}$$

such that it annihilates all polynomials Q of degree $\leq p - 1$, and is 1 on the special polynomial x^p, i.e.

$$L(Q) = 0 \quad \text{for } \deg Q \leq p - 1,$$
$$L(x^p) = 1.$$

Exercise 11.10 Note that $\alpha_{i,l_i} \neq 0$ for all $i = 1, \ldots, d$. Prove the uniqueness of the functional L.

Exercise 11.11 Using the equalities of Definition 11.9 write the coefficients α_{ij} in (11.11) as Vandermonde determinants. The explicit coefficients in (11.10) are a special case.

Now let us determine precisely the coefficients α_{ij}. Let us denote the polynomial

$$\Omega(z) = \prod_{i=1}^{s} (z - t_i)^{\nu_i}.$$

It has the following representation through elementary fractions:[2]

$$\frac{1}{\Omega(z)} = \sum_{i=1}^{s} \sum_{j=0}^{l_i - 1} \frac{\beta_{ij}}{(z - t_i)^{j+1}}.$$

We have the following result [4, p. 7].

---

[2] This representation is frequently used in integral calculus.

Chebyshev splines 193

Theorem 11.12 *The divided difference operator is given by*

$$[t_1,\ldots,t_p]f = \sum_{i=1}^{s}\sum_{j=0}^{l_i-1} \frac{\beta_{ij}}{j!} f^{(j)}(t_i),$$

where the coefficients β_{ij} are those in the representation of the function $1/\Omega(z)$ through elementary fractions.

The proof of this theorem will follow from the proof of the following *residuum representation* of the divided difference operator which was established by *Frobenius* [4, p. 8].

Theorem 11.13 *Let the function $f(z)$ be analytic in a closed simply connected domain containing the rectifiable curve Γ which encircles the points $\{t_1, t_2, \ldots, t_d\}$. We have the following equality:*

$$[t_1,\ldots,t_p]f = \frac{1}{2\pi i}\int_\Gamma \frac{f(z)}{\Omega(z)} dz. \tag{11.12}$$

Proof Evidently, the functional given by the integral over Γ is a linear one, and due to the residuum theorem [8], we obtain

$$\frac{1}{2\pi i}\int_\Gamma \frac{f(z)}{\Omega(z)} dz = \sum_{i=1}^{s}\sum_{j=0}^{l_i-1} \frac{\beta_{ij}}{j!} f^{(j)}(t_i).$$

We see that this functional annihilates all polynomials of degree $\leq p - 1$. Indeed, it is clear from Cauchy's residuum theorem that the integral over Γ has the same value for all simple closed contours which contain the points t_j, $j = 1,\ldots, d$. In particular, we choose

$$\Gamma_R = \{Re^{i\theta} : 0 \leq \theta < 2\pi\},$$

for $R \longrightarrow \infty$. By $l_1 + \cdots + l_d = p + 1$ it follows that $\deg \Omega = p + 1$. Then for every polynomial f of degree $\leq p - 1$ we obtain the estimate

$$\left|\int_{\Gamma_R} \frac{f(z)}{\Omega(z)} dz\right| \leq \int_0^{2\pi}\left|\frac{f(z)}{\Omega(z)}\right| R\, d\theta \leq \int_0^{2\pi} \frac{C}{R^2} R\, d\theta$$

$$\leq \frac{2\pi C}{R} \stackrel{R\to\infty}{\longrightarrow} 0,$$

where $C > 0$ is a constant, which proves the statement. On the other hand, if we take $f(x) = x^p$, since the polynomial $\Omega(z) - z^{p+1}$ has degree $\leq p$, we obtain in a similar way

$$\left|\int_{\Gamma_R} \left(\frac{1}{z} - \frac{z^p}{\Omega(z)}\right) dz\right| \leq \int_0^{2\pi}\left|\frac{\Omega(z) - z^{p+1}}{z\Omega(z)}\right| R\, d\theta$$

$$= \int_0^{2\pi} \frac{C}{R^2} R\, d\theta \stackrel{R\to\infty}{\longrightarrow} 0,$$

194 *Multivariate polysplines*

where $C > 0$ is a constant. Since

$$\int_{\Gamma_R} \frac{1}{z} dz = 2\pi i$$

this completes the proof of the theorem. ∎

We have the following classical result [9, p. 4].

Corollary 11.14 *Let us denote by $P(x)$ the polynomial which interpolates the function $f(x)$ at the points t_i with multiplicity l_i for $i = 1, 2, \ldots, d$. It has degree $\leq p + 1$. Then the divided difference of $f(x)$ coincides with the leading coefficient a_{p+1} in front of the term x^{p+1} of the polynomial $P(x)$, i.e.*

$$[t_1, t_2, \ldots, t_p]f = a_{p+1}.$$

Proof It is evident that

$$[t_1, t_2, \ldots, t_p]f = [t_1, t_2, \ldots, t_p]P,$$

since all derivatives coincide. On the other hand if $P(x) = a_{p+1} x^{p+1} + P_1(x)$, where $P_1(x)$ is a polynomial of degree $\leq p$, we obtain by Theorem 11.13 the equalities

$$[t_1, t_2, \ldots, t_p]P = \frac{1}{2\pi i} \int_{\Gamma_R} \frac{P(z)}{\Omega(z)} dz = \frac{1}{2\pi i} \int_{\Gamma_R} \frac{a_{p+1} z^{p+1} + P_1(z)}{\Omega(z)} dz$$

$$= \frac{1}{2\pi i} \int_{\Gamma_R} \frac{a_{p+1} z^{p+1}}{\Omega(z)} dz = a_{p+1}.$$

∎

Exercise 11.15 *Compute the coefficients α_{ij} in the case of double multiplicity for t_1.*

Exercise 11.16 *In the case of equidistant points $t_i = t_1 + (i-1)h$, we have the formula for the one-dimensional forward divided difference operator δ_h^k of order k and step $h > 0$ if we put $x = t_1$, as follows [50, p. 53]:*

$$\delta_h^p f(x) = \sum_{j=0}^{p} (-1)^{p-j} \binom{p}{j} f(x + jh) \tag{11.13}$$

$$= p! h^p \cdot [t_1, \ldots, t_{p+1}] f(\cdot + x).$$

11.2.2 Divided difference operators for Chebyshev systems

Now let us consider a generalization of the notion of divided difference operator for the case of the *Chebyshev system*. The basis for that generalization is an analog to Definition 11.9, p. 192. Since the polynomials Q_p are solutions to $d^p Q_p / dt^p = 0$, we will consider as its generalization the elements of the space $\mathcal{U}_p$.

We will be looking for linear combinations, or rather functionals, defined on $\mathcal{U}_p$, such that

$$l(f) = \sum_{j=1}^{p+1} \alpha_j f(t_j)$$

is zero for all $f \in \mathcal{U}_p$. So far we need the analog of the polynomial $Q_p(t) = t^p$. The main "nonuniqueness" point here is that a Chebyshev system $U_p = \{u_i\}_{i=1}^{p}$ has a nonunique extension $U_{p+1} = \{u_i\}_{i=1}^{p+1}$. Different choices of the function u_{p+1} provide different divided difference operators.

Let us also note that these functionals $l(f)$ may be considered as *mean value properties*. Indeed, if we fix some of the points t_j with $\alpha_j \neq 0$, for simplicity say t_1 has $\alpha_1 \neq 0$, then for every $f \in [\mathcal{U}_p]_{\text{lin}}$ we will have

$$f(t_1) = \sum_{j=2}^{p+1} -\frac{\alpha_j}{\alpha_1} f(t_j).$$

First we need to embed the system of order p into a system of order $p+1$, defined on a larger interval [50, p. 364].

Theorem 11.17 *If $U_p = \{u_i\}_{i=1}^{p}$ is an ECT-system on the interval I then we can extend all functions on any larger interval containing I so that U_p remains an ECT-system.*

If $U_p = \{u_i\}_{i=1}^{p}$ is in canonical form (11.3) then there exists a function u_{p+1} such that the new system $U_{p+1} = \{u_i\}_{i=1}^{p+1}$ is also a canonical ECT-system of $p+1$ functions on the interval I.

In particular, in the case of an operator of the type $L(d/dt)$ where L is a polynomial with only real zeros the first part of the theorem is evident; the second will be obtained by considering the operator $L(d/dt)((d/dt) - \tau_{p+1})$ for some τ_{p+1} by means of the transition formula in (11.8), p. 190.

We consider the extension of the *ECT*-system U_p to an *ECT*-system U_{p+1}. For a sufficiently smooth function f we define the *divided difference of order p with respect to U_{p+1}* by putting [50, p. 368],

$$[t_1, t_2, \ldots, t_{p+1}]_{U_{p+1}} f = D\begin{pmatrix} t_1, \ldots, t_p, t_{p+1} \\ u_1, \ldots, u_p, f \end{pmatrix} \bigg/ D\begin{pmatrix} t_1, \ldots, t_{p+1} \\ u_1, \ldots, u_{p+1} \end{pmatrix}.$$

The main property of the so-defined operator is that it *annihilates* the functions in the space U_p.

Theorem 11.18 *Let $f \in \mathcal{U}_p$. Then*

$$[t_1, t_2, \ldots, t_{p+1}]_{U_{p+1}} f = 0$$

and

$$[t_1, t_2, \ldots, t_{p+1}]_{U_{p+1}} u_{p+1} = 1.$$

196 Multivariate polysplines

If there are coincidences among some points t_i, i.e.

$$\{t_1 \leq t_2 \leq \cdots \leq t_{p+1}\} = \{\overbrace{\tau_1 = \cdots = \tau_1}^{l_1} < \cdots < \overbrace{\tau_d = \cdots = \tau_d}^{l_d}\}$$

with $l_1 + \cdots + l_d = p+1$ then we have the representation

$$[t_1, t_2, \ldots, t_{p+1}]_{U_{p+1}} f = \sum_{i=1}^{d} \sum_{j=1}^{l_i} \alpha_{ij} L_{j-1} f(\tau_i)$$

where $\alpha_{i l_i} \neq 0$, $i = 1, \ldots, d$. This may be reduced to ordinary derivatives

$$[t_1, t_2, \ldots, t_{p+1}]_{U_{p+1}} f = \sum_{i=1}^{d} \sum_{j=1}^{l_i} \beta_{ij} \frac{d^{j-1}}{dt^{j-1}} f(\tau_i)$$

where $\beta_{i l_i} \neq 0$, $i = 1, \ldots, d$. If the values of two functions f and g coincide in the sense of

$$\frac{d^{j-1}}{dt^{j-1}} f(\tau_i) = \frac{d^{j-1}}{dt^{j-1}} g(\tau_i) \quad \text{for } 1 \leq i \leq d, \ 1 \leq j \leq l_i,$$

then the divided differences also coincide:

$$[t_1, t_2, \ldots, t_{p+1}]_{U_{p+1}} f = [t_1, t_2, \ldots, t_{p+1}]_{U_{p+1}} g.$$

We have the following *recursion* formula.

Theorem 11.19 *Suppose that $t_1 \neq t_{p+1}$. Then*

$$[t_1, t_2, \ldots, t_{p+1}]_{U_{p+1}} f = \frac{[t_2, \ldots, t_{p+1}]_{U_p} f - [t_1, t_2, \ldots, t_p]_{U_p} f}{[t_2, \ldots, t_{p+1}]_{U_{p+1}} u_{p+1} - [t_1, t_2, \ldots, t_p]_{U_p} u_{p+1}}.$$

11.2.3 Lagrange–Hermite interpolation formula for Chebyshev systems

A beautiful and natural application of the above generalized divided difference operator is the following **Lagrange–Hermite interpolation** formula with remainder, which is written as in the usual *Taylor formula*, with the remainder in a *differential form*.

Theorem 11.20 *Let $\mathcal{U}_p$ be an ECT-system on $I = [a, b]$, and the interpolation points $\tau_1 < \tau_2 < \cdots < \tau_d$ be given in I. Let the positive integers l_i satisfy $l_1 + \cdots + l_d = p$. Then for every given set of real numbers $\{z_{ij} : i = 1, \ldots, d; \ j = 1, \ldots, l_i\}$ there exists a unique element $u \in \mathcal{U}_p$ such that the interpolation property holds*

$$\frac{d^{j-1}}{dt^{j-1}} u(\tau_i) = z_{ij} \quad (i = 1, \ldots, d; \ j = 1, \ldots, l_i).$$

In the case of
$$z_{ij} = \frac{d^{j-1}}{dt^{j-1}} f(\tau_i) \quad (i = 1, \ldots, d; \ j = 1, \ldots, l_i)$$
for some function f then
$$f(t) - u(t) = \varphi(t)[t_1, t_2, \ldots, t_p, t]_{U_{p+1}} f$$
where
$$\varphi(t) = \frac{D_{U_{p+1}}(t_1, t_2, \ldots, t_p, t)}{D_{U_p}(t_1, t_2, \ldots, t_p)}.$$

Exercise 11.21 *Write the above for the Lagrange interpolation.*

11.3 Dual operator and ECT-system

Assuming that we have the positive functions w_i, $i = 1, \ldots, p$, on the interval I, we first define the adjoint operators to the differentiation operators D_i

$$D_0^* f = f, \tag{11.14}$$

$$D_i^* f = \frac{1}{w_{p-i+1}} \frac{d}{dt} f(t) \quad \text{for } i = 1, \ldots, p.$$

Then we define the *formal adjoint* operator L_i^* to the operator L_i by putting

$$L_i^* = D_i^* \ldots D_0^* \quad (0 \leq i \leq p). \tag{11.15}$$

Now it is clear that the following set of functions are solutions to the ordinary differential equation $L_p^* u(t) = 0$ in I:

$$\left.\begin{array}{ll} u_1^*(t; a) &= 1, \\ u_2^*(t; a) &= \int_a^t w_p(t_p)\, dt_p, \\ \quad \vdots & \\ u_p^*(t; a) &= \int_a^t w_p(t_p) \int_a^{t_p} \ldots \int_a^{t_2} w_2(t_2)\, dt_2 \ldots dt_p, \end{array}\right\} \tag{11.16}$$

i.e.
$$u_i^*(t; a) = \int_a^t w_{i-1}(t) u_{i-1}^*(t)\, dt \quad \text{for } i = 2, 3, \ldots, p+1.$$

Evidently, we also have the equality
$$L_j^* u_i^*(a) = 0 \quad \text{for } 0 \leq j \leq i-2, \ 1 \leq i \leq p.$$

The system of functions $U_p^* = \{u_i^*\}_{i=1}^p$, which is obviously an *ECT-system*, is known as the *dual canonical ECT-system*.

Theorem 11.22 *The space spanned by the system $U_p^* = \{u_i^*\}_{i=1}^p$ is the null space for the operator L_p^*.*

11.3.1 Green's function and Taylor formula

In order to write a generalization to the Taylor formula we will need the analogs to the Green function which in the case of algebraic polynomials is $(x - y)_+^{j-1}$. Let us put [50, p. 373]

$$h_j(x; y) = w_1(x) \int_y^x w_2(t_2) \int_y^{t_2} \cdots \int_y^{s_{j-1}} w_j(t_j) \, ds_j \cdots dt_2 \tag{11.17}$$

and for the Green function

$$g_j(x; y) = \begin{cases} h_j(x) & \text{for } x \geq y, \\ 0 & \text{otherwise.} \end{cases} \tag{11.18}$$

Now applying Theorem 11.20, p. 196, on Hermite interpolation, we obtain the solubility of the following initial value problem [50, p. 376], where the remainder is written in the form of a **Peano kernel**.

Theorem 11.23 *The following initial-value problem has a unique solution:*

$$L_p f(t) = h(t) \quad \text{for } t \in I,$$
$$L_i f(a) = f_i \quad \text{for } 0 \leq i \leq p - 1,$$

where the function $h \in L^1(I)$ *and* f_i *are given real numbers. The solution* f *is represented in the form*

$$f(t) = u(t) + \int_a^b g_p(t; y) h(y) \, dy,$$

where $u \in \mathcal{U}_p$ *is the unique solution to equalities*

$$L_i u(a) = f_i \quad \text{for } 0 \leq i \leq p - 1,$$

by Theorem 11.20, p. 196.

The following relation holds:

$$g_j(x; y) = (-1)^{m-j} L_{m-j,y}^* g_m(x; y) \quad \text{for } j = 1, 2, \ldots, m - 1.$$

As a corollary, for every function f which satisfies $L_p f \in L^1(I)$ we obtain the generalized Taylor expansion [50, p. 376]

$$f(t) = u_f(t) + \int_a^b g_p(t; y) h(y) \, dy \quad \text{for } t \text{ in } I, \tag{11.19}$$

where $u_f \in \mathcal{U}_p$ is the unique element satisfying

$$L_i u(a) = L_i f(a) \quad \text{for } 0 \leq i \leq p - 1.$$

This result may be embedded in a more general multivariate framework. We have a similar result for the general elliptic boundary value problems, which are Green formulas for a set of boundary conditions, see Section 23.3, p. 475.

We also have the *dual generalized Taylor expansion*. For every function f such that $L_p^* f \in L^1(I)$ the following representation holds [50, p. 378]:

$$f(y) = u_f^*(y) + \int_a^b g_p(t; y)(-1)^p L_p^* f(t) dt \quad \text{for } t \text{ in } I, \tag{11.20}$$

where u_f^* is the unique element in $\mathcal{U}_p^*$ which satisfies

$$L_i^* u_f^*(b) = L_i^* f(b) \quad \text{for } 0 \le i \le p-1.$$

The last result is based on the observation that $g_p(x; y)$ is also the Green function for the operator L_p^*.

11.4 Chebyshev splines and one-sided basis

The *Chebyshev spline* of order $m - 1$ is defined as a function which is a piecewise element of a system $\mathcal{U}_m$ of order m. Recall the polynomial splines where the degree of the polynomials is $m - 1$ and the basis is $\{1, t, t^2, \ldots, t^{m-1}\}$. It may also have different smoothness at every knot. For simplicity we will consider the simplest case of splines where the smoothness is of order $m - 2$. This will be the case that is most studied in the present book.

Let us provide a rigorous definition [50, p. 378]. We consider an interval $[a, b]$ and knots x_i, $i = 0, \ldots, k+1$, such that

$$a = x_0 < x_1 < \cdots < x_k < x_{k+1} = b. \tag{11.21}$$

Definition 11.24 *The function $u(t)$, for $t \in [a, b]$, is a Chebyshev spline (with multiplicity $m_1 = m_2 = \cdots = m_k = 1$) if and only if, for the restrictions of $u(t)$ to every interval (x_i, x_{i+1}) given by*

$$u_i(t) = u(t)|_{(x_i, x_{i+1})} \quad \text{for } i = 0, \ldots, k,$$

we have $u_i \in \mathcal{U}_m$, for $i = 0, \ldots, k$, and the smoothness conditions also hold

$$\frac{d^j}{dt^j} u_i(x_{i+1}) = \frac{d^j}{dt^j} u_{i+1}(x_{i+1}) \tag{11.22}$$

for $j = 0, 1, \ldots, m-2$, and $i = 0, 1, \ldots, k-1$.

These last conditions show that $u(t)$ belongs to $C^{m-2}([a, b])$. We denote the space of all such splines by $\mathcal{S}_m = \mathcal{S}_m(x_0, x_1, \ldots, x_{k+1})$.

It is easy to compute the dimension of $\mathcal{S}_m$. It is equal to the dimension of all piecewise $\mathcal{U}_m$-functions, which is equal to $(k+1)m$ minus the number of restrictions (11.22), which in turn are $(m-1)k$, i.e.

$$\dim(\mathcal{S}_m) = m + k.$$

As in the case of polynomial splines one has a natural system of **one-sided basis functions**. It is constructed through the functions $g_s(x; y)$ which are the analog to the Green functions $(x - y)_+^{s-1}$ [50, p. 379].

Theorem 11.25 *The functions*

$$\rho_{i,1}(t) = \{g_m(t; x_i): \text{ for } i = 1, \ldots, k\}$$

$$\rho_{0,j} = \{g_{m-j+1}(t; x_0): \text{ for } j = 1, \ldots, m\}$$

form a basis of $\mathcal{S}_m$. We have the representation

$$\rho_{i,j}(t) = (-1)^{j-1} L_{j-1}^* g_m(t; x_i)$$

for the above range of the indices i and j.

Due to the lack of boundary conditions at the point $x_0 = a$ we need all functions $g_s(t; x_0)$, $s = 1, \ldots, m$, which have a singularity at x_0 for the basis of $\mathcal{S}_m$.

An important property of the Chebyshev splines is that we may count the number of *zeros of the splines having compact support*, see Powell [42, p. 230], the result being the same as in the case of polynomial splines.

Theorem 11.26 *Let $h \in \mathcal{S}_m(x_0, x_1, \ldots, x_{k+1})$ and $h(t)$ be identically zero on the intervals $[x_0, x_u]$ and $[x_v, x_{k+1}]$ for some integers $0 < u < v < k + 1$. Let us denote by Z the number of zeros of $h(t)$ in the open interval (x_u, x_v). Then if $h(t)$ is not identically zero in any subinterval of (x_u, x_v) it follows that:*

$$Z \leq v - u - m,$$

i.e. Z is less than or equal to the length of the chain of knots minus the dimension of the space $\mathcal{U}_m$.

An important consequence may be drawn about the minimal length of the support of Chebyshev splines, and here the result is the same as for the polynomial splines.

Corollary 11.27 *If a spline $u \in \mathcal{S}_m(x_0, x_1, \ldots, x_{k+1})$ has a compact support equal to the interval $[x_u, x_v]$ then its length satisfies*

$$m \leq v - u.$$

Now let us fix our attention to the L-splines with $L = L_{(k)}^{2p}$ which are of even order. We recall the results of Section 10.15, p. 177. Further we refer to Ahlberg et al. [1, Chapter VI], where generalized L-splines are considered with operators $L = L_1 L_1^*$. The operator L_1 may have variable coefficients and L_1^* is formally self-adjoint. All considerations there are also valid in the case of a positive weight function which is the case for function r^{n-1} for $r > 0$. Thus we obtain the following result about existence and uniqueness of interpolation Chebyshev splines for the operator $L = L_{(k)}^p$.

Theorem 11.28 *Let us consider for an arbitrary integer $p \geq 1$ the Chebyshev splines with respect to the operator $L = L_{(k)}^{2p}$ on an arbitrary interval $[a, b]$, where $a > 0$. The set of knots $x_1, x_2, \ldots, x_k$, satisfies $a = x_0 < x_1 < \ldots x_{k+1} = b$. For every set of* interpolation data $\{c_j\}_{j=0}^{k+1}$ there exists a unique interpolation Chebyshev spline $u(r)$, i.e. satisfying

$$u(x_j) = c_j \quad for \ j = 0, 1, \ldots, k+1,$$

which satisfies the boundary conditions

$$u^{(\ell)}(a) = \alpha_\ell \quad for \ \ell = 1, 2, \ldots, p-1,$$
$$u^{(\ell)}(a) = \beta_\ell \quad for \ \ell = 1, 2, \ldots, p-1,$$

where α_ℓ and β_ℓ are arbitrary constants.

The proof is left as an exercise for the reader. One has to prove the first integral identity as in [1, Chapter VI.3]. As we have already stated in Section 10.15, p. 177, we can use the ready-made results for the polysplines. Since our case is simpler than the general case considered in [1] the above theorem is completely analogous to Theorem 5.8.1 in [1], where the polynomial case is provided.

11.4.1 TB-splines, or the Chebyshev B-splines as a Peano kernel for the divided difference

A classical fact of one-dimensional spline theory is that the basis ρ_{ij} defined above does not provide a good tool for numerical computations owing to its instability [9, 42]. Such a tool is provided by the basis of splines having compact support.

Let us introduce the Chebyshev splines having compact support. As we will see below in formula (11.26), p. 202, they arise, just like the polynomial splines, as Peano kernels of the divided difference operator.

We assume that a set of $m+1$ knots is given

$$y_i \leq y_{i+1} \leq \cdots \leq y_{i+m} \tag{11.23}$$

such that $y_i < y_{i+m}$. The index i is auxiliary and will be used when the above points y are a subset of a larger set of knots. The associated *Chebyshev spline of order m* (or TB-spline) is given by

$$Q_i^m(t) = (-1)^m [y_i, y_{i+1}, \ldots, y_{i+m}]_{U_{m+1}^*} g_m(t; y) \tag{11.24}$$

where we have used the divided difference operator arising from the dual ECT-system U_{m+1}^* associated with U_{m+1} and defined in (11.16), p. 195.

The following result holds [50, p. 380].

Theorem 11.29 *The TB-spline $Q_i^m(t)$ has the property*

$$Q_i^m(t) = 0 \quad for \ t \leq y_i \ or \ t \geq y_{i+m};$$
$$Q_i^m(t) > 0 \quad for \ y_i < t < y_{i+m}.$$

The most important property of the TB-splines is that they form a basis which is "stable". In order to have sufficient TB-splines we need some points from the interval

$[a, b]$, see [50, p. 116]. Actually, we need m points to the left of a and m points to the right of b in order to be able to cover all subintervals of $[a, b]$ with sufficient Q_is.

Let the knots y_j, $i = 1, \ldots, 2m + k$, be given with

$$y_1 \leq y_2 \leq \cdots \leq y_{2m+k}.$$

These knots are called the *extended partition of* $[a, b]$ if they satisfy

$$\left.\begin{array}{l} a < y_{m+1} < \cdots < y_{m+k} < b, \\ y_1 \leq \cdots \leq y_m \leq a, \\ b \leq y_{m+k+1} \leq \cdots \leq y_{2m+k}. \end{array}\right\} \quad (11.25)$$

Theorem 11.30 *For the extended partition y_j, $j = 1, \ldots, 2m + k$, of $[a, b]$ we assume that $b < y_{2m+k}$. We assume that the ECT-system U_m is defined on the interval $[y_1, y_{2m+k}]$. If the TB-spline Q_i is given by formula (11.24) then the set of functions $\{Q_i(t)\}_{i=1}^{m+k}$ is a basis of the set of all Chebyshev splines $\mathcal{S}_m$ on $[a, b]$.*

Let us note that the proof uses essentially the result on the zeros of the compactly supported splines, Theorem 11.26, p. 200.

Corollary 11.31 *Every Chebyshev spline $u(t)$ with this property satisfies*

$$u(t) = C Q_i^m(t) \quad for \ t \in [a, b],$$

for some constant C.

According to Theorem 11.17, p. 195, the system U_m, which is an *ECT-system* on $[a, b]$, can be extended to an *ECT-system* on every bigger interval, hence, on $[y_1, y_{2m+k}]$.

We have the **Peano kernel** representation of the divided difference operator which motivates the definition of the TB-splines in (11.24). Indeed, applying the operator $[y_i, \ldots, y_{i+m}]_{U_{m+1}^*}$ to the Taylor expansion (11.20), p. 199, we obtain

$$[y_i, \ldots, y_{i+m}]_{U_{m+1}^*} f = \int_{y_i}^{y_{i+m}} Q_i(t) L_m^* f(t) \, dt \quad (11.26)$$

which holds for every function satisfying $L_m^* f \in L_1[y_i, y_{i+m}]$. For $f = u_{m+1}^*$ we obtain by the definition of u_{m+1}^* that

$$1 = \int_{y_i}^{y_{i+m}} Q_i(t) \, dt$$

which shows that the TB-splines are naturally normalized. So far we will need another normalization with respect to the sup norm. It is given by

$$N_i(t) = \alpha_i Q_i(t)$$

where the constants

$$\alpha_i = \frac{D_{U_{m+1}^*}(y_i, \ldots, y_{i+m}) D_{U_{m-1}^*}(y_{i+1}, \ldots, y_{i+m-1})}{D_{U_m^*}(y_{i+1}, \ldots, y_{i+m}) D_{U_m^*}(y_i, \ldots, y_{i+m-1})}. \quad (11.27)$$

We have $\alpha_i > 0$ since U_{m+1}^* is an *ECT*-system.

The following **Marsden identity** for normalized TB-splines shows how the basis of the set U_m may be expanded in terms of the normalized TB-splines N_i, cf. [50, p. 383].

Theorem 11.32 *Let us put*

$$\varphi_i(t) = \frac{D_{U_m^*}(y_{i+1}, \ldots, y_{i+m-1}, t)}{D_{U_{m-1}^*}(y_{i+1}, \ldots, y_{i+m-1})} \quad \text{for } i = 1, 2, \ldots, m+k. \tag{11.28}$$

Then the function h_m defined in (11.17), p. 198, satisfies

$$h_m(x; y) = \sum_{i=1}^{m=k} (-1)^{m-1} \varphi_i(y) N_i(x).$$

Also

$$u_1(t) = \sum_{i=1}^{m+k} N_i(t), \tag{11.29}$$

$$u_j(t) = \sum_{i=1}^{m+k} \eta_i^j N_i(t) \quad \text{for } j = 2, 3, \ldots, m,$$

where

$$\eta_i^j = (-1)^{j-1} L_{m-j}^* \varphi_i(y)|_{y=a}.$$

From (11.29) and the Markov-type inequality for ECT-systems, it follows that [50, p. 385]

$$|D^k N_i(x)| \le \frac{C}{\Delta^k} \quad (k = 0, 1, \ldots, m-1)$$

where C depends only on the weights $w_1, \ldots, w_m$ defining the system U_m.

11.4.2 Dual basis and Riesz basis property for the TB-splines

Let us define the set $\{\lambda_i\}_{i=1}^{m+k}$ of dual functionals on the space $L_p[a, b]$ such that

$$(\lambda_i, N_j) = \delta_{ij} \quad (i, j = 1, 2, \ldots, m+k)$$

where $(\cdot, \cdot)$ is the integral scalar product. We define

$$(\lambda_i, f) = \int_{y_i}^{y_{i+m}} f(t) L_m^* \psi_i(t) \, dt$$

$$\psi_i(t) = \alpha_i \varphi_i(t) G_i(t)$$

where the function $\varphi_i(t)$ is defined in (11.28) and α_i is defined in (11.27). The functions G_j are defined through the following *perfect spline* [50, p. 139]:

$$B_m^*(t) = m(-1)^m [v_0, v_1, \ldots, v_m](t-v)_+^{m-1}$$

$$v_i = \cos\left(\frac{m-i}{m}\right)\pi \quad (i = 0, 1, \ldots, m)$$

and the transition function g is given by [50, p. 141],

$$g(t) = \begin{cases} 0 & \text{for } t < -1, \\ \int_{-1}^{t} B_m^*(s)\,ds & \text{for } -1 \leq t < 1, \\ 1 & \text{for } 1 \leq t, \end{cases}$$

respectively, [50, p. 145],

$$G_j(t) = g\left(\frac{2t - y_j - y_{j+m}}{y_{j+m} - y_j}\right) \quad \text{for } j = 1, \ldots, m+k.$$

The norms of the functionals λ_j are estimated in Theorem 11.32.

Theorem 11.33 *The functionals λ_j, $j = 1, \ldots, m+k$, form a dual basis for the TB-splines N_i, $i = 1, \ldots, m+k$, and*

$$\|\lambda_i\| = \sup_f \frac{|(\lambda_i, f)|}{\|f\|_{L_p[y_i, y_{i+m}]}} \leq C_{m,w} \underline{\Delta}^{-1/p} \quad (i = 1, \ldots, m+k)$$

where the constant $C_{m,w}$ depends only on m and the weights w_i.

Let us recall the constants M_i defined in (11.9), p. 191. The most fundamental property of the TB-splines which is the "stability" or the bounded "conditioning number" is expressed in Theorem 11.33.

Theorem 11.34 *Let $1 \leq p \leq \infty$. For $i = 1, 2, \ldots, m+k$ let us put*

$$B_{i,p}(t) = \overline{\Delta}^{-1/p} N_i(t).$$

Then there exist constants C_1 and C_2 satisfying $0 < C_1, C_2 < \infty$ which depend only on m and $\underline{M}_i$, $\overline{M}_i$, $i = 1, \ldots, m$, and such that

$$\left(\sum_{i=1}^{m+k} |c_i|^p\right) \leq C_1 \left\|\sum_{i=1}^{m+k} c_i B_{i,p}\right\|_{L_p[a,b]}^p \leq C_2 \left(\sum_{i=1}^{m+k} |c_i|^p\right)$$

for every sequence $c_1, c_2, \ldots, c_{m+k}$.

11.5 Natural Chebyshev splines

Here we provide briefly the results on **natural L-splines** which are very similar to those for natural polynomial splines. We follow Schumaker [50, Chapter 9.8, p. 396]. They will be necessary for the spherical polysplines supported in the ball $B(0; r_N)$.

The so-called *natural* polynomial splines are well known for their beautiful extremal properties related to the Holladay theorem. They are polynomials only of odd degree. In the same manner we will define *natural Chebyshev splines* for an even-order *ECT-system*.

Let U_{2p} be an *ECT-system* of order $2p$ which is in the *canonical* form as defined in (11.3), p. 188. It generates as in (11.7), p. 189, the *ECT-space* $\mathcal{U}_{2p}$. For every $j = 1, 2, \ldots, 2p$, we have the *ECT-spaces* $\mathcal{U}_j$ given by

$$\mathcal{U}_j = \lin\{u_i,\ i = 1, 2, \ldots, j\}.$$

Let us assume that a partition of the interval $[a, b]$ be provided by the points x_i satisfying

$$a = x_0 < x_1 < \cdots < x_k < x_{k+1} = b,$$

where $x_1, x_2, \ldots, x_k$ will be the knots of the splines. We will work only with knots with simple multiplicity.

Definition 11.35 *Let the function $u(t)$, $t \in [a, b]$ be a Chebyshev spline with respect to U_{2p} in the sense of Definition 11.24, p. 199, i.e. it belongs to $C^{2p-2}([a, b])$ and its restriction to every interval $[x_j, x_{j+1}]$, denoted by $u_{|[x_j, x_{j+1}]}$, belongs to the space $\mathcal{U}_{2p}$, for $j = 0, 1, \ldots, k$. Then $u(t)$ is called,*

- *a **left-sided natural Chebyshev spline** with respect to the ECT-space $\mathcal{U}_{2p}$ if the first piece*

$$u_0(t) = u_{|[a, x_1]} \quad \text{belongs to } \mathcal{U}_p,$$

- *a **right-sided natural Chebyshev spline** with respect to the ECT-space $\mathcal{U}_{2p}$ if the last piece*

$$u_k(t) = u_{|[x_k, b]} \quad \text{belongs to } \mathcal{U}_p,$$

- *a **natural Chebyshev spline** with respect to the ECT-space $\mathcal{U}_{2p}$ if it is both left-sided and right-sided.*

Let us denote these spaces by $\mathcal{LS}_{2p} = \mathcal{LS}_{2p}(x_0, x_1, \ldots, x_{k+1})$, $\mathcal{RS}_{2p} = \mathcal{RS}_{2p}(x_0, x_1, \ldots, x_{k+1})$, and $\mathcal{NS}_{2p} = \mathcal{NS}_{2p}(x_0, x_1, \ldots, x_{k+1})$, respectively.

Remark 11.36 Obviously, referring to Corollary 11.3, p. 188, if we have an operator $L = L_1 L_2$ where L_1 and L_2 are of order p then the system $\mathcal{U}_p$ is generated by the operator L_1 and it is a subsystem, as above, of the system $\mathcal{U}_{2p}$ generated by the operator L.

Proposition 11.37 *The spaces of natural Chebyshev splines have the following dimension:*

$$\dim(\mathcal{LS}_{2p}(x_0, x_1, \ldots, x_{k+1})) = p + k,$$
$$\dim(\mathcal{RS}_{2p}(x_0, x_1, \ldots, x_{k+1})) = p + k,$$
$$\dim(\mathcal{NS}_{2p}(x_0, x_1, \ldots, x_{k+1})) = k.$$

The proof follows by a direct count of the dimensions.

The main point of our discussion will be the construction of a full set of (left-sided, right-sided and natural) Chebyshev splines with "small support". We assume that the partition $\{x_i\}$ is extended through the *left-oriented* set $\{y_i\}_{i=1}^{3p+k}$ similar to (11.25), p. 202.

It has p points to the left of the interval $[a, b]$, $2p$ points to the right of the interval $[a, b]$, and k points inside $[a, b]$ coinciding with the x_is. Thus it satisfies

$$y_{p+1} = x_1, \ y_{p+2} = x_2, \ldots, y_{p+k} = x_k,$$

and

$$a < y_{p+1} < \cdots < y_{p+k} < b,$$
$$y_1 \leq \cdots \leq y_p \leq a,$$
$$b \leq y_{p+k+1} \leq \cdots \leq y_{3p+k}.$$

Let us denote by $g_{2p}(x; y)$ and $g_{2p}^*(x; y)$ the Green functions associated with the *ECT*-systems U_{2p} and U_{2p}^*, respectively. We put

$$L_{i,j}^{2p}(x) := [y_i, y_{i+1}, \ldots, y_{i+j}]_{U_{j+1}^*} g_{2p}^*(y; x),$$

$$R_{i,j}^{2p}(x) := (-1)^j [y_i, y_{i+1}, \ldots, y_{i+j}]_{U_{j+1}^*} g_{2p}(x; y).$$

Using these functions we may construct a basis for the spaces $\mathcal{L}S_{2p}(x_0, x_1, \ldots, x_{k+1})$, $\mathcal{R}S_{2p}(x_0, x_1, \ldots, x_{k+1})$, and $\mathcal{N}S_{2p}(x_0, x_1, \ldots, x_{k+1})$. We will be interested mainly in the space $\mathcal{L}S_{2p}(x_0, x_1, \ldots, x_{k+1})$. Let us note that for working with natural splines one needs the minimum degrees of freedom which are provided by the condition $k \geq 2p$. This allows us to solve the interpolation problem. In the case of *left-sided natural splines* we need the condition $k \geq p$.

Theorem 11.38 *Let $k \geq p$, and let the left-oriented extended partition $\{y_i\}_{i=1}^{3p+k}$ be given. We define the splines*

$$B_i(x) = \begin{cases} L_{p+1, p+i-1}^{2p}(x) & \text{for } i = 1, 2, \ldots, p, \\ N_i^{2p}(x) & \text{for } i = p+1, p+2, \ldots, p+k, \end{cases}$$

where $N_i^{2p}(x)$ is the normalized T B-spline associated with the knots $y_i, y_{i+1}, \ldots, y_{i+2p}$. Then the functions $\{B_i(x)\}_{i=1}^{p+k}$ form a basis for the space of left-sided natural splines $\mathcal{L}S_{2p}(x_0, x_1, \ldots, x_{k+1})$.

Needless to say the above basis is by no means a unique construction. Its main support properties are summarized in Proposition 11.38.

Proposition 11.39 *Let us have two points satisfying $y_i < y_{i+j}$ and $0 \leq j < 2p$. Then*

$$\operatorname{supp} L_{i,j}^{2p} \subset [-\infty, y_{i+j}],$$

$$L_{i,j}^{2p}(x) > 0 \quad \text{for } x < y_{i+j}.$$

For all $x < y_i$ the following representation holds:

$$L_{i,j}^{2p}(x) = (-1)^{2p-j} u_{2p-j}(x) + \sum_{\nu=1}^{2p-j-1} \alpha_\nu u_\nu(x).$$

In a similar way,

$$\operatorname{supp} R_{i,j}^{2p} \subset [y_i, \infty),$$

$$R_{i,j}^{2p}(x) > 0 \quad \text{for } x > y_i,$$

and for $x \geq y_{i+j}$ the representation

$$R_{i,j}^{2p}(x) = u_{2p-j}(x) + \sum_{\nu=1}^{2p-j-1} \beta_\nu u_\nu(x)$$

holds.

The following interpolation problem is solved by the natural splines.

Theorem 11.40 *Let the operator L be equal to $L_{(k)}^{2p}$. We consider the Chebyshev splines generated through it on an interval $[a, b]$ for $a > 0$, and which are also left-sided natural in the case of $a = 0$. The system $\mathcal{U}_p$ contains the elements $r^k, r^{k+2}, \ldots, r^{k+2p-2}$ which are analytic at $r = 0$. Let the numbers c_i, $i = 1, 2, \ldots, k+1$, be given. Then there exists a unique left-sided natural Chebyshev spline $u(t)$ such that*

$$u(x_i) = c_i \quad \text{for } i = 1, 2, \ldots, k+1,$$

and it satisfies $p - 1$ additional boundary conditions,

$$u^{(\ell)}(x_{k+1}) = b_\ell \quad \text{for } \ell = 1, \ldots, p-1.$$

A similar result holds for the right-sided natural splines.

The proof is left to the reader as an exercise. It is very similar to the standard one [50] in the case of natural Chebyshev splines and is also based on their arguments provided after the interpolation Theorem (11.28), p. 200, for the usual Chebyshev splines. The results of the present Section are essentially used for the construction of polysplines on annuli with a compact support, see Theorem 9.7, p. 124.

Chapter 12

Appendix on Fourier series and Fourier transform

If we denote by $L_2(0, 2\pi)$ the set of all measurable complex-valued functions f defined on the interval $(0, 2\pi)$ and satisfying

$$\int_0^{2\pi} |f(x)|^2 dx < \infty,$$

then we have the Fourier series representation, and even more. We have combined the classical *Riesz–Fisher* theorem and others [52, 65, Chapter 9].

Theorem 12.1 *For every $f \in L_2(0, 2\pi)$ we have the representation*

$$f(x) = \sum_{j=-\infty}^{\infty} c_j e^{ijx} \quad \text{for } x \in (0, 2\pi),$$

with Fourier coefficients

$$c_j := \frac{1}{2\pi} \int_0^{2\pi} f(x) e^{-ijx} dx,$$

and convergence of the series in L_2 sense, i.e.

$$\lim_{N \to \infty} \int_0^{2\pi} \left| f(x) - \sum_{j=-N}^{N} c_j e^{ijx} \right|^2 dx = 0.$$

The second important property of the Fourier series is that the sequence of coefficients $\{c_j\}$ belongs to the space ℓ_2 and has the same norm as the function given below.

210   *Multivariate polysplines*

Theorem 12.2 *If $f \in L_2(0, 2\pi)$ then*

$$\frac{1}{2\pi} \int_0^{2\pi} |f(x)|^2 \, dx = \sum_{j=-\infty}^{\infty} |c_j|^2, \tag{12.1}$$

and this identity establishes an isometric isomorphism between the space of functions $f \in L_2(0, 2\pi)$ and the space of all infinite sequences $\{c_j\} \in \ell_2$.

We also have the so-called expansion in Fourier series by trigonometric functions provided by Theorem 12.3.

Theorem 12.3 *For every real-valued function $f \in L_2(0, 2\pi)$ we have representation*

$$f(x) = \frac{u_0}{2} + \sum_{k=1}^{\infty} (u_k \cos kx + v_k \sin kx), \tag{12.2}$$

where, as usual, the coefficients u_k and v_k are given by

$$u_k = \frac{1}{\pi} \int_0^{2\pi} f(y) \cos ky \, dy \quad \text{for } k = 0, 1, 2, \ldots,$$

$$v_k = \frac{1}{\pi} \int_0^{2\pi} f(y) \sin ky \, dy \quad \text{for } k = 1, 2, \ldots,$$

and the convergence is in the L_2-sense.

This result easily follows from Theorem 12.1, p. 209, [64]. Indeed, for a real-valued function f, from the equalities

$$\cos k\varphi = \frac{e^{ik\varphi} + e^{-ik\varphi}}{2},$$

$$\sin k\varphi = \frac{e^{ik\varphi} - e^{-ik\varphi}}{2i},$$

we see that (12.2) is equivalent to the representation

$$f(\varphi) = \sum_{k=-\infty}^{\infty} c_k e^{ik\varphi},$$

where the coefficients c_k are given by the equalities

$$c_k = \frac{1}{2}\left(a_k + \frac{b_k}{i}\right) \quad \text{for } k \in \mathbb{Z},$$

and obviously satisfy

$$c_{-k} = \overline{c_k} \quad \text{for } k \in \mathbb{Z}.$$

If we now have a complex-valued function $f(\varphi) = f_1(\varphi) + i f_2(\varphi)$, since $f \in L_2(\mathbb{S}^1)$ is equivalent to $f_1, f_2 \in L_2(\mathbb{S}^1)$, we have the same representation for the real and the imaginary parts $f_1(\varphi)$ and $f_2(\varphi)$.

Fourier series and Fourier transform

The convergence in the above theorem is in L_2-sense, i.e. for $N \to \infty$ we have

$$\int_0^{2\pi} \left| f(x) - \frac{u_0}{2} - \sum_{k=1}^{N}(u_k \cos kx + v_k \sin kx) \right|^2 dx \longrightarrow 0.$$

Note that the series on the right-hand side is 2π-periodic but we have said nothing about the periodicity of the function f itself. In order to obtain uniform convergence we need stronger conditions [65, Chapter 9.44].

Theorem 12.4 *Let the function $f(x)$ be continuous for $0 \le x \le 2\pi$, periodic, i.e. $f(0) = f(2\pi)$ and the integral $\int_0^{2\pi} |f'(x)| dx$ be finite. Then the trigonometric Fourier series for $f(x)$ converges uniformly to $f(x)$.*

Example 12.5 *The function $f(x) = |x - 1|$ satisfies all conditions of Theorem 12.4 although the derivative at $x = 1$ does not exist.*

It is very useful for the Dirichlet problem for harmonic functions to consider **sine** trigonometric Fourier series. They appear naturally as Fourier series of odd functions. Indeed, let the function $f(x)$ be 2π-periodic and assume that it is also odd, i.e. it satisfies

$$f(-x) = -f(x).$$

Since

$$u_k = \frac{1}{\pi} \int_0^{2\pi} f(y) \cos ky \, dy$$

$$= \frac{1}{\pi} \int_{-\pi}^{\pi} f(y) \cos ky \, dy \quad \text{for } k = 0, 1, 2, \ldots,$$

it follows that $u_k = 0$ for $k = 0, 1, 2, \ldots$, and similarly

$$v_k = \frac{2}{\pi} \int_0^{\pi} f(y) \sin ky \, dy \quad \text{for } k = 1, 2, 3, \ldots.$$

Corollary 12.6 *Let the function $f(x)$ defined on the whole real axis be π-periodic, continuous and the integral $\int_0^{\pi} |f'(x)|^2 dx$ be finite. Then the sine trigonometric Fourier series converges uniformly to $f(x)$.*

Further we list some elementary properties of the Fourier transform. We assume that the functions are complex valued.

For every function $f \in L_1(\mathbb{R}^n)$ we define the **Fourier transform** by

$$\widehat{f}(\xi) := \mathcal{F}[f](\xi) := \int_{\mathbb{R}^n} f(x) e^{-i\xi x} dx. \tag{12.3}$$

It is remarkable that this transform can be extended to the space $L_2(\mathbb{R}^n)$. For functions $f \in L_2(\mathbb{R}^n)$ one puts

$$\widehat{f}(\xi) := \mathcal{F}[f](\xi) := \lim_{N \to \infty} \int_{|x| < N} f(x) e^{-i\xi x} dx,$$

see Rudin [49].

The **inverse Fourier transform** is defined by

$$\mathcal{F}^{-1}[g](x) = \frac{1}{(2\pi)^n} \cdot \int_{\mathbb{R}^n} f(\xi) e^{i\xi x} \, d\xi. \tag{12.4}$$

For every two functions $f, g \in L_2(\mathbb{R})$ we have the scalar product

$$\langle f, g \rangle := \int_{\mathbb{R}^n} f(x) \cdot \overline{g(x)} \, dx,$$

where the bar over a quantity denotes its complex conjugate.

Theorem 12.7 *Let $f, g \in L_2(\mathbb{R}^n)$. Then the following identity holds:*

1. *The **Parseval** identity*

$$\langle f, g \rangle = \langle \widehat{f}, \widehat{g} \rangle,$$

i.e.

$$\int_{\mathbb{R}^n} f(x) \cdot \overline{g(x)} \, dx = \frac{1}{(2\pi)^n} \cdot \int_{\mathbb{R}^n} \widehat{f}(\xi) \cdot \overline{\widehat{g}(\xi)} \, d\xi. \tag{12.5}$$

2. *For every function $f \in L_2(\mathbb{R}^n)$ the **inversion** formula holds*

$$\mathcal{F}^{-1}\mathcal{F}f = f \tag{12.6}$$

and $\mathcal{F}$ is a one-to-one map of $L_2(\mathbb{R}^n)$ onto itself.

The convolution of two functions is defined by

$$f(x) * g(x) = \int_{\mathbb{R}^n} f(x-y)g(y) \, dy.$$

Theorem 12.8 *Let $f, g \in L_2(\mathbb{R}^n)$. Then the Fourier transform of the convolution of f and g is the product of their Fourier transforms, i.e.*

$$\mathcal{F}[f * g](\xi) = \mathcal{F}[f](\xi) \cdot \mathcal{F}[g](\xi). \tag{12.7}$$

12.1 Bibliographical notes

We have added a number of useful references below in the Bibliography which we have not used directly in the text. They are related to RBFs, scattered data in geophysics, and to properties of the polyharmonic functions.

Bibliography to Part I

[1] Ahlberg, J., Nilson, E. and Walsh, J. *The Theory of Splines and their Applications*, Academic Press, New York, 1967. (Russian transl. "Mir", Moscow, 1972)

[2] Axler, Sh., Bourdon, P. and Ramey, W. *Harmonic Function Theory*, Springer-Verlag, New York, 1992.

[3] Aronszajn, N., Creese, T.M. and Lipkin, L.J. *Polyharmonic Functions*, Clarendon Press, Oxford, 1983.

[4] Bojanov, B., Hakopian, H.A. and Sahakian, A.A. *Spline Functions and Multivariate Interpolation*, Kluwer Academic Publishers, Dordrecht, 1993.

[5] Bos, L.N., Levenberg, P., Milman, B.A. and Taylor. Tangential Markov inequalities characterize algebraic submanifolds of $\mathbb{R}^N$. *Indiana Univ. Math., J.* 44 (1995), pp. 115–138.

[6] Briggs, J.C. Machine contouring using minimum curvature. *Geophysics*, 39 (1974), pp. 39–48.

[7] Budak, B., Samarskii, A. and Tikhonov, A. *A Collection of Problems on Mathematical Physics*, Pergamon Press, Oxford, London, 1964.

[8] Conway, J.B. *Functions of One Complex Variable*, 2nd edition, Springer-Verlag, New York, 1978.

[9] de Boor, C. *A Practical Guide to Splines*, Springer-Verlag, New York, 1978.

[10] de Boor, C. Topics in multivariate approximation theory. In: *Topics in Numerical Analysis, Lecture Notes in Mathematics, 965*, Turner, P. R. (Ed.), 1981, pp. 39–78.

[11] de Boor C., Höllig, K. and Riemenschneider, S. *Box Splines*, Springer-Verlag, Berlin, 1993.

[12] Duchon, J. Interpolation des fonctions de deux variables suivant le principe de la flexion des plaques minces, R.A.I.R.O. *Sr. Anal. numer.* 10, (1976), No. R-3, pp. 5–12.

[13] Dyn, N. Interpolation and approximation by radial and related functions. In: *Approximation Theory VI*, C.K. Chui, L.L. Schumaker and J.D. Ward (Eds), Academic Press, New York, 1989, pp. 211–234.

[14] Dyn, N. and Ron, A. Recurrence relations for Tchebycheffian B-splines. *J. Anal. Mathem.* 51, (1988), pp. 118–138.

[15] Dyn, N. and Ron, A. Cardinal translation invariant Tchebycheffian B-splines. *Approx. Theory Appl.* 6, (1990), No. 2, pp. 1–12.

[16] Farin, G. *Curves and Surfaces for Computer-Aided Geometric Design. A Practical Guide*. Academic Press, Inc., San Diego, CA, 1997.

[17] Floater, M.S. and Iske, A. Thinning and approximation of large Sets of Scattered data. In: *Advanced Topics in Multivariate Approximation, (Montecatini Terme, 1995)* F. Fontanella, K. Jetter and P.-J. Laurent (Eds), pp. 87–96, World Sci. Publishing, River Edge, NJ, 1996.

[18] Freeden, W., Gervens, T. and Schreiner, M. *Constructive Approximation on the Sphere*, Oxford Science Publications, Clarendon Press, Oxford, 1998.
[19] *Function Estimates,* Contemporary Mathematics, vol. 59, J.S. Marron Ed., AMS Providence, RI, 1986.
[20] Gonzalez-Casanova, P. and Alvarez, R. Splines in geophysics. *Geophysics*, 50, (1985) No. 12, pp. 2831–2848.
[21] Hayman, W. K. and Korenblum, B. Representation and uniqueness theorems for polyharmonic functions. *J. Anal. Math.* 60 (1993), pp. 113–133.
[22] Helms, L.L. *Introduction to Potential Theory*, Wiley-Interscience, New York, 1969.
[23] Hobson, E.W. *The Theory of Spherical and Ellipsoidal Harmonics*, Cambridge University Press, 1955 reprinted from the first edition of 1931.
[24] Hoschek, J. and Lasser, D. *Fundamentals of Computer Aided Geometric Design*. Translated from the 1992 German edition by Lany L. Schumaker. A.K. Peters, Ltd., Wellesley, MA, 1993.
[25] Kalf, H. On the expansion of a function in terms of spherical harmonics in arbitrary dimensions. *Bull. Belg. Math. Soc. – Simon Stevin* 2, (1995) No. 4, pp. 361–380.
[26] Karlin, S. *Total Positivity*, Stanford University Press, 1968.
[27] Karlin, S. Generalized Markov–Bernstein type inequalities for spline functions. In: *Studies in Spline Functions and Approximation Theory*, S. Karlin *et al.* (Eds), Academic Press, New York, 1976, pp. 461–484.
[28] Karlin, S. and Studden, W. J., *Tchebycheff Systems: with Applications in Analysis and Statistics*, Interscience Publishers, New York, 1966.
[29] Lancaster, P. and Salkauskas, K. *Curve and Surface Fitting*, Academic Press, London, 1986.
[30] Laurent, P.–J. *Approximation et Optimisation*, Hermann, Paris, 1972.
[31] Light, W. Some aspects of radial basis function approximation. In: *Approximation Theory, Spline Functions and Applications*, S.P. Singh (Ed.), Kluwer Academic Publishers, Amsterdam, 1992, pp. 163–190.
[32] Lyche, T. A recurrence relation for Chebyshevian B-splines. *Constructive Approximation* 1, (1985), pp. 155–173.
[33] Madych, W.R. Polyharmonic cardinal splines. *J. Approx. Theory* 60 (1990), No. 2, pp. 141–156.
[34] Madych, W.R. and Nelson, S.A. Polyharmonic cardinal splines. *J. Approx. Theory* 60, (1990), pp. 141–156.
[35] Magnus, W., Oberhettinger, F. and Soni, R. *Formulas and Theorems for the Special Functions of Mathematical Physics*, 3rd edition. Springer–Verlag, Berlin, 1966.
[36] Mikhlin, S.G. *Mathematical Physics, an Advanced Course*, North-Holland Publ. Co., Amsterdam, 1970.
[37] Muskhelishvili, N.I. *Some Basic Problems of the Mathematical Theory of Elasticity*. Noordhoff International Publishing, Leiden, 1977.
[38] Narasimhan, R. *Analysis on Real and Complex Manifolds*. Advanced Studies in Pure Mathematics. Vol. 1, North-Holland Publishing Company, Amsterdam, 1968.
[39] Otis, A. and Barnett, M. Surface harmonic of products of Cartesian coordinates. *Mathemation Comp.* 18, (1964), No. 88, pp. 635–643.
[40] Parker, R.L. Harmonic splines in geomagnetism. In: *Function Estimates*, American Mathematical Society, Providence, Rhode Island, 1986, pp. 63–76

[41] Pontryagin, L.S. *Ordinary Differential Equations*, Addison-Wesley, Reading, Mass.–Palo Alto, Calif.–London, 1962.
[42] Powell, M.J.D., *Approximation Theory and Methods*, Cambridge University Press, Cambridge, 1981.
[43] Powell, M.J.D. The theory of radial basis function approximation in 1990. In: *Advances in Numerical Analysis. Vol. 2: Wavelets, Subdivision Algorithms, and Radial Basis Functions*, Proceedings of the 4th Summer School, Lancaster, UK, 1990, pp. 105–210, Oxford University Press, New York, 1992.
[44] Powell, M.J.D. A review of algorithms for thin plate spline interpolation in two dimensions. In: *Advanced Topics in Multivariate Approximation*, (*Montecatini Terme, 1995*) F. Fontanella, K. Jetter and P.-J. Laurent (Eds), pp. 303–322, World Sci. Publishing, River Edge, NJ, 1996.
[45] Ragozin, D.L. Polynomial approximation on compact manifolds and homogeneous spaces, *Trans. Am. Math. Soc.* 150 (1970), pp. 41–53.
[46] Ragozin, D.L. Constructive polynomial approximation on spheres and projective spaces. *Trans. Am. Math. Soc.* 162 (1971), pp. 157–170.
[47] Ragozin, D.L., Uniform convergence of spherical harmonic expansions. *Math. Ann.* 195 (1972), pp. 87–94.
[48] Rudin, W. *Functional Analysis*, McGraw-Hill, New York, 1973.
[49] Rudin, W. *Real and Complex Analysis*, McGraw-Hill, New York, 1976.
[50] Schumaker, L.L. *Spline Functions: Basic Theory*, J. Wiley and Sons, New York, Chichester–Brisbane–Toronto, 1981.
[51] Seeley, R. Spherical harmonics. *Amer. Math. Monthly*, 73 (1966), pp. 115–121.
[52] Smirnov, V.I. *A Course of Higher Mathematics. Vol. III: Part 2*, Pergamon Press, Oxford–London–New York–Paris, 1964, (Available in German where the name of the author is given as W.I. Smirnow).
[53] Smith, M.H.F. and Wessel, P. Gridding with continuous curvature splines in tension. *Geophysics* 55 (1990), No. 3, pp. 293–305.
[54] Sobolev, S.L. *Introduction to the Theory of Cubature Formulas*, Nauka, Moscow, 1974 (in Russian).
[55] Sobolev, S.L. *Cubature Formulas and Modern Analysis*, Gordon and Breach Science Publishers, Montreux, 1992.
[56] Stein, E. *Singular Integrals and Differentiability Properties of Functions*, Princeton University Press, Princeton, New Jersey, 1970.
[57] Stein, E. and Weiss, G. *Introduction to Fourier Analysis on Euclidean Spaces*, Princeton University Press, Princeton, New Jersey, 1971.
[58] SURFER – Golden Software, Golden, Colorado, 1997.
[59] Timoshenko, S.P. and Goodier, J.N. *Theory of Elasticity*, 3rd edition, McGraw-Hill, New York, 1970.
[60] Triebel, H. *Higher Analysis*. Johann Ambrosius Barth, Leipzig, 1992. (available in German).
[61] Treves, F. *Basic Linear Partial Differential Equations*, Academic Press, New York, 1975.
[62] Vekua, I.N. *New methods for Solving Elliptic Equations*, New York: NH Publ. Co., John Wiley and Sons, Inc., 1967.

[63] Wahba, G. *Spline Models for Observational Data*, SIAM, Philadelphia, Pennsylvania, 1990.
[64] Weinberger, H.F. *A First Course in Partial Differential Equations*, Blaisdell Publishing Company, New York–Toronto–London, 1965.
[65] Whittaker, E.T. and Watson, G.N. *A Course of Modern Analysis*, Cambridge University Press, London, 1965.
[66] Zweiling, K. *Grundlagen einer Theorie der Biharmonischen Polynome*, Verlag Technik, Berlin, 1952.
[67] Zygmund, A. *Trigonometric Series*, Vols. I and II, Cambridge University Press, London, 1977.

Part II

Cardinal polysplines in $\mathbb{R}^n$

The main purpose of Part II is to find and develop a proper polyspline analog to the notion of the cardinal splines. By definition, in the one-dimensional case the cardinal splines are those having knots at the integer points, or somewhat more generally, at the points $\alpha + \beta j$ where $j \in \mathbb{Z}$, for some fixed numbers α and β. There is a very beautiful theory, which was mainly developed by Schoenberg, the main results being summarized in his short monograph [18].[2] The results in this theory may be considered as a part of *harmonic analysis* due to the fact that the basic cardinal splines may be viewed as a Fourier transform of the function

$$\left(\frac{\sin \xi/2}{\xi/2}\right)^k.$$

Let us start with the *polysplines on strips*. Now trying to invent our *polyspline Ansatz*[1] let us imagine that we have polysplines on infinitely many strips, i.e. the *knot-surfaces* are infinitely many parallel hyperplanes. It seems very natural to term "cardinal" those polysplines which have equidistant hyperplanes. It is not very difficult to see that this is indeed a proper *Ansatz* and one may obtain many results by generalizing the one-dimensional case.[2]

For *polysplines on annuli*, i.e. when the knot-surfaces are infinitely many concentric spheres $S(0; r_j)$, finding the proper Ansatz is a real intellectual challenge. Its answer is far from evident but it is interesting that it is unique! The hint to the answer is hidden in the representation of polyharmonic functions in the annulus in Corollary 10.38, p. 173. By this corollary if $h(x)$ satisfies $\Delta^p h(x) = 0$ in the annulus $A_{r_j, r_{j+1}}$ and belongs to L_2 then $h(x)$ has the representation

$$h(x) = \sum_{k=0}^{\infty} \sum_{\ell=1}^{d_k} f_{k,\ell}(\log r) Y_{k,\ell}(\theta) \quad \text{for } r_j < r < r_{j+1},$$

where the one-dimensional function $f_{k,\ell}(v)$ is a solution to the equation

$$M_{k,p}\left(\frac{d}{dv}\right) f_{k,\ell}(v) = 0 \quad \text{for } r_j < e^v < r_{j+1}.$$

Recall that by formula (10.26), p. 169, the operator $M_{k,p}$ has constant coefficients. Furthermore, we have seen in Part I and more specially in Theorem 9.7, p. 124, that h is a polyspline if and only if for every two indexes k and ℓ the function $f_{k,\ell}(v)$ is an L-spline for the operator $L = M_{k,p}$! Now the question is whether we have a reasonable "cardinal" theory of such L-splines? Yes, we do! It has been developed by Micchelli [12, 13]. Some of the results have been given concise and elementary proofs by Schoenberg [19] in the same volume.

Eureka! We will call h a "cardinal polyspline on annuli" if all components $f_{k,\ell}$ are cardinal L-splines with knots at $j \in \mathbb{Z}$. Thus we see that the "break-radii" have to satisfy $r_j = e^j$, hence the break-surfaces for the polyspline h will be the spheres $S(0; e^j)$.

---

[1] The meaning of *Ansatz* was discussed in the footnote on p. 32.

[2] Due to the lack of space we omit the consideration of the cardinal polysplines on strips. We treat in detail only the case of the technically more complicated cardinal polysplines on annuli. The reader will be able to follow the same scheme and produce similar results for the cardinal polysplines on strips.

It now becomes clear to the reader what the motivation was for the compendium on representation of polyharmonic functions in the annulus, and further what the motivation is to have an exposition of the results of Micchelli on cardinal L-splines coming in the next chapter.

Last but not least the one-dimensional cardinal splines serve as a basic example for the wavelet analysis. We plan to mimic this construction by using *cardinal polysplines*. Thus a major motivation for the detailed study of the cardinal theory of polysplines in the present Part is their application to "polyharmonic wavelet analysis" in Part III.

Finally, we want to warn the reader that there will be some weak overlapping of the notations in some Chapters of the present Part. Following the tradition by $L(x)$ we will sometimes denote the fundamental spline function of Schoenberg and this may be mixed with the operator L for the L-spline. This overlapping is indeed very weak and we prefer to retain the original notations of Schoenberg. We will eventually repeat this warning at the proper place.

Chapter 13

Cardinal L-splines according to Micchelli

In the present chapter we provide an extended study of the cardinal L-splines following the approach of Ch. Micchelli, including results by I. Schoenberg, Dyn and Ron.[1]

13.1 Cardinal L-splines and the interpolation problem

The theory of cardinal splines and more specifically cardinal L-splines is a beautiful area of spline analysis which deserves much attention in view of its recent applications to wavelet analysis.

Within the general theory of splines the *theory of cardinal splines*, or the splines having only integers as knots, plays a very important and specific role. First, technically it may be considered more as a subset of harmonic analysis than of the general spline theory. Indeed, one may view the whole theory as study of Fourier inverse of functions of the type

$$\left(\frac{\sin \xi/2}{\xi/2}\right)^m = \left(\frac{e^{i\xi/2} - e^{-i\xi/2}}{i\xi}\right)^m = e^{i\xi m/2}\left(\frac{1 - e^{-i\xi}}{i\xi}\right)^m.$$

Let us denote by Q_m the usual polynomial B-spline of degree m with knots at the points $\{0, 1, \ldots, m\}$ and with support coinciding with the interval $[0, m]$, and let us introduce the "centralized" spline $M_m(x) = Q_m(x + m/2)$, having support $[-m/2, m/2]$ and knots at $\{-m/2, -m/2+1, \ldots, m/2\}$. Then by the properties of the Fourier transform

---

[1] In view of the terminology that has been established, see Chapter 11 and *Schumaker* [22], it would be more appropriate to use the name "cardinal Chebyshev splines" since the theory of Micchlli only concerns operators L having constant coefficients, and the corresponding splines are Chebyshev splines.

222 Multivariate polysplines

we obtain the following equality [18, pp. 11,12]:

$$\widehat{Q_m}(\xi) = \left(\frac{1-e^{-i\xi}}{i\xi}\right)^m = \left(\widehat{Q_1}(\xi)\right)^m,$$

$$\widehat{M_m}(\xi) = e^{i\xi m/2}\widehat{Q_m}(\xi) = \left(\frac{2\sin(\xi/2)}{\xi}\right)^m = \left(\widehat{M_1}(\xi)\right)^m.$$

By taking the inverse Fourier transform we see that

$$Q_m(x) = Q_{m-1}(x) * Q_1(x) = \int_0^1 Q_{m-1}(x-y)\,dy, \qquad (13.1)$$

hence, we have a simple constructive and inductive definition of the compactly supported spline $Q_m(x)$.

The cardinal L-splines with compact support (TB-splines) which we will study may also be considered as a part of harmonic analysis since their Fourier transforms are given by

$$\widehat{Q_m}(x) = \frac{\prod_{j=1}^m (e^{-\lambda_j} - e^{-i\xi})}{\prod_{j=1}^m (i\xi - \lambda_j)},$$

where λ_j are real constants. Thus we have in a similar way

$$Q_{m-1}(x) = Q_{m-1}(x) * Q_1(x),$$

which provides a simple method to generate the most important function of the whole theory. So far this visual simplicity is only superficial.

The reader should be aware that one may start reading the present chapter from Section 13.10, where the compactly supported splines are introduced, since for the majority of standard numerical work one does not need much more. However, as will become clear in Chapter 14, p. 267, in order to understand the deeper properties of the functions Q_m, which are further necessary for the *wavelet analysis* in Part III, one really needs the whole theory developed in the present chapter. In particular, one needs the notions of Euler polynomials $A_m(x; \lambda)$ and the Euler–Frobenius polynomials $\Pi_m(\lambda) = A_m(0; \lambda)$, the location of their zeros etc.

The theory of polynomial cardinal splines, including the theory of the *Euler–Frobenius* and *Euler polynomials* related to them, was developed mainly by Schoenberg till the mid-1970s. He has summarized almost all the results in his fascinating book [18].[2]

During the last decade there has been renewed interest in cardinal splines in view of their applications to *wavelet theory*. One may even say that the cardinal splines were reborn in wavelet analysis in the works of Chui, [3], who generalized such a fundamental notion as the Euler–Frobenius polynomial for an arbitrary scaling function $\phi(x)$ generating a *multiresolution analysis*.

---

[2] One has to mention also the initiating work of Quade and Collatz [16], and that of Tchakaloff [35] of which Schoenberg was apparently not aware. More about the beautiful analytic work of Tchakaloff, which has been published in Bulgarian with a French summary, [2, p. 39].

As we have already said the theory of *cardinal L-splines* has been developed by Micchelli and Schoenberg.[3] The cardinal L-splines possess most of the advantageous properties of the usual (polynomial) cardinal splines. However, one important property which distinguishes the L-splines from the polynomial splines is that they are not scale-invariant, i.e. if $f(x)$ is an L-spline for some operator L then $f(\alpha x)$ is not an L-spline for the same operator. In the case of constant coefficient operators L the function $f(\alpha x)$ is still an L-spline, but for another operator. This is essentially used in the theory of *non-stationary wavelets* developed by de Boor et al. ([5] p. 150). The last is a construction which we will need for the wavelet analysis using *cardinal polysplines on annuli* to be treated in Part III of the present book, and for that reason all details of the one-dimensional construction will be studied here.

Cardinal L-splines are a basic tool for our study. For that reason we prefer to give an independent exposition of the theory, which does not refer to the fundamental theory of L-splines (and Chebyshev splines) with general knots developed by Schumaker [22], which we have already used in Part I. Such an exposition will give an opportunity for a reader who is mainly interested in wavelet analysis to have a complete and logically closed understanding of the subject. Let us note that the same results may be obtained [6, 7] by following the approach to L-splines of [22].

In the present section we will follow closely the approach and most of Micchelli's notations [12, 13]. Let us give some basic definitions and notations. Let the real numbers $\lambda_1, \ldots, \lambda_{Z+1}$ be given. We will consider the *nonordered* vector

$$\Lambda := \Lambda_{Z+1} := [\lambda_1, \lambda_2, \ldots, \lambda_{Z+1}], \tag{13.1a}$$

where some of the numbers λ_j may have repetitions. The number of repetitions of a number λ in Λ will be termed the **multiplicity** of λ.

There are different ways to give a good representation of such vectors Λ but they are all overburdened with indices. For example, we might write [22, p. 20]

$$t_1 \le t_2 \le \cdots \le t_m = \overbrace{\tau_1, \ldots, \tau_1}^{l_1}, \ldots, \overbrace{\tau_d, \ldots, \tau_d}^{l_d},$$

where $\sum_{i=1}^{d} l_i = m$. Another possibility is to put [2, p. 5]

$$(t_1, t_2, \ldots, t_m) = ((\tau_1, l_1), \ldots, (\tau_d, l_d)).$$

By using the notation [·] for such a vector we avoid having to describe the multiplicity of the entries every time. For almost all our purposes the representation by a nonordered vector Λ will be adequate.[4]

---

[3] Micchelli constructs his theory in a way close to the meditative approach to cardinal splines developed by Schoenberg. This is based mainly on the Euler exponential spline.

[4] We note that a large part of the theory in the present chapter holds for complex numbers λ_j. In such a case the so-called W-property, associated with the name of Polya, holds only for intervals with bounded length, i.e. the set U_{Z+1} is not Chebyshev over arbitrary large intervals and one has to keep this in mind. See for examples, the comment of Schoenberg [19, p. 251]. The results which we need for the cardinal polysplines require no such generality, while the last would overburden some proofs.

Further we introduce the polynomial

$$q_{Z+1}(z) := q_{Z+1}[\Lambda](z) := \prod_{j=1}^{Z+1}(z - \lambda_j) \qquad (13.2)$$

and the operator $\mathcal{L}_{Z+1}$ defined by

$$\mathcal{L}_{Z+1}[\Lambda]f(x) := q_{Z+1}\left(\frac{d}{dx}\right)f(x) = \prod_{j=1}^{Z+1}\left(\frac{d}{dx} - \lambda_j\right)f(x) \qquad (13.3)$$

where, if it is clear from the context, we will drop the dependence on the set Λ and simply write $\mathcal{L}_{Z+1}f$ or q_{Z+1}.[5]

Let us introduce the set of solutions, sometimes called **L-polynomials**, over the whole real axis:

$$U_{Z+1} := U_{Z+1}[\Lambda] = \{u \text{ in } C^\infty(\mathbb{R}): \mathcal{L}_{Z+1}[\Lambda]u(x) = 0 \text{ for } x \text{ in } \mathbb{R}\}.$$

As will be discussed in the sections below the fact that λ_j are real constants provides the following important properties:

1. The set U_{Z+1} is Chebyshev over the whole real axis, i.e. every $\varphi \in U_{Z+1}$ has no more than Z real zeros.

2. The set U_{Z+1} is translation invariant, i.e. if $\varphi \in U_{Z+1}$ then for every real number α we have $\varphi(x - \alpha) \in U_{Z+1}$.

3. The classical polynomial case is obtained as a special case, when $\lambda_1 = \lambda_2 = \cdots = \lambda_{Z+1} = 0$. In this case we have the following:

$$q_{Z+1}(z) = z^{Z+1}, \quad \mathcal{L}_{Z+1}[\Lambda]f(x) = \frac{d^{Z+1}}{dx^{Z+1}}f(x),$$

and U_{Z+1} is the set of all polynomials of degree $\leq Z$.

We are using the notation $Z + 1$ in order to make our notation consistent with the standard one-dimensional polynomial case. As is known, the dimension of the space U_{Z+1} is

$$\dim U_{Z+1} = Z + 1,$$

but in the polynomial case the degree of the polynomials is $\leq Z$.

Definition 13.1 *The class of* **cardinal L-splines** *for the operator* $\mathcal{L}_{Z+1}[\Lambda]$ *is defined as the set of those functions* $u(x) \in C^{Z-1}(\mathbb{R})$ *which on every interval* $(j, j + 1)$ *is a solution of* $\mathcal{L}_{Z+1}[\Lambda]u(x) = 0$, *i.e.*

$$\mathcal{S}_{Z+1} := \mathcal{S}_{Z+1}[\Lambda] := \{u \text{ in } C^{Z-1}(\mathbb{R}): u_{|(j,j+1)} \text{ in } U_{Z+1} \text{ for all } j \in \mathbb{Z}\}. \qquad (13.4)$$

---

[5] As we will see below, the global C^∞ solutions of $\mathcal{L}_{P+1}f(x) = 0$ are linear combinations of expressions $R_j(x) \cdot e^{\lambda_j x}$, where $R_j(x)$ is a polynomial with $\deg R_j \leq$ (multiplicity of t_j) -1. For that reason these splines are sometimes called *exponential*. This terminology should not be mixed with the so-called "exponential Euler spline" of Schoenberg which we will meet below.

In this definition by $g_{|(j,j+1)}$ we have, as usual, denoted the restriction of the function g to the interval $(j, j+1)$. In order to save notation, by $g_{|(j,j+1)} \in U_{Z+1}$ we mean that $g_{|(j,j+1)}$ is a restriction of an element of U_{Z+1}.

(We use q_{Z+1} for the polynomial instead of p_{Z+1} of Micchelli [12, 13]. We also write $\mathcal{S}_{Z+1}$ for the space of splines instead of $\mathcal{S}_Z$. Let us note again that his notation [12, 13] tends to preserve the tradition of the polynomial splines where the *degree* of the polynomials is Z and the dimension of the space is $Z + 1$. We put as a central index $Z + 1$ instead.)

The main problem solved by Schoenberg for polynomial splines and by Micchelli for the above introduced L-splines is the so-called *cardinal interpolation problem*. They have found the conditions within which the problem:

$$u(j+\alpha) = y_j \quad \text{for all } j \text{ in } \mathbb{Z}, \tag{13.5}$$

has a solution u in $\mathcal{S}_{Z+1}$. Here α is a constant such that $0 \le \alpha < 1$.[6] In order to formulate the complete solution for (13.5) we need the class of null L-splines, which is defined as

$$\mathcal{S}^0_{Z+1} := \{u \text{ in } \mathcal{S}_{Z+1} : u(j+\alpha) = 0 \text{ for all } j \text{ in } \mathbb{Z}\}.$$

We always assume that α is fixed. The following result is basic in Micchelli [12, 13, p. 204], and Schoenberg [19]. It generalizes the classical result of Schoenberg about cardinal interpolation through polynomial splines from his book [18].

Theorem 13.2 *1. The space $\mathcal{S}^0_{Z+1}$ has dimension*

$$\dim(\mathcal{S}^0_{Z+1}) = \begin{cases} m = Z - 1 & \text{for } \alpha = 0, \\ m = Z & \text{for } 0 < \alpha < 1. \end{cases}$$

2. The space $\mathcal{S}^0_{Z+1}$ is spanned by m eigensplines $S_1, S_2, \ldots, S_m$ which satisfy the equation

$$S_i(x+1) = \tau_i S_i(x) \quad \text{for } i = 1, \ldots, m,$$

where the constants τ_i (called the eigenvalues of the problem) satisfy

$$\tau_1 < \tau_2 < \cdots < \tau_m < 0.$$

3. Let $\tau_i \ne -1$ for $i = 1, \ldots, m$. Then there exists a fundamental cardinal L-spline $L(x) \in \mathcal{S}_{Z+1}$, i.e. a spline such that[7]

$$L(j+\alpha) = \begin{cases} 0 & \text{for } j \ne 0, \\ 1 & \text{for } j = 0. \end{cases}$$

---

[6] The results about solubility of this problem are extended by Schoenberg [19] without any extra effort to the cardinal grid $\{jh + \alpha : \text{for all } j \text{ in } \mathbb{Z}\}$, where $h > 0$ is an arbitrary constant.

[7] As we said already in the Introduction to this Part, the reader does not have to mix this L with the operator L. We have preserved Schoenberg's original notation which was also used by Micchelli and Chui.

There exist positive constants A, B such that

$$|L(x)| \leq Ae^{-B|x|} \quad \text{for all } x \text{ in } \mathbb{R}.$$

4. *Let $\tau_i \neq -1$ for $i = 1, \ldots, m$. Let the sequence y_j be of power growth, i.e. for some $\gamma \geq 0$ it satisfies $y_j = O(|j|^\gamma)$. Then there exists a unique $u \in \mathcal{S}_{Z+1}$ which has a power growth, i.e.*

$$|u(x)| = O(|x|^\gamma) \quad \text{for all } x \text{ in } \mathbb{R},$$

and which interpolates the data y_j, i.e.

$$u(j + \alpha) = y_j \quad \text{for all } j \text{ in } \mathbb{Z}.$$

We have the representation

$$u(x) = \sum_{i=-\infty}^{\infty} y_j L(x - j).$$

Later, in Theorem 13.33, p. 238, we will provide another criterion for solving the cardinal interpolation problem. One of our main purposes in Part II will be to find an analog to the above theorem for cardinal polysplines.

13.2 Differential operators and their solution sets U_{Z+1}

Let us introduce some operators decomposing the operator $\mathcal{L}_{Z+1}$ of Section 13.1.

When the nonordered vector $\Lambda = [\lambda_1, \lambda_2, \ldots, \lambda_{Z+1}]$ is given we define the following operators:

$$\mathcal{D}_j f(x) := \left(\frac{d}{dx} - \lambda_j\right) f(x) = e^{\lambda_j x} \frac{d}{dx} e^{-\lambda_j x} f(x) \quad \text{for } j = 1, \ldots, Z+1, \quad (13.6)$$

$$\mathcal{D}_0 f(x) := f(x).$$

Evidently, for every integer $s \geq 1$ we have

$$[\mathcal{D}_j]^s f(x) = e^{\lambda_j x} \frac{d^s}{dx^s} e^{-\lambda_j x} f(x).$$

For every integer $s \geq 1$ we will define the following differential operators:

$$\mathcal{L}_s f(x) := \mathcal{D}_1 \cdots \mathcal{D}_s f(x),$$
$$\mathcal{L}_0 f(x) := f(x).$$

As was said above, the space of C^∞ solutions of the equation

$$\mathcal{L}_{Z+1} f(x) = 0 \quad \text{for } x \text{ in } \mathbb{R},$$

which we have denoted by $U_{Z+1}[\Lambda]$ will be important.

In order to develop some intuition in the reader who is not experienced in differential equations, we provide the following simple, standard facts from the theory of ODEs concerning the space $U_{Z+1}[\Lambda]$, see Pontryagin [15].

Example 13.3 $\dim U_{Z+1} = Z + 1$.

Example 13.4 *If*
$$\lambda_1 = \lambda_2 = \cdots = \lambda_{Z+1} = 0$$
then
$$\mathcal{L}_{Z+1} f(x) = \frac{d^{Z+1}}{dx^{Z+1}} f(x)$$
and U_{Z+1} is the set of all algebraic polynomials of degree $\leq Z$, i.e.
$$U_{Z+1} = \{1, x, x^2, \ldots, x^Z\}_{\text{lin}}.$$

Here $\{\cdot\}_{\text{lin}}$ denotes the linear hull of the set of functions inside the brackets.

Example 13.5 *If all λ_j are pairwise different, i.e. $\lambda_i \neq \lambda_j$ for $i \neq j$, then*
$$U_{Z+1} = \{e^{\lambda_1 x}, e^{\lambda_2 x}, \ldots, e^{\lambda_{Z+1} x}\}_{\text{lin}}.$$

Example 13.6 *The constants belong to the set $U_{Z+1}[\Lambda]$ if and only if there exists an index j for which $\lambda_j = 0$.*

Example 13.7 *If*
$$\lambda_1 = \lambda_2 = \cdots = \lambda_{Z+1}$$
then the set $U_{Z+1}[\Lambda]$ coincides with all algebraic polynomials of degree $\leq Z$, times $e^{\lambda_1 x}$, i.e.
$$U_{Z+1}[\Lambda] = \{e^{\lambda_1 x}, xe^{\lambda_1 x}, \ldots, x^Z e^{\lambda_1 x}\}_{\text{lin}}$$
$$= \{R(x)e^{\lambda_1 x} : R \text{ is a polynomial of } \deg R \leq Z\}.$$

Example 13.8 *More generally, let the set Λ be given by*
$$\Lambda = \Big[\underbrace{\tau_1, \tau_1, \ldots, \tau_1}_{m_1}, \underbrace{\tau_2, \tau_2, \ldots, \tau_2}_{m_2}, \ldots, \ldots, \ldots, \underbrace{\tau_\ell, \tau_\ell, \ldots, \tau_\ell}_{m_\ell}\Big],$$
where $m_1 + m_2 + \cdots + m_\ell = Z + 1$. Then
$$U_{Z+1}[\Lambda] = \{R_1(x)e^{\tau_1 x}, R_2(x)e^{\tau_2 x}, \ldots, R_\ell(x)e^{\tau_\ell x}\}_{\text{lin}}$$
where the polynomials R_j satisfy
$$\deg R_1(x) \leq m_1 - 1, \ \deg R_2(x) \leq m_2 - 1, \ldots, \ \deg R_\ell(x) \leq m_\ell - 1.$$

13.3 Variation of the set $U_{Z+1}[\Lambda]$ with Λ and other properties

Let us see how the set $U_{Z+1}[\Lambda]$ changes with the variation of Λ. If all values of λ_j are pairwise different, we have seen in Example 13.5, p. 227, that $U_{Z+1}[\Lambda] = \{e^{\lambda_1 x}, e^{\lambda_2 x}, \ldots, e^{\lambda_{Z+1} x}\}_{\text{lin}}$. Now let $\lambda_2 \to \lambda_1$. Obviously, for $\lambda_2 \neq \lambda_1$ we have

$$U_{Z+1}[\lambda_1, \lambda_2, \lambda_3, \ldots, \lambda_{Z+1}] = \left\{ e^{\lambda_1 x}, \frac{e^{\lambda_1 x} - e^{\lambda_2 x}}{\lambda_1 - \lambda_2}, e^{\lambda_3 x}, \ldots, e^{\lambda_{Z+1} x} \right\}_{\text{lin}},$$

which in the limit $\lambda_2 \to \lambda_1$ gives

$$U_{Z+1}[\lambda_1, \lambda_1, \lambda_3, \ldots, \lambda_{Z+1}] = \{e^{\lambda_1 x}, x e^{\lambda_1 x}, e^{\lambda_3 x}, \ldots, e^{\lambda_{Z+1} x}\}_{\text{lin}}.$$

In this way we obtain the sets U_{Z+1} in Examples 13.7 and 13.8.

This kind of limiting process will often be used below – the reason is that several formulas are much simpler to write in the case of pairwise different λ_js. Then, using the above limiting argument, we will also obtain the result in the case of arbitrary λ_js.

Theorem 13.9 *The space $U_{Z+1}[\Lambda]$ is translation invariant, i.e. if $\varphi(x)$ belongs to $U_{Z+1}[\Lambda]$ then $\varphi(x-c)$ belongs to $U_{Z+1}[\Lambda]$ for every real number c. The space $U_{Z+1}[\Lambda]$ is not scaling invariant, i.e. if $\varphi(x)$ belongs to $U_{Z+1}[\Lambda]$ then in general it does not follow that $\varphi(hx)$ belongs to $U_{Z+1}[\Lambda]$ for arbitrary real number h.*

The first statement is due to the fact that

$$e^{\lambda_j(x-c)} = e^{-\lambda_j c} \cdot e^{\lambda_j x} \text{ belongs to } U_{Z+1}[\Lambda],$$

and the last is true since only for $h = \lambda_i/\lambda_j$, we have

$$e^{h \lambda_j x} \text{ belongs to } U_{Z+1}[\Lambda].$$

So far, if we consider another operator, namely

$$\mathcal{L}_{Z+1}[hT] = \prod_{j=1}^{Z+1} \left(\frac{d}{dx} - ht_j \right),$$

then evidently

$$e^{h t_j x} \text{ belongs to } U_{Z+1}[hT].$$

Here we have used the notation for the nonordered vector

$$hT := [ht_1, \ldots, ht_{Z+1}].$$

This simple fact will be used further in the wavelet analysis.

Theorem 13.10 *The space $U_{Z+1}[\Lambda]$ is Chebyshev on the whole real line, i.e. if $\varphi(x)$ belongs to $U_{Z+1}[\Lambda]$ then φ has no more than Z real zeros.*

Theorem 13.10 is in fact a reformulation of Theorem 11.4, p. 188, and is important to us.

Exercise 13.11 *Prove Theorem 13.10 in the case of pairwise different $\lambda_j s$.*

Hint: *If some $\varphi(x) \in U_{Z+1}[\Lambda]$ has $Z + 1$ different zeros $x_1, \ldots, x_{Z+1}$, then on every interval (x_j, x_{j+1}) we have a point ξ_j where $\varphi'(\xi_j) = \lambda_1 \varphi(\xi_j)$. Indeed, on the interval (x_j, x_{j+1}) the continuous function $\varphi'(x)$ changes its sign. However, on the same interval the continuous function $\lambda_1 \varphi(x)$ is zero at both endpoints. It follows that the function $\varphi'(x) - \lambda_1 \varphi(x)$ changes sign on the interval (x_j, x_{j+1}), hence, by Rolle's theorem there exists a $\xi_j \in (x_j, x_{j+1})$ such that $\varphi'(\xi_j) = \lambda_1 \varphi(\xi_j)$. Now the function $\varphi'(x) - \lambda_1 \varphi(x)$ belongs to $U_{Z+1}[\lambda_2, \lambda_3, \ldots, \lambda_{Z+1}]$. Proceed further using inductive reasoning.*

13.4 The Green function $\phi_Z^+(x)$ of the operator $\mathcal{L}_{Z+1}$

Here we introduce the so-called Green function associated with the operator $\mathcal{L}_{Z+1}$. This function is the analog to the function $(x - t)_+^Z$ in the polynomial case. We put

$$\phi_Z(x) := [\lambda_1, \lambda_2, \ldots, \lambda_{Z+1}]_z e^{xz},$$

where the index of the *divided difference* $[\lambda_1, \lambda_2, \ldots, \lambda_{Z+1}]_z$ means that it is taken with respect to the variable z. Let us note that, using the equivalent definition of divided difference through residuum, see Chapter 11, formula (11.12), p. 193, we have

$$\phi_Z(x) = \int_\Gamma \frac{e^{xz}}{q_{Z+1}(z)} dz, \qquad (13.7)$$

where the contour Γ in the complex plane surrounds the zeros of the polynomial $q_{Z+1}(z)$. In particular, in the case of pairwise different $\lambda_j s$ we obtain

$$\phi_Z(x) = \sum_{j=1}^{Z+1} \frac{e^{\lambda_j x}}{q'_{Z+1}(\lambda_j)}. \qquad (13.8)$$

We define the function $\phi_Z^+(x)$ as follows:

$$\phi_Z^+(x) := \begin{cases} \phi_Z(x) & \text{for } x \geq 0, \\ 0 & \text{for } x < 0. \end{cases}$$

The following result shows that the function $\phi_Z^+(x)$ is the Green function for the operator $\mathcal{L}_{Z+1}$.

Proposition 13.12 *The function $\phi_Z^+(x)$ is the Green function for the operator $\mathcal{L}_{Z+1}$, i.e. it satisfies the following three equivalent properties.*

1. $\phi_Z^+(x)$ belongs to $C^{Z-1}(\mathbb{R})$.
2. The following equalities hold:
$$\begin{cases} \mathcal{D}_0\mathcal{D}_1\cdots\mathcal{D}_\ell\phi_Z(x)|_{x=0+} = 0 & \text{for } \ell = 0,\ldots,Z-1, \\ \mathcal{D}_0\mathcal{D}_1\cdots\mathcal{D}_Z\phi_Z(x)|_{x=0+} = 1, \end{cases} \quad (13.9)$$

where the operators $\mathcal{D}_j$ were defined in (13.6), p. 226.

3. The equalities in (13.9) are equivalent to the following:
$$\frac{d^\ell}{dx^\ell}\phi_Z(x)|_{x=0+} = 0 \quad \text{for } \ell = 0,\ldots,Z-1,$$
$$\frac{d^Z}{dx^Z}\phi_Z(x)|_{x=0+} = 1.$$

The function $\phi_Z(x)$ is also the unique element in U_{Z+1} which satisfies these equalities.

Proof By the residuum representation (13.7), p. 229, we obtain, for $\ell = 0,\ldots,Z-1$, the equalities
$$\mathcal{D}_0\mathcal{D}_1\cdots\mathcal{D}_\ell\phi_Z(x)|_{x=0+} = \left[\int_\Gamma \frac{\mathcal{D}_0\mathcal{D}_1\cdots\mathcal{D}_\ell e^{xz}}{q_{Z+1}(z)}dz\right]_{|x=0+}$$
$$= \int_\Gamma \prod_{j=1}^\ell (z-\lambda_j)\frac{1}{q_{Z+1}(z)}dz.$$

Taking for Γ the large circle $\Gamma_R = \{R\cdot e^{i\theta} : 0 \leq \theta < \theta\}$, we obtain the estimate
$$|\mathcal{D}_0\mathcal{D}_1\cdots\mathcal{D}_\ell\phi_Z(x)|_{x=0+}| \leq \int_0^{2\pi}\left|\prod_{j=1}^\ell(z-\lambda_j)\frac{1}{q_{Z+1}(z)}\right|R\,d\theta$$
$$\leq \int_0^{2\pi}\frac{C}{R^{Z+1-\ell}}R\,d\theta = \frac{2\pi C}{R^{Z-\ell}}$$

which after letting $R \longrightarrow \infty$ proves the first part of (2).

Since
$$\left[\frac{\mathcal{D}_0\mathcal{D}_1\cdots\mathcal{D}_Z e^{xz}}{q_{Z+1}(z)}\right]_{|x=0+} = \frac{1}{z-\lambda_{Z+1}},$$

and by the Cauchy residuum theorem
$$\int_\Gamma \frac{1}{z-\lambda_{Z+1}}dz = 2\pi i,$$

we obtain the second equality in (2).

Point (3) follows easily by induction in s since
$$\prod_{j=1}^s\left(\frac{d}{dx}-\lambda_j\right) = \frac{d^s}{dx^s}+\sum_{j=0}^{s-1}c_j\frac{d^j}{dx^j}.$$

The uniqueness as stated follows since the dimension of the space $U_{Z+1}(\Lambda)$ is $Z+1$. ∎

Let us denote by $\phi_Z^+[\lambda_1 + \gamma, \ldots, \lambda_{Z+1} + \gamma](x)$ the Green function corresponding to the nonordered vector

$$\Lambda + \gamma := [\lambda_1 + \gamma, \lambda_2 + \gamma, \ldots, \lambda_{Z+1} + \gamma].$$

Proposition 13.13 *The Green function ϕ_Z^+ satisfies the following identity:*

$$\phi_Z^+[\Lambda + \gamma](x) = e^{\gamma x} \phi_Z^+[\Lambda](x).$$

Proof Since the function $e^{\gamma x}$ is C^∞ it follows that $e^{\gamma x} \phi_Z^+[\lambda_1, \lambda_2, \ldots, \lambda_{Z+1}](x) \in C^{Z-1}(R)$, hence, the function $e^{\gamma x} \phi_Z^+[\lambda_1, \lambda_2, \ldots, \lambda_{Z+1}](x)$ satisfies the first row of conditions in (13.9), p. 230. The last condition in (13.9) is satisfied owing to the Leibnitz formula for differentiation of a product

$$\left\{\prod_{j=1}^{Z}\left(\frac{d}{dx} - \lambda_j\right)\right\} [e^{\gamma x} \phi_Z^+[\lambda_1, \lambda_2, \ldots, \lambda_{Z+1}](x)]|_{x=0+}$$

$$= \left\{\frac{d^Z}{dx^Z} + \sum_{j=0}^{Z-1} c_j \frac{d^j}{dx^j}\right\} [e^{\gamma x} \phi_Z^+[\lambda_1, \lambda_2, \ldots, \lambda_{Z+1}](x)]|_{x=0+}$$

$$= \frac{d^Z}{dx^Z} [e^{\gamma x} \phi_Z^+[\lambda_1, \lambda_2, \ldots, \lambda_{Z+1}](x)]|_{x=0+}$$

$$= \left[e^{\gamma x} \frac{d^Z}{dx^Z} \phi_Z^+[\lambda_1, \lambda_2, \ldots, \lambda_{Z+1}](x)\right]_{|x=0+}$$

$$= 1,$$

which completes the proof. ∎

Exercise 13.14 *Prove the above theorem without residuum, assuming for simplicity that all λ_j are pairwise different.*
  Hint: Then

$$\phi_Z(x) = \sum_{j=1}^{Z+1} \frac{1}{q_Z'(\lambda_j)} e^{\lambda_j x}.$$

and it follows that

$$\mathcal{D}_0 \mathcal{D}_1 \cdots \mathcal{D}_Z \phi_Z(x)|_{x=0+} = \prod_{j=1}^{Z}(\lambda_{Z+1} - \lambda_j) \cdot \frac{1}{q_Z'(\lambda_{Z+1})} \cdot [e^{\lambda_{Z+1} x}]|_{x=0+}$$

$$= \prod_{j=1}^{Z}(\lambda_{Z+1} - \lambda_j) \cdot \frac{1}{q_Z'(\lambda_{Z+1})}$$

$$= 1.$$

In Theorem 11.25, p. 200, we provided the basic result about the one-sided basis generated by the Green function for general Chebyshev splines. Here we specify that general result for the case of the cardinal splines $\mathcal{S}_{Z+1}$ defined in (13.4), p. 224.

Theorem 13.15 *Let us denote by $\mathcal{S}_{Z+1}[a, b]$ the L-splines in $\mathcal{S}_{Z+1}$ having break-points in the interval $[a, b]$ where a, b are integers. Then the set of shifts $\{\phi_Z^+[\Lambda](x - j) : j = a - 1, \ldots, b\}$ is a linear basis for the space $\mathcal{S}_{Z+1}[a, b]$.*

This is a classic result in spline theory and is proved simply by counting dimensions.

Corollary 13.16 *The function $\phi_Z^+(x) \in L_2(\mathbb{R})$ if and only if $\lambda_j < 0$ for all $j = 1, \ldots, Z + 1$.*

This will be used later in wavelet analysis using L-splines.

Exercise 13.17 *Prove Corollary 13.16. Hint: Use representation (13.8), p. 229, of $\phi_Z(x)$ in the case of pairwise different λ_j. Another possibility is to use representation (13.7).*

Remark 13.18 1. Let us note that the function $\phi_Z^+(x)$ may be considered as an L-spline with the only knot the point 0. Since the dimension of U_{Z+1} is $Z + 1$ (as we have seen in Proposition 13.12, p. 229) this is the only L-spline with a knot at 0. This corollary means, roughly speaking, that the space of splines in $L_2(\mathbb{R})$ with the only knot 0 has dimension zero or one, and both cases are described.

2. Let us note that due to the translation invariance of the space U_{Z+1} it follows that the function $\phi_Z^+[\Lambda](x - y)$ is the Green function associated with the operator $\mathcal{L}_{Z+1}$ in the most general sense of this notion, see Section 11.3.1, especially formula (11.18) on p. 198.

13.5 The dictionary: L-polynomial case

In order to make the transition from the classical polynomial case to the case of solutions of the operator $\mathcal{L}_{Z+1}$ called L-polynomials we provide the following dictionary of notions:

$$\frac{d^{Z+1}}{dx^{Z+1}} \longrightarrow \prod_{j=1}^{Z+1} \left(\frac{d}{dx} - \lambda_j\right),$$

$$\pi_Z \longrightarrow U_{Z+1},$$

$$(x - t)_+^Z \longrightarrow \phi_Z^+(x - t),$$

where, as usual, π_Z denotes the set of polynomials of degree $\leq Z$.

13.6 The generalized Euler polynomials $A_Z(x; \lambda)$

In the classical theory of cardinal splines the so-called *Euler* and *Euler–Frobenius* polynomials, see Schoenberg [18, p. 21], play a major role. These two notions may be generalized to the case of the differential operators $\mathcal{L}_{Z+1}$, and they also play an analogous role in the theory of cardinal L-splines.

Lemma 13.19 *Let $\lambda \neq e^{\lambda_i}$ for $i = 1, \ldots, Z + 1$. If for some function u in U_{Z+1} we have $u^{(j)}(1) = \lambda u^{(j)}(0)$ for $j = 0, 1, \ldots, Z$, then $u \equiv 0$.*

Proof For $Z = 0$ we have $\Lambda = [\lambda_1]$ and the space $U_1[\Lambda]$ is one-dimensional with elements $Ce^{\lambda_1 x}$ for an arbitrary constant C. Hence, $u(1) = u(0)$ implies $\lambda = e^{\lambda_1}$ if $C \neq 0$.

For $Z \geq 1$ we obtain from the first two conditions

$$u(1) = u(0), \quad u'(1) = u'(0),$$

that

$$\left(\frac{d}{dx} - \lambda_{Z+1}\right) u(1) = \left(\frac{d}{dx} - \lambda_{Z+1}\right) u(0).$$

However, $((d/dx) - \lambda_{Z+1})u(x)$ belongs to $U_Z[\lambda_1, \ldots, \lambda_Z]$, which shows that we may proceed inductively. ■

Let us note that the above proof is similar to the proof of the equivalence of points (2) and (3) in Proposition 13.12, p. 229, and we can see that conditions $u^{(j)}(1) = \lambda u^{(j)}(0)$ are equivalent to conditions $\mathcal{D}_0 \mathcal{D}_1 \cdots \mathcal{D}_\ell u(1) = \lambda \mathcal{D}_0 \mathcal{D}_1 \cdots \mathcal{D}_\ell u(0)$.

Thanks to the above lemma we may introduce the very important function $A(x; \lambda)$ of the theory developed by Micchelli which is a generalization of the classical *Euler polynomial* considered by Schoenberg [18, p. 21].[8] Let us first put

$$G(z; x, \lambda) := \frac{e^{xz}}{e^z - \lambda}.$$

Definition 13.20 *Let $\lambda \neq e^{\lambda_i}$ for $i = 1, \ldots, Z + 1$. Then the function $A_Z(x; \lambda) = A_Z[\Lambda](x; \lambda)$ is defined as the unique element in $U_{Z+1} = U_{Z+1}[\Lambda]$ which is a solution of the boundary value problem*

$$A_Z^{(j)}(1; \lambda) = \lambda A_Z^{(j)}(0; \lambda) \quad \text{for } j = 0, \ldots, Z - 1,$$

$$A_Z^{(Z)}(1; \lambda) = \lambda A_Z^{(Z)}(0; \lambda) + 1.$$

*It will be called the **Euler polynomial**.*

In fact $A_Z(x; \lambda)$ is L-polynomial. Where necessary we will write $A_Z[\Lambda](x; \lambda)$ or $A_Z[\lambda_1, \lambda_2, \ldots, \lambda_{Z+1}](x; \lambda)$ in order to stress the dependence in the nonordered vector Λ.

Recalling the operators $\mathcal{D}_j$ defined in (13.6), p. 226, we can state the most important properties of the function $A(x; \lambda)$.

---

[8] We have taken the notation $A(x; \lambda)$ from Schoenberg.

Theorem 13.21 *The function $A_Z[\Lambda](x;\lambda)$ satisfies the following properties:*

1. $\begin{cases} \mathcal{L}_i A_Z(1;\lambda) = \lambda \mathcal{L}_i A_Z(0;\lambda) & \text{for } i = 0, 1, \ldots, Z-1, \text{ and} \\ \mathcal{L}_Z A_Z(1;\lambda) = \lambda \mathcal{L}_Z A_Z(0;\lambda) + 1. \end{cases}$
2. $\mathcal{D}_{Z+1} A_Z[\lambda_1, \lambda_2, \ldots, \lambda_{Z+1}](x;\lambda) = A_{Z-1}[\lambda_1, \lambda_2, \ldots, \lambda_Z](x;\lambda)$.
3. $A_Z(x;\lambda) = [\lambda_1, \lambda_2, \ldots, \lambda_{Z+1}]_z G(z; x, \lambda)$, the last being the divided difference with respect to the variable z.
4. $A_Z(x+1;\lambda) - \lambda A_Z(x;\lambda) = \phi_Z(x)$.

Exercise 13.22 *Prove properties (1) and (2).*
Hint: *Use induction as in the proof of Lemma 13.19, p. 233.*

Proof Let us prove properties (3) and (4).
Assuming for simplicity that $\lambda_j \neq \lambda_i$ for $j \neq i$, we see that the space U_{Z+1} is spanned by the exponentials $e^{\lambda_j x}$, hence, for some constants σ_j we have

$$A_Z(x;\lambda) = \sum_{j=1}^{Z+1} \sigma_j e^{\lambda_j x}.$$

Now the conditions in Definition 13.20 give

$$\sum_{j=1}^{Z+1} \sigma_j \lambda_j^i (e^{\lambda_j} - \lambda) = 0 \quad \text{for } i = 0, \ldots, Z-1,$$

$$\sum_{j=1}^{Z+1} \sigma_j \lambda_j^Z (e^{\lambda_j} - \lambda) = 1,$$

which is a linear system with respect to σ_j which has a determinant, multiple of the Vandermonde

$$\prod_{j=1}^{Z+1} (e^{\lambda_j} - \lambda) \cdot \det[\lambda_j^i]_{i=0, j=1}^{Z, Z+1} = \prod_{j=1}^{Z+1} (e^{\lambda_j} - \lambda) \cdot \prod_{i<j} (\lambda_i - \lambda_j).$$

The solution is given by

$$\sigma_j = \frac{1}{(e^{\lambda_j} - \lambda)} \cdot \frac{1}{q'_{Z+1}(\lambda_j)},$$

which proves

$$A_Z(x;\lambda) = \sum_{j=1}^{Z+1} \frac{1}{q'_{Z+1}(\lambda_j)} \cdot \frac{e^{\lambda_j x}}{(e^{\lambda_j} - \lambda)} \quad (13.10)$$

which is exactly (3). From this formula property (4) follows directly. ∎

Exercise 13.23 *Prove (3) by checking directly that all conditions of Definition 13.20, p. 233, are satisfied by the function $[\lambda_1, \lambda_2, \ldots, \lambda_{Z+1}]_z G(z; x, \lambda)$.*

We have the following useful corollary.

Corollary 13.24 *For every λ such that $\lambda \neq e^{\lambda_j}$ for all $j = 1, 2, \ldots, Z+1$, the function $A_Z(x; \lambda)$ permits the residuum representation*

$$A_Z(x; \lambda) = \frac{1}{2\pi i} \int_\Gamma \frac{1}{q_{Z+1}(z)} \frac{e^{xz}}{e^z - \lambda} dz, \tag{13.11}$$

where the closed contour Γ surrounds the zeros of $q_{Z+1}(z)$, i.e. all elements of $\Lambda = [\lambda_1, \lambda_2, \ldots, \lambda_{Z+1}]$, and excludes the zeros of the function $\exp(xz)/(e^z - \lambda)$.

The proof is due to the Frobenius representation of divided difference as residuum, see formula (11.12), p. 193.

Let us put

$$r(\lambda) := \prod_{j=1}^{Z+1}(e^{\lambda_j} - \lambda) = \sum_{j=0}^{Z+1} r_j \lambda^j, \tag{13.12}$$

$$s(\lambda) := \prod_{j=1}^{Z+1}(e^{-\lambda_j} - \lambda) = \sum_{j=0}^{Z+1} s_j \lambda^j. \tag{13.13}$$

We have the following important representation of the function $A_Z(x; \lambda)$.

Corollary 13.25 *The expression*

$$\Pi_Z(\lambda; x) := r(\lambda) A_Z(x; \lambda) \tag{13.14}$$

is a polynomial of degree $\leq Z$ in λ. The polynomial $\Pi_Z(\lambda; 0)$ has degree $\leq Z - 1$.

The polynomial $\Pi_Z(\lambda) := \Pi_Z(\lambda; 0)$ is known as the **Euler–Frobenius polynomial**.

Corollary 13.25 follows directly from formula (13.10). We obtain the representation

$$\Pi_Z(\lambda; x) = \phi_Z(x)\lambda^Z + \cdots + e^{\lambda_1 + \cdots + \lambda_{Z+1}} \phi_Z(x - 1).$$

We see that the points $\lambda = e^{\lambda_j}$ are singular for the function $A_Z(x; \lambda)$. However, the polynomials $\Pi_Z(\lambda; x)$ also make sense for such values of the parameter λ.

Proposition 13.26 *Assume that all $\lambda_s \in \Lambda$ are pairwise different. Then for every $x \in \mathbb{R}$ and for every $\lambda_s \in \Lambda$ we have the following equality:*

$$\Pi_Z(e^{\lambda_s}; x) = -r'(e^{\lambda_s}) \cdot \frac{e^{\lambda_s x}}{q'_{Z+1}(\lambda_s)}. \tag{13.15}$$

Proof By the definition of the function in formula (13.14) we obtain

$$\Pi_Z(\lambda; x) = r(\lambda) A_Z(x; \lambda) = \prod_{j=1}^{Z+1}(e^{\lambda_j} - \lambda) \cdot \sum_{j=1}^{Z+1} \frac{1}{q'_{Z+1}(\lambda_j)} \cdot \frac{e^{\lambda_j x}}{e^{\lambda_j} - \lambda}$$

$$= \sum_{j=1}^{Z+1} \frac{e^{\lambda_j x}}{q'_{Z+1}(\lambda_j)} \cdot \prod_{\ell=1, \ell \neq j}^{Z+1}(e^{\lambda_\ell} - \lambda). \tag{13.16}$$

Hence,
$$\Pi_Z(e^{\lambda_s}; x) = -r'(e^{\lambda_s}) \cdot \frac{e^{\lambda_s x}}{q'_{Z+1}(\lambda_s)},$$
which completes the proof. ∎

Exercise 13.27 *Find an expression for $\Pi_Z(\lambda; x)$ when the values of λ_j are not pairwise different as in equality (13.16). Consider the case when $Z+1 = 2p$ with $\lambda_1 = \cdots = \lambda_p$ and $\lambda_{p+1} = \cdots = \lambda_{2p}$.*

13.7 Generalized divided difference operator

In the general theory of Chebyshev splines presented in Section 11.2, p. 191, we have a divided difference operator which is not uniquely determined. It is important for the cardinal L-splines that the coefficients of the polynomials $r(\lambda)$ and $s(\lambda)$ determine an elegant expression for a *divided difference operator*.

Let us consider the polynomial
$$q^*_{Z+1}[\Lambda](z) := q^*_{Z+1}(z) := q_{Z+1}[-\Lambda](z) = \prod_{j=1}^{Z+1}(z + \lambda_j),$$
which evidently satisfies $q^*_{Z+1}(z) = (-1)^{Z+1} q_{Z+1}(-z)$ by the definition of $q^*_{Z+1}(z)$ in (13.2), p. 224.

Naturally, we will define by $U^*_{Z+1}[\Lambda]$ the space of C^∞-solutions of the equation
$$\mathcal{L}^*_{Z+1}[\Lambda] f := q^*_{Z+1}[\Lambda]\left(\frac{d}{dx}\right) f(x) = 0 \quad \text{for } x \text{ in } \mathbb{R}.$$

We have the following *divided difference operators* for cardinal L-splines.

Proposition 13.28 *If the coefficients r_j and s_j are those defined, respectively, in (13.12) and (13.13), p. 235, then for every function $f(x)$ in $U_{Z+1}[\Lambda]$*
$$\sum_{j=1}^{Z+1} r_j f(j) = 0. \tag{13.17}$$

Also, for every f in $U^*_{Z+1}[\Lambda]$
$$\sum_{j=1}^{Z+1} s_j f(j) = 0. \tag{13.18}$$

Proof For simplicity, we first assume that all λ_j are pairwise different. In such a case every solution to $\mathcal{L}_{Z+1} f(x) = 0$ is a linear combination of simple exponents
$$f(x) = \sum_{l=1}^{Z+1} \sigma_l e^{\lambda_l x}.$$

But for every $l = 1, \ldots, Z+1$

$$\sum_{j=1}^{Z+1} r_j e^{\lambda_l j} = \sum_{j=1}^{Z+1} r_j (e^{\lambda_l})^j = r(e^{\lambda_l}) = 0$$

holds, hence $\sum_{j=1}^{Z+1} r_j f(j) = 0$.

For arbitrary values of λ_j let us note that all the equalities above depend continuously on the parameters λ_j, $j = 1, \ldots, Z+1$. We proceed by perturbing the coinciding λ_js so that the perturbed values do not coincide and then we apply a limiting argument in all the above equalities. In a similar way we prove the second part, since the solutions of $\mathcal{L}^*_{Z+1}(d/dx) f(x) = 0$ are linear combinations of exponents of the type $e^{-\lambda_j x}$. ∎

13.8 Zeros of the Euler–Frobenius polynomial $\Pi_Z(\lambda)$

Recall that we have termed the polynomial

$$\Pi_Z(\lambda) := \Pi_Z(\lambda; 0)$$

Euler–Frobenius polynomial

From formula (13.10), p. 234, we see that the values $\lambda = e^{\lambda_i}$ are generally speaking singular for the function $A_Z(x; \lambda)$. Lemma 13.29 gives an answer to what happens if $\lambda = e^{\lambda_i}$ for some i. It plays a central role in solving the *cardinal interpolation problem* (13.5), p. 225.

Lemma 13.29 *Let $U_{Z+1}[\Lambda] = \{u_1, \ldots, u_{Z+1}\}_{\text{lin}}$.*
For any α with $0 \leq \alpha < 1$ the system of equations

$$y^{(i)}(1) = \lambda y^{(i)}(0) \quad \text{for } i = 0, \ldots, Z-1,$$
$$y(\alpha) = 0,$$

has a nontrivial solution y in U_{Z+1} if and only if $\lambda \neq e^{\lambda_i}$ for $i = 1, \ldots, Z+1$ and $A_Z(\alpha; \lambda) = 0$.

More precisely, the determinant of the above linear system with respect to the variables c_j, where $y(x) = \sum_{j=1}^{Z+1} u_j(x)$, is proportional to $A_Z(\alpha; \lambda)$.

Exercise 13.30 *Prove the above lemma. Hint: Follow a way similar to the one we used to obtain formula (13.10), p. 234.*

We will not prove the following fundamental theorem since its proof will not be necessary later in our study.

Theorem 13.31 *1. If $\lambda \geq 0$ and $\lambda \neq e^{\lambda_i}$ for $i = 1, \ldots, Z+1$, then as a function of x, $A_Z(x; \lambda)$ has no zeros in the interval $(0, 1)$. If $\lambda < 0$ then $A_Z(x; \lambda)$ has exactly one simple zero in the interval $[0, 1)$.*

2. Let us fix α with $0 < \alpha < 1$. Then as a function of λ, $A_Z(\alpha; \lambda)$ has exactly Z different zeros

$$\tau_1(\alpha) < \cdots < \tau_Z(\alpha) < 0$$

which interlace the zeros of $A_{Z-1}(\alpha, \lambda) = A_{Z-1}[\lambda_1, \ldots, \lambda_Z](\alpha, \lambda)$.

3. For $Z \geq 2$ the polynomial $\Pi_Z(\lambda) = r(\lambda)A_Z(0;\lambda)$ has exactly $Z-1$ negative zeros which interlace the $Z-2$ zeros

$$\tau_1(0) < \cdots < \tau_{Z-1}(0) < 0$$

of $A_{Z-1}(0;\lambda) = A_{Z-1}[\lambda_1,\ldots,\lambda_Z](0,\lambda)$.

Micchelli proves this theorem by applying a generalized Budan–Fourier-type result for the zeros of L-polynomials [13, pp. 210–211]. Schoenberg has provided a more elementary proof of the above result [19, p. 256, Theorems 1 and 2, pp. 258, Lemma 1].

13.9 The cardinal interpolation problem for L-splines

In view of the above results we see that for every α with $0 < \alpha < 1$ there exist precisely Z solutions of the zero interpolation problem (13.5), p. 225, i.e. elements of the space S_Z^0. They correspond to the different solutions of equation $A_Z(\alpha,\lambda) = 0$.

Proposition 13.32 *For every α satisfying $0 < \alpha < 1$ the dimension of S_Z^0 is exactly Z, while for $\alpha = 0$ it is $Z-1$.*

For the proof see the illuminating explanation by Schoenberg either in his book [18, Lecture 4, pp. 35, 36], or in his paper [19, p. 269].
We put

$$S_j(x) := A_Z(x, \tau_j(\alpha)) \quad \text{for } 0 \leq x \leq 1$$

and extend it for every x in $\mathbb{R}$ by means of the functional equation

$$S_j(x+1) = \tau_j(\alpha)S_j(x).$$

We proceed in a similar way for $\alpha = 0$ but there we use the $Z-1$ zeros of $A_Z(0;\lambda)$.

Let us note that all these elements have an exponential growth. Indeed, if $-1 < \tau_j(\alpha) < 0$ then due to

$$S_j(m) = \tau_j^m(\alpha)S_j(0) \quad \text{for all } m \text{ in } \mathbb{Z},$$

for all $m < 0$ we have an exponential growth for $m \to \infty$. If $\tau_j(\alpha) < -1$ then we obtain exponential growth for all $m > 0$ for $m \to -\infty$.

Let us denote by ξ the unique simple zero of $A_Z(x;-1)$ satisfying $0 \leq \xi < 1$. Thus, if $\alpha \neq \xi$ an element S_Z^0 of power growth does not exist.

This obtains the main result of the cardinal interpolation.

Theorem 13.33 *Let ξ be, as above, the unique zero of $A_Z(x;-1)$ in the interval $[0,1)$. Then for every α with $0 \leq \alpha < 1$ such that $\alpha \neq \xi$ and any bi-infinite sequence of power growth $\{y_j\}_{j=-\infty}^{\infty}$ there exists a unique spline $u(x)$ in S_Z of power growth for which*

$$u(\alpha + j) = y_j \quad \text{for all } j \text{ in } \mathbb{Z},$$

and $u(x)$ is given by the cardinal series

$$u(x) = \sum_{j=-\infty}^{\infty} y_j L(x-j).$$

Here $L(\cdot)$ is the fundamental cardinal L-spline of Theorem 13.2, p. 225.

13.10 The cardinal compactly supported L-splines Q_{Z+1}

One of the important features of Micchelli's approach to cardinal L-splines is the simple and natural way in which the cardinal TB-spline[9] functions are obtained, compared with the general construction of the TB-splines in Section 11.4.1, p. 201.

We will consider a more general situation by taking the mesh $h\mathbb{Z}$ instead of $\mathbb{Z}$. This will be particularly important when we study L-spline wavelets.

We assume as usual that the nonordered vector $\Lambda = [\lambda_1, \lambda_2, \ldots, \lambda_{Z+1}]$ of the real numbers is fixed. We will consider the *cardinal mesh*

$$h\mathbb{Z} := \{jh : \text{ all } j \text{ in } \mathbb{Z}\}$$

where we have taken some fixed number $h > 0$.[10] The reader may simplify the results below by putting $h = 1$.

Definition 13.34 *The (forward) TB-spline for the cardinal L-spline space $\mathcal{S}_Z(\Lambda)$ is defined by*

$$Q_{Z+1}(x) := Q_{Z+1}[\Lambda; h](x) := \sum_{j=0}^{Z+1} \phi_Z^+(x - jh) s_{j,h} \tag{13.19}$$

where

$$s_h(x) := s_h[\Lambda](x) := \prod_{j=1}^{Z+1} (e^{-\lambda_j h} - x) = \sum_{j=0}^{Z+1} s_{j,h} x^j. \tag{13.20}$$

The notation $Q_{Z+1} = Q_{Z+1}[\Lambda; h] = Q[\Lambda; h]$ will be used on equal rights depending on what we want to emphasize. Obviously the notation $Q_{Z+1}[\Lambda; h]$ is redundant since Λ will normally have $Z+1$ elements, but if this is not the case we will use this notation.

In the case $Z = 0$ we obtain

$$s_h(x) = e^{-\lambda_1 h} - x = s_0 + s_1 x, \quad \text{with}$$

$$s_0 = e^{-\lambda_1 h}, \quad s_1 = -1.$$

---

[9] The notion TB-spline is the generalization of the polynomial B-spline. It means an L-spline with a minimal compact support.

[10] Such a cardinal mesh is considered by Schoenberg [19]. The case considered by Micchelli [12, 13] is that of $h = 1$.

240 Multivariate polysplines

By the properties of $\phi_Z(x)$ in Proposition 13.12, p. 229, we have
$$\phi_0(x) = e^{\lambda_1 x},$$
hence,
$$Q_1(x) = e^{-\lambda_1 h} e^{\lambda_1 x} \chi_{[0,h]}(x), \qquad (13.21)$$
where $\chi_{[0,h]}(x)$ is the characteristic function of the interval $[0, h]$, i.e. by definition
$$\chi_{[0,h]}(x) := \begin{cases} 1 & \text{for } 0 \leq x \leq h, \\ 0 & \text{elsewhere.} \end{cases} \qquad (13.22)$$

Here is the most central result of classical spline theory but formulated in our L-spline setting (see the case of general Chebyshev splines in Theorem 11.29, p. 201):

Proposition 13.35 *1. The spline $Q_{Z+1}(x) = Q_{Z+1}[\Lambda; h](x)$ defined by formula (13.19) is a T B-spline for the operator $\mathcal{L}_{Z+1}[\Lambda]$ on the mesh $h\mathbb{Z}$, i.e. it is a nonnegative function, has a minimal compact support in the sense that no (nonzero) L-spline with smaller support exists, and it is the unique L-spline up to a multiplicative constant with support $[0, Ph + h]$. The above means that*
$$\begin{cases} Q_{Z+1}(x) > 0 & \text{for } 0 < x < Ph + h, \\ Q_{Z+1}(x) = 0 & \text{for } x \leq 0 \quad \text{or} \quad x \geq Ph + h. \end{cases}$$

Proof We prove only that the support of $Q_{Z+1}(x)$ is contained in the interval $[0, Ph + h]$.

Assuming for simplicity that all values of λ_j are pairwise different, by the definition of the function $\phi_Z(x)$ in formula (13.8), p. 229, we obtain
$$\phi_Z(x - jh) = \sum_{l=1}^{Z+1} \frac{1}{q'_{Z+1}(\lambda_l)} e^{\lambda_l (x - jh)},$$
hence, for $x \geq Z + 1$ we have
$$Q_{Z+1}(x) = \sum_{j=0}^{Z+1} \phi_Z(x - jh) s_j$$
$$= \sum_{j=0}^{Z+1} \left(\sum_{l=1}^{Z+1} \frac{1}{q'_{Z+1}(\lambda_l)} e^{\lambda_l (x - jh)} \right) s_j$$
$$= \sum_{l=1}^{Z+1} \frac{1}{q'_{Z+1}(\lambda_l)} e^{\lambda_l x} \left(\sum_{j=0}^{Z+1} e^{-\lambda_l jh} s_j \right)$$
$$= \sum_{l=1}^{Z+1} \frac{1}{q'_{Z+1}(\lambda_l)} e^{\lambda_l x} \cdot s_h(e^{-\lambda_l h}) = 0.$$

Thus for $x \leq 0$, $Q_{Z+1}(x) = 0$ follows immediately from the definition of the function $\phi_Z^+(x)$. ∎

Exercise 13.36 *Prove the above result by using the residuum representation of $\phi_Z^+(x)$ in formula (13.7), p. 229.*

Exercise 13.37 *This is another classic result in spline theory. Prove the minimality of the support for $Q_{Z+1}(x)$ stated in Proposition 13.35.*

Hint: Recall that the dimension of U_{Z+1} is $Z+1$ and check that the number of smoothness conditions on $Q_{Z+1}(x)$ (note $Q_{Z+1}(x)$ belongs to $C^{Z-1}(\mathbb{R})$) is $(Z+2)Z$. Count the dimensions.

Theorem 13.38 is the most basic of all results about compactly supported spline, (for the general Chebyshev splines see Theorem 11.30, p. 202).

Theorem 13.38 *Take for simplicity $h = 1$.*
1. No element of the set of shifts

$$\{Q_{Z+1}(x-j): \text{ for all } j \text{ in } \mathbb{Z}\}$$

is a finite linear combination of the others.
2. Denote by $\mathcal{S}_Z(\Lambda)[a,b]$ the space of cardinal L-splines in $\mathcal{S}_Z(\Lambda)$ which have their support only in the interval $[a,b]$ where a, b are two integers. Then the set of shifts

$$\{Q_{Z+1}(x-j): \text{ for } j = a-Z, \ldots, b+Z\}$$

forms a linear basis of $\mathcal{S}_Z(\Lambda)[a,b]$. All elements in this set of shifts are linearly independent.

13.11 Laplace and Fourier transform of the cardinal TB-spline Q_{Z+1}

Since the function $\phi_Z^+(x)$ has no compact support we may not consider its Fourier transform in the classical sense. On the other hand the function $Q_{Z+1}(x) = Q_{Z+1}[\Lambda; h](x)$ is a linear combination of shifts (integer translates) of $\phi_Z^+(x)$ but has a compact support and for that reason its Fourier transform is defined in a classical sense. Because of this we first compute the *Laplace transform* $\mathfrak{L}[\phi_Z^+](z)$ which makes sense for some subdomain of the complex plane and after that we extend by analytical argument the formula obtained for $\mathfrak{L}[Q_{Z+1}](z)$. Then we use the fact that the *Fourier transform* is obtained through the *Laplace transform* at the point $z = i\xi$.

Proposition 13.39 *The Laplace transform of the function Q_{Z+1} is given by*

$$\mathfrak{L}[Q_{Z+1}[\Lambda; h]](z) = \int_{-\infty}^{\infty} Q_{Z+1}(x) e^{-xz} dx = \frac{s_h(e^{-zh})}{q_{Z+1}(z)} = \frac{\prod_{j=1}^{Z+1}(e^{-\lambda_j h} - e^{-zh})}{\prod_{j=1}^{Z+1}(z - \lambda_j)}$$

for every complex number $z \in \mathbb{C}$. The Fourier transform is, respectively,

$$\widehat{Q_{Z+1}[\Lambda; h]}(\xi) = \mathfrak{L}[Q_{Z+1}](i\xi) = \frac{s_h(e^{-i\xi h})}{q_{Z+1}(i\xi)} = \frac{\prod_{j=1}^{Z+1}(e^{-\lambda_j h} - e^{-i\xi h})}{\prod_{j=1}^{Z+1}(i\xi - \lambda_j)}. \quad (13.23)$$

Proof Assuming for simplicity that all λ_js are pairwise different, we can easily see that

$$\mathcal{L}[\phi_Z^+(x)](z) = \int_{-\infty}^{\infty} \phi_Z^+(x) e^{-xz}\, dx \tag{13.24}$$

$$= \frac{1}{q_{Z+1}(z)} \quad \text{for Re } z > \max_{j=1,\ldots,Z+1} \lambda_j,$$

which follows directly from the representation of the function $\phi_Z^+(x)$, formula (13.8), p. 229. Indeed, we have

$$\int e^{\lambda_j x} e^{-zx}\, dx = \frac{-1}{\lambda_j - z} \quad \text{for Re } z > \max_{j=1,\ldots,Z+1} \lambda_j$$

and

$$\int_{-\infty}^{\infty} \phi_Z^+(x) e^{-xz}\, dx = \sum_{j=1}^{Z+1} \frac{1}{q'_{Z+1}(\lambda_j)} \frac{1}{z - \lambda_j} = \frac{1}{q_{Z+1}(z)}.$$

The last equality is a standard result in the representation of a rational polynomial through simple fractions, easily checked by multiplying with $(z - \lambda_j)$ and substituting $z = \lambda_j$ thereafter.

Hence, by the standard properties of the *Laplace* transform we obtain

$$\int_{-\infty}^{\infty} Q_{Z+1}(x) e^{-xz}\, dx = \left(\sum_{j=0}^{Z+1} s_j e^{-zjh} \right) \cdot \int_{-\infty}^{\infty} \phi_Z^+(x) e^{-xz}\, dx = \frac{s_h(e^{-zh})}{q_{Z+1}(z)}$$

which is now true for every complex number $z \in \mathbb{C}$ since $Q_{Z+1}(x)$ has a compact support and we extend the right-hand side analytically. For coinciding λ_js the result follows by a continuity argument. ∎

From the above there immediately follows a relationship between the TB-spline for different $h\mathbb{Z}$, which shows that we may reduce the study of $Q_{Z+1}[\Lambda; h]$ to the case of $h = 1$. For that reason it makes sense to introduce a simplified notation for $h = 1$, namely

$$Q_{Z+1}[\Lambda](x) := Q_{Z+1}[\Lambda; 1](x). \tag{13.24a}$$

By formula (13.23) we obtain, through simple transformations,

$$\widehat{Q_{Z+1}[\Lambda; h]}(\xi) = \frac{\prod_{j=1}^{Z+1}(e^{-\lambda_j h} - e^{-i\xi h})}{\prod_{j=1}^{Z+1}(i\xi - \lambda_j)} \tag{13.25}$$

$$= h^{Z+1} \frac{\prod_{j=1}^{Z+1}(e^{-\lambda_j h} - e^{-i\xi h})}{\prod_{j=1}^{Z+1}(ih\xi - h\lambda_j)}$$

$$= h^{Z+1} \cdot \widehat{Q_{Z+1}[h\Lambda]}(h\xi).$$

Taking the inverse Fourier transform we obtain Proposition 13.40.

Proposition 13.40 *The TB-spline $Q_{\mathbb{Z}+1}[h\Lambda]$ on the mesh $\mathbb{Z}$ and the TB-spline $Q_{\mathbb{Z}+1}[\Lambda; h](x)$ on the mesh $h\mathbb{Z}$ are related by the equality:*

$$Q_{\mathbb{Z}+1}[\Lambda; h](x) = h^Z \cdot Q_{\mathbb{Z}+1}[h\Lambda]\left(\frac{x}{h}\right). \tag{13.26}$$

It should be noted that the last function has singularities at hj, $j \in \mathbb{Z}$, since the function $Q_{\mathbb{Z}+1}[h\Lambda](y)$ has singularities at $j \in \mathbb{Z}$.

Exercise 13.41 *Prove Proposition 13.40 by using the residuum representation of $\phi_{\mathbb{Z}}(x)$ in formula (13.7), p. 229.*

13.12 Convolution formula for cardinal *TB*-splines

Here we will prove an important generalization of the inductive convolution formula known for the polynomial cardinal splines, see Schoenberg [18, p. 12, formula (1.9)], which we have mentioned in the introduction as formula (13.1), p. 220.

Assume that we are given the nonordered vector $\Lambda = [\lambda_1, \ldots, \lambda_N]$ and let us denote by m_j the number of entries in Λ for the number λ_j, i.e. m_j is the multiplicity of λ_j.[11] As above we denote by $Q[\Lambda](x) = Q_{N+1}[\Lambda](x)$ the L–spline on $\mathbb{Z}$ which corresponds, according to formula (13.19), p. 239, to the set Λ. ". We denote by $Q[\Lambda](x) = Q_{N+1}[\Lambda](x)$ the L-spline which corresponds according to formula (13.19), p. 239, to the set Λ. Here we drop the subindex $N + 1$ of Q as inessential for the present consideration.

The Fourier transform of $Q[\Lambda](x)$ which we have by formula (13.23), p. 241, is equal to

$$\widehat{Q[\Lambda]}(\xi) = \frac{\prod_{j=1}^{N}(e^{-\lambda_j} - e^{-i\xi})}{\prod_{j=1}^{N}(i\xi - \lambda_j)}.$$

Assume that a *subdivision* of the nonordered vector Λ be given, i.e. two other nonordered vectors Λ_1 and Λ_2 are determined by $\Lambda_1 = [u_1, \ldots, u_{N_1}]$ and $\Lambda_2 = [v_1, \ldots, v_{N_2}]$ with $N = N_1 + N_2$, and the number of entries of λ_j in Λ_1 plus the number of entries of λ_j in Λ_2 is equal to m_j. Evidently, we have

$$\widehat{Q[\Lambda]}(\xi) = \widehat{Q[\Lambda_1]}(\xi) \cdot \widehat{Q[\Lambda_2]}(\xi).$$

By taking the inverse Fourier transform and using a basic property of the Fourier transform, namely to convert the convolution between two functions into their product, see (12.7), p. 212, we obtain

$$\widehat{Q[\Lambda_1] * Q[\Lambda_2]}(\xi) = \widehat{Q[\Lambda_1]}(\xi) \cdot \widehat{Q[\Lambda_2]}(\xi).$$

This completes the proof.

---

[11] See the conventions about non–ordered vectors Λ on p. 223.

Proposition 13.42 *If the sets Λ_1, Λ_2 and Λ are defined as above then the corresponding TB-splines satisfy the following equality:*

$$Q[\Lambda](x) = Q[\Lambda_1](x) * Q[\Lambda_2](x).$$

In particular,

$$Q[\mu, \lambda_1, \ldots, \lambda_s](x) = e^{-\mu h} \cdot \int_0^h Q[\lambda_1, \ldots, \lambda_s](x-y) \cdot e^{\mu y} dy, \quad (13.27)$$

and

$$Q[\lambda_1, \ldots, \lambda_s](x) = Q[\lambda_1](x) * Q[\lambda_2](x) * \cdots * Q[\lambda_N](x). \quad (13.28)$$

Let us recall that in the last equality $Q[\lambda_j](x)$ is the TB-spline corresponding to the vector $\Lambda = [\lambda_j]$ which has a unique element and which by formula (13.21), p. 240, is given by

$$Q[\lambda_j](x) = e^{-\lambda_j h} e^{\lambda_j x} \chi_{[0,h]}(x)$$

or in the case of the mesh $\mathbb{Z}$ is given by $Q[\lambda_j](x) = e^{-\lambda_j} e^{\lambda_j x} \chi_{[0,1]}(x)$. It has the Fourier transform

$$\widehat{Q[\lambda_j]}(\xi) = e^{-\lambda_j h} \int_0^h e^{\lambda_j x} e^{-i\xi x} dx = e^{-\lambda_j h} \cdot \frac{e^{(\lambda_j - i\xi)h} - 1}{\lambda_j - i\xi}$$

which coincides with the general formula (13.23), p. 241.

13.13 Differentiation of cardinal TB-splines

We now prove Theorem 13.43 by means of the convolution formula for different order TB-splines.

Theorem 13.43 *If we use the notation for the TB-spline as in (13.19), p. 239, for the mesh $h\mathbb{Z}$, then the following formula holds:*

$$\left(\frac{d}{dx} - \mu\right) Q[\mu, \lambda_1, \ldots, \lambda_s](x)$$

$$= -e^{-\mu h}(Q[\lambda_1, \ldots, \lambda_s](x-h) \cdot e^{\mu h} - Q[\lambda_1, \ldots, \lambda_s](x))$$

$$= e^{-\mu h} Q[\lambda_1, \ldots, \lambda_s](x) + Q[\lambda_1, \ldots, \lambda_s](x-h).$$

Proof In formula (13.27), p. 244, we have proved that

$$Q[\mu, \lambda_1, \ldots, \lambda_s](x) = e^{-\mu h} \cdot \int_0^h Q[\lambda_1, \ldots, \lambda_s](x-y) \cdot e^{\mu y} dy.$$

Let us differentiate it. We obtain after integration by parts, and using the fact that $(d/dx)g(x - y) = -(d/dy)g(x - y)$, the following equality:

$$\frac{d}{dx}Q[\mu, \lambda_1, \ldots, \lambda_s](x)$$

$$= -e^{-\mu h} \cdot \int_0^h \frac{d}{dy} Q[\lambda_1, \ldots, \lambda_s](x - y) \cdot e^{\mu y}\, dy$$

$$= -e^{-\mu h} \left(Q[\lambda_1, \ldots, \lambda_s](x - y) \cdot e^{\mu y}\Big|_{y=0}^{y=h} - \mu \int_0^h Q[\lambda_1, \ldots, \lambda_s](x - y) \cdot e^{\mu y}\, dy \right)$$

$$= -e^{-\mu h}(Q[\lambda_1, \ldots, \lambda_s](x - h) \cdot e^{\mu h} - Q[\lambda_1, \ldots, \lambda_s](x))$$

$$+ \mu \int_0^h Q[\mu, \lambda_1, \ldots, \lambda_s](x - y) \cdot e^{\mu y}\, dy,$$

which completes the proof. ∎

13.14 Hermite–Gennocchi-type formula

We may easily derive an analog to the classical Hermite–Gennocchi formula [2, p. 9].

In order to be able to apply the Fourier transform we have to work at least temporarily with functions in $L_2(\mathbb{R})$. For an arbitrary function $f \in L_{1,\text{loc}}(\mathbb{R}) \cap L_2(\mathbb{R})$ let us consider

$$I := \int_{-\infty}^{\infty} f(x) Q[\Lambda](x)\, dx.$$

Using the Parseval identity (12.5), p. 212, and the convolution formula (13.28), p. 244, we obtain

$$I = \frac{1}{2\pi} \int_{-\infty}^{\infty} \widehat{f}(\xi) \cdot \overline{\widehat{Q[\Lambda]}(\xi)}\, d\xi$$

$$= \frac{1}{2\pi} \int_{-\infty}^{\infty} \widehat{f}(\xi) \cdot \overline{\widehat{Q[\lambda_1]}(\xi)} \cdot \overline{\widehat{Q[\lambda_2]}(\xi)} \cdot \ldots \cdot \overline{\widehat{Q[\lambda_N]}(\xi)}\, d\xi$$

$$= \frac{1}{2\pi} \int_{-\infty}^{\infty} \widehat{f}(\xi) \cdot \left\{ \prod_{j=1}^{N} e^{-\lambda_j h} \right\} \cdot \left\{ \prod_{j=1}^{N} \int_0^h e^{\lambda_j x} e^{i\xi x}\, dx \right\} d\xi$$

$$= \frac{1}{2\pi} \left\{ \prod_{j=1}^{N} e^{-\lambda_j h} \right\} \cdot \underbrace{\int_0^h \cdots \int_0^h}_{N} \int_{-\infty}^{\infty} \widehat{f}(\xi) \cdot \prod_{j=1}^{N} e^{\lambda_j x_j} e^{i\xi x_j}\, dx_1 \cdots dx_N\, d\xi$$

$$= \frac{1}{2\pi} \left\{ \prod_{j=1}^{N} e^{-\lambda_j h} \right\} \cdot \underbrace{\int_0^h \cdots \int_0^h}_{N} \prod_{j=1}^{N} e^{\lambda_j x_j} \cdot \int_{-\infty}^{\infty} \widehat{f}(\xi) \cdot e^{i\xi(x_1 + \cdots + x_N)}\, d\xi\, dx_1 \cdots dx_N.$$

246   *Multivariate polysplines*

Further we apply the inverse Fourier transform $\mathcal{F}^{-1}$, see (12.4), p. 212. By formula (12.6), p. 212, it has the property that $\mathcal{F}^{-1}\mathcal{F} = id$, which implies the equality:

$$I = \left\{ \prod_{j=1}^{N} e^{-\lambda_j h} \right\} \cdot \underbrace{\int_0^h \cdots \int_0^h}_{N} \prod_{j=1}^{N} e^{\lambda_j x_j} \cdot f(x_1 + \cdots + x_N) \, dx_1 \cdots dx_N.$$

Let us note that both sides also make sense for functions in $L_{1,\mathrm{loc}}(\mathbb{R})$, and by the approximation argument we may prove it for all functions in $L_{1,\mathrm{loc}}(\mathbb{R})$.

Thus we have proved the *generalized Hermite–Gennocchi formula*[12].

Theorem 13.44 *If the nonordered vector* $\Lambda = [\lambda_1, \ldots, \lambda_N]$ *is given and the function* f *belongs to* $L_{1,\mathrm{loc}}(\mathbb{R})$ *(i.e. f belongs to $L_1(a,b)$ for every finite interval (a,b)). Then the corresponding TB-spline $Q[\Lambda](x)$ defined on the mesh $h\mathbb{Z}$ satisfies the identity*

$$\int_{-\infty}^{\infty} f(x) Q[\Lambda](x) \, dx$$

$$= \left\{ \prod_{j=1}^{N} e^{-\lambda_j h} \right\} \cdot \underbrace{\int_0^h \cdots \int_0^h}_{N} \prod_{j=1}^{N} e^{\lambda_j x_j} \cdot f(x_1 + \cdots + x_N) \, dx_1 \cdots dx_N. \quad (13.29)$$

Exercise 13.45 *Recall that the left-hand side of equality (13.29) is equal to the divided difference of a function g for which $\mathcal{L}^*_{P+1} g = f$, which will be proved in Theorem 13.59, p. 258. Combining both results we obtain the equality which is usually known as the* Hermite–Gennocchi *formula. Prove the above result for noncardinal L-splines when $Q[\Lambda](x)$ is the corresponding compactly supported TB-spline without using the Fourier transform.*

13.15   Recurrence relation for the *TB*-spline

As an application of the Fourier transform of the TB-spline $Q[\Lambda](x)$ we may easily prove a recurrence relation which expresses the values of $Q[\Lambda]$ through values of lower order TB-splines.[14][13]

---

[12] This result has been proved by Dyn and Ron [7].
[13] This result was first proved by Dyn and Ron [6, 7].

Theorem 13.46 *If $\lambda_1 \neq \lambda_{Z+1}$ then the following recurrence relation holds:*

$$Q[\lambda_1, \lambda_2, \ldots, \lambda_{Z+1}](x) = \frac{e^{-\lambda_{Z+1}}}{\lambda_1 - \lambda_{Z+1}} Q[\lambda_2, \ldots, \lambda_{Z+1}](x)$$

$$+ \frac{-e^{-\lambda_1}}{\lambda_1 - \lambda_{Z+1}} Q[\lambda_1, \ldots, \lambda_Z](x)$$

$$+ \frac{-1}{\lambda_1 - \lambda_{Z+1}} Q[\lambda_2, \ldots, \lambda_{Z+1}](x-1)$$

$$+ \frac{1}{\lambda_1 - \lambda_{Z+1}} Q[\lambda_1, \ldots, \lambda_Z](x-1). \quad (13.30)$$

Proof By assumption $\lambda_1 \neq \lambda_{Z+1}$. We will be looking for the constants in the equality

$$Q[\lambda_1, \lambda_2, \ldots, \lambda_{Z+1}](x) = C_1 Q[\lambda_2, \ldots, \lambda_{Z+1}](x) + C_2 Q[\lambda_1, \ldots, \lambda_Z](x)$$
$$+ C_3 Q[\lambda_2, \ldots, \lambda_{Z+1}](x-1) + C_4 Q[\lambda_1, \ldots, \lambda_Z](x-1).$$

We carry out some algebraic operations. First, we take the Fourier transform on both sides and obtain

$$\frac{\prod_{j=1}^{Z+1}(e^{-\lambda_j} - e^{-i\xi})}{\prod_{j=1}^{Z+1}(i\xi - \lambda_j)}$$

$$= C_1 \frac{\prod_{j=2}^{Z+1}(e^{-\lambda_j} - e^{-i\xi})}{\prod_{j=2}^{Z+1}(i\xi - \lambda_j)} + C_2 \frac{\prod_{j=1}^{Z}(e^{-\lambda_j} - e^{-i\xi})}{\prod_{j=1}^{Z}(i\xi - \lambda_j)}$$

$$+ C_3 e^{-i\xi} \frac{\prod_{j=2}^{Z+1}(e^{-\lambda_j} - e^{-i\xi})}{\prod_{j=2}^{Z+1}(i\xi - \lambda_j)} + C_4 e^{-i\xi} \frac{\prod_{j=1}^{Z}(e^{-\lambda_j} - e^{-i\xi})}{\prod_{j=1}^{Z}(i\xi - \lambda_j)}.$$

Then we divide the last by $\left(\prod_{j=2}^{Z}(e^{-\lambda_j} - e^{-i\xi})\right) \Big/ \left(\prod_{j=2}^{Z}(i\xi - \lambda_j)\right)$, after putting

$$z = e^{-i\xi}, \quad T_1 = e^{-\lambda_1}, \quad T_2 = e^{-\lambda_{Z+1}},$$

we obtain

$$\frac{(T_1 - z)(T_2 - z)}{(i\xi - \lambda_1)(i\xi - \lambda_{Z+1})}$$
$$= \frac{(C_1 + C_3 z)(i\xi - \lambda_{Z+1})(T_1 - z) + (C_2 + C_4 z)(i\xi - \lambda_1)(T_2 - z)}{(i\xi - \lambda_1)(i\xi - \lambda_{Z+1})}.$$

By comparing both sides as polynomials of z we obtain the system:

$$C_1(i\xi - \lambda_{Z+1})T_1 + C_2(i\xi - \lambda_1)T_2 = T_1 T_2,$$
$$-C_1(i\xi - \lambda_{Z+1}) + C_3(i\xi - \lambda_{Z+1})T_1 - C_2(i\xi - \lambda_1) + C_4(i\xi - \lambda_1)T_2 = -(T_1 + T_2),$$
$$-C_3(i\xi - \lambda_{Z+1}) - C_4(i\xi - \lambda_1) = 1.$$

Now we compare the coefficients in front of the variable ξ in the first and the third equations which gives

$$C_1 T_1 + C_2 T_2 = 0,$$
$$C_3 + C_4 = 0,$$

which again gives, using the first and the third equations, the solution

$$C_1 = \frac{T_2}{\lambda_1 - \lambda_{Z+1}} = \frac{e^{-\lambda_{Z+1}}}{\lambda_1 - \lambda_{Z+1}},$$

$$C_2 = -\frac{C_1 T_1}{T_2} = -\frac{-T_1}{\lambda_1 - \lambda_{Z+1}} = \frac{-e^{-\lambda_1}}{\lambda_1 - \lambda_{Z+1}},$$

$$C_3 = \frac{-1}{\lambda_1 - \lambda_{Z+1}},$$

$$C_4 = \frac{1}{\lambda_1 - \lambda_{Z+1}}.$$

One checks directly that these constants also satisfy the second equation, hence they solve the above system. ∎

13.16 The adjoint operator $\mathcal{L}^*_{Z+1}$ and the TB-spline $Q^*_{Z+1}(x)$

In Section 13.7, p. 236, we introduced the adjoint polynomial q^*_{Z+1} and the adjoint operator $\mathcal{L}^*_{Z+1}$ for the purposes of the generalized divided difference operators. Here we will need them again for defining the adjoint TB-spline. It is helpful to work with the formally adjoint operator times $(-1)^{Z+1}$

$$\mathcal{L}^*_{Z+1}[\Lambda]\left(\frac{d}{dx}\right) := \mathcal{L}_{Z+1}[-\Lambda]\left(\frac{d}{dx}\right) := \prod_{j=1}^{Z+1}\left(\frac{d}{dx} + \lambda_j\right) \tag{13.31}$$

with the polynomial

$$q^*_{Z+1}(\lambda) := \prod_{j=1}^{Z+1}(\lambda + \lambda_j) = (-1)^{Z+1} q_{Z+1}(-\lambda). \tag{13.32}$$

The corresponding TB-spline on the mesh $h\mathbb{Z}$ is given by

$$Q^*_{Z+1}(x) := (-1)^{Z+1} \sum_{j=0}^{Z} \phi^+_Z(jh - x) r_j \tag{13.33}$$

where we have put, as in (13.12), p. 235,

$$r_h(x) := \sum_{j=0}^{Z+1} r_{j,h} x^j = \prod_{j=1}^{Z+1} (e^{\lambda_j h} - x).$$

We will drop the second index and write r_j instead of $r_{j,h}$ if the context allows.

Proposition 13.47 *The polynomials $r_h(x)$ and $s_h(x)$ are related through the equality*

$$x^{Z+1} r_h\left(\frac{1}{x}\right) = (-1)^{Z+1} e^{h(\lambda_1 + \cdots + \lambda_{Z+1})} \cdot s_h(x). \tag{13.34}$$

Proof We have evidently

$$x^{Z+1} r_h\left(\frac{1}{x}\right) = x^{Z+1} \prod_{j=1}^{Z+1} \left(e^{\lambda_j h} - 1/x\right) = \prod_{j=1}^{Z+1} (x e^{\lambda_j h} - 1)$$

$$= \exp\left(h\left(\sum_{j=1}^{Z+1} \lambda_j\right)\right) \cdot \prod_{j=1}^{Z+1} (x - e^{-\lambda_j h})$$

which proves the statement. ∎

Due to the properties of ϕ_Z^+ proved in Proposition 13.12, p. 229, one may prove Proposition 13.48.

Proposition 13.48 *The following equality holds:*

$$Q_{Z+1}^*(Zh + h - x) = e^{(\lambda_1 + \cdots + \lambda_{Z+1})h} \cdot Q_{Z+1}(x). \tag{13.35}$$

Proof It is clear from

$$Q_{Z+1}^*(Zh + h - x) = (-1)^{Z+1} \sum_{j=0}^{Z+1} \phi_Z^+(x - (Z + 1 - j)h) r_j$$

that $Q_{Z+1}^*(Zh + h - x)$ is an L-spline for the operator $\mathcal{L}_{Z+1}$ with a support in the half-axis $x \geq 0$. By the definition of the polynomial $r(x)$, and by Proposition 13.35, p. 240, applied to the operator $\mathcal{L}_{Z+1}^*[\Lambda] = \mathcal{L}_{Z+1}[-\Lambda]$, it follows that the support of $Q_{Z+1}^*(Zh + h - x)$ coincides with the interval $[0, Zh + h]$. By the uniqueness of such a TB-spline, which we proved in Proposition 13.35 it follows that $Q_{Z+1}^*(Zh + h - x)$ and $Q_{Z+1}(x)$ are proportional, i.e. for some constant

$$Q_{Z+1}^*(Zh + h - x) = C Q_{Z+1}(x).$$

In order to obtain this constant it suffices to check this equality for $x = h$.
By the definition of the functions Q_{Z+1} and ϕ_Z^+ we have

$$Q_{Z+1}(h) = \phi_Z^+(h) s_{0,h},$$

$$Q_{Z+1}^* = (-1)^{Z+1} \phi_Z^+(h) r_{Z+1,h}.$$

It follows that
$$C = \frac{(-1)^{Z+1} r_{Z+1,h}}{s_{0,h}}.$$

From the definition of the polynomials $s_h(\lambda)$ and $r_h(\lambda)$ we see directly that
$$s_{0,h} = e^{-(\lambda_1 + \cdots + \lambda_{Z+1})h},$$
$$r_{Z+1,h} = (-1)^{Z+1},$$

which completes the proof. ∎

13.17 The Euler polynomial $A_Z(x; \lambda)$ and the TB-spline $Q_{Z+1}(x)$

For simplicity we consider only the case $h = 1$. The function $A_Z(x; \lambda)$ and the TB-splines Q^*_{Z+1} are related by Proposition 13.49.

Proposition 13.49 *The following equality holds:*
$$A_Z(x; \lambda) = (-1)^Z \frac{\sum_{j=0}^{Z} Q^*_{Z+1}(j+1-x)\lambda^j}{r(\lambda)}. \qquad (13.36)$$

Proof Let us make the direct expansion
$$G(z) = \frac{e^{xz}}{e^z - \lambda} = \sum_{j=0}^{\infty} \lambda^j e^{-(j+1-x)z},$$

hence, by the definition of the function ϕ_Z it follows that:
$$A_Z(x; \lambda) = [\lambda_1, \lambda_2, \ldots, \lambda_{Z+1}]_z G(z)$$
$$= \sum_{j=0}^{\infty} \lambda^j \phi_Z(x - j - 1)$$
$$= \frac{\left(\sum_{j=0}^{\infty} \lambda^j \phi_Z(x - j - 1)\right) \cdot \left(\sum_{j=0}^{Z+1} r_j \lambda^j\right)}{r(\lambda)}$$
$$= (-1)^Z \cdot \frac{\sum_{j=0}^{Z} Q^*_{Z+1}(j+1-x)\lambda^j}{r(\lambda)}$$

which completes the proof. ∎

We now obtain an important representation of the *Euler–Frobenius* polynomials $\Pi_Z(\lambda; 0)$.

Proposition 13.50 We have the following symmetric representation of the polynomial $\Pi_Z(\lambda; 0)$:

$$\Pi_Z(\lambda; 0) = (-1)^Z \cdot \lambda^{(Z-1)/2} \cdot e^{\lambda_1 + \cdots + \lambda_{Z+1}} \cdot \sum_{l=-(Z-1)/2}^{(Z-1)/2} Q_{Z+1}\left(\frac{Z+1}{2} - l\right) \lambda^l. \tag{13.37}$$

Proof By formula (13.14), p. 235, we obtain

$$\Pi_Z(\lambda; x) := r(\lambda) A_Z(x; \lambda) = (-1)^Z \cdot \sum_{j=0}^{Z} Q^*_{Z+1}(j + 1 - x) \lambda^j.$$

Further we use formula (13.35), i.e., $Q^*_{Z+1}(Zh + h - x) = e^{(\lambda_1 + \cdots + \lambda_{Z+1})h} \cdot Q_{Z+1}(x)$. After putting

$$\frac{Z-1}{2} - j = l$$

we obtain the following equalities:

$$\Pi_Z(\lambda; 0) = (-1)^Z \cdot \sum_{j=0}^{Z} Q^*_{Z+1}(j+1) \lambda^j$$

$$= (-1)^Z \cdot e^{\lambda_1 + \cdots + \lambda_{Z+1}} \cdot \sum_{j=0}^{Z-1} Q_{Z+1}(Z - j) \lambda^j$$

$$= (-1)^Z \cdot e^{\lambda_1 + \cdots + \lambda_{Z+1}} \sum_{j=0}^{Z-1} Q_{Z+1}\left(\frac{Z-1}{2} - j + \frac{Z+1}{2}\right) \lambda^j$$

$$= (-1)^Z \cdot \lambda^{Z-1/2} \cdot e^{\lambda_1 + \cdots + \lambda_{Z+1}} \cdot \sum_{l=-(Z-1)/2}^{(Z-1)/2} Q_{Z+1}\left(\frac{Z+1}{2} + l\right) \lambda^{-l}. \tag{13.38}$$

Since the function $Q_{Z+1}(x + (Z+1)/2)$ is symmetrized around zero in the sense that its support is the interval $[-(Z+1)/2, (Z+1)/2]$ it follows that

$$\Pi_Z(\lambda; 0) = (-1)^Z \cdot \lambda^{(Z-1)/2} \cdot e^{\lambda_1 + \cdots + \lambda_{Z+1}} \cdot \sum_{l=-(Z-1)/2}^{(Z-1)/2} Q_{Z+1}((Z+1)/2 - l) \lambda^l,$$

which completes the proof. ∎

Note that in the polynomial case we have $Q_{Z+1}((Z+1)/2 - \ell) = Q_{Z+1}((Z+1)/2 + \ell)$ for every $\ell \in \mathbb{Z}$, which plays a key role for the symmetry of the zeros of the polynomial $\Pi_Z(\lambda)$. The above result will be used in Section 13.23, p. 261, to prove a remarkable symmetry property of the zeros of $\Pi_Z(\lambda; 0)$ in the special case when the numbers λ_j arise through the spherical operators. We have the same symmetry so far in the case of a nonordered vector Λ which is symmetric.

Theorem 13.51 *Let the nonordered vector Λ be symmetric, i.e. $\Lambda = -\Lambda$. Then for every number $x \in \mathbb{R}$ we have*

$$Q_{Z+1}[\Lambda]\left(\frac{Z+1}{2} - x\right) = Q_{Z+1}[\Lambda]\left(\frac{Z+1}{2} + x\right), \tag{13.39}$$

or, equivalently,

$$Q_{Z+1}[\Lambda](Z+1-x) = Q_{Z+1}[\Lambda](x).$$

Proof Assuming for simplicity that all λ_js are different, on every interval $[\ell, \ell+1]$ we have the representation

$$Q_{Z+1}[\Lambda](x) = \sum_{j=1}^{Z+1} \alpha_j e^{\lambda_j x},$$

which implies that the function $Q_{Z+1}[\Lambda](Z+1-x)$ is again L-spline since

$$Q_{Z+1}[\Lambda](Z+1-x) = \sum_{j=1}^{Z+1} \alpha_j e^{\lambda_j(Z+1-x)}$$

$$= \sum_{j=1}^{Z+1} e^{\lambda_j(Z+1)} \alpha_j e^{-\lambda_j x}$$

on every interval $[\ell, \ell+1]$. The function $Q_{Z+1}[\Lambda](Z+1-x)$ has the same support $[0, Z+1]$. Due to the uniqueness of the compactly supported L-spline $Q_{Z+1}[\Lambda](x)$ it follows that

$$Q_{Z+1}[\Lambda](Z+1-x) = C Q_{Z+1}[\Lambda](x)$$

for some constant C. After putting $x = (Z+1)/2$ we obtain $C = 1$, since $Q_{Z+1}[\Lambda]((Z+1)/2) \neq 0$. ∎

Proposition 13.52 *Let $A_Z^*(x; \lambda)$ be the function corresponding by Definition 13.20, p. 233, to the polynomial q_{Z+1}^*. Then*

$$A_Z\left(1-x; \frac{1}{\lambda}\right) = (-1)^{Z-1} \lambda A_Z^*(x; \lambda) \quad \text{for } 0 \leq x \leq 1. \tag{13.40}$$

If the nonordered vector $\Lambda = [\lambda_1, \lambda_2, \ldots, \lambda_{Z+1}]$ is symmetric with respect to zero, i.e. $\Lambda = -\Lambda$, then the unique zero of equation $A_Z(x; -1) = 0$ satisfying $0 \leq x < 1$ is equal to $1/2$ for odd Z, and is equal to 0 for even Z.[14]

---

[14] In Micchelli's paper [13, p. 216, Remark 2.3], the last statement is obviously wrong since the formula is wrong; it is correct on p. 213 and on p. 224 of the paper. It is correct in Schoenberg's paper on L-splines [19, p. 268].

Micchelli's cardinal L-splines

Proof (1) Assuming for simplicity that all λ_js are different we apply formula (13.10), p. 234, and obtain

$$A_Z^*(x;\lambda) = \sum_{j=1}^{Z+1} \frac{1}{q_{Z+1}^{*\prime}(-\lambda_j)} \frac{e^{-\lambda_j x}}{e^{-\lambda_j}-\lambda} = \frac{(-1)^Z}{\lambda} \cdot \sum_{j=1}^{Z+1} \frac{1}{q_{Z+1}'(\lambda_j)} \frac{e^{(1-x)\lambda_j}}{1/\lambda - e^{\lambda_j x}}$$

$$= \frac{(-1)^{Z-1}}{\lambda} A_Z\left(1-x; \frac{1}{\lambda}\right).$$

(2) Now if $\Lambda = -\Lambda$ it follows that $A_Z^*(x;\lambda) = A_Z(x;\lambda)$, hence

$$A_Z(1-x;-1) = (-1)^Z A_Z(x;-1).$$

If Z is even it follows that $A_Z(1-x;-1) = A_Z(x;-1)$. For $x = 0$ this gives $A_Z(1;-1) = A_Z(0;-1)$. On the other hand by the very definition (Definition 13.20, p. 233) of $A_Z(x;\lambda)$ we have $A_Z(1;-1) = -A_Z(0;-1)$, which implies $A_Z(0;-1) = 0$.

(3) Let Z be odd. It follows that $A_Z(1-x;-1) = -A_Z(x;-1)$. For $x = 1/2$ we obtain $A_Z(1/2;-1) = -A_Z(1/2;-1)$ which implies that $A_Z(1/2;-1) = 0$. ∎

We immediately obtain the following useful corollary.

Corollary 13.53 *If the nonordered vector Λ is symmetric, i. e. $\Lambda = -\Lambda$, then the zeros of the equation $\Pi_Z(\lambda) = 0$, which have been defined in Theorem 13.31, p. 237, are all different from -1, i.e. $\Pi_Z(-1) \neq 0$.*

Proof By Proposition 13.52, p. 252, we find that $\xi = 1/2$ is the only zero of the equation $A_Z(x;-1) = 0$. Let some τ_i be a solution to $\Pi_Z(\tau_i) = 0$, and $\tau_i = -1$. Then by $\Pi_Z(\lambda) = r(\lambda)A_Z(0;\lambda)$ it follows that $A_Z(0;-1) = 0$. This contradiction proves the corollary. ∎

Remark 13.54 Since $r(-1) = \prod(e^{\lambda_j}+1) > 0$ it follows that the zeros of A_Z and Π_Z are the same.

13.18 The leading coefficient of the Euler–Frobenius polynomial $\Pi_Z(\lambda)$

We will now compute the leading coefficient of the Euler–Frobenius polynomial $\Pi_Z(\lambda) = \Pi_Z(\lambda; 0)$.

Since $\Pi_Z(\lambda; 0) = r(\lambda)A_Z(0;\lambda)$ and $Q_{Z+1}^*(Z+1) = 0$ we find from formula (13.36), p. 250, that the leading coefficient of $\Pi_Z(\lambda; 0)$ is

$$(-1)^Z Q_{Z+1}^*(Z)\lambda^{Z-1}.$$

By formula (13.35), p. 249, we obtain the equalities:

$$Q^*_{Z+1}(Z) = e^{\lambda_1 + \cdots + \lambda_{Z+1}} Q_{Z+1}(1),$$
$$Q_{Z+1}(1) = \phi_Z^+(1) \cdot s_0,$$
$$s_0 = e^{-\lambda_1 - \cdots - \lambda_{Z+1}},$$
$$\phi_Z^+(1) = \sum_{j=1}^{Z+1} \frac{e^{\lambda_j}}{q'_{Z+1}(\lambda_j)},$$

which imply

$$Q^*_{Z+1}(Z) = \sum_{j=1}^{Z+1} \frac{e^{\lambda_j}}{q'_{Z+1}(\lambda_j)}.$$

Hence,

$$\Pi_Z(\lambda; 0) = (-1)^Z \sum_{j=1}^{Z+1} \frac{e^{\lambda_j}}{q'_{Z+1}(\lambda_j)} \cdot \prod_{j=1}^{Z-1} (\lambda - v_j), \tag{13.41}$$

where v_j are the zeros of the polynomial $\Pi_Z(\lambda; 0)$ which, we will see, are all real and negative.

13.19 Schoenberg's "exponential" Euler L-spline $\Phi_Z(x; \lambda)$ and $A_Z(x; \lambda)$

The word *exponential* is used in a different sense by Schoenberg [20, 21, p. 256] where he introduces "exponential L-splines of basis λ". This sense has nothing to do with the exponential splines used by other authors [17]. For that reason we have put it in quotation marks and used the expression "exponential" Euler.

Now we will obtain an expression for the *"exponential" Euler L-spline*, through the basic function $A_Z(x; \lambda)$.

The *"exponential" Euler L-spline* is defined in a natural way, generalizing the polynomial case of Schoenberg by putting

$$\Phi_Z(x; \lambda) = \sum_{j=-\infty}^{\infty} \lambda^j Q_{Z+1}(x - j). \tag{13.42}$$

It is always a convergent series since only a finite number of terms are nonzero. It evidently has the remarkable *"exponential property"* (and for that reason Schoenberg has called it exponential):

$$\Phi_Z(x+1;\lambda) = \sum_{j=-\infty}^{\infty} \lambda^j Q_{Z+1}(x+1-j) \qquad (13.43)$$

$$= \lambda \sum_{j=-\infty}^{\infty} \lambda^{j-1} Q_{Z+1}(x-(j-1))$$

$$= \lambda \Phi_Z(x;\lambda);$$

it reminds us of the property of the exponential function λ^x which satisfies the same equation

$$\lambda^{x+1} = \lambda \cdot \lambda^x.$$

Since $Q_{Z+1}(x)$ is differentiable $Z-1$ times, if we differentiate the above equality (13.43) l times where $l \leq Z-1$, and put $x=0$, it follows that:

$$\Phi_Z^{(l)}(1;\lambda) = \lambda \Phi_Z^{(l)}(0;\lambda) \quad \text{for } l = 0, 1, \ldots, Z-1.$$

Hence, by Definition 13.20, p. 233, and Lemma 13.19, p. 233, for $0 \leq x \leq 1$ the function $\Phi_Z(x;\lambda)$ is proportional to $A_Z(x;\lambda)$.

Now we establish a link between these two fundamental functions, $\Phi_Z(x;\lambda)$ and $A_Z(x;\lambda)$.

Proposition 13.55 *The following relation holds for $0 \leq x \leq 1$:*

$$\Phi_Z(x;\lambda) = \frac{(-1)^Z}{\lambda^Z} e^{-\lambda_1 - \cdots - \lambda_{Z+1}} \cdot \Pi_Z(x;\lambda) \qquad (13.44)$$

$$= \frac{(-1)^Z}{\lambda^Z} e^{-\lambda_1 - \cdots - \lambda_{Z+1}} r(\lambda) A_Z(x;\lambda)$$

$$= -\lambda s \left(\lambda^{-1}\right) A_Z(x;\lambda).$$

Proof Now let us use equality (13.35), p. 249, namely $Q_{Z+1}^*(Zh + h - x) = e^{(\lambda_1 + \cdots + \lambda_{Z+1})h} \cdot Q_{Z+1}(x)$ for $h = 1$. In (13.36), p. 250, we have obtained the equality

$$A_Z(x;\lambda) = (-1)^Z \cdot [r(\lambda)]^{-1} \cdot \sum_{j=0}^{Z} Q_{Z+1}^*(j+1-x)\lambda^j.$$

Hence for every x satisfying $0 \leq x \leq 1$, we obtain

$$A_Z(x;\lambda) = \frac{(-1)^Z e^{\lambda_1 + \cdots + \lambda_{Z+1}}}{r(\lambda)} \cdot \lambda^Z \cdot \sum_{j=0}^{Z} Q_{Z+1}(x+Z-j)\lambda^{j-Z}$$

$$= \frac{(-1)^Z e^{\lambda_1 + \cdots + \lambda_{Z+1}}}{r(\lambda)} \cdot \lambda^Z \cdot \sum_{j=-\infty}^{\infty} \lambda^j Q_{Z+1}(x-j)$$

$$= \frac{(-1)^Z e^{\lambda_1 + \cdots + \lambda_{Z+1}}}{r(\lambda)} \cdot \lambda^Z \cdot \Phi_Z(x;\lambda). \qquad (13.45)$$

By the definition of the polynomial Π_Z in formula (13.51), p. 261, we obtain

$$\Phi_Z(x;\lambda) = \frac{(-1)^Z}{\lambda^Z} e^{-\lambda_1-\cdots-\lambda_{Z+1}} \cdot \Pi_Z(x;\lambda).$$

∎

We have the following symmetry property.

Theorem 13.56 *If the nonordered vector Λ is symmetric, i.e. $\Lambda = -\Lambda$, then*

$$\Phi_Z\left(\frac{Z+1}{2}; \frac{1}{z}\right) = \Phi_Z\left(\frac{Z+1}{2}; z\right) \quad \text{for all } z \text{ in } \mathbb{C}. \tag{13.46}$$

Proof In Theorem 13.51, p. 252, we have proved

$$Q_{Z+1}(Z+1-x) = Q_{Z+1}(x) \quad \text{for all } x \text{ in } \mathbb{R},$$

and in equality (13.42), p. 254, we have

$$\Phi_Z(0; z) = \sum_{j=-\infty}^{\infty} z^j Q_{Z+1}(-j)$$

$$= \sum_{j=0}^{Z+1} z^{-j} Q_{Z+1}(j).$$

These imply by the exponential property of Φ_Z the following:

$$\Phi_Z\left(\frac{Z+1}{2}; z\right) = z^{(Z+1)/2} \sum_{j=0}^{Z+1} Q_{Z+1}(j) z^{-j}$$

$$= z^{-(Z+1)/2} \sum_{j=0}^{Z+1} Q_{Z+1}(j) z^j,$$

which completes the proof. ∎

We immediately obtain the following useful corollary.

Corollary 13.57 *If the nonordered vector Λ is symmetric, i.e. $\Lambda = -\Lambda$, then $\Phi_Z(0; z) \neq 0$ for all complex numbers z, with $|z| = 1$.*

The proof follows directly from Corollary 13.53, p. 253, and the relation between Π_Z and Φ_Z given by formula (13.44), p. 255, above.

Let us apply formula (13.44). We use the relation for $\lambda = e^{\lambda_s}$, $1 \leq s \leq Z+1$. This gives the equality

$$\Phi_Z(x; e^{\lambda_s}) = (-1)^Z e^{-(\lambda_1+\cdots+\lambda_{Z+1})} e^{-\lambda_s Z} \cdot \Pi_Z(x; e^{\lambda_s}).$$

By formula (13.15), p. 235, we see that in the case of pairwise different λ_j we obtain:

$$\Pi_Z(x; e^{\lambda_s}) = r(e^{\lambda_s}) \cdot A_Z(x; e^{\lambda_s}) = \frac{-r'(e^{\lambda_s})}{q'_{Z+1}(\lambda_s)} \cdot e^{\lambda_s x}.$$

Hence,

$$\Phi_Z(x; e^{\lambda_s}) = (-1)^Z e^{-(\lambda_1 + \cdots + \lambda_{Z+1})} e^{-\lambda_s Z} \cdot \frac{-r'(e^{\lambda_s})}{q'_{Z+1}(\lambda_s)} \cdot e^{\lambda_s x}. \tag{13.47}$$

13.20 Marsden's identity for cardinal L-splines

There is an important *normalization* property which is analogous to the classical *Marsden identity* for polynomial splines.

Proposition 13.58 *Assume that all λ_j are pairwise different. Then for every λ_s in Λ and for every x in $\mathbb{R}$ we have the following identity:*

$$\sum_{j=-\infty}^{\infty} e^{j\lambda_s} \cdot Q_{Z+1}(x - j) = \Phi_Z(x; e^{\lambda_s}) \tag{13.48}$$

$$= (-1)^{Z+1} e^{-(\lambda_1 + \cdots + \lambda_{Z+1})} e^{-\lambda_s Z} \cdot \frac{r'(e^{\lambda_s})}{q'_{Z+1}(\lambda_s)} \cdot e^{\lambda_s x}.$$

It is clear that the sum on the left-hand side is finite over j satisfying $0 < x - j < Z + 1$. The proof is obtained by applying the above formula for $\Phi_Z(x; \lambda)$ in (13.47), p. 257. This result is useful for estimating the norm of Q_{Z+1}.

13.21 Peano kernel and the divided difference operator in the cardinal case

Here we provide a direct proof that the *TB*-spline $Q_{Z+1}(x)$ is indeed the Peano kernel for the divided difference operator defined in formula (13.18), p. 236, through the polynomial $s(\lambda)$.

We compute the divided difference in the case of different λ_js. First, we recall the adjoint operator of formula (13.31), p. 248,

$$\mathcal{L}^*_{Z+1}\left(\frac{d}{dx}\right) := (-1)^{Z+1} \prod_{j=1}^{Z+1} \left(-\frac{d}{dx} - \lambda_j\right) = (-1)^{Z+1} \prod_{j=1}^{Z+1} \mathcal{D}^*_j$$

where $\mathcal{D}^*_j = -(d/dx) - \lambda_j$ is the operator formally adjoint to the operator $\mathcal{D}_j = d/dx - \lambda_j$ defined in formula (13.6), p. 226.[15]

---

[15] These operators differ from those of Dyn and Ron [7, p. 5]. However, the difference between the operators $\mathcal{L}^*_{P+1}$ is not large.

258 Multivariate polysplines

Recalling the properties of the functions $\phi_Z(x)$ in Proposition 13.12, p. 229, and the definition of the TB-spline Q_{Z+1} in (13.19), p. 239, now we have the following *Peano identity* for the *generalized divided difference operator* given by formula (13.18), p. 236:

Theorem 13.59 *We assume that the function f is C^∞. Then the following Peano-type identity holds:*

$$\int_{-\infty}^{\infty} Q_{Z+1}(x) \mathcal{L}_{Z+1}^* f(x) dx = (-1)^{Z+1} \sum_{j=0}^{Z+1} s_j \cdot f(j). \tag{13.49}$$

Proof First, recall the properties of the function $\phi_Z(x)$ which are stated in Proposition 13.12, p. 229. We assume without restricting the generality that f has a compact support. By the definition of Q_{Z+1} in (13.19), p. 239, we obtain

$$I := \int_{-\infty}^{\infty} Q_{Z+1}(x) \mathcal{L}_{Z+1}^* f(x) dx$$

$$= (-1)^{Z+1} \int_{-\infty}^{\infty} Q_{Z+1}(x) \mathcal{D}_1^* \cdots \mathcal{D}_{Z+1}^* f(x) dx$$

$$= (-1)^{Z+1} \sum_{j=0}^{Z+1} s_j \int_j^{\infty} \mathcal{D}_Z \cdots \mathcal{D}_1 \phi_Z(x-j) \mathcal{D}_{Z+1}^* f(x) dx$$

$$= (-1)^Z \sum_{j=0}^{Z+1} s_j \int_j^{\infty} \mathcal{D}_Z \cdots \mathcal{D}_1 \phi_Z(x-j) \left(\frac{d}{dx} + \lambda_{Z+1} \right) f(x) dx.$$

Further we integrate by parts and apply the properties of the function ϕ_Z in Proposition 13.12, p. 229,

$$I = (-1)^Z \sum_{j=0}^{Z+1} s_j \mathcal{D}_Z \cdots \mathcal{D}_1 \phi_Z(x-j) \cdot f(x) \Big|_{x=j+}^{x=\infty}$$

$$+ (-1)^Z \sum_{j=0}^{Z+1} s_j \left(- \int_j^{\infty} \frac{d}{dx} \mathcal{D}_Z \cdots \mathcal{D}_1 \phi_Z(x-j) f(x) dx \right.$$

$$\left. + \int_j^{\infty} \mathcal{D}_Z \cdots \mathcal{D}_1 \phi_Z(x-j) \lambda_{Z+1} f(x) dx \right)$$

$$= (-1)^{Z+1} \sum_{j=0}^{Z+1} s_j \cdot f(j) + (-1)^{Z+1} \sum_{j=0}^{Z+1} s_j \int_j^\infty \mathcal{D}_{Z+1} \mathcal{D}_Z \cdots \mathcal{D}_1 \phi_Z(x-j) f(x)\, dx$$

$$= (-1)^{Z+1} \sum_{j=0}^{Z+1} s_j \cdot f(j),$$

which completes the proof. ∎

13.22 Two-scale relation (refinement equation) for the TB-splines $Q_{Z+1}[\Lambda; h]$

Assuming the nonordered vector Λ given, we denote by $Q_{Z+1}[\Lambda](x)$ the TB-spline defined according to formula (13.19), p. 239, for the mesh $\mathbb{Z}$. As before we denote, by $Q_{Z+1}[t\Lambda; h](x)$ the TB-spline for the mesh $h\mathbb{Z}$ and for the nonordered vector $t\Lambda = [t\lambda_1, \ldots, t\lambda_{Z+1}]$. Up to now we have mainly used the notation

$$Q_{Z+1}(x) = Q_{Z+1}[\Lambda](x)$$

without indicating the dependence on h.[16] We note again that the index $Z+1$ is redundant but useful to have.

It is important for *wavelet analysis* to consider the relation between the TB-spline Q_{Z+1} for the cardinal L-splines on the mesh $h\mathbb{Z} := \{jh \colon \text{for } j \text{ in } \mathbb{Z}\}$ and the TB-spline on the mesh $2h\mathbb{Z} := \{2jh \colon \text{for } j \text{ in } \mathbb{Z}\}$, where as above h is a fixed positive number. One says that $h\mathbb{Z}$ is a *refinement* of $2h\mathbb{Z}$. We have seen in Section 13.10, p. 239, that the TB-spline $Q_{Z+1}[\Lambda; h]$ has support on the interval $[0, Zh + h]$ and break-points jh for $j = 0, 1, \ldots, Z+1$. In a similar way on the mesh $2h\mathbb{Z}$ the compactly supported TB-spline $Q_{Z+1}[\Lambda; 2h]$ has a support $[0, (Z+1)2h]$ with break-points $j2h$ for all $j = 0, 1, \ldots, Z+1$. On the other hand, obviously $Q_{Z+1}[\Lambda; 2h](x)$ is also an L-spline on the mesh $h\mathbb{Z}$. According to Theorem 13.38, p. 241, the integer shifts $Q_{Z+1}[\Lambda; h](x - \ell h)$ form a basis for all compactly supported splines on $\mathbb{R}$, hence it is possible to express $Q_{Z+1}[\Lambda; 2h]$ as a linear combination of the shifts $Q_{Z+1}[\Lambda; h](x - \ell h)$. Theorem 13.60 provides the exact linear combination.

Theorem 13.60 *We have the representation, called the* two-scale relation *or* refinement equation

$$Q_{Z+1}[\Lambda; 2h](x) = \sum_{\ell=0}^{Z+1} \gamma_\ell Q_{Z+1}[\Lambda; h](x - \ell h), \qquad (13.50)$$

where the two-scale sequence *is*

$$\gamma_\ell = (-1)^\ell s_\ell \quad \text{for } \ell = 0, 1, \ldots, Z+1,$$

and the two-scale symbol *is* $Z(e^{-i\xi h}) = s_h(-e^{-i\xi h})$.[17]

---

[16] In the notation of de Boor et al. [5], we have $Q_{Z+1}[\Lambda](x) = N_\Lambda(x)$.

[17] See Part III for this terminology.

Proof Let us take the Fourier transform on both sides of the equality (13.50). Due to

$$\widehat{Q_{Z+1}[\Lambda; h](x - \ell h)}(\xi) = \int_{-\infty}^{\infty} Q_{Z+1}[\Lambda; h](x - \ell h) e^{-i\xi x} \, dx$$

$$= e^{-i\xi \ell h} \widehat{Q_{Z+1}[\Lambda; h]}(\xi),$$

we obtain

$$\widehat{Q_{Z+1}[\Lambda; 2h]}(\xi) = \sum_{\ell=0}^{Z+1} \gamma_\ell e^{-i\xi \ell h} \widehat{Q_{Z+1}[\Lambda; h]}(\xi).$$

We obtain from formula (13.25), p. 242,

$$\widehat{Q_{Z+1}[\Lambda; 2h]}(\xi) = \prod_{j=1}^{Z+1} (e^{-\lambda_j h} + e^{-i\xi h}) \widehat{Q_{Z+1}[\Lambda; h]}(\xi)$$

$$= s_h(-e^{-i\xi h}) \widehat{Q_{Z+1}[\Lambda; h]}(\xi).$$

Since

$$s_h(-e^{-i\xi h}) = \sum_{\ell=0}^{Z+1} s_\ell (-e^{-i\xi h})^\ell = \sum_{\ell=0}^{Z+1} s_\ell (-1)^\ell e^{-i\xi \ell h},$$

the proof will be completed by taking the inverse Fourier transform. ∎

Theorem 13.60 is another interpretation of Proposition 13.40, p. 243, where we have established a relation between the Fourier transforms of $Q_{Z+1}[\Lambda; 2h]$ and of $Q_{Z+1}[\Lambda; h]$.

This relation is quite close to being understood as a *generalized two-scale relation*. Anyway, we have a simple transition from one level to the other in the wavelet spaces, which will be much exploited in Part III.

Remark 13.61 *Due to the translation invariance we have the same coefficients for all shifts $Q_{Z+1}(x - 2\ell h)$.*

Remark 13.62 *If $\Lambda = [0, \ldots, 0]$, which corresponds to the usual polynomial case, we see that due to $h\Lambda = \Lambda$ it follows that:*

$$Q_{Z+1}[\Lambda; h](x) = h^Z \cdot Q_{Z+1}[\Lambda] \left(\frac{x}{h}\right),$$

which provides us with a scale invariant set of compactly supported functions. Chui [3] uses this in his cardinal spline wavelet analysis. For the nonzero vector Λ we have the nonstationary wavelet analysis of de Boor et al. [5].

13.23 Symmetry of the zeros of the Euler–Frobenius polynomial $\Pi_Z(\lambda)$

We now consider the special case of the nonordered vector Λ which is generating the spherical operator $M_{k,p}$, see formula (10.26), p. 169. We will prove a remarkable symmetry property of the compactly supported spline Q_{Z+1} and of the *Euler–Frobenius polynomial* $\Pi_Z(\lambda) = \Pi_Z(\lambda; 0)$ which are available due to the "almost" symmetry properties of the corresponding vector $\Lambda = [\lambda_1, \lambda_2, \ldots, \lambda_{2p}]$.

We consider the operator $L = M_{k,p}$. We have

$$Z = 2p - 1$$

and the nonordered vector $\Lambda = [\lambda_1, \lambda_2, \ldots, \lambda_{2p}]$ is given by

$$\begin{aligned}&\lambda_1 = -n-k+2, \quad \lambda_2 = -n-k+4, \quad \ldots, \quad \lambda_p = -n-k+2p,\\&\lambda_{p+1} = k, \quad \lambda_{p+2} = k+2, \quad \ldots, \quad \lambda_{2p} = k+2p-2.\end{aligned} \quad (13.51)$$

By the definition of the Euler–Frobenius polynomial and by the proof of Proposition 13.50, p. 250, namely equality (13.38) we have

$$\Pi_Z(\lambda) = e^{\lambda_1 + \cdots + \lambda_{Z+1}} \cdot (-1)^Z \cdot \sum_{j=0}^{Z} Q_{Z+1}(j) \lambda^{Z-j} \quad (13.52)$$

$$= e^{\lambda_1 + \cdots + \lambda_{Z+1}} \cdot (-1)^Z \cdot \sum_{j=0}^{Z-1} Q_{Z+1}(Z-j) \lambda^j.$$

Let us note that in the case of arbitrary symmetric set $\Lambda = -\Lambda$ we will have $Q_{Z+1}(j) = Q_{Z+1}(Z+1-j)$. Indeed, in such a case the function

$$Q_{Z+1}(Z+1-x)$$

is a piecewise linear combination of

$$\{e^{-\lambda_1 x}, \ldots, e^{-\lambda_{Z+1} x}\} = \{e^{\lambda_1 x}, \ldots, e^{\lambda_{Z+1} x}\},$$

hence, due to the uniqueness of the compactly supported TB-spline Q_{Z+1} with support $[0, Z+1]$ it follows that:

$$Q_{Z+1}(Z+1-x) = C \cdot Q_{Z+1}(x)$$

for some constant $C > 0$. But for $x = (Z+1)/2$ we obtain $Q_{Z+1}((Z+1)/2) = C \cdot Q_{Z+1}((Z+1)/2)$, hence $C = 1$. Thus by Proposition 13.50, p. 250, we obtain

$$\Pi_Z(\lambda) = \lambda^{Z-1} \Pi_Z\left(\frac{1}{\lambda}\right).$$

Hence

$$\Pi_Z(\lambda) = 0$$

262 *Multivariate polysplines*

implies
$$\Pi_Z\left(\frac{1}{\lambda}\right) = 0.$$

We will see that for the above special choice of the vector Λ in (13.51) we have a rather similar picture since the set Λ "symmetrizes" for $k \to \infty$. We know that the function $Q_{Z+1}(Z+1-x)$ is a piecewise linear combination of the functions
$$\{e^{-\lambda_1 x}, e^{-\lambda_2 x}, \ldots, e^{-\lambda_{2p} x}\}.$$

Due to the "almost" symmetry of the vector Λ we see that after multiplying with $e^{(\lambda_1+\lambda_{2p})x}$ the basis for $-\Lambda$ changes into the basis for Λ, namely
$$e^{(\lambda_1+\lambda_{2p})x} \cdot \{e^{-\lambda_1 x}, e^{-\lambda_2 x}, \ldots, e^{-\lambda_{2p} x}\} = \{e^{\lambda_1 x}, e^{\lambda_2 x}, \ldots, e^{\lambda_{2p} x}\}.$$

We have used the equalities
$$\lambda_1 + \lambda_{2p} = -n - k + 2 + k + 2p - 2 = -n + 2p,$$

$$-\lambda_j + \lambda_1 + \lambda_{2p} = k + 2(p-j)$$
$$= \lambda_{p+p-j} \quad \text{for } j = 1, \ldots, p,$$

$$-\lambda_{p+j} + \lambda_1 + \lambda_{2p} = -n - k + 2(p+1-j)$$
$$= \lambda_{p+1-j} \quad \text{for } j = 1, \ldots, p.$$

Thus by the uniqueness of the compactly supported spline we obtain
$$e^{(\lambda_1+\lambda_{2p})x} Q_{Z+1}(Z+1-x) = C \cdot Q_{Z+1}(x).$$

By putting $x = (Z+1)/2$ it follows that
$$C = e^{(\lambda_1+\lambda_{2p})(Z+1)/2} = e^{(\lambda_1+\lambda_{2p})p} = e^{(-n+2p)p}.$$

Thus we have proved the following result about the symmetry of the compactly supported *TB*-spline.

Theorem 13.63 *For the special choice of the set Λ given by (13.51), p. 261, we have*
$$Q_{Z+1}(Z+1-x) = e^{(-n+2p)p} \cdot e^{-(-n+2p)x} \cdot Q_{Z+1}(x) \tag{13.53}$$
$$= e^{(-n+2p)(p-x)} \cdot Q_{Z+1}(x).$$

It should be noted that this result is independent of k.

Now we will draw some consequences about the symmetry of the polynomial Π_Z and its zeros. We obtain from (13.52) the equalities
$$\Pi_Z(\lambda) = e^{\lambda_1+\cdots+\lambda_{Z+1}} \cdot (-1)^Z \cdot \sum_{j=0}^{Z} Q_{Z+1}(j)\lambda^{Z-j}$$
$$= e^{\lambda_1+\cdots+\lambda_{Z+1}} \cdot (-1)^Z \cdot e^{-(-n+2p)p} \cdot \sum_{j=0}^{Z} e^{(-n+2p)j} Q_{Z+1}(Z+1-j)\lambda^{Z-j}.$$

Let us recall that since $Q_{Z+1}(Z+1) = 0$ the term with $j = 0$ is zero. If we put $i = Z - j$ or $j = Z - i$ we see that

$$\Pi_Z(\lambda; 0) = e^{\lambda_1 + \cdots + \lambda_{Z+1}} \cdot (-1)^Z \cdot e^{-(-n+2p)p} \cdot \sum_{i=0}^{Z} e^{(-n+2p)(Z-i)} Q_{Z+1}(i+1) \lambda^i$$

$$= \left(\frac{\lambda}{e^{-n+2p}}\right)^{-1} e^{\lambda_1 + \cdots + \lambda_{Z+1}} \cdot (-1)^Z \cdot e^{-(-n+2p)p} \cdot e^{(-n+2p)Z}$$

$$\times \sum_{i=0}^{Z} Q_{Z+1}(i+1) \left(\frac{\lambda}{e^{-n+2p}}\right)^{i+1}$$

$$= \left(\frac{\lambda}{e^{-n+2p}}\right)^{-1+Z} e^{\lambda_1 + \cdots + \lambda_{Z+1}} \cdot (-1)^Z \cdot e^{-(-n+2p)p} \cdot e^{(-n+2p)Z}$$

$$\times \sum_{i=0}^{Z} Q_{Z+1}(i+1) \left(\frac{e^{-n+2p}}{\lambda}\right)^{Z-(i+1)}$$

$$= \lambda^{Z-1} \cdot C \cdot \Pi_Z\left(\frac{e^{-n+2p}}{\lambda}; 0\right)$$

for a constant C which may be defined by the above and it is clear that $C \neq 0$. We find this constant by putting $\tilde{\lambda} = \sqrt{e^{-n+2p}}$. This gives

$$\Pi_Z(\tilde{\lambda}; 0) = \tilde{\lambda}^{Z-1} \cdot C \cdot \Pi_Z(\tilde{\lambda}; 0),$$

hence since $\Pi_Z(\lambda; 0)$ has only negative zeros we obtain

$$C = e^{-p(-n+2p)/2}.$$

By the general theory, see Theorem 13.31, p. 237, we know that all $Z - 1 = 2p - 2$ zeros of $\Pi_Z(\lambda; 0)$ satisfy

$$\mu_{Z-1} < \cdots < \mu_1 < 0,$$

hence we see that all zeros separate into two groups. Thus we have proved Theorem 13.64.

Theorem 13.64 *For the special choice of Λ given by (13.51), p. 261, we have the symmetry*

$$\Pi_Z(\lambda; 0) = \lambda^{Z-1} \cdot e^{-p(-n+2p)/2} \cdot \Pi_Z\left(\frac{e^{-n+2p}}{\lambda}; 0\right).$$

If for some $\lambda \neq 0$ we have

$$\Pi_Z(\lambda; 0) = 0$$

then also

$$\Pi_Z\left(\frac{e^{-n+2p}}{\lambda}; 0\right) = 0.$$

Hence, the $Z - 1 = 2p - 2$ zeros of the equation $\Pi_Z(\lambda; 0) = 0$ satisfy

$$\mu_j \mu_{2p-2-j+1} = e^{-n+2p} \quad \text{for } j = 1, \ldots, p-1,$$

and

$$\mu_{2p-2} < \cdots < \mu_p < -\sqrt[2]{e^{(-n+2p)}} < \mu_{p-1} < \cdots < \mu_1 < 0.$$

We see again the *remarkable fact that this symmetry is completely independent of k*, in particular the constant $-\sqrt[2]{e^{(-n+2p)}}$ is independent of k.

These results will be used in the cardinal interpolation with polysplines in Section 15.7.

13.24 Estimates of the functions $A_Z(x; \lambda)$ and $Q_{Z+1}(x)$

We will provide some important estimates of the function $A(x; \lambda)$ for the special choice of the set Λ above in (13.51), p. 261, and the somewhat more general cases considered in [9].

Using the residuum representation (13.11), p. 235, of the function $A(x; \lambda)$ we prove the following.

Theorem 13.65 *Let the vector Λ be the one given by (13.51), p. 261. Let K be a compact subset of the complex plane, $0 \notin K$ and hence $e^{\lambda_j} \notin K$ for large k. Then for every $\varepsilon > 0$ there exist a constant $C > 0$ and an integer k_0 such that for all $k \geq k_0$, for all $\lambda \in K$, and for all x satisfying $0 \leq x \leq 1$, the following estimate holds:*

$$|A_Z(x; \lambda)| \leq \frac{C}{k^Z}. \tag{13.54}$$

Proof We will prove the estimate first for all x satisfying $0 \leq x \leq 1 - \delta$ for every small $\delta > 0$. Then it will follow for all $0 \leq x \leq 1$ by the symmetry property (13.40), p. 252, i.e.

$$A_Z\left(1 - x; \frac{1}{\lambda}\right) = (-1)^{Z-1} \lambda A_Z^*(x; \lambda) \quad \text{for } 0 \leq x \leq 1.$$

For simplicity we consider the case $p = 2$, $Z = 2p - 1 = 3$, and $K = \{|\lambda| = 1\}$. By formula (13.11), p. 235, we have

$$A_3(0; \lambda) = \frac{1}{2\pi i} \int_\Gamma \frac{dz}{q_4(z)(e^z - \lambda)},$$

where Γ is a contour (or a sum of contours with the same orientation) in the complex plane which surrounds all points $\{t_1, \ldots, t_8\}$ and does not surround the points $i\varphi$ for real φ such that $e^{i\varphi} = \lambda$.

We will choose $\Gamma = \Gamma_1 \cup \Gamma_2$, where Γ_1 is a circle which surrounds the points λ_1, λ_2, and Γ_2 is a circle which surrounds the points λ_3, λ_4, namely we put

$$\Gamma_j := \{z \in \mathbb{C} : |z - z_j| = R_j\} \quad \text{for } j = 1, 2,$$

where
$$z_1 := \frac{\lambda_1 + \lambda_2}{2} = -n - k + 3,$$
$$R_1 := |z_1| - 2 = +n + k + 1,$$

and
$$z_2 := \frac{\lambda_3 + \lambda_4}{2} = k + 1,$$
$$R_2 := |z_2 - \lambda_3| + 1 = 2.$$

As will become clear, we have chosen these circles in order to obtain the best possible estimate.

Indeed, for large k and some constant $C_1 > 0$ we have the inequality
$$|q_4(z)| \geq C_1 |2k - 1|^2 \quad \text{for } z \in \Gamma_2.$$

On the other hand, $|e^z - \lambda| \geq |e^z| - |\lambda|$ implies for large k the inequality
$$|e^z - \lambda| \geq e^{k-1} - 2 \quad \text{for } z \in \Gamma_2 \text{ and } \lambda \in K.$$

The above inequalities imply
$$I_2 := \left| \int_{\Gamma_2} \frac{dz}{q_4(z)(e^z - \lambda)} \right| \leq \frac{R_2}{C_1 |2k - 1|^2 (e^{k-1} - 2)}.$$

This estimate provides exponential decay for the integral over Γ_2 for $k \to \infty$.

On the other hand for the integral over the circle Γ_1 for an appropriate constant $C_2 > 0$ we have
$$|q_4(z)| \geq C_2 R_1^2 \lambda_3^2 \quad \text{for } z \in \Gamma_1,$$
$$|e^z - \lambda| \geq 1 - e^{-2} \quad \text{for } z \in \Gamma_1 \text{ and } \lambda \in K,$$

and obtain for an appropriate $C_2' > 0$ the estimate
$$I_1 := \left| \int_{\Gamma_1} \frac{dz}{q_4(z)(e^z - \lambda)} \right| \leq C_2' \frac{R_1}{k^4 |e^{-2} - 1|} = C_2' \frac{n + k - 1}{k^4 |e^{-2} - 1|}.$$

Since $A_3(x; \lambda)$ is the sum of the two integrals the statement of the theorem follows. ∎

We can now provide an optimal estimate for the compactly supported spline. According to formulas (13.42) and (13.44) we obtain for $0 \leq x \leq 1$ the representation
$$\sum_{j=-\infty}^{\infty} Q_{Z+1}(x - j) = (-1)^Z e^{-\lambda_1 - \lambda_2 - \cdots - \lambda_{Z+1}} r_{Z+1}(1) A_Z(x; 1).$$

Taking all terms in the sum we see that

$$\max_{x \in \mathbb{R}} Q_{Z+1}(x) \leq e^{-\lambda_1 - \lambda_2 - \cdots - \lambda_{Z+1}} |r_{Z+1}(1)| \max_{x \in [0,1]} |A_Z(x;1)|.$$

Theorem 13.66 *Let the compactly supported spline $Q_{Z+1}(x)$ correspond to the vector Λ of (13.51), p. 261. Then for $k \to \infty$ it satisfies the asymptotic order*

$$\max_{x \in \mathbb{R}} Q_{Z+1}(x) \approx \frac{e^{pk}}{k^Z}. \qquad (13.55)$$

Proof The estimate of $\max_{x \in [0,1]} |A_Z(x;1)|$ comes from the above theorem. Since $\lambda_j \to 0$ for $j = 1, 2, \ldots, p$, and $\lambda_j \to \infty$ for $j = p+1, \ldots, 2p$, the estimate of the asymptotic order of $r_{Z+1}(1)$ is

$$|r_{Z+1}(1)| \leq \prod_{j=1}^{Z+1} |e^{\lambda_j} - 1| \leq C e^{pk}.$$

This completes the proof. ∎

Chapter 14

Riesz bounds for the cardinal L-splines Q_{Z+1}

The main purpose of this chapter is to study the *Riesz bounds* for the set of shifts of the TB-spline $Q_{Z+1}(x) = Q_{Z+1}[\Lambda](x)$, namely the constants A and B in the inequality

$$A\|c\|_{\ell_2} \le \left\|\sum c_j Q_{Z+1}(x-j)\right\|_{L_2(\mathbb{R})} \le B\|c\|_{\ell_2}.$$

The point of our analysis will be to understand their dependence on the vector Λ. There are two cases that are particularly interesting.

- The vector Λ is obtained by the solutions of the spherical operators $L_{(k)}^p$, see the formula (10.28), p. 170; i.e. $\Lambda_k = [\lambda_1, \lambda_2, \ldots, \lambda_{2p}]$ is given by

$$\begin{aligned}&\lambda_1 = -n-k+2, \quad \lambda_2 = -n-k+4, \quad \ldots, \quad \lambda_p = -n-k+2p,\\ &\lambda_{p+1} = k, \qquad\qquad \lambda_{p+2} = k+2, \qquad \ldots, \quad \lambda_{2p} = k+2p-2.\end{aligned} \quad (14.1)$$

We will let $k \to \infty$.

- The vector Λ is obtained from the Fourier transform of the operator Δ^p in the strip, given by (9.6), p. 120; i.e. $\Lambda_\eta = [\lambda_1, \lambda_2, \ldots, \lambda_{2p}]$ is given by

$$\begin{aligned}\lambda_1 = \lambda_2 = \cdots = \lambda_p = -\eta,\\ \lambda_{Z+1} = \lambda_{p+2} = \cdots = \lambda_{2p} = \eta.\end{aligned} \quad (14.2)$$

We will let $\eta \to \infty$.

We will concentrate our efforts on the first case which is more difficult and will be further applied to the wavelet analysis through polysplines on annuli in Part III. We will see that A and B have the same asymptotics for $k \to \infty$ and satisfy

$$\frac{B}{A} \xrightarrow{k \to \infty} 1.$$

268 Multivariate polysplines

Our analysis is inspired by the analysis carried out by Chui [3, Chapter 4.2] in the case of polynomial cardinal splines. It is interesting that many of the techniques he used [3] for the estimation of the Riesz bounds can be conveyed in a nontrivial way to the present case of L-splines although other technical problems also have to be solved.

Let us recall some results which are useful for the analysis of the Riesz basis.

For a bi-infinite sequence $c = \{c_j\}_{j=-\infty}^{\infty}$ we put

$$\|c\|_{\ell_2} = \left(\sum_{j=-\infty}^{\infty} |c_j|^2 \right)^{1/2}.$$

We have the following, see again [3, p. 76, Theorem 3.24].

Theorem 14.1 *Let $\varphi \in L_2(\mathbb{R})$. If the two constants A, B satisfy $0 < A \leq B < \infty$ then the following are equivalent:*

(i) The translates of the function $\varphi(x)$ satisfy the Riesz condition

$$A\|c\|_{\ell_2}^2 \leq \left\| \sum_{j=-\infty}^{\infty} c_j \varphi(x-j) \right\|_{L_2(R)}^2 \leq B\|c\|_{\ell_2}^2.$$

(ii) The Fourier transform $\widehat{\varphi}(\xi)$ of $\varphi(x)$ satisfies

$$A \leq \sum_{j=-\infty}^{\infty} |\widehat{\varphi}(\xi + 2\pi j)|^2 \leq B$$

for almost every $\xi \in \mathbb{R}$.

The constants A, B are called **Riesz bounds.**

It is important for the wavelet analysis to consider the mesh $h\mathbb{Z}$. There we have the linear combinations

$$\sum_{\nu=-\infty}^{\infty} c_\nu \phi(x - h\nu).$$

We have the following more general version of Theorem 14.1.

Theorem 14.2 *Let $\phi \in L_2(\mathbb{R})$. If the Riesz bounds A, B satisfy $0 < A \leq B < \infty$, then for every sequence $\{c_\nu\} \in \ell_2$ the inequality*

$$A\|c\|_{\ell_2}^2 \leq \int_{-\infty}^{\infty} \left| \sum_{\nu=-\infty}^{\infty} c_\nu \phi(x - h\nu) \right|^2 dx \leq B\|c\|_{\ell_2}^2$$

holds if and only if

$$hA \leq \sum_{\nu=-\infty}^{\infty} \left| \widehat{\phi}\left(\frac{\xi + 2\pi \ell}{h} \right) \right|^2 \leq hB \quad \text{a.e. on } \mathbb{R}.$$

Proof Indeed, after the change of variables $(x/h) \longrightarrow x$, we obtain

$$\int_{-\infty}^{\infty} \left| \sum_{v=-\infty}^{\infty} c_v \phi(x-hv) \right|^2 dx = \int_{-\infty}^{\infty} \left| \sum_{v=-\infty}^{\infty} c_v \phi \left(h \left(\frac{x}{h} - v \right) \right) \right|^2 dx$$

$$= h \int_{-\infty}^{\infty} \left| \sum_{v=-\infty}^{\infty} c_v \phi(h(x-v)) \right|^2 dx.$$

For the last we have by the Riesz criterion of Theorem 14.1, p. 268, that

$$A \sum_{v=-\infty}^{\infty} |c_v|^2 \leq h \int_{-\infty}^{\infty} \left| \sum_{v=-\infty}^{\infty} c_v \phi(h(x-v)) \right|^2 dx \leq B \sum_{v=-\infty}^{\infty} |c_v|^2$$

if and only if

$$\frac{A}{h} \leq \sum_{v=-\infty}^{\infty} \left| \widehat{\phi(hx)}(\xi + 2\pi\ell) \right|^2 \leq \frac{B}{h}.$$

Since

$$\widehat{\phi(hx)}(\xi) = \int_{-\infty}^{\infty} e^{-i\xi x} \phi(hx) \, dx = \frac{1}{h} \int_{-\infty}^{\infty} e^{-i\frac{\xi}{h}y} \phi(y) \, dy$$

$$= \frac{1}{h} \widehat{\phi} \left(\frac{\xi}{h} \right),$$

the above criterion will become

$$\frac{A}{h} \leq \frac{1}{h^2} \sum_{v=-\infty}^{\infty} \left| \widehat{\phi} \left(\frac{\xi + 2\pi\ell}{h} \right) \right|^2 \leq \frac{B}{h},$$

which proves the statement of the proposition. ∎

Theorem 14.3 is another useful result which helps to compute the above sum of translates of the Fourier transforms [3, p. 48, Theorem 2.28].

Theorem 14.3 *Let one of the following two conditions hold for the function $\varphi \in L_2(\mathbb{R})$, which may be complex-valued:*

(i) $\varphi(x) = O(|x|^{-\beta})$, for $\beta > 1$, and $\widehat{\varphi}(x) = O(|x|^{-\alpha})$, for $\alpha > \frac{1}{2}$, both for $|x| \to \infty$.

(ii) The function $\widehat{\varphi}(x)$ is continuous, has a compact support, and has a bounded variation on its support.[1]

Then the following equality holds:

$$\sum_{j=-\infty}^{\infty} |\widehat{\varphi}(\xi + 2\pi j)|^2 = \sum_{j=-\infty}^{\infty} \left\{ \int_{-\infty}^{\infty} \varphi(y+j) \overline{\varphi(y)} \, dy \right\} e^{-ij\xi}$$

for all $\xi \in \mathbb{R}$.

---

[1] For a function to have a "bounded variation" is equivalent to it being the difference of two nondecreasing functions.

14.1 Summary of necessary results for cardinal L-splines

For the reader's convenience we have summarized the results which we will use below. This summary is similar to Chui [3, Theorem 4.3, p. 85] and the reader is thus facilitated to check the analogy with the polynomial case.[2]

Remark 14.4 *We have the following properties of the TB-splines, which we formulate for simplicity on the mesh $\mathbb{Z}$:*

1. *Using the Theorem on convolution of TB-splines (formula (13.27), p. 244) for an arbitrary real number μ we may define the TB-spline $Q_1[\mu](x)$ by putting (see formula (13.21), p. 240)*

$$Q_1[\mu](x) = e^{-\mu} e^{\mu x} \chi_{[0,1]}(x),$$

and inductively, we may define for every nonordered vector of real constants $[\lambda_1, \lambda_2, \ldots, \lambda_m]$ for every integer $m \geq 1$, by putting

$$Q_m[\lambda_1, \lambda_2, \ldots, \lambda_m](x) = Q_{m-1}[\lambda_1, \lambda_2, \ldots, \lambda_{m-1}](x) * Q_1[\lambda_m](x)$$

$$= \int_0^1 Q_{m-1}[\lambda_1, \lambda_2, \ldots, \lambda_{m-1}](x - y) \cdot Q_1[\lambda_m](y) \, dy.$$

2. *For every continuous function $f \in C(\mathbb{R})$ we have the **Hermite–Gennocchi identity** (13.29), p. 246,*

$$\int_{-\infty}^{\infty} f(x) Q[\Lambda](x) \, dx = \left\{ \prod_{j=1}^{N} e^{-\lambda_j h} \right\}$$

$$\cdot \underbrace{\int_0^h \cdots \int_0^h}_{N} \prod_{j=1}^{N} e^{\lambda_j x_j} \cdot f(x_1 + \cdots + x_N) \, dx_1 \cdots dx_N.$$

3. *For every function $g \in C^{Z+1}$ we have the **Peano kernel** representation formula (13.49), p. 258:*

$$\int_{-\infty}^{\infty} Q_{Z+1}(x) \mathcal{L}_{Z+1}^* f(x) \, dx = (-1)^{Z+1} \sum_{j=0}^{Z+1} s_j \cdot f(j),$$

where the adjoint operator $\mathcal{L}_{Z+1}^*$ is defined in Chapter 13.

4. $\operatorname{supp} Q_{Z+1} = [0, Z+1]$.

5. $Q_{Z+1}(x) > 0$ for $0 < x < Z + 1$.

---

[2] Let us note that m in the last reference is equal to $Z + 1$ in our notation.

6. The **Marsden identity** (13.48), p. 257, holds for Q_{Z+1} and for every integer s with $1 \le s \le Z+1$, with the second equality holding if λ_s has multiplicity one (i.e. $q'_{Z+1}(\lambda_s) \ne 0$):

$$\sum_{j=-\infty}^{\infty} e^{j\lambda_s} \cdot Q_{Z+1}(x-j) = \Phi_Z(x; e^{\lambda_s})$$

$$= (-1)^{Z+1} e^{-(\lambda_1 + \cdots + \lambda_{Z+1})} e^{-\lambda_s Z} \cdot \frac{r'(e^{\lambda_s})}{q'_{Z+1}(\lambda_s)} \cdot e^{\lambda_s x}.$$

7. The **differentiation** formula holds for the TB-splines Q_{Z+1} by Theorem 13.43, p. 244,

$$\left(\frac{d}{dx} - \mu\right) Q[\mu, \lambda_1, \ldots, \lambda_Z](x) = \frac{1}{e^{\mu}} \{Q[\lambda_1, \ldots, \lambda_Z](x) - Q[\lambda_1, \ldots, \lambda_Z](x-1) e^{\mu}\}.$$

8. The **recurrence** relation for Q_{Z+1} from (13.30), p. 247: if $\lambda_1 \ne \lambda_{Z+1}$ then

$$Q[\lambda_1, \lambda_2, \ldots, \lambda_{Z+1}](x)$$

$$= \frac{e^{-\lambda_{Z+1}}}{\lambda_1 - \lambda_{Z+1}} Q[\lambda_2, \ldots, \lambda_{Z+1}](x) + \frac{-e^{-\lambda_1}}{\lambda_1 - \lambda_{Z+1}} Q[\lambda_1, \ldots, \lambda_Z](x)$$

$$+ \frac{-1}{\lambda_1 - \lambda_{Z+1}} Q[\lambda_2, \ldots, \lambda_{Z+1}](x-1) + \frac{1}{\lambda_1 - \lambda_{Z+1}} Q[\lambda_1, \ldots, \lambda_Z](x-1).$$

9. The **symmetry** property relates $Q_{Z+1}(((Z+1)/2) + x)$ and $Q_{Z+1}(((Z+1)/2) - x)$, namely

$$Q_{Z+1}(Z+1-x) = e^{(-n+2p)p} \cdot e^{-(-n+2p)x} \cdot Q_{Z+1}(x)$$

$$= e^{(-n+2p)(p-x)} \cdot Q_{Z+1}(x)$$

which was proved in Theorem 13.63, p. 262, for the special case of vector Λ of (10.28), p. 170.

Let us remark that, in the polynomial case, the result of point (7) is used by Chui [3, p. 179] essentially for the construction of the wavelet.

14.2 Riesz bounds

Let us proceed to find the Riesz bounds for the set of shifts of the function $Q_{Z+1}(x)$. We will need the notion of the symmetrized nonordered vector.

Definition 14.5 *Let the nonordered vector* $\Lambda = [\mu_1, \ldots, \mu_s]$ *be given. The* **symmetrized vector** $\widetilde{\Lambda}$ *is a nonordered vector which consists of $2s$ elements, namely all μ_j and all $-\mu_j$, and we put*

$$\widetilde{\Lambda} := [\mu_1, \ldots, \mu_s, -\mu_1, \ldots, -\mu_s].$$

We now have the following fundamental theorem.

Theorem 14.6 Let the nonordered vector $\Lambda = [\lambda_1, \lambda_2, \ldots, \lambda_{Z+1}]$ be given. Let

$$S(\xi) := S[\Lambda](\xi) := \sum_{j=-\infty}^{\infty} |\widehat{Q_{Z+1}[\Lambda]}(\xi + 2\pi j)|^2. \quad (14.3)$$

Then for every $\xi \in \mathbb{R}$ the equality

$$S(\xi) = e^{-\lambda_1 - \cdots - \lambda_{Z+1}} \Phi_{2Z+1}[\widetilde{\Lambda}](Z+1, e^{i\xi}) \quad (14.4)$$
$$= e^{-\lambda_1 - \cdots - \lambda_{Z+1}} \Phi_{2Z+1}[\widetilde{\Lambda}](Z+1, e^{-i\xi})$$

holds where the nonordered vector $\widetilde{\Lambda}$ is the symmetrized vector of Λ, and the function $\Phi_{2Z+1}[\widetilde{\Lambda}](x; \lambda)$ is the Schoenberg "exponential Euler L-spline" defined in (13.42), p. 254, by

$$\Phi_{2Z+1}[\widetilde{\Lambda}](x; \lambda) = \sum_{j=-\infty}^{\infty} Q_{2Z+2}[\widetilde{\Lambda}](x-j) \lambda^j.$$

The set of shifts
$$\{Q_{Z+1}[\Lambda](x-j): \text{ for all } j \in \mathbb{Z}\}$$
is a **Riesz basis** for the linear space it generates
$$\operatorname*{clos}_{L_2(\mathbb{R})} \{Q_{Z+1}(x-j): \text{ for all } j \in \mathbb{Z}\},$$

in other words, there exist two constants A, B with $0 < A \leq B < \infty$ called **Riesz bounds** such that for every bi-infinite sequence $\{c_j\}_{j=-\infty}^{\infty}$ in ℓ_2 the following inequality holds:

$$A\|c\|_{\ell_2}^2 \leq \left\| \sum_{j=-\infty}^{\infty} c_j Q_{Z+1}(x-j) \right\|_{L_2(\mathbb{R})}^2 \leq B\|c\|_{\ell_2}^2. \quad (14.5)$$

The **Riesz bounds** are given by

$$A = e^{-\lambda_1 - \cdots - \lambda_{Z+1}} |\Pi_{2Z+1}(-1)|,$$
$$B = e^{-\lambda_1 - \cdots - \lambda_{Z+1}} |\Pi_{2Z+1}(1)|.$$

Here the function $\Pi_{2Z+1}(\lambda) = \Pi_{2Z+1}[\widetilde{\Lambda}](\lambda)$ is the Euler–Frobenius polynomial related to Φ_{2Z+1} by formula (13.44), p. 255.

If we prove the first equality in (14.4) then the second holds since $S(\xi)$ is a real number by its definition. On the other hand the second equality follows directly from the fact that $\widetilde{\Lambda}$ is a symmetric nonordered vector and by (13.46), p. 256.

In order to make the proof of the above theorem more transparent we will carry it out in the special case $Z = 3$. The general case does not differ essentially.

Proof Let $Z = 3$. We have the nonordered vector $\Lambda = [\lambda_1, \lambda_2, \lambda_3, \lambda_4]$. Recall that in formula (13.19), p. 239, we have defined the compactly supported TB-spline $Q_4[\Lambda](x)$ and it has a support coinciding with the interval $[0, 4]$.

The linear independence of the set $\{Q_4(x - k): \text{ for all } k \in \mathbb{Z}\}$ follows by the fundamental Theorem 13.38, p. 241. We shall prove that this set is a Riesz basis.

By Theorem 14.1, p. 268, we see that inequality (14.5) holds for some constants $A, B > 0$ if and only if the inequality

$$A \leq S(\xi) \leq B, \qquad \text{a.e. in } \mathbb{R},$$

holds, where we have put

$$S(\xi) = \sum_{j=-\infty}^{\infty} |\widehat{Q_4(x)}(\xi + 2\pi j)|^2. \tag{14.6}$$

We would like to apply Theorem 14.3, p. 269, to $S(\xi)$. By formula (13.23), p. 241, for the Fourier transform of Q_4 we find that, for large $|\xi| \to \infty$ and for some constant $C > 0$, the following inequality holds:

$$|\widehat{Q_4}(\xi)| = \left| \prod_{j=1}^{4} \frac{(e^{-\lambda_j} - e^{-i\xi})}{i\xi - \lambda_j} \right| \leq \frac{C}{|\xi|^4}.$$

Since $Q_4(x)$ has a compact support (hence decays arbitrarily fast in x) we may indeed apply Theorem 14.3, (i), p. 269, and obtain the equality

$$S(\xi) = \sum_{j=-\infty}^{\infty} \left\{ \int_{-\infty}^{\infty} Q_4(y+j)Q_4(y)\,dy \right\} e^{-ij\xi}.$$

Further we apply the *generalized Hermite–Gennocchi formula* provided in Remark 14.4, (2), on p. 270, to every integral inside the sum, and change the variables, by putting $\chi_j = 1 - \tau_j$, to obtain the equality

$$\int_{-\infty}^{\infty} Q_4(y+j)Q_4(y)\,dy$$

$$= \int_{[0,1]^4} Q_4(\chi_1 + \cdots + \chi_4 + j) e^{\lambda_1(\chi_1-1)+\cdots+\lambda_4(\chi_4-1)} \, d\chi_1 \ldots d\chi_4$$

$$= \int_{[0,1]^4} Q_4(4 - \tau_1 - \cdots - \tau_4 + j) e^{(-\lambda_1)\tau_1+\cdots+(-\lambda_4)\tau_4} \, d\tau_1 \ldots d\tau_4.$$

To the last integral we now apply *four* times the convolution formula for TB-splines provided above in Remark 14.4, (1), p. 270. This gives the equality

$$\int_{-\infty}^{\infty} Q_4(y+j)Q_4(y)\,dy = e^{-\lambda_1-\cdots-\lambda_4} \cdot Q_8[\widetilde{\Lambda}](4+j),$$

where $\widetilde{\Lambda}$ is the *nonordered vector* which is symmetrization of the vector Λ, i.e.

$$\widetilde{\Lambda} = [\lambda_1, \lambda_2, \lambda_3, \lambda_4, -\lambda_1, -\lambda_2, -\lambda_3, -\lambda_4]$$
$$= [t_1, t_2, t_3, t_4, t_5, t_6, t_7, t_8],$$

where in the last we assume that

$$t_1 \leq t_2 \leq t_3 \leq t_4 \leq t_5 \leq t_6 \leq t_7 \leq t_8,$$

and $Q_8(x) = Q_8[\widetilde{\Lambda}](x)$ is the TB-spline which, by formula (13.19), p. 239, corresponds to the nonordered vector $\widetilde{\Lambda}$. It corresponds to the polynomial q_8 and the operator $\mathcal{L}_8$ defined in (13.3), p. 224, given by

$$\mathcal{L}_8 = q_8\left(\frac{d}{dx}\right) = \prod_{j=1}^{8}\left(\frac{d}{dx} - t_j\right),$$

where we have put $Z = 7$. Further we apply all definitions to the case of $Z = 7$ and the L-splines generated through the operator $\mathcal{L}_8$. Let us return to the expression for $S(\xi)$ given now by

$$S(\xi) = e^{-\lambda_1 - \lambda_2 - \cdots - \lambda_4} \cdot \sum_{j=-\infty}^{\infty} Q_8(4+j)e^{-ij\xi} \qquad (14.7)$$

which we have to analyze.

Let us note the important circumstance that the set $\widetilde{\Lambda}$ is symmetric, i.e. $\lambda \in \widetilde{\Lambda}$ implies $-\lambda \in \widetilde{\Lambda}$.

Let us recall the definition of the Schoenberg's *exponential Euler L-spline*, the function Φ_7 in formula (13.42), p. 254. We have

$$\Phi_7\left(4; \frac{1}{\lambda}\right) = \sum_{j=-\infty}^{\infty} Q_8(4+j)\lambda^j = \sum_{j=-3}^{3} Q_8(4+j)\lambda^j$$

(recall that the support of Q_8 coincides with the interval $[0, 8]$, $Q_8(8) = Q_8(0) = 0$), which gives for $\lambda = e^{-i\xi}$ the representation

$$S(\xi) = C_S \cdot \Phi_7(4; e^{i\xi}),$$

where we have put

$$C_S := e^{-\lambda_1 - \cdots - \lambda_4}.$$

Let us put, as before in formula (13.12), p. 235,

$$r_8(\lambda) = \prod_{j=1}^{8}(e^{t_j} - \lambda). \qquad (14.8)$$

According to the theory developed by Micchelli we have the following relation between the *generalized Euler–Frobenius polynomial* Π_7 and the TB-spline Q_8, see formula (13.14), and formula (13.45), p. 255,

$$\Pi_7(\lambda) = r_8(\lambda)A_7(0; \lambda). \qquad (14.9)$$

Further by the basic properties of Schoenberg's *exponential Euler L-spline*, formulas (13.43), and (13.45), p. 255, it follows that

$$\Phi_7(4;\lambda) = \lambda^4 \Phi_7(0;\lambda) = \lambda^4 \lambda s_8\left(\frac{1}{\lambda}\right) A_7(0;\lambda)$$

$$= \lambda^5 s_8\left(\frac{1}{\lambda}\right) \frac{\Pi_7(\lambda)}{r_8(\lambda)}.$$

The relation between the functions $s_8(\lambda)$ and $r_8(\lambda)$ in formula (13.34), p. 249, implies

$$\Phi_7(4;\lambda) = \lambda^{-3} \Pi_7(\lambda) \quad \text{for all } \lambda \text{ in } \mathbb{C}.$$

Hence, for $\lambda = e^{i\xi}$ we obtain

$$S(\xi) = C_S \cdot e^{-3i\xi} \cdot \Pi_7(e^{i\xi}).$$

According to Theorem 13.31, p. 237, the polynomial $\Pi_7(\lambda)$ has precisely six negative zeros $v_1, \ldots, v_6$ such that

$$0 > v_1 > \cdots > v_6,$$

$$v_1 v_6 = v_2 v_5 = v_3 v_4 = 1.$$

Now we return to the techniques used by Chui [3, pp. 89–90] and we will find the estimates from above and from below by using these properties of the roots of the polynomial $\Pi_7(\lambda)$. According to formula (13.41), p. 254, we have the representation

$$\Pi_7(\lambda) = D \cdot \prod_{j=1}^{6} (\lambda - v_j),$$

where D is a nonzero constant which will no longer play an important role and so we do not specify its value. Hence, since the roots v_j are real we obtain the equalities

$$|\Pi_7(e^{-i\xi})| = |D| \cdot \prod_{j=1}^{6} |e^{-i\xi} - v_j|$$

$$= |D| \cdot \prod_{j=1}^{3} |e^{-i\xi} - v_j| \left|e^{-i\xi} - \frac{1}{v_j}\right|$$

$$= |D| \cdot \prod_{j=1}^{3} \frac{|e^{i\xi} - v_j|^2}{|v_j|}$$

$$= |D| \cdot \prod_{j=1}^{3} \frac{1 - 2v_j \cos\xi + v_j^2}{|v_j|}.$$

Since all v_j satisfy $v_j < 0$, we have the evident estimates

$$|\Pi_7(e^{-i\xi})| \leq \prod_{j=1}^{3} \frac{1 - 2v_j + v_j^2}{|v_j|} = |\Pi_7(1)|$$

$$= \prod_{j=1}^{3} \frac{(1-v_j)^2}{|v_j|},$$

and

$$|\Pi_7(e^{-i\xi})| \geq |D| \cdot \prod_{j=1}^{3} \frac{1 + 2v_j + v_j^2}{|v_j|} = |\Pi_7(-1)|$$

$$= |D| \cdot \prod_{j=1}^{3} \frac{(1+v_j)^2}{|v_j|}.$$

Finally, we obtain the estimate from above:

$$|S(\xi)| = C_S \cdot |\Pi_7(e^{i\xi})|$$
$$\leq C_S \cdot |\Pi_7(1)| = B,$$

and in a similar way the estimate from below:

$$|S(\xi)| \geq C_S \cdot |\Pi_7(-1)| = A.$$

Due to the above mentioned properties of the zeros of the polynomial Π_7 we see that

$$\Pi_7(1) \neq \infty, \qquad \Pi_7(-1) \neq 0,$$
$$r_8(1) \neq 0, \qquad r_8(-1) \neq 0.$$

It follows that

$$0 < A \leq B < \infty$$

which proves the theorem. ∎

We now obtain Corollary 14.7 about the shifts on the mesh $h\mathbb{Z}$.

Corollary 14.7 *Using the notation of Theorem 14.6, p. 272. let us put*

$$S[\Lambda; h](\xi) := \sum_{j=-\infty}^{\infty} \left| \widehat{\varrho_{Z+1}[\Lambda; h]} \left(\frac{\xi + 2\pi j}{h} \right) \right|^2.$$

Then for every $\xi \in \mathbb{R}$, we obtain the equality

$$S[\Lambda; h](\xi) = h^{2Z+2} e^{-(\lambda_1 + \cdots + \lambda_{Z+1})h} \cdot \Phi_{2Z+1}[h\widetilde{\Lambda}](Z+1, e^{i\xi}).$$

Consequently, the set of shifts

$$\{Q_{Z+1}[\Lambda; h](x - jh): \text{ for all } j \in \mathbb{Z}\}$$

forms a Riesz basis *with* Riesz bounds

$$A_1 = h^{2Z+1} A, \qquad B_1 = h^{2Z+1} B,$$

where A, B are Riesz bounds for the set of shifts

$$\{Q_{Z+1}[h\Lambda](x - j): \text{ for all } j \in \mathbb{Z}\}.$$

Proof By formula (13.25), p. 242, we have

$$\widehat{Q_{Z+1}[\Lambda; h]}(\xi) = h^{Z+1} \cdot \widehat{Q_{Z+1}[h\Lambda]}(h\xi),$$

which implies that

$$S[\Lambda; h](\xi) = h^{2Z+2} \cdot \sum_{j=-\infty}^{\infty} |\widehat{Q_{Z+1}[h\Lambda]}(\xi + 2\pi j)|^2$$

$$= h^{2Z+2} e^{-(\lambda_1 + \cdots + \lambda_{Z+1})h} \cdot \Phi_{2Z+1}[h\widetilde{\Lambda}](Z+1, e^{i\xi}).$$

To prove the second statement we use formula (13.26), p. 243, namely

$$Q_{Z+1}[\Lambda; h](x) = h^Z \cdot Q_{Z+1}[h\Lambda]\left(\frac{x}{h}\right),$$

which implies that

$$\sum_{j=-\infty}^{\infty} c_j Q_{Z+1}[\Lambda; h](x - jh) = h^Z \sum_{j=-\infty}^{\infty} c_j Q_{Z+1}[h\Lambda]\left(\frac{x}{h} - j\right).$$

By the substitution $(x/h) \to x$ we obtain

$$\left\| h^Z \sum_{j=-\infty}^{\infty} c_j Q_{Z+1}[h\Lambda]\left(\frac{x}{h} - j\right) \right\|_{L_2(\mathbb{R})}^2$$

$$= \int_{-\infty}^{\infty} \left| h^Z \sum_{j=-\infty}^{\infty} c_j Q_{Z+1}[h\Lambda]\left(\frac{x}{h} - j\right) \right|^2 dx$$

$$= h^{2Z+1} \left\| \sum_{j=-\infty}^{\infty} c_j Q_{Z+1}[h\Lambda](x - j) \right\|_{L_2(\mathbb{R})}^2.$$

By Theorem 14.6 the set of shifts $\{Q_{Z+1}[h\Lambda](x - jh): \text{ for all } j \in \mathbb{Z}\}$ has the Riesz basis property with bounds A, B. Now using the equivalence Theorem 14.1, p. 268, we see that the set of shifts $\{Q_{Z+1}[\Lambda; h](x - jh): j \in \mathbb{Z}\}$ has the Riesz basis property with Riesz bounds A, B if and only if

$$h^{2Z+1} A \|c\|_{\ell_2} \leq h^{2Z+1} \sum_{j=-\infty}^{\infty} |\widehat{Q_{Z+1}[h\Lambda]}(\xi + 2\pi j)|^2 \leq h^{2Z+1} B \|c\|_{\ell_2}.$$

Hence $A_1 = h^{2Z+1} A$ and $B_1 h^{2Z+1} B$. This completes the proof. ∎

Let us prepare the reader for the fact that later on we will consider shifts of the set Λ_k for some constant β and all the above results are effortlessly carried over to this case. We define the shifted nonordered vector $\Lambda_k + \beta$ by

$$\Lambda_k + \beta := [\lambda_1 + \beta, \ldots, \lambda_{2p} + \beta]. \tag{14.10}$$

14.3 The asymptotic of $A_Z(0; \lambda)$ in k

The special choice of the nonordered vector Λ of the form (14.1), p. 267, arising through the spherical operators $L_{(k)}^p$ and the related constant coefficients operators $M_{k,p}$, see (10.26), p. 169, is of central interest for us.

For simplicity we will often specify the case $p = 2$, where $Z = 2p - 1 = 3$, and the operator $M_{k,2}$ is given by formula (13.3), p. 224, namely

$$M_{k,2}\left(\frac{d}{dv}\right) = \prod_{j=1}^{4}\left(\frac{d}{dv} - \lambda_j\right)$$

where

$$\Lambda_k = \begin{cases} \lambda_1 = -n - k + 2, \\ \lambda_2 = -n - k + 4, \\ \lambda_3 = k, \\ \lambda_4 = k + 2. \end{cases} \tag{14.11}$$

This corresponds to the biharmonic polysplines. The corresponding symmetrized vector is $\widetilde{\Lambda} = [t_1, t_2, \ldots, t_8]$. When the dimension n of the space satisfies $n \geq 6$ or is odd then for large k we have only pairwise different t_js given by

$$\begin{aligned} &t_1 = -n - k + 2; \quad t_2 = -n - k + 4; \quad t_3 = -k - 2; \quad t_4 = -k; \\ &t_5 = k; \quad\quad\quad\quad\; t_6 = k + 2; \quad\quad\;\; t_7 = n + k - 4; \quad t_8 = n + k - 2. \end{aligned} \tag{14.12}$$

For us it is essential to see that asymptotically for $k \longrightarrow \infty$

$$\frac{B}{A} \longrightarrow \text{const}.$$

First we have to analyze the asymptotics of the function $A_Z(0; \lambda)$.

Theorem 14.8 *Let the vector $\widetilde{\Lambda}$ be the symmetrized vector of the vector $\Lambda_k = [\lambda_1, \lambda_2, \ldots, \lambda_{2p}]$, where λ_j are those coming from the spherical operators $L_{(k)}$, and are given by (14.1), p. 267. Then the function $A_{4p-1}(x; \lambda)$ which corresponds to the vector $\widetilde{\Lambda}_k$ satisfies the estimates*

$$\frac{C_1}{k^{4p-1}} \leq |A_{4p-1}(0; \lambda)| \leq \frac{C_2}{k^{4p-1}} \quad \text{for all } |\lambda| = 1, \tag{14.13}$$

for some constants $C_1, C_2 > 0$, i.e. $A_{4p-1}(0; \lambda) \approx 1/(k^{4p-1})$ for $k \to \infty$.[3]

Proof For simplicity we prove the case $p = 2$.

Assume also for simplicity that all t_js are pairwise different. According to formula (13.10), p. 234, we have the representation

$$A_7(0; \lambda) = \sum_{j=1}^{8} \frac{1}{q_8'(t_j)} \cdot \frac{1}{e^{t_j} - \lambda},$$

where the polynomial $q_8(\lambda)$ is given by $q_8(\lambda) = \prod_{j=1}^{8}(\lambda - t_j)$. In order to compute the asymptotics for $k \to \infty$, we will use the special structure of the vector $\widetilde{\Lambda} = [t_1, \ldots, t_8]$.

Let us consider $A_7(0; \lambda)$ as a function of k for all $|\lambda| = 1$. In the above expression for $A_7(0; \lambda)$ it is evident that the second half of the terms satisfy

$$\sum_{j=5}^{8} \frac{1}{q_8'(t_j)} \cdot \frac{1}{e^{t_j} - \lambda} \xrightarrow{k \to \infty} 0,$$

since $t_j > 0$ for $j = 5, \ldots, 8$. With regard to the other terms, we see that

$$\frac{1}{e^{t_j} - \lambda} \xrightarrow{k \to \infty} \frac{1}{-\lambda} \quad \text{for } j = 1, \ldots, 4,$$

since $t_j < 0$, for $j = 1, \ldots, 4$. Hence, asymptotically, for $k \to \infty$

$$A_7(0; \lambda) \approx \frac{1}{-\lambda} \cdot \sum_{j=1}^{4} \frac{1}{q_8'(t_j)}.$$

However, $q_8'(t_j)$ are polynomials of k whereas, for all $j \leq 4$, e^{t_j} contains $\exp(-k)$. Thus for large k we have

$$\sum_{j=5}^{8} \frac{1}{q_8'(t_j)} \cdot \frac{1}{e^{t_j} - \lambda} = \frac{T_1(k, e^{-k})}{T_2(k, e^{-k})},$$

and

$$\frac{1}{-\lambda} \sum_{j=5}^{8} \frac{1}{q_8'(t_j)} = \frac{T_1(k, 0)}{T_2(k, 0)},$$

---

[3] We will say that two sequences a_k and b_k have the same asymptotic order for $k \to \infty$ if and only if there exist two positive constants C_1 and C_2 such that $C_1|a_k| \leq |b_k| \leq C_2|a_k|$ for sufficiently large k. In such a case we will write $a_k \approx b_k$.

where $T_1(\cdot, \cdot)$ and $T_2(\cdot, \cdot)$ are polynomials of two variables. However, in Lemma 14.9, p. 280, we will prove that there exist two constants $D_1, D_2 > 0$ such that

$$\frac{D_1}{k^7} \leq \sum_{j=1}^{4} \frac{1}{q_8'(t_j)} \leq \frac{D_2}{k^7} \quad \text{for all } k \geq 1.$$

It is easy to see from the above that there exist two constants $C_1', C_2' > 0$, independent of k and l, such that for $k \to \infty$ the following two inequalities hold:

$$C_1' \left| \frac{1}{\lambda} \right| \left| \sum_{j=1}^{4} \frac{1}{q_8'(t_j)} \right| \leq |A_7(0; \lambda)| \leq C_2' \left| \frac{1}{\lambda} \right| \left| \sum_{j=1}^{4} \frac{1}{q_8'(t_j)} \right|.$$

Hence we obtain

$$\frac{C_1}{k^7} \leq |A_7(0; \lambda)| \leq \frac{C_2}{k^7}.$$

■

Lemma 14.9 *There exist two constants $C_1, C_2 > 0$ such that for $k \to \infty$ the following estimate holds:*

$$\frac{C_1}{k^{4p-1}} \leq \left| \sum_{j=1}^{2p} \frac{1}{q_{4p}'(t_j)} \right| \leq \frac{C_2}{k^{4p-1}}.$$

Proof We will use a residuum representation. Let us put $\gamma(\theta) = e^{i\theta}$ for $\theta \in [0, 2\pi]$ and $\Gamma = k(\gamma(\theta) - 1)$. Then for k large enough the contour Γ will encompass the negative points $t_1, t_2, \ldots, t_{2p}$ but not the points t_j with $j \geq 2p + 1$. Thus we have

$$U_k := \sum_{j=1}^{2p} \frac{1}{q_{4p}'(t_j)} = \frac{1}{2\pi i} \int_\Gamma \frac{dz}{q_{4p}(z)}.$$

Since

$$q_{4p}(z) = \prod_{j=1}^{4p} (z - t_j) = \prod_{j=1}^{p} (z - (-n - k + 2j)) \prod_{j=0}^{p-1} (z - (k + 2j)),$$

we see that for larger k

$$k^{4p-1} U_k = \frac{1}{2\pi i} \int_0^{2\pi} \prod_{j=1}^{p} \left(\gamma + \frac{n - 2j}{k} \right)^{-1} \prod_{j=0}^{p-1} \left(\gamma - 2 - \frac{2j}{k} \right)^{-1} \gamma' \, d\theta.$$

For $k \to \infty$ the last integral obviously converges to

$$\int_\gamma z^{-p} (z - 2)^{-p} \, dz,$$

which is easily seen to be nonzero. This proves the lemma. ■

Remark 14.10 1. A more subtle study of the behavior of the function $A_{4p-1}(x;\lambda)$ is carried out in [9], where the estimate from above is proved in the same way.

2. From the proof it is clear that Theorem 14.8 holds for $\lambda \in K \subset \mathbb{C}$ where K is a compact set satisfying $\{0\} \notin K$.

14.4 Asymptotic of the Riesz bounds A, B

We now provide the asymptotic of A and B with respect to the parameter k.

Theorem 14.11 As in Theorem 14.6, p. 272, for arbitrary integer $p \geq 1$ we denote by A, B the Riesz bounds for the set of shifts $\{Q_{2p}[\Lambda](x-j)\colon \text{for all } j \in \mathbb{Z}\}$, where $Q_{2p}[\Lambda](x)$ is the T B-spline defined through the vector $\Lambda = \Lambda_k = [\lambda_1, \lambda_2, \ldots, \lambda_{2p}]$ given by (14.1), p. 267. Then for $k \to \infty$ the asymptotic order of A and B is given by

$$A \approx \frac{e^{2pk}}{k^{4p-1}}, \qquad B \approx \frac{e^{2pk}}{k^{4p-1}},$$

i.e. there exist two constants $C_1, C_2 > 0$ such that

$$C_1 \frac{e^{2pk}}{k^{4p-1}} \leq A \leq C_2 \frac{e^{2pk}}{k^{4p-1}}, \qquad C_1 \frac{e^{2pk}}{k^{4p-1}} \leq B \leq C_2 \frac{e^{2pk}}{k^{4p-1}},$$

for all $k \geq 1$.

Now let us take $\Lambda = \Lambda_k + (n/2)$. The Riesz bounds A, B in Theorem 14.6 for this vector Λ have the same asymptotic order as those for the vector Λ_k.

Proof We will use the notation and results of Theorem 14.6, p. 272, where we saw that we can take the Riesz bounds

$$A := C_S \cdot |\Pi_{4p-1}(-1)|,$$
$$B := C_S \cdot |\Pi_{4p-1}(1)|,$$

where

$$C_S = e^{-\lambda_1 - \cdots - \lambda_{2p}}.$$

As before by $\widetilde{\Lambda} = [t_1, t_2, \ldots, t_{4p}]$ we denote the vector which is the symmetrized vector of the vector $\Lambda_k = [\lambda_1, \lambda_2, \ldots, \lambda_{2p}]$. From the definition of the function $r_{4p}[\widetilde{\Lambda}]$ in (13.12), p. 235, we obtain asymptotically for $k \to \infty$ the following order:

$$r_{4p}(\lambda) \approx e^{t_{2p+1} + \cdots + t_{4p}} \cdot (-\lambda)^{2p} \quad \text{for all } |\lambda| = 1.$$

We apply Theorem 14.8, p. 279, about the asymptotic order of $A_{4p-1}(0;\lambda)$, which gives

$$\Pi_{4p-1}(1) = r_{4p}(1) A_{4p-1}(0;1) \approx e^{t_{2p+1}+\cdots+t_{4p}} \cdot \frac{1}{k^{4p-1}},$$

$$\Pi_{4p-1}(-1) = r_{4p}(-1) A_{4p-1}(0;-1) \approx e^{t_{2p+1}+\cdots+t_{4p}} \cdot \frac{1}{k^{4p-1}}.$$

Due to $\sum_{j=1}^{2p} \lambda_j = \text{const}$ and $\sum_{j=2p+1}^{4p} t_j = 2pk + \text{const}$ the last implies

$$A = C_S |\Pi_{4p-1}(-1)| \approx C_S \cdot e^{t_{2p+1}+\cdots+t_{4p}} \cdot \frac{1}{k^{4p-1}}$$

$$\approx \frac{e^{2pk}}{k^{4p-1}} \quad \text{for } k \geq 1.$$

In a similar way we prove the asymptotic order of B. ∎

Remark 14.12 *It is easy to see by the same method of proof that the result remains true if we consider the nonordered vector $\Lambda = \Lambda_k + \beta$ for some constant β. The essence here is that the set Λ_k is "almost symmetric"; and one might use larger generalizations but we will not need them further.*

14.4.1 Asymptotic for TB-splines Q_{Z+1} on the mesh $h\mathbb{Z}$

Using the transformation formula (13.26), p. 243, namely

$$Q[\Lambda, h](x) = h^Z Q_{Z+1}[h\Lambda]\left(\frac{x}{h}\right),$$

we see that we may reduce the L-splines on the grid $h\mathbb{Z}$ to L-splines on the grid $\mathbb{Z}$, which will be important for the wavelet analysis.

Corollary 14.13 *Let for $p \geq 1$ the nonordered vector Λ be given by (14.1), p. 267, and its symmetrization $\widetilde{\Lambda}$ be given by $\widetilde{\Lambda} = [t_1, t_2, \ldots, t_{4p}]$. Let us denote the constants*

$$D_1 := \lambda_1 + \lambda_2 + \ldots + \lambda_{2p},$$
$$D_2 + 2pk := t_{2p+1} + t_{2p+2} + \ldots + t_{4p}.$$

Then the asymptotic order of the Riesz bounds A, B for the set of shifts

$$\{Q[\Lambda; h](x - jh) : \text{ for all } j \in \mathbb{Z}\}$$

is given for all $k \geq 1$ by

$$A \approx e^{h(D_2 - D_1)} h^{2Z} \frac{e^{2pkh}}{k^{4p-1}}, \quad B \approx e^{h(D_2 - D_1)} h^{2Z} \frac{e^{2pkh}}{k^{4p-1}}.$$

Proof The proof follows by replacing Λ by $h\Lambda$ in Theorem 14.11, p. 281. Indeed, as in the proof of this theorem we obtain the Riesz bounds for the set of shifts $\{Q[h\Lambda](x - h) : \text{ for all } j \in \mathbb{Z}\}$, given by

$$A_1 := C_S \cdot |\Pi_{4p-1}(-1)| \approx C_S \cdot e^{h(t_{2p+1}+\cdots+t_{4p})} \cdot \frac{1}{h \cdot k^{4p-1}}$$

$$= e^{-h(\lambda_1+\lambda_2+\cdots+\lambda_{2p})} \cdot e^{h(t_{2p+1}+\cdots+t_{4p})} \cdot \frac{1}{k^{4p-1}},$$

as is the case for B_1. Note that we have also reconsidered the asymptotic of $A_{4p-1}[h\Lambda](0;1)$, which is now – by the proof of Theorem 14.8, p. 279 equal to $1/(h \cdot k^{4p-1})$. Further we apply Corollary 14.7, p. 276. This ends the proof. ∎

Note also that for $k = 0$ we have $A, B > 0$ by Theorem 14.6, p. 272.

14.5 Synthesis of compactly supported polysplines on annuli

An immediate application of the above results is the construction of compactly supported cardinal polysplines on annuli, i.e. of polysplines having as break-surfaces the spheres $S(0; e^j)$ and having compact support. They are the analog to the TB-spline $Q_{\mathbb{Z}+1}$ in the one-dimensional case. We will provide the main arguments for such a construction but will not pay much attention to the details.

Again we assume that the vector $\Lambda = \Lambda_k$ is given by (14.1), p. 267. As we have said in Theorem 14.11, p. 281, the Riesz bounds A, B for the set of functions

$$\left\{ Q_{\mathbb{Z}+1}\left[\Lambda_k + \frac{n}{2}, h\right](v - vh) \colon \text{ for all } v \in \mathbb{Z} \right\}$$

have the same sharp asymptotic in k, i.e. if we put

$$\varphi(k) = \frac{e^{2pkh}}{k^{4p-1}} \quad \text{for } k \geq 1,$$

$$\varphi(0) = 1,$$

then there exist constants C_1 and C_2 such that

$$0 < C - C_1 \leq \frac{A}{\varphi(k)} \leq \frac{B}{\varphi(k)} \leq C + C_2 < \infty.$$

Remark 14.14 *Let us observe that we have the above results for every nonordered vector $\Lambda_k + \beta$, where $\beta \in \mathbb{R}$, i.e. not only for $\beta = n/2$.*

As a result, we obtain Definition 14.15.

Definition 14.15 *Let β be an arbitrary real number. The normed TB-spline $\widetilde{Q}$ is defined by the equality*

$$\widetilde{Q}[\Lambda_k + \beta; h](v) := \frac{1}{\sqrt{\varphi(k)}} Q[\Lambda_k + \beta; h](v). \tag{14.14}$$

For $\widetilde{Q}[\Lambda_k + \beta; h]$ we obviously have the Riesz bounds $A = C - C_1$ and $B = C + C_2$ which are independent of $k \geq 0$. It means that for every integer $k \geq 0$, and for every sequence $\{c_j\} \in \ell_2$, the following Riesz inequality holds:

$$(C - C_1) \sum_{j=-\infty}^{\infty} |c_j|^2 \leq \left\| \sum_{j=-\infty}^{\infty} c_j \widetilde{Q}[\Lambda_k + \beta; h](v - j) \right\|^2 \leq (C + C_2) \sum_{j=-\infty}^{\infty} |c_j|^2.$$

$$\tag{14.15}$$

284 Multivariate polysplines

For simplicity let us now take the special case $p = 2$ when the TB-spline is given by
$$Q_4(v) = Q_4\left[\Lambda_k + \frac{n}{2}\right](v).$$

The *normed* TB-spline is given by
$$\tilde{Q}_4(v) := \sqrt{\frac{k^7}{e^{4k}}} \cdot Q_4(v).$$

The importance of the vector $\Lambda_k + (n/2)$ will become clear from the next explanation.

Let the polyspline f of order $p \geq 1$ have as break-surfaces the spheres $S(0; e^j)$ and belong to L_2. Let its expansion in spherical harmonics be
$$f(x) = \sum_{k=0}^{\infty} \sum_{\ell=1}^{d_k} f^{k,\ell}(\log r) Y_{k,\ell}(\theta), \quad \text{where} \quad x = r\theta.$$

Here the functions $f^{k,\ell}(v)$ are L-splines for $L = M_{k,p} = \mathcal{L}[\Lambda_k]$. Since $f \in L_2(\mathbb{R}^n)$ it follows that
$$\|f\|^2_{L_2(\mathbb{R}^n)} = \int_{\mathbb{R}^n} \left| \sum_{k=0}^{\infty} \sum_{\ell=1}^{d_k} f^{k,\ell}(\log r) Y_{k,\ell}(\theta) \right|^2 dx$$
$$= \int_0^{\infty} \int_{\mathbb{S}^{n-1}} \left| \sum_{k=0}^{\infty} \sum_{\ell=1}^{d_k} f^{k,\ell}(\log r) Y_{k,\ell}(\theta) \right|^2 r^{n-1} dr\, d\theta < \infty.$$

After the change $v = \log r$ we obtain
$$\|f\|^2_{L_2(\mathbb{R}^n)} = \sum_{k=0}^{\infty} \sum_{\ell=1}^{d_k} \int_{-\infty}^{\infty} \left| e^{(n/2)v} f^{k,\ell}(v) \right|^2 dv < \infty.$$

It is important to note that the functions $e^{(n/2)v} f^{k,\ell}(v)$ are L-splines for the operator $L = \mathcal{L}[\Lambda_k + (n/2)]$.

Now let us again specify $p = 2$. We have the representation
$$e^{(n/2)v} f^{k,\ell}(v) = \sum_{\nu=-\infty}^{\infty} f_\nu^{k,\ell} \cdot \tilde{Q}_4(v - \nu).$$

By applying the above inequality (14.15), p. 283, we obtain
$$\|f\|^2_{L_2(\mathbb{R}^n)} = \sum_{k=0}^{\infty} \sum_{\ell=1}^{d_k} \left\| \sum_{\nu=-\infty}^{\infty} f_\nu^{k,\ell} \cdot \tilde{Q}_4(v - \nu) \right\|^2_{L_2}$$
$$\geq (C - C_1) \sum_{k=0}^{\infty} \sum_{\ell=1}^{d_k} \sum_{\nu=-\infty}^{\infty} \left| f_\nu^{k,\ell} \right|^2,$$

which shows that
$$\sum_{k=0}^{\infty}\sum_{\ell=1}^{d_k}\left|f_v^{k,\ell}\right|^2 < \infty.$$

Hence for every $v \in \mathbb{Z}$ the series
$$f_v(x) := \sum_{k=0}^{\infty}\sum_{\ell=1}^{d_k} f_v^{k,\ell} \cdot \widetilde{Q}_4(\log r - v) Y_{k,\ell}(\theta)$$

is convergent. It has a compact support coinciding with the set of x which is such that $r = |x|$ is in the support of the function $Q_4(\log r - v)$, i.e. in the interval
$$r \in [e^v, e^{v+4}].$$

Thus the function
$$Q_4(x) = \sum_{k=0}^{\infty}\sum_{\ell=1}^{d_k} g^{k,\ell} \cdot \widetilde{Q}_4(\log r) Y_{k,\ell}(\theta)$$

is a polyspline which is an analog to the one-dimensional spline Q_4. The choice of the coefficients $g^{k,\ell}$ gives many such polysplines.

Remark 14.16 *Let us note that the smoothness of $Q_4(x)$ is related to the smoothness of the function*
$$\sum_{k=0}^{\infty}\sum_{\ell=1}^{d_k} g^{k,\ell} \cdot Y_{k,\ell}(\theta)$$

on the sphere. We refer to Section (23.1.3), p. 464, which indicates when a function on the sphere belongs to the Sobolev space $H^s(\mathbb{S}^{n-1})$.

This is a general way to construct polysplines with a compact support which is analogous to the one-dimensional case where the support of $Q_4(v)$ is $[0, 4]$. So this polyspline might be considered as the "polyspline version" of the TB-spline $Q_4(\log r - v)$. Let us state the precise result, although we have not provided a rigorous proof of all the points.

Theorem 14.17 *Every cardinal biharmonic polyspline f which is in $C^2(\mathbb{R}^n)$ and in $L_2(\mathbb{R}^n)$ permits the following representation:*
$$f(x) = \sum_{v=-\infty}^{\infty} f_v(e^{-j}x), \qquad (14.16)$$

where $f_v(x)$ is a biharmonic polyspline in $C^2(\mathbb{R}^n)$ having compact support coinciding with the closure of the annulus $A(1, e^4)$ enclosed by the radii 1 and e^4.

If f has a compact support and $0 \notin \mathrm{Supp}(f)$ then the sum in (14.16) is finite.

From the above reasoning it is clear that this result also holds for an arbitrary integer $p \geq 1$.

Remark 14.18 *The above Theorem may be considered as a sampling result.*

Chapter 15

Cardinal interpolation polysplines on annuli

15.1 Introduction

The main purpose of this chapter is to find and develop a proper Ansatz for the interpolation of *polysplines on annuli* which would correspond to the *one-dimensional cardinal interpolation* with splines, or *L*-splines, studied by Schoenberg and Micchelli.

In Theorem 9.7, p. 124, we proved that every polyspline of order p on annuli, i.e. a polyspline $h(x)$ with concentric spheres as break-surfaces, is reduced through expansion in spherical harmonics to infinitely many L-splines. Namely, if the break-surfaces are the spheres $S(0, r_j)$ for $j = 1, 2, \ldots, N$, and the expansion of $h(x)$ in spherical harmonics for the basis $\{Y_{k,\ell}(\theta)\}$ is given by

$$h(x) = \sum_{k=0}^{\infty} \sum_{\ell=1}^{d_k} h_{k,\ell}(\log r) Y_{k,\ell}(\theta) \quad \text{for } r_j < r < r_{j+1},$$

then for every pair of indexes $k = 0, 1, 2, \ldots; \ell = 1, 2, \ldots, d_k$, the one-dimensional function $f_{k,\ell}(v)$ is an L-spline and satisfies

$$M_{k,p}\left(\frac{d}{dv}\right) h_{k,\ell}(v) = 0 \quad \text{for } r_j < e^v < r_{j+1}.$$

In the same theorem we have seen that for every fixed k and ℓ the interpolation conditions satisfied by every $f_{k,\ell}(v)$ are actually

$$h_{k,\ell}(\log r_j) = \psi_j^{k,\ell} \quad \text{for } j = 1, 2, \ldots, N,$$

where the data are taken from the expansion of the function $h(x)$ on the sphere $S(0; r_j)$, i.e. from

$$h(r_j\theta) = f_j(\theta) = \sum_{k=0}^{\infty} \sum_{\ell=1}^{d_k} \psi_j^{k,\ell} Y_{k,\ell}(\theta).$$

288 *Multivariate polysplines*

However, due to the results of Micchelli we have seen in Chapter 13 that the interpolation problem
$$h_{k,\ell}(\alpha + j) = c_j \quad \text{for } j \in \mathbb{Z}, \tag{15.1}$$
makes sense and is solvable for data $\{c_j\}_{j \in \mathbb{Z}}$ having power growth if and only if the constant α satisfies some condition arising through the operator $M_{k,p}$.

The above determines in a unique way our **Ansatz** and the notion of **cardinal polysplines on annuli**. Indeed, if the data points have to be at
$$v = \alpha + j$$
by the inverse transform $r = e^v$ it follows that the break-spheres $S(0; r_j)$ have to have radii
$$r_j = e^{\alpha + j}.$$

By Micchelli's theory for every index k there exists one value ξ_k for which the above problem is not uniquely solvable, see Theorem 13.33, p. 238. An important point of our analysis will be to show that the interpolation problem (15.1) is solvable for all indexes k and ℓ with parameter $\alpha = 0$. Note that in such a case the interpolation holds again on the same spheres, which are the break-surfaces.

We will see in the present chapter that the polyspline $h(x)$ has every right to be called the **cardinal interpolation polyspline**. Of course, one might think about a more general problem where
$$r_j = ab^j.$$
In that case the break-points of the one-dimensional L-splines for the operators $M_{k,p}$ are
$$\log r_j = \log a + j \log b.$$
This case has also been considered in Chapter 13.[1]

It is interesting that if one applies the so-called Mellin transform to the function $h(x)$ (i.e. the sequence of two transforms $(r, \theta) \to (v = \log r, \theta) \to (\xi, \theta)$ where the last is Fourier transform only in the variable v) then we will indeed have a cardinal analysis.

In order to set the above ideas into motion we will need the whole machinery of cardinal L-splines which we have developed in Chapter 13.

15.2 Formulation of the cardinal interpolation problem for polysplines

Now let us assume that a cardinal polyspline h on annuli is given. We assume that the interpolation data functions f_j are prescribed on the spheres $S_j = S(0; e^{j+\alpha})$ of radii $r_j = e^{j+\alpha}$ and have the expansion in spherical harmonics
$$f_j(\theta) = \sum_{k=0}^{\infty} \sum_{\ell=1}^{d_k} \psi_j^{k,\ell} Y_{k,\ell}(\theta) \quad \text{for all } \theta \text{ in } \mathbb{S}^{n-1}.$$

---
[1] This is exactly the case of cardinal L-splines studied by Schoenberg [19].

When we consider the problem for interpolation with cardinal polysplines we have to do the interpolation for the data $\psi_j^{k,\ell}$ for every $k \geq 0$ and $\ell = 1, \ldots, d_k$. Actually, we have different problems only for different ks. We will see that for $\alpha = 0$ the interpolation problem is solvable for data satisfying some power growth.

There is also nonuniqueness in choosing the *Ansatz* for "power growth data".

1. We may consider *power growth in the mean*

$$\left(\int |f_j(\theta)|^2 d\theta\right)^{1/2} \leq C \cdot |j|^\gamma \quad \text{for } |j| \longrightarrow \infty \tag{15.2}$$

for some $\gamma > 0$. Thus if we consider a function $f(x)$ defined for every $x = r\theta \in R^n$, where the spherical coordinates are given by $r = |x|, \theta \in \mathbb{S}^{n-1}$, then

$$f_j(\theta) = f(e^{j+\alpha}\theta)$$

and the above condition may be rewritten as

$$\left(\int |f(x)|^2 d\theta\right)^{1/2} \leq C \cdot |\log |x||^\gamma \quad \text{for } |x| = e^j \longrightarrow 0 \text{ or } \infty$$

since for $|j| \longrightarrow \infty$ the relation $|j + \alpha|^\gamma \approx |j|^\gamma$ holds, i.e. there exist two positive constants C_1, C_2 and a number j_0 such that

$$C_1 |j|^\gamma \leq |j + \alpha|^\gamma \leq C_2 |j|^\gamma \quad \text{for } j \geq j_0.$$

2. We may consider *pointwise power growth*

$$|f_j(\theta)| \leq C \cdot |j|^\gamma \tag{15.3}$$

$$= C \cdot |\log |x||^\gamma \quad \text{for } |x| = e^j \text{ and } j \longrightarrow \pm\infty.$$

3. We may consider *power growth for all spherical components*, i.e.

$$|\psi_j^{k,\ell}| \leq C \cdot |j|^\gamma \tag{15.4}$$

$$= C \cdot |\log |x||^\gamma \quad \text{for } |x| = e^j \text{ and } j \longrightarrow \pm\infty.$$

4. We may specify the growth along every component k, ℓ, namely for some constants $D_{k,\ell} > 0$ we have

$$|\psi_j^{k,\ell}| \leq D_{k,\ell} \cdot |j|^\gamma \tag{15.5}$$

$$= D_{k,\ell} \cdot |\log |x||^\gamma \quad \text{for } |x| = e^j \text{ and } j \longrightarrow \pm\infty.$$

We will see that we may obtain definite results in case (4). On the other hand, all possible definitions of "power growth" provided above are contained in the last case. Indeed, it is evident, due to

$$\int |f_j(\theta)|^2 d\theta = \sum_{k=0}^{\infty} \sum_{\ell=1}^{d_k} |\psi_j^{k,\ell}|^2,$$

that (2) $\Rightarrow$ (1) $\Rightarrow$ (3) $\Rightarrow$ (4).

The most remarkable result of the present chapter is that all components $h^{k,\ell}(\cdot)$ of the cardinal polyspline $h(x)$ which interpolates the above data satisfy the inequality

$$\left|h^{k,\ell}(v)\right| \leq \widetilde{C} D_{k,\ell} |v|^{\gamma} \quad \text{for all } v \in \mathbb{R}$$

with a constant $\widetilde{C} > 0$ independent of the index k![2]

15.3 $\alpha = 0$ is good for all L-splines with $L = M_{k,p}$

As we have already explained, in view of the theory developed by Micchelli and Schoenberg, not every α is good for cardinal interpolation with L-splines for all $L = M_{k,p}$ with $k \geq 0$ and $\ell = 1, 2, \ldots, d_k$. We will prove that the value $\alpha = 0$ is good. It will mean that it is possible to solve the cardinal interpolation problem

$$u_{(k)}(j) = y_j^{(k)} \quad \text{for all } j \in \mathbb{Z}, \tag{15.6}$$

where $u_{(k)}(v)$ is a cardinal L-spline with $L = M_{k,p}$.

Let us recall that every operator $M_{k,p}$ which is given by formula (10.26), on p. 169, is an operator of order $Z = 2p - 1$, and we have, respectively, the polynomial q_{Z+1} defined by formula (13.3), p. 224, here given by $M_{k,p}(\cdot)$, i.e.

$$q_{2p}(z) = M_{k,p}(z) = \prod_{j=1}^{2p} (z - \lambda_j).$$

In order to emphasize the role of the polynomial $M_{k,p}(z)$ we will use its notation instead of q_{Z+1}. We have the function $A_{2p-1}(x; \lambda)$ defined above in (13.10), p. 234. Here we will denote it by $A_{2p-1}^{(k)}(x; \lambda)$ in order to emphasize the role of the parameter k.

In order to study the cardinal interpolation problem (15.6), p. 290, for every k, according to Theorem 13.33, p. 238, we have to consider the unique root of the equation $A_{2p-1}^{(k)}(x; -1) = 0$ for $0 \leq x < 1$. We have to show that 0 is not a concentration point of the zeros for varying k. For that reason we need the asymptotic behavior of the root of the equation for $k \longrightarrow \infty$.

As in Chapter 14, we will restrict ourselves to the case $p = 2$ of the biharmonic operator Δ^2 which is the simplest nontrivial case of a polyspline. Thus we have

$$p = 2,$$
$$Z = 2p - 1 = 3.$$

This will make things technically simpler and it will be clear that the result holds for an arbitrary integer $p \geq 1$.

---

[2] We prove this inequality for the L-splines with respect to the operators $L = M_{k,p}$. Obviously such an estimate does not involve the index ℓ.

Then we have as in (14.11), p. 278,

$$\lambda_1 = -n-k+2; \quad \lambda_2 = -n-k+4; \quad \lambda_3 = k; \quad \lambda_4 = k+2. \tag{15.7}$$

We will compute the function $A^{(k)}_{2p-1}(x;\lambda) = A^{(k)}_3(x;\lambda)$ in the case $p=2$. When k is large enough then all λ_j are pairwise different and

$$\lambda_1 < \lambda_2 < \lambda_3 < \lambda_4,$$

hence,

$$A^{(k)}_3(x;\lambda) = \sum_{j=1}^{4} \frac{1}{(M_{k,2})'(\lambda_j)} \cdot \frac{e^{\lambda_j x}}{e^{\lambda_j} - \lambda}. \tag{15.8}$$

In the case of coinciding λ_js, say $\lambda_1 = \lambda_2$, we simply take the limit of the above formula for $\lambda_2 = \lambda_1 + \varepsilon$ and $\varepsilon \longrightarrow 0$, or apply the residuum representation of A_Z in (13.11), p. 235.

According to Micchelli's theorem 13.31, p. 237, equation

$$0 = A^{(k)}_3(x;-1) = \sum_{j=1}^{4} \frac{1}{(M_{k,2})'(\lambda_j)} \cdot \frac{e^{\lambda_j x}}{e^{\lambda_j} + 1} \tag{15.9}$$

has a unique solution which satisfies $0 \le x < 1$. We will prove that $x = 0$ is not a root.

Now let $k \ge 0$ be fixed. Assume that $x = 0$ is a solution of (15.9) then, after multiplying with $\prod_{j=1}^{4}(e^{\lambda_j} + 1)$ we obtain

$$\sum_{i=1}^{l} \beta_i e^{\rho_i} = 0;$$

here β_i are non-zero integers and ρ_i are integers obtained by sums of some of the constants $\lambda_j(k)$. Due to the special form of the constants $\lambda_j(k)$ provided in (15.7) above we see that at least one of the ρ_is is non-zero. Thus we may apply the classical theorem of *Lindemann* on transcendental numbers which states that the above equality is impossible, see e.g. K. Mahler [11, p. 213] or A. Baker [1, p. 6]. Thus we see that a number-theoretical argument proves that for every individual k we have $A^{(k)}_3(0;-1) \ne 0$. We need to see that 0 is not a concentration point of the zeros of $A^{(k)}_3(x;-1)$. This is proved in Theorem 15.1.

Theorem 15.1 *Consider the function*

$$f_j(x) = k^2 e^{kx} A^{(k)}_3(x;\lambda).$$

Then for every number ε with $0 < \varepsilon < 1/2$ the sequence f_k converges uniformly to the function

$$f(x,\lambda) = -C\frac{1}{\lambda}e^{(2-n)x}(e^{2x} - 1)$$

for all $x \in [0, 1/2 - \varepsilon)$ and $\lambda < -\varepsilon$, where the constant C is not equal to zero.

292 Multivariate polysplines

Proof We put
$$I := k^2 e^{kx} A_3^{(k)}(x; \lambda)$$

and for large k we obtain the representation
$$I = e^{kx} \sum_{j=1}^{4} \frac{k^2}{(M_{k,2})'(\lambda_j)} \cdot \frac{e^{\lambda_j x}}{e^{\lambda_j} - \lambda}.$$

It is easily seen that for some constants $d_j \neq 0$ for $j = 1, \ldots, 4$, we have
$$\frac{k^2}{(M_{k,2})'(\lambda_j)} \xrightarrow{k \to \infty} d_j. \tag{15.10}$$

Now if, for uniformity, we put $\lambda_1 = -k + C_1$, $\lambda_2 = -k + C_2$, $\lambda_3 = k + C_3$ and $\lambda_4 = k + C_4$, we obtain for $D = \max d_j$ the following estimate:
$$\left| e^{kx} \sum_{j=3}^{4} \frac{k^2}{(M_{k,2})'(\lambda_j)} \cdot \frac{e^{\lambda_j x}}{e^{\lambda_j} - \lambda} \right| \leq D \sum_{j=3}^{4} \frac{e^{C_j x} e^{2xk}}{|e^{k+C_j} - \lambda|}.$$

Since for large k we have $|e^{k+C_j} - \lambda| \geq e^k$, it follows that
$$|I| \leq C e^{(2x-1)k}.$$

The last converges uniformly to zero for $k \to \infty$ and $x \in [0, 1/2 - \varepsilon)$.

It follows that the function I converges uniformly to the function
$$f(x, \lambda) = -\frac{1}{\lambda} \sum_{j=1}^{2} d_j e^{C_j x}.$$

A straightforward computation completes the proof. ∎

Let us note that the case of arbitrary $p \geq 1$ does not differ essentially, we leave the details to the reader, or see [9], and from the above theorem we obtain the following corollary.

Corollary 15.2 *Let us denote by ξ_k the unique solution to equation*
$$A_{2p-1}^{(k)}(x; -1) = 0$$

satisfying $0 \leq x_k < 1$. Then
$$\xi_k \xrightarrow{k \to \infty} \frac{1}{2}.$$

15.4 Explaining the problem

Thus, by the previous section, we are free to consider the cardinal interpolation polysplines which have as break-surfaces the spheres $S(0; e^j)$, satisfying the interpolation conditions

$$h(e^j \theta) = f_j(\theta) \quad \text{for } \theta \in \mathbb{S}^{n-1} \text{ and } j \in \mathbb{Z}.$$

So far we will need some more subtle estimates in order to prove the power growth of the polyspline $h(x)$ itself. Indeed, the power growth of the data f_j, condition (15.5), p. 289, is

$$|\psi_j^{k,\ell}| \leq D_{k,\ell} \cdot |j|^\gamma \quad \text{for all } j \in \mathbb{Z},$$

and with $\alpha = 0$ by the Micchelli theory for every $k \geq 0$ and $\ell = 1, 2, \ldots, d_k$, there exists an $M_{k,2}$-spline (L-spline with operator $L = M_{k,2}$) $u^{k,\ell}(v)$ which satisfies

$$\begin{cases} u^{k,\ell}(j) = \psi_j^{k,\ell} & \text{for all } j \in \mathbb{Z}, \\ |u^{k,\ell}(v)| \leq C_k \cdot D_{k,\ell} \cdot |v|^\gamma & \text{for all } v \in \mathbb{R}. \end{cases}$$

We have here the constants $C_k > 0$ which depend on the operator $M_{k,2}$.

We will prove below (at least in the case $p = 2$; for the general case see [9]) that one may choose

$$C_k = \widetilde{C} = \text{const}$$

independent of the index $k \geq 0$. This is the most remarkable feature of the cardinal polyspline interpolation.[3]

Now we investigate the power growth of the *polysplines*

$$h(x) = \sum_{k=0}^{\infty} \sum_{\ell=1}^{d_k} h^{k,\ell}(\log r) Y_{k,\ell}(\theta).$$

We would like to have at least a bounded L_2-norm on every sphere of radius $r > 0$, and we also expect an estimate of the growth of this L_2-norm. By the above we have an estimate of the form

$$\int |h(r\theta)|^2 \, d\theta = \sum_{k=0}^{\infty} \sum_{\ell=1}^{d_k} |h^{k,\ell}(\log r)|^2$$

$$\leq \sum_{k=0}^{\infty} \sum_{\ell=1}^{d_k} \widetilde{C}^2 \cdot D_{k,\ell}^2 \cdot |\log r|^{2\gamma}$$

$$= |\log r|^{2\gamma} \cdot \sum_{k=0}^{\infty} d_k \cdot \widetilde{C}^2 \cdot D_{k,\ell}^2,$$

---
[3] Note that this "stability" of the constant C_k might be expected owing to the "almost symmetry" of the vector Λ_k for large k.

where d_k is the number of linearly independent spherical harmonics in the space $\mathcal{H}_k$ as defined in (10.9), p. 145. Its asymptotic is given by formula (10.11), p. 145,

$$d_k \approx \frac{2}{(n-2)!} k^{n-2} \quad \text{for } k \longrightarrow \infty.$$

Hence $\int |h(r\theta)|^2 \, d\theta < \infty$ will be ensured for every r if the inequality

$$\sum_{k=0}^{\infty} k^{n-2} \cdot \widetilde{C}^2 \cdot \sum_{\ell=1}^{d_k} D_{k,\ell}^2 < \infty \tag{15.11}$$

holds.

15.5 Schoenberg's results on the fundamental spline $L(X)$ in the polynomial case

The *fundamental cardinal spline function* $L(X)$ plays a basic role in interpolation. By definition it is a spline which satisfies

$$L(\nu) = \delta_{\nu 0} := \begin{cases} 1 & \text{for } \nu = 0, \\ 0 & \text{for } \nu \neq 0, \end{cases}$$

where δ_{ij} denotes the usual Kronecker symbol. It has been studied thoroughly by Schoenberg in [18–20], although none of these references contains all the details in one place.[4] We take only the case of *odd-degree* polynomials, namely the $(2m-1)$-degree case. Let us recall that there are some essential differences between the odd-degree and the even-degree cases. Specifically, the interpolation points in the odd-degree case may coincide with the break-points of the splines in order to have a well-defined interpolation problem. The main point is that the $M_{k,p}$-splines, which we will consider further correspond to the odd-degree case, since the operator $M_{k,p}$ is of even order.

All the necessary properties of the Euler–Frobenius polynomials $\Pi_{2m-1}(z)$ are provided in [18, Theorem 2.1, p. 22].[5] From [18, p. 37], we know that the polynomial

$$\Pi_{2m-1}(z) = (2m-1)! z^{m-1} \sum_{j=-m+1}^{m-1} M_{2m}(j) z^j$$

has $2m-2$ zeros satisfying[6]

$$\mu_{2m-2} < \cdots < \mu_m < -1 < \mu_{m-1} < \cdots < \mu_1 < 0$$

$$\mu_1 \mu_{2m-2} = \mu_2 \mu_{2m-3} = \cdots = 1.$$

---

[4] We remind the reader again that the function $L(x)$ has been studied by Schoenberg and we have retained his notation. This function is also widely used in the Wavelet Analysis by *Chui* [3]. This notation has nothing to do with the operator L used for the L-splines.

[5] Or consult the more general case of L-splines in Theorem 13.31, p. 237, in Chapter 14 which will be particularly useful for the generalizations that follow in Section 15.6.

[6] Micchelli uses reversed ordering, loc. cit., p. 228.

Recall that the polynomial $\Pi_{2m-1}(z)$ has coefficient 1 for the highest-order term and

$$\Pi_{2m-1}(z) = \prod_{j=1}^{2m-2}(z-\mu_j).$$

By the above it follows that the polynomial $\Pi_{2m-1}(z)$ is not zero in the ring $\{z : |\mu_{m-1}| < |z| < |\mu_m|\}$ which contains the unit circle $|z| = 1$. Let us recall that we have a symmetric B-spline M_{2m}, i.e.

$$M_{2m}(j) = M_{2m}(-j),$$

and it has a compact support coinciding with the interval $[-m, m]$. Let us note that from this symmetry directly follows:

$$\sum_{j=-m+1}^{m-1} M_{2m}(j)z^j = \sum_{j=-m+1}^{m-1} M_{2m}(j)\frac{1}{z^j}.$$

It follows that for z on the circle $\{|z| = 1\}$ we have the expansion [20, pp. 181, 182]:

$$\frac{1}{\sum_{j=-m+1}^{m-1} M_{2m}(j)z^j} = \frac{(2m-1)!z^{m-1}}{\Pi_{2m-1}(z)}$$

$$= \frac{((2m-1)!}{(1-(\mu_1/z))\cdots(1-(\mu_{m-1}/z))((z/\mu_m)-1)\cdots((z/\mu_{2m-2})-1)\mu_m\cdots\mu_{2m-2}}$$

$$= \sum_{j=-\infty}^{\infty} \omega_j z^j.$$

Now the symmetry of M_{2m} implies

$$\omega_{-j} = \omega_j \quad \text{for all } j \in \mathbb{Z}.$$

A simple way to obtain this is by considering the representation in simple fractions and using the fact that $\mu_{2m-1-j} \cdot \mu_j = 1$ for $j = 1, 2, \ldots, m-1$, namely

$$\frac{1}{(1-(\mu_1/z))\cdots(1-(\mu_{m-1}/z))((z/\mu_m)-1)\cdots((z/\mu_{2m-2})-1)}$$

$$= \sum_{l=1}^{m-1} \frac{A_l}{(1-(\mu_l/z))} + \sum_{l=m}^{2m-2} \frac{B_l}{(1-(z/\mu_l))}$$

$$= \sum_{l=1}^{m-1} A_l \sum_{j=0}^{\infty}\left(\frac{\mu_l}{z}\right)^j + \sum_{l=m}^{2m-2} B_l \sum_{j=0}^{\infty}(\mu_{2m-1-l}\cdot z)^j$$

$$= \sum_{l=1}^{m-1} A_l \sum_{j=0}^{\infty}\left(\frac{\mu_l}{z}\right)^j + \sum_{l=1}^{m-1} B_{2m-1-l}\sum_{j=0}^{\infty}(\mu_l \cdot z)^j.$$

We obtain the expression

$$\omega_j = \frac{(2m-1)!}{\mu_m \cdots \mu_{2m-2}} \cdot \sum_{l=1}^{m-1} B_{2m-1-l} \cdot \mu_l^j$$

$$= (2m-1)! \cdot \mu_1 \cdots \mu_{m-1} \cdot \sum_{l=1}^{m-1} B_{2m-1-l} \cdot \mu_l^j \quad \text{for } j \geq 0.$$

Due to $\omega_{-j} = \omega_j$ the last implies,

$$|\omega_j| = O(|\mu_{m-1}|^{|j|}) \quad \text{for } |j| \longrightarrow \infty.$$

The last relation is usually written as, [21, p. 414],

$$|\omega_j| \leq C e^{-\gamma j} \quad \text{for all } j \in \mathbb{Z}$$

for some constant $C > 0$, where we have put

$$e^{-\gamma} = |\mu_{m-1}| \quad \text{or} \quad \gamma = -\log |\mu_{m-1}|.$$

Obviously $\gamma > 0$.[7]

The fundamental spline function $L_{2m-1}(x)$ is now given by

$$L_{2m-1}(x) = \sum_{j=-\infty}^{\infty} \omega_j M_{2m}(x - j).$$

It is involved in the following identity:

$$1 = \left(\sum_{j=-m+1}^{m-1} M_{2m}(j) z^j \right) \cdot \left(\sum_{j=-\infty}^{\infty} \omega_j z^j \right) = \sum_{\nu=-\infty}^{\infty} L_{2m-1}(\nu) z^\nu,$$

which implies its basic property

$$L_{2m-1}(j) = \delta_{j0} \quad \text{for all } j \in \mathbb{Z}.$$

The estimate for $L_{2m-1}(x)$ follows directly from its definition [21, p. 415]

$$|L_{2m-1}(x)| \leq M_{2m}(0) \cdot \sum_{x+m>j>x-m} |\omega_j|$$

$$\leq M_{2m}(0) \cdot C \cdot \sum_{x+m>j>x-m} e^{-\gamma j}$$

---

[7] Let us note that in 1973 [18, p. 38], Schoenberg defined the fundamental function L_{2m-1} as a linear combination of shifts of M_{2m} but in his earlier work of 1972, [21, p. 414], he used L_m as a linear combination of shifts of M_m, so one must be careful not to mix the notations.

where C and γ are as above. Since for $|x| \geq m$ we have $M_{2m}(x) = 0$, it easily follows by subtracting that

$$|L_{2m-1}(x)| \leq M_{2m}(0) \cdot C \cdot e^{-\gamma(x-m)} \sum_{j=0}^{\infty} e^{-\gamma j}$$

$$= M_{2m}(0) \cdot C \cdot e^{\gamma m} \frac{1}{1 - e^{-\gamma}} \cdot e^{-\gamma x}.$$

For this constant we put

$$C_{2m-1} = M_{2m}(0) \cdot C \cdot e^{\gamma m} \frac{1}{1 - e^{-\gamma}}.$$

Let us proceed to the cardinal interpolation. Assume that we are given some data sequence $\{y_j\}$ having power growth, i.e. for some constant $A > 0$ [21, p. 415],

$$|y_j| \leq A(|j|^s + 1) \quad \text{for all } j \in \mathbb{Z}$$

holds. Then for the cardinal interpolation spline $S(x)$ we put

$$S(x) := \sum y_j L(x - j).$$

We immediately obtain the estimate

$$|S(x)| \leq \sum_{j=-\infty}^{\infty} AC_{2m-1}(|j|^s + 1)e^{-\gamma|x-j|}.$$

Lemma 15.3 *For sufficiently large $|x|$ the inequality*

$$|S(x)| \leq F|x|^s$$

holds.

Proof Due to the symmetry we see that, indeed,

$$\frac{1}{x^s} \sum_{j=-\infty}^{\infty} (|j|^s + 1)e^{-\gamma|x-j|} \leq F_1 \quad \text{for } x \longrightarrow +\infty.$$

For $j \leq 0$ the sum is bounded

$$\sum_{j=-\infty}^{0} (|j|^s + 1)e^{-\gamma|x-j|} = \sum_{j=-\infty}^{0} (|j|^s + 1)e^{-\gamma(x-j)}$$

$$= e^{-\gamma x} \sum_{j=-\infty}^{0} (|j|^s + 1)e^{\gamma j}$$

$$\leq 1 \cdot \sum_{j=0}^{\infty} (j^s + 1)e^{-\gamma j}$$

$$\leq 1 + \int_0^{\infty} (t^s + 1)e^{-\gamma t} dt.$$

It follows that we need to estimate only the second part of the series which splits into two parts:

$$\sum_{j=1}^{\infty}\left(\frac{j}{x}\right)^s e^{-\gamma|x-j|} = \sum_{j \leq x+1}^{\infty} + \sum_{j > x+1}^{\infty}$$

$$< \sum_{j \leq x+1}^{\infty} e^{-\gamma(x-j)} + \sum_{j > x+1}^{\infty} j^s x^{-s} e^{-\gamma|x-j|}$$

$$= O(1) + x^{-s} \sum_{j > x+1}^{\infty} j^s e^{-\gamma(j-x)}$$

$$= O(1) + x^{-s} e^{\gamma x} \sum_{j > x+1}^{\infty} j^s e^{-\gamma j}.$$

If we consider only x in the interval (ξ, ∞) where the function $x^s e^{-\gamma x}$ is decreasing and convex, then we estimate from above by integral the last sum and obtain

$$x^{-s} e^{\gamma x} \sum_{j > x+1}^{\infty} j^s e^{-\gamma j} < x^{-s} e^{\gamma x} \int_x^{\infty} t^s e^{-\gamma t} dt$$

$$= \int_x^{\infty} \left(\frac{t}{x}\right)^s e^{-\gamma(t-x)} dt$$

$$= \int_0^{\infty} \left(1 + \frac{u}{x}\right)^s e^{-\gamma u} du. \quad \blacksquare$$

15.6 Asymptotic of the zeros of $\Pi_Z(\lambda; 0)$

We return to the polysplines. We restrict ourselves to the case $p = 2$ as stated above.
In formula (15.8), p. 291, for large k we have

$$A_3^{(k)}(0; \lambda) := \sum_{j=1}^4 \frac{1}{(M_{k,2})'(\lambda_j)} \cdot \frac{1}{e^{\lambda_j} - \lambda}.$$

As we saw in (15.10), p. 292, there exist constants $d_j \neq 0$ such that

$$\frac{(M_{k,2})'(\lambda_j)}{k^2} \longrightarrow d_j \quad \text{for } j = 1, \ldots, 4.$$

This gives the polynomial Π_3 by

$$\Pi_3(\lambda) = r_3(\lambda) A_3^{(k)}(0; \lambda)$$

$$= C_k[(e^{\lambda_1} - e^{\lambda_4})(n + 2k - 4)(e^{\lambda_2} - \lambda)(e^{\lambda_3} - \lambda)$$
$$+ (e^{\lambda_3} - e^{\lambda_2})(n + 2k)(e^{\lambda_1} - \lambda)(e^{\lambda_4} - \lambda)],$$

where $C_k \neq 0$.

We are looking for the solutions μ_1 and μ_2 of
$$\Pi_3(\lambda) = 0.$$

We know by the zero properties of the function Π_Z in Theorem 13.64, p. 263, that they satisfy
$$\mu_1\mu_2 = e^{-n+4},$$
and
$$0 > \mu_1 > -\sqrt{e^{-n+4}} > \mu_2.$$

Proposition 15.4 *The following limit relations hold:*
$$\mu_2 \xrightarrow{k\to\infty} -\infty, \quad \mu_1 \xrightarrow{k\to\infty} 0.$$

Proof Let us assume that $|\mu_2|$ has a bounded subsequence $\mu_2(k)$, i.e. for some $C' > 0$ we assume that
$$\sqrt{e^{-n+4}} < |\mu_2(k)| \leq C' \quad \text{for } k \longrightarrow \infty.$$

We divide both sides of the equality $\Pi_3(\lambda) = 0$ by $e^{\lambda_3+\lambda_4}$ and obtain for $\lambda = \mu_2(k)$ the following:
$$(n+2k-4)(e^{\lambda_2}-\lambda)\left(\frac{e^{\lambda_1}}{e^{\lambda_4}}-1\right)\left(1-\frac{\lambda}{e^{\lambda_3}}\right) = (n+2k)(e^{\lambda_1}-\lambda)\left(\frac{e^{\lambda_2}}{e^{\lambda_3}}-1\right)\left(1-\frac{\lambda}{e^{\lambda_4}}\right).$$

Obviously on both sides we have expressions of the form
$$(n+2k-4)\left(\lambda + \frac{C_1(k)}{e^k}\right) = (n+2k)\left(\lambda + \frac{C_2(k)}{e^k}\right),$$
where $C_1(k)$ and $C_2(k)$ are bounded for $k \to \infty$. We subtract the right-hand side from the left and obtain
$$(n+2k)\left(\frac{C_1(k)-C_2(k)}{e^k}\right) = 4\left(\lambda + \frac{C_1(k)}{e^k}\right)$$
which implies $4 = 0$. This contradiction proves the proposition. ∎

We are interested in the explicit values of the roots μ_1 and μ_2. Let us compute them. Let us put
$$\gamma_k = \frac{(e^{\lambda_1}-e^{\lambda_4})(n+2k-4)}{(e^{\lambda_3}-e^{\lambda_2})(n+2k)}.$$

We obtain
$$\mu_{1,2} = \frac{-B \pm \sqrt[2]{B^2-4AC}}{2A}$$

where

$$A = (1 + \gamma_k)$$
$$B = -[\gamma_k(e^{\lambda_2} + e^{\lambda_3}) + (e^{\lambda_1} + e^{\lambda_4})]$$
$$C = \gamma_k e^{-n+4} + e^{-n+4}.$$

Thus

$$\mu_{1,2} \approx \frac{(2e^{k+2})/(k) \pm \sqrt[2]{(2e^{k+2}/k)^2 - 4(1-e^2)^2 e^{-n+4}}}{2(1-e^2)}.$$

We can easily see that

$$\gamma_k \xrightarrow{k \to \infty} -e^2.$$

Let us compute the asymptotic of μ_1 and μ_2. For $k \longrightarrow \infty$ we have

$$e^{\lambda_1} \longrightarrow 0$$
$$\gamma_k e^{\lambda_2} \longrightarrow 0,$$

hence, for $k \longrightarrow \infty$ we have the two asymptotics

$$\mu_2 \approx \frac{e^k}{k} \frac{2e^2}{1-e^2}$$
$$\mu_1 \approx \frac{k}{e^k} \frac{(1-e^2)}{2e^{n-2}}.$$

Evidently, $\mu_2 < \mu_1 < 0$. We see that if

$$c = n - 2p,$$

then we may write

$$\mu_1 \mu_2 = e^{-c},$$

which is precisely the relation of Theorem 13.64, p. 263.

15.7 The fundamental spline function $L(X)$ for the spherical operators $M_{k,p}$

We now consider the fundamental spline function L for the operators $M_{k,p}$. We will obtain precise estimates for the constants related to the function $L(X)$ analogous to Section 15.5. We will confine ourselves to the case $p = 2$.

Since $p = 2$, we have $Z = 2p - 1 = 3$. Then we have as in (15.7), p. 291, the vector Λ given by

$$\lambda_1 = -n - k + 2; \quad \lambda_2 = -n - k + 4; \quad \lambda_3 = k; \quad \lambda_4 = k + 2. \quad (15.12)$$

Let us recall that the polynomials Π_Z are not monic, i.e. their leading coefficients are not equal to 1. We have already obtained a representation of their leading coefficient in (13.41), p. 254. As we saw above the two zeros of $\Pi_3(\lambda)$ satisfy

$$\mu_1 \mu_2 = e^{-n+4},$$

$$\mu_2 < -\sqrt[2]{e^{-n+4}} < \mu_1 < 0,$$

$$\mu_2 \xrightarrow{k \to \infty} -\infty.$$

We will follow the scheme for constructing the fundamental interpolation spline function introduced by Schoenberg, which we have explained in Section 15.5, p. 294.

We consider the expansion of the reciprocal to the function

$$\sum_{l=-(Z-1)/2}^{(Z-1)/2} Q_{Z+1}\left(\frac{Z+1}{2} + l\right) z^l$$

which, by Theorem 13.64, p. 263, about symmetry and by the leading coefficient formula (13.41), p. 254, is equal to

$$\frac{1}{\sum_{l=-(Z-1)/2}^{(Z-1)/2} Q_{Z+1}((Z+1)/2 + l) z^l} = \frac{(-1)^Z \cdot z^{-(Z-1)/2} \cdot e^{\lambda_1 + \cdots + \lambda_{Z+1}}}{\Pi_Z(z^{-1}; 0)}$$

$$= \frac{(-1)^Z \cdot e^{p(-n+2p)/2} \cdot z^{(Z-1)/2} \cdot e^{\lambda_1 + \cdots + \lambda_{Z+1}}}{\Pi_Z(e^{-c}z; 0)}$$

$$= \frac{e^{p((-n+2p)/2)} \cdot z^{(Z-1)/2} \cdot e^{\lambda_1 + \cdots + \lambda_{Z+1}}}{\sum_{j=1}^{Z+1} \frac{e^{\lambda_j}}{M'_{k,p}(\lambda_j)} \cdot \prod_{j=1}^{Z-1}(e^{-c}z - \mu_j)}$$

$$= C_Z \frac{z^{(Z-1)/2}}{\prod_{j=1}^{Z-1}(e^{-c}z - \mu_j)},$$

where recalling that $c = n - 2p$, for simplicity we have also put

$$C_Z = \frac{e^{pc/2} e^{\lambda_1 + \cdots + \lambda_{Z+1}}}{\sum_{j=1}^{Z+1} \frac{e^{\lambda_j}}{M'_{k,p}(\lambda_j)}}.$$

We also use the notation $M_{k,p}(z) = q_{Z+1}(\cdot)$.

Thus for $p = 2$ and $Z = 3$ we obtain the following identity:

$$C_Z \cdot \frac{z^{(Z-1)/2}}{\prod_{j=1}^{Z-1}(e^{-c}z - \mu_j)} = C_3 \cdot \frac{z}{\prod_{j=1}^{2}(e^{-c}z - \mu_j)} = \sum_{j=-\infty}^{\infty} \omega_j z^j$$

with the constant C_3 given by

$$C_3 = \frac{e^{p(-n+4)/2} e^{-2n+8}}{\sum_{j=1}^{4} \frac{e^{\lambda_j}}{M'_{k,2}(\lambda_j)}},$$

where the constant C_3' is independent of k. As we have seen in (15.10), p. 292, for $k \to \infty$ we have $q_4'(\lambda_j)/k^2 \to d_j \neq 0$. It follows that the asymptotic order of C_3 for $k \to \infty$ is given by:

$$C_3 \approx \frac{q_4'(\lambda_4)}{e^{\lambda_4}} = \frac{k^2}{e^k}. \tag{15.13}$$

By using the constants ω_j we will define the fundamental interpolation spline $L(X)$. We have

$$\frac{z}{(e^{-c}z - \mu_1)(e^{-c}z - \mu_2)} = \frac{z}{\mu_2 - \mu_1} \left(\frac{1}{e^{-c}z - \mu_2} - \frac{1}{e^{-c}z - \mu_1} \right)$$

$$= \frac{-z}{\mu_2 - \mu_1} \left(\frac{1}{\mu_2} \frac{1}{1 - (e^{-c}z/\mu_2)} + \frac{1}{e^{-c}z(1 - (\mu_1/e^{-c}z))} \right)$$

$$= \frac{-z}{\mu_2 - \mu_1} \left[\frac{1}{\mu_2} \sum_{l=0}^{\infty} \left(\frac{e^{-c}z}{\mu_2} \right)^l + \frac{1}{e^{-c}z} \sum_{l=0}^{\infty} \left(\frac{\mu_1}{e^{-c}z} \right)^l \right]$$

$$= \frac{1}{\mu_1 - \mu_2} \left[e^c \sum_{l=0}^{\infty} \left(\frac{e^{-c}z}{\mu_2} \right)^{l+1} + e^c \sum_{l=0}^{\infty} \left(\frac{\mu_1}{e^{-c}z} \right)^l \right],$$

hence, if we put

$$C_3' = C_3 e^c,$$

we obtain

$$\omega_j = C_3' \frac{1}{\mu_1 - \mu_2} \left(\frac{e^{-c}}{\mu_2} \right)^j \quad \text{for } j = 1, 2, \ldots,$$

$$\omega_{-j} = C_3' \frac{1}{\mu_1 - \mu_2} \left(\frac{\mu_1}{e^{-c}} \right)^j = C_3' \frac{1}{\mu_1 - \mu_2} \left(\frac{1}{\mu_2} \right)^j \quad \text{for } j = 0, 1, \ldots.$$

We see that we have a somewhat more complicated relationship between ω_j and ω_{-j}, namely

$$\omega_{-j} = \omega_j e^{cj} \quad \text{for } j = 1, 2, 3, \ldots,$$

compared with the polynomial case where $\omega_j = \omega_{-j}$. Obviously, the sign of ω_j alternates with j for large k and

$$\omega_j = C_3' O \left(\frac{1}{|\mu_2|^{|j|+1}} \right) \quad \text{for } j \in \mathbb{Z}.$$

Since we only have the above expressions for large values of k we see that in such a case $|1/\mu_2| < 1$ which guarantees

$$|\omega_j| \xrightarrow{j \to \infty} 0, \quad |\omega_{-j}| \xrightarrow{j \to \infty} 0.$$

Further, following Schoenberg [19, p. 272], as in Section 15.5 we define the fundamental function $L(x)$ by the equality

$$L(x) = \sum_{j=-\infty}^{\infty} \omega_j Q_4(x-j).$$

A simple computation shows that

$$\frac{1}{\sum_{l=-(Z-1)/2}^{(Z-1)/2} Q_{Z+1}((Z+1)/2+l)z^l} \sum_{j=-\infty}^{\infty} \omega_j z^j = \sum_{\nu=-\infty}^{\infty} L(\nu)z^{\nu}$$

which shows that $L(x)$ is indeed a "fundamental spline function".

Remark 15.5 *There is an elegant formula in Micchelli [13, p. 224, formula (26)] for the function $L(x)$, namely*

$$L(x) = \frac{1}{2\pi i} \int_{\Gamma} \frac{1}{z} \frac{A_Z(x;z)}{A_Z(\alpha;z)} dz,$$

where the contour Γ is the parametrization of the unit circle $\Gamma = \{e^{i\phi} : \text{for } 0 \leq \phi \leq 2\pi\}$. The function $L(x)$ satisfies $L(\alpha + j) = \delta_{j0}$ for all $j \in \mathbb{Z}$.

We proceed in a manner similar to that in Section 15.5.

15.7.1 Estimate of the fundamental spline $L(x)$

Since supp Q_4, is equal to the interval $[0, 4]$, for $x \geq 4$ we obtain the estimate

$$|L(x)| \leq \max_x(Q_4(x)) \cdot \sum_{x-4<j<x} |\omega_j|$$

$$\leq \max_x(Q_4(x)) \cdot C_3' \cdot \sum_{j>x-4} \left|\frac{1}{\mu_2}\right|^j \cdot \frac{1}{|\mu_1 - \mu_2|}$$

$$\leq \max_x(Q_4(x)) \cdot C_3' \cdot \frac{1}{|\mu_1 - \mu_2|} \cdot \left|\frac{1}{\mu_2}\right|^{x-4} \sum_{j=0}^{\infty} \left|\frac{1}{\mu_2}\right|^j.$$

Since we take only large k for which $|\mu_2| \geq 1$ we obtain

$$|L(x)| \leq \max_x(Q_4(x)) \cdot C_3' \cdot \frac{1}{|\mu_1 - \mu_2|} \cdot \left|\frac{1}{\mu_2}\right|^x \sum_{j=0}^{\infty} \left|\frac{1}{\mu_2}\right|^j$$

$$\leq \max_x(Q_4(x)) \cdot C_3' \cdot \frac{1}{|\mu_1 - \mu_2|} \cdot \left|\frac{1}{\mu_2}\right|^x.$$

In a similar way for all x satisfying $x \leq 0$ we obtain the estimate

$$|L(x)| \leq \max_x(Q_4(x)) \cdot C_3' \cdot \frac{1}{|\mu_1 - \mu_2|} \cdot \left|\frac{1}{\mu_2}\right|^{|x|}.$$

For $0 \leq x \leq 4$ we obtain the estimate

$$|L(x)| \leq \max_x(Q_4(x)) \cdot C_3' \cdot \frac{1}{|\mu_1 - \mu_2|}$$
$$\times \left(\frac{1}{|\mu_2|^4} + \frac{1}{|\mu_2|^3} + \frac{1}{|\mu_2|^2} + \frac{1}{|\mu_2|^1} + 1 \right).$$

This last case is crucial since we see that for sufficiently large k the very last sum is bounded by 5. However, we are looking for an upper bound of $L(x)$ of the type $CD^{|x|}$ for some constant $D < 1$ and a constant $C > 0$. For that reason we put

$$D := \max_{k \geq k_0} \left| \frac{1}{\mu_2} \right|,$$

where we have taken k_0 large enough, as above, which also guarantees that μ_2 satisfies $|1/\mu_2| < 1$. In the cases $x < 0$ and $x \geq 4$ we see that $|1/\mu_2| \leq D$ implies

$$|L(x)| \leq \max_x(Q_4(x)) \cdot C_3' \cdot \frac{1}{|\mu_1 - \mu_2|} \cdot D^{|x|}.$$

For $0 \leq x \leq 4$ we obtain

$$|L(x)| \leq \max_x(Q_4(x)) \cdot C_3' \cdot \frac{1}{|\mu_1 - \mu_2|} \frac{5}{D^4} D^{|x|}.$$

Hence, since $5/D^4 > 1$ the estimate we finally obtain is

$$|L(x)| \leq \max_x(Q_4(x)) \cdot C_3' \frac{1}{|\mu_1 - \mu_2|} \frac{5}{D^4} D^{|x|} \quad \text{for all } x \in \mathbb{R}. \tag{15.14}$$

15.7.2 Estimate of the cardinal spline $S(x)$

First, we assume that k is fixed and large enough.

Now let us assume that we have the power growth data, i.e. for some constant $A > 0$ we have a sequence $\{y_j\}$ satisfying

$$|y_j| \leq A(|j|^s + 1) \quad \text{for all } j \in \mathbb{Z}.$$

We assume as in [21] that $s \geq 0$.

We consider the interpolation cardinal L-spline (for the operator $\mathcal{L}[\Lambda_k]$) as constructed in the previous section:

$$S(x) = \sum_{j=-\infty}^{\infty} y_j L(x - j).$$

We see that for sufficiently large $|x|$ by means of estimate (15.14), p. 304, we obtain

$$|S(x)| \leq \max_x(Q_4(x)) \cdot C_3' \cdot \frac{1}{|\mu_1 - \mu_2|} \cdot A \cdot \sum_{j=-\infty}^{\infty} (|j|^s + 1) D^{|x-j|}.$$

Let us note that the inequality

$$S_1 = \sum_{j=-\infty}^{\infty} (|j|^s + 1) D^{|x-j|} \leq F|x|^s$$

holds, as in the one-dimensional case, by Lemma 15.3, p. 297. And the constant F is obviously independent of $k \geq k_0$.

It follows that for $k \geq k_0$ the following estimate holds:

$$|S(x)| \leq \max_x (Q_4(x)) \cdot C_3' \cdot \frac{1}{|\mu_1 - \mu_2|} \cdot A \cdot F \cdot |x|^s. \tag{15.15}$$

According to Theorem 13.66, p. 266, we have the following estimate of the asymptotic order of Q_{Z+1} for $k \to \infty$:

$$\max_{x \in \mathbb{R}} Q_4(x) \approx \frac{e^{2k}}{k^3}.$$

As we have seen the asymptotic order of the root μ_2 for $k \to \infty$ is given by

$$\mu_2 \approx \frac{e^k}{k} \frac{2e^2}{1 - e^2}.$$

For the constant C_3' we see from (15.13), p. 302, that

$$C_3' \approx \frac{q_4'(\lambda_4)}{e^{\lambda_4}} = \frac{k^2}{e^k}.$$

Let us return to the cardinal spline $S(x)$. We obtain the *final* estimate

$$|S(x)| \leq CC' \frac{e^{2k}}{k^3} \cdot \frac{k^2}{e^k} \cdot \frac{k}{e^k} \cdot A \cdot F \cdot |x|^s = C''A|x|^s$$

where the constant C'' is independent of k.

Thus we have proved that the constants C_k defined in (15.11), p. 294, may be chosen independently of $k \geq 0$ by choosing the maximum of all corresponding constants for $k = 0, 1, \ldots, k_0 - 1$ and the above constant C''. We will use $\widetilde{C}$ for that constant.

15.8 Synthesis of the interpolation cardinal polyspline

We have done all the preparatory work and now we can synthesize the interpolation cardinal polyspline.

As we have explained in Section 15.4, p. 293, we have to choose the constants $D_{k,l}$ in order to obtain convergent series for the polysplines. From the above estimates we obtain the following result which we formulate only in the case $p = 2$.

Theorem 15.6 *Let the vector Λ be given by (15.12), p. 300. Let the functions $f_j(\theta)$, for $j \in \mathbb{Z}$, be given data on the spheres $S(0; e^j)$ and have there the Fourier–Laplace expansion*

$$f_j(\theta) = \sum_{k=0}^{\infty} \sum_{\ell=1}^{d_k} \psi_j^{k,\ell} Y_{k,\ell}(\theta) \quad \text{for all } \theta \text{ in } \mathbb{S}^{n-1}.$$

We assume that all f_j belong to the Sobolev class $H^{3\frac{1}{2}}(\mathbb{S}^{n-1})$. We assume that they satisfy the condition of power growth, i.e. there exist constants $D_{k,\ell}$ such that for some constant $\gamma > 0$ and for all pairs k and ℓ

$$|\psi_j^{k,\ell}| \leq D_{k,\ell} \cdot |j|^\gamma \quad \text{for all } j \in \mathbb{Z}$$

holds, and the constants $D_{k,\ell}$ satisfy the inequality

$$C := \tilde{C}^2 \sum_{k=0}^{\infty} k^{n-2} \left(\sum_{\ell=1}^{d_k} D_{k,\ell}^2 \right) < \infty.$$

Then the interpolation biharmonic polyspline h exists in the Sobolev class $H^4(\mathbb{R}^n \setminus \cup S(0; e^j))$ and satisfies the power growth

$$\int_{\mathbb{S}^{n-1}} |h(r\theta)|^2 \, d\theta \leq C \cdot |\log r|^{2\gamma}.$$

Proof The convergence in L_2 of the Fourier–Laplace series of $h(r\theta)$ follows by the inequalities established above. The power growth of h follows from the inequality for the function $S(x)$ established in the previous section.

Since the data functions f_j belong to the Sobolev space $H^{3\frac{1}{2}}$ we may apply the general regularity Theorem 21.3, p. 435, for polysplines, which implies that in every annulus f belongs to the Sobolev class H^4. This finishes the proof. ∎

The general case is given in [9].

15.9 Bibliographical notes

In the area of cardinal L-splines, except for the results of *Ch. Micchelli, I. Schoenberg,* and *N. Dyn* and *A. Ron* one has to mention the recurrence relations obtained for the first time by *Lyche*, see [10]. See also the results of *ter Morsche* [14].

We have added also a reference to a number of papers of *S. L. Sobolev*, [23]–[33], about some interesting properties of the zeros of the Euler polynomial. He applied these results to obtain optimal quadrature formulas.

Bibliography to Part II

[1] Baker, A., *Transcendental Number Theory*, Cambridge University Press, Cambridge, 1975.

[2] Bojanov, B., Hakopian, H.A. and Sahakian, A.A. *Spline Functions and Multivariate Interpolation*, Kluwer Academic Publishers, Dordrecht, 1993.

[3] Chui, Ch. *An Introduction to Wavelets*, Academic Press, Boston, 1992.

[4] Curry, H.B. and Schoenberg, I.J. On Polya frequency functions, IV: The fundamental spline functions and their limits. *J. Analyse Math.*, 17 (1966), pp. 71–107.

[5] de Boor, C., DeVore, R. and Ron, A. On the construction of multivariate (pre)wavelets. *Constructive Approximation*, 9 (1993), pp. 123–166.

[6] Dyn N. and Ron, A. Recurrence relations for Tchebycheffian B-splines. *J. d'anal. Math.*, 51 (1988), pp. 118–138.

[7] Dyn N. and Ron, A. Cardinal translation invariant Tchebycheffian B-splines, *Approx. Theory Appl.* 6, (1990), No. 2, pp. 1–12.

[8] Jetter K., Multivariate approximation from the cardinal interpolation point of view. *Approximation Theory VII* (Austin, TX, 1992), pp. 131–161, Academic Press, Boston, MA, 1993.

[9] Kounchev, O. and Render, H. Multivariate cardinal splines through spherical harmonics, submitted, 2000.

[10] Lyche, T. A recurrence relation for Chebyshevian B-splines, *Constructive Approximation*, 1 (1985), pp. 155–173.

[11] Mahler, K., *Lectures on Transcendental Numbers*, Lecture Notes in Mathematics, 546, Springer–Verlag, 1976.

[12] Micchelli, Ch. Oscillation matrices and cardinal spline interpolation. In: *Studies in Spline Functions and Approximation Theory*, S. Karlin *et al.* (Eds), Academic Press, NY, 1976, pp. 163–202.

[13] Micchelli, Ch. Cardinal L-splines, In: *Studies in Spline Functions and Approximation Theory*, S. Karlin *et al.* (Eds), Academic Press, NY, 1976, pp. 203–250.

[14] ter Morsche, H.G. Interpolation and extremal properties of L-spline functions, thesis, Eindhoven University of Technology, Eindhoven, 1982.

[15] Pontryagin, L.S. *Ordinary Differential Equations*, Addison-Wesley, 1962.

[16] Quade, W. and Collatz, L. Zur Interpolationstheorie der reellen periodischen Funktionen. *Sitzungsber. der Preuss. Akad. der Wiss. Phys.–Math. Kl.*, XXX (1938), pp. 383–429.

[17] Ron, A. Exponential box splines. *Constructive Approximation*, 4 (1988), pp. 357–378.

[18] Schoenberg, I.J. *Cardinal Spline Interpolation*, SIAM, Philadelphia, Pennsylvania, 1973.

[19] Schoenberg, I.J. On Micchelli's theory of cardinal L-splines. In: *Studies in Spline Functions and Approximation Theory*, S. Karlin et al. (Eds), Academic Press, NY, 1976, pp. 251–276.

[20] Schoenberg, I.J. Cardinal interpolation and spline functions. *J. Approx. Theory* 2 (1969), pp. 167–206.

[21] Schoenberg, I.J. Cardinal interpolation and spline functions: II. Interpolation of data of power growth. *J. Approx. Theory*, 6 (1972), pp. 404–420.

[22] Schumaker, L.L. *Spline Functions: Basic Theory*, J. Wiley and Sons, NY, Chichester–Brisbane–Toronto, 1981.

[23] Sobolev, S.L. Les coefficients optimaux des formules d'integration approximative. *Ann. Scuola Norm. Sup. Pisa Cl. Sci.*, (4) 5 (1978), No. 3, pp. 455–469. and Errata: (French) *Ann. Scuola Norm. Sup. Pisa Cl. Sci.*, (4) 6 (1979), No. 4, p. 729.

[24] Sobolev, S.L. On the algebric order of exactness of approximate integration formulas. (Russian) *Partial differential equations (Novosibirsk, 1983)*, pp. 4–11, 219, "Nauka" Sibirsk. Otdel., Novosibirsk, 1986.

[25] Sobolev, S.L. Comportement asymptotique des racines des polynomes d'Euler. (French. English, Italian summary) [Asymptotic behavior of the roots of Euler polynomials.] *Rend. Sem. Mat. Fis. Milano*, 52 (1982), pp. 221–243.

[26] Sobolev, S.L. On extreme roots of Euler polynomials. (Russian) *Dokl. Akad. Nauk SSSR*, 242 (1978) No. 5, pp. 1016–1019.

[27] Sobolev, S.L. Convergence of cubature formulas on various classes of periodic functions. (Russian) *Theory of cubature formulas and the application of functional analysis to problems of mathematical physics (Russian)*, pp. 122–140, 167, Trudy Sem. S. L. Soboleva, No. 1, 1976, Akad. Nauk SSSR Sibirsk. Otdel., Inst. Mat., Novosibirsk, 1976.

[28] Sobolev, S.L. On the asymptotic behavior of Euler polynomials. (Russian) *Dokl. Akad. Nauk, SSSR*, 245 (1979), No. 2, pp. 304–308.

[29] Sobolev, S.L. More on roots of Euler polynomials. (Russian) *Dokl. Akad. Nauk SSSR*, 245 (1979), No. 4, pp. 801–804.

[30] Sobolev, S.L. Roots of Euler polynomials. (Russian) *Dokl. Akad. Nauk SSSR*, 235 (1977), No. 2, pp. 277–280.

[31] Sobolev, S.L. Coefficients of best quadrature formulas. (Russian) *Dokl. Akad. Nauk SSSR*, 235 (1977), No. 1, pp. 34–37.

[32] Sobolev, S.L. Convergence of cubature formulas on elements of $L_2^{(m)}$ (Russian) *Dokl. Akad. Nauk SSSR*, 228 (1976), No. 1, pp. 45–47.

[33] Sobolev, S.L. Convergence of cubature formulae on infinitely differentiable functions. (Russian) *Dokl. Akad. Nauk SSSR*, 223 (1975), No. 4, pp. 793–796.

[34] *Studies in Spline Functions and Approximation Theory*, Eds. S. Karlin et al., Academic Press, New York, 1976.

[35] Tschakaloff, L. On a certain representation of the Newton divided differences in interpolation theory and its applications, (Bulgarian. French summary) *Annuaire de l'Univ. de Sofia, Fiz. Mat. Fakultet*, v. 34 (1938), pp. 353–405.

Part III

Wavelet analysis

In the Introduction to Part II we explained how naturally the cardinal polysplines appear. This is especially non-trivial in the case of polysplines with break-surfaces on concentric spheres. Similarly the notion of *polyharmonic multiresolution analysis* appears in a natural manner. Let us explain how it happens in the case of polysplines on annuli.

First, recall that every cardinal polyspline on annuli has break-surfaces the spheres $S(0; e^v)$ for all integers $v \in \mathbb{Z}$, and possesses an expansion in spherical harmonics of the form

$$h(x) = \sum_{k=0}^{\infty} \sum_{\ell=1}^{d_k} h^{k,\ell}(\log r) Y_{k,\ell}(\theta).$$

For every couple of indexes k and ℓ the function $h^{k,\ell}(v)$ is a cardinal L-spline for the constant coefficients operator $L = M_{k,p}$ with knots at $v \in \mathbb{Z}$.

What would be more natural than to make multiresolution analysis (MRA) of every one-dimensional component $h^{k,\ell}(v)$, i.e. of the L-splines for the operator $L = M_{k,p}$? And the refinement of the knot set $\mathbb{Z}$ will be $(1/2)\mathbb{Z}$, etc. Furthermore we are lucky that the basic elements of wavelet analysis for cardinal L-splines has been created by de Boor *et al.* [9].[1]

Having obtained MRA for every component $h^{k,\ell}$ it remains only to assemble the puzzle by means of the above formula (III).[2]

The above program is easy to describe in general terms but it takes a good deal of work to accomplish. First, let us recall that in the polynomial case the detailed cardinal spline wavelet analysis has been carried out by Chui [4], [5]. Accordingly, in Chapter 16 we provide a brief review of his results. This review will be our compass when studying the cardinal L-spline wavelet analysis. The transition from the polynomial spline case to the L-spline case is highly non-trivial and we were aware of that in the proof of the *Riesz inequalities* for the shifts of the TB-spline $Q_{\mathbb{Z}+1}$ in Chapter 14, p. 267.

The Cardinal L-spline wavelet analysis which we develop in Chapter 17 uses the whole machinery of cardinal L-splines which we have developed in Part II. It is amazing that all the main results of Chui's approach permit non-trivial generalizations for the cardinal L-spline wavelet analysis. The dependence on the vector Λ is essential and is emphasized in these results. The Chui's results are reduced to the special case of the vector $\Lambda = [0, ..., 0]$.

Finally in Chapter 18 we obtain the assembled "polyharmonic wavelet". It has some interesting properties. Needless to say, it does not satisfy the axioms of the MRA as established by Y. Meyer and S. Mallat, see [14]. So far it satisfies some of them and also some other properties which we provide in Theorem 18.9, p. 380. Put in a proper framework these properties may be considered as the axioms of what we call *Polyharmonic Multiresolution Analysis*.

---

[1] The Bibliography is at the end of the present Part.

[2] As we already mentioned in the Introduction to Part II, due to the lack of space we do not consider in detail the cardinal Polysplines on strips. For that reason we do not consider here the Wavelet Analysis generated by Polysplines on strips. We note only that its formulas are simpler than the annular case, in particular the refinement is by considering parallel hyperplanes having say first coordinate $t \in 2^{-j}\mathbb{Z}$, see Definition 9.1, p. 118.

It is remarkable that we may preserve the basic scheme of the usual MRA but if we introduce some proper substitutes to the basic notions:

1. There is no *refinement equation* but we have a *refinement operator,* see Theorem 18.9, p. 380. This refinement operator is generated by the *non-stationary scaling operator* for the L-splines defined in formula (17.6), p. 328.
2. In Section 18.5, p. 379, we see that there is a unique function which we call *father wavelet* and it generates the spaces PV_j of the *polyharmonic MRA* in a "non-stationary" way. In a similar way, in Section 18.6, p. 384, we see that there is a unique function which we call *mother wavelet* which generates the wavelet spaces PW_j in a "non-stationary" way.
3. To have the whole picture completed let us recall that we have also the *sampling operator* provided by formula (14.16), p. 285, of Part II.

Hence, the main conclusion of the present Part is that the attempt to make a reasonable Multiresolution Analysis by means of a refining sequence of spaces of *cardinal polysplines (on annuli or strips)* leads to a considerable reconsideration of the whole store of basic notions of MRA.

The present Part provides a detailed study of only one example of "spherical polyharmonic wavelet analysis". Similar wavelet analysis may be carried out for other elliptic differential operators of the form

$$A(r) \frac{\partial^2}{\partial r^2} + B(r) \frac{\partial}{\partial r} + \frac{1}{r^2} \Delta_\theta$$

which are possibly degenerate at the origin but split into infinitely many one-dimensional operators with constant coefficients.

What might be the area of application of such wavelets which have *by definition* singularities on whole $(n-1)$-dimensional surfaces? Let us point out to a possible application for analyzing, e.g. in $\mathbb{R}^2$, images having singularities on curves. The problem of efficient computational analysis of such images has been indicated by Meyer and Mallat – the standard wavelet paradigm is not efficient for analyzing images having $(n-1)$-dimensional singularities. This problem has been given a thorough consideration in a series of papers of D. Donoho with coauthors. In particular, the curvelets by D. Donoho and E. Candes [2] have been created with the main purpose to solve this problem. The polyharmonic wavelets may be considered as an alternative approach to this problem.

Finally, let us note that much as we do not like it many of the formulas in the present Part are overburdened with indexes and arguments which makes it somewhat heavy to read. On the other hand this detailed exposition would provide the reader with the opportunity to check the correctness of all formulas.

Chapter 16

Chui's cardinal spline wavelet analysis

As is usual in wavelet analysis, we will be working in the space $L_2(\mathbb{R})$ of square summable complex valued functions with the scalar product defined by

$$\langle f, g \rangle := \langle f, g \rangle_{L_2(\mathbb{R})} := \int_{-\infty}^{\infty} f(x)\overline{g(x)}\,dx \qquad (16.1)$$

for every two functions $f, g \in L_2(\mathbb{R})$. We have the norm

$$\|f\|^2 := \|f\|_{L_2(\mathbb{R})}^2 := \langle f, f \rangle. \qquad (16.2)$$

16.1 Cardinal splines and the sets V_j

Denote by V_j the closure in $L_2(\mathbb{R})$ of the space of all **cardinal polynomial splines** (which are in $L_2(\mathbb{R})$) of polynomial degree m having knots at the points

$$\mathbb{Z} \cdot 2^{-j} = \left\{ \frac{\ell}{2^j} \colon \text{ for all } \ell \in \mathbb{Z} \right\}. \qquad (16.3)$$

By definition their smoothness is C^{m-1}. Evidently, since $\mathbb{Z} \cdot 2^{-j_1} \subset \mathbb{Z} \cdot 2^{-j_2}$ for every two integers j_1 and j_2 satisfying $j_1 < j_2$, we have the inclusions

$$\cdots \subset V_{-2} \subset V_{-1} \subset V_0 \subset V_1 \subset V_2 \subset \cdots.$$

We use the term *cardinal* in the wider sense, understanding splines with knots at the set $\mathbb{Z} \cdot h = \{\ell h \colon \text{ for all } \ell \in \mathbb{Z}\}$ for some number $h > 0$.

The most important function in the theory of cardinal splines and also of spline wavelet analysis is the compactly supported B-spline $N_m(x) \in V_0$ with support coinciding with the interval $[0, m]$ and with knots at the integers. Following the tradition in

MRA we will denote it through $\phi(x)$, i.e.

$$\phi(x) := N_m(x), \qquad (16.4)$$

and we will call it the **scaling function**. We know that N_m is symmetric around the point $m/2$ which is the center of the interval $[0, m]$, i.e.

$$N_m(x) = N_m(m - x) \quad \text{for } x \in \mathbb{R}.$$

An important formula is the one providing the **Fourier transform** of N_m

$$\widehat{N_m}(\xi) = \left(\frac{1 - e^{-i\xi}}{i\xi}\right)^m, \qquad (16.5)$$

[5, p. 53]. It is remarkable that the cardinal spline $N_m(x)$ which has knots at the integer points $\mathbb{Z}$ generates through shifts not only the space V_0, i.e.[3]

$$V_0 = \underset{L_2(\mathbb{R})}{\text{clos}}\,\{\phi(x - \ell): \text{ for all } \ell \in \mathbb{Z}\}, \qquad (16.6)$$

but also the spaces V_j. For that purpose one forms the 2^j-dilates of the function $N_m(x)$, namely $N_m(2^j x)$, and considers its shifts, thus obtaining for all $j \in \mathbb{Z}$, the equality

$$V_j = \underset{L_2(\mathbb{R})}{\text{clos}}\,\{\phi(2^j x - \ell): \text{ for all } \ell \in \mathbb{Z}\}. \qquad (16.7)$$

Actually, this is due to the fact that the function $\phi(2^j x) = N_m(2^j x)$ is again piecewise polynomial of degree $\leq m$ but with knots on the mesh $\mathbb{Z}2^{-j} = \{\ell/2^j: \text{ for all } \ell \in \mathbb{Z}\}$.

Since $V_0 \subset V_1$ we have, [4, p. 91, formula (4.3.2)] the central relation in MRA called the **two-scale relation** or **refinement equation**[4] for $\phi(x)$, namely

$$\phi(x) = \sum_{j=-\infty}^{\infty} p_j \phi(2x - j), \qquad (16.8)$$

where in fact the sequence $\{p_j\}$ is finite and is given by the coefficients of the polynomial

$$P(z) := \frac{1}{2}\sum_{j=0}^{m} p_j z^j := \left(\frac{1+z}{2}\right)^m. \qquad (16.9)$$

In order to make clear this *most fundamental relation* in wavelet theory, let us show that it is easy to prove. After taking the Fourier transform of (16.8), we obtain

$$\widehat{\phi}(\xi) = \frac{1}{2}\sum_{j=0}^{m} p_j e^{-ij(\xi/2)} \cdot \widehat{\phi}\left(\frac{\xi}{2}\right), \qquad (16.10)$$

---

[3] The notation $\text{clos}_{L_2(\mathbb{R})}$ means the linear and topological hull in the norm of $L_2(\mathbb{R})$.

[4] In the setting of some authors these are *prewavelets* (de Boor et al. [9]), in the setting of Chui these are *semi-orthogonal wavelets*..

and by substituting the explicit formula for the Fourier transform (16.5) above we obtain

$$\left(\frac{1-e^{-i\xi}}{i\xi}\right)^m = P(z)\left(\frac{1-e^{-i\xi/2}}{i\xi/2}\right)^m.$$

Now the identity

$$(1-e^{-i\xi})^m = \left(1-e^{-i\xi/2}\right)^m \left(1+e^{-i\xi/2}\right)^m$$

implies (16.9).

The principal property of the scaling function ϕ is that the set of shifts $\{\phi(x-\ell): \text{ for all } \ell \in \mathbb{Z}\}$ is a **Riesz basis** of V_0. The notion of Riesz basis has become a replacement condition for orthogonal basis since every Riesz basis may be orthonormalized and many properties of the orthogonal basis are preserved. By the definition of a Riesz basis there exist two constants A, B with $0 < A \le B < \infty$, and for every sequence $\{c_j\} \in \ell_2$ holds

$$A \sum_{j=-\infty}^{\infty} |c_j|^2 \le \left\| \sum_{j=-\infty}^{\infty} c_j \phi(x-j) \right\|_{L_2(\mathbb{R})} \le B \sum_{j=-\infty}^{\infty} |c_j|^2. \tag{16.11}$$

The constants A, B are called **Riesz bounds**. An *equivalent condition* for a basis to be Riesz is that

$$\lfloor \phi \rfloor := \sum_{j=-\infty}^{\infty} |\widehat{\phi}(\xi + 2\pi j)|^2 < \infty \quad \text{a.e. in } \mathbb{R}. \tag{16.12}$$

Below we will find an explicit expression for this infinite sum.

16.2 The wavelet spaces W_j

The **wavelet spaces** W_j are defined uniquely through the properties holding for all j in $\mathbb{Z}$, namely:

$$V_{j+1} = V_j \oplus W_j,$$
$$W_j \subset V_{j+1},$$

or, equivalently,

$$W_j := V_{j+1} \ominus V_j \quad \text{for all } j \text{ in } \mathbb{Z}. \tag{16.13}$$

Here $\oplus$ means the orthogonal sum of two linear spaces, and in the context above we have two mutually orthogonal subspaces, V_j and W_j; their usual sum gives V_{j+1}. The last means that in $V_{j+1} = V_j \oplus W_j$ we have a sum but not simply isomorphism!

Since

$$\operatorname*{clos}_{L_2(\mathbb{R})} \left\{ \bigcup_{j=-\infty}^{\infty} V_j \right\} = L_2(\mathbb{R}) \tag{16.14}$$

we obtain the expansion

$$L_2(\mathbb{R}) = \bigoplus_{j=-\infty}^{\infty} W_j. \tag{16.15}$$

Any function may be expanded in a **generalized Fourier series**

$$f(x) = \sum_{j=-\infty}^{\infty} w_j \quad \text{with } w_j \in W_j \text{ for all } j \in \mathbb{Z}. \tag{16.16}$$

The major factor contributing to the charm of the *classical cardinal spline wavelets*[5] is that, roughly speaking, one has only one generating function ψ for all spaces W_j. More precisely, ψ is called the mother **wavelet** if its shifts generate W_0, i.e.

$$W_0 = \operatorname*{clos}_{L_2(\mathbb{R})} \{\psi(x - \ell) : \text{ for all } \ell \in \mathbb{Z}\} \tag{16.17}$$

and the 2^j-dilates

$$\psi_{j,\ell}(x) := 2^{j/2} \psi(2^j x - \ell)$$

generate the spaces W_j, i.e.

$$W_j = \operatorname*{clos}_{L_2(\mathbb{R})} \{\psi_{j,\ell}(x) : \text{ for all } \ell \in \mathbb{Z}\}. \tag{16.18}$$

The norming of the functions $\psi_{j,\ell}(x)$ is important since it preserves the L_2-norm of linear combinations at every level j for $j \in \mathbb{Z}$, namely

$$\int_{-\infty}^{\infty} \left| \sum_{\ell=-\infty}^{\infty} c_\ell 2^{j/2} \psi(2^j x - \ell) \right|^2 dx = \int_{-\infty}^{\infty} \left| \sum_{\ell=-\infty}^{\infty} c_\ell \psi(x - \ell) \right|^2 dx,$$

hence the set of shifts $\{\psi_{j,\ell}(x)\}_{\ell \in \mathbb{Z}}$ has the same Riesz constants A, B as the shifts of $\psi(x)$.

Let us note that the polynomial spline wavelets owe their algorithmic effectiveness to the fact that the polynomial splines are scale-invariant. That is: if $f(x)$ is a polynomial spline, then for every number h the function $f(hx)$ is also a polynomial spline. In some sense, dilations do not change the physical nature of the basic space of functions used to construct MRA. It is important to note that in a similar way all classical wavelet constructions and the axioms of MRA rely upon this principle [8, 14], namely that the *physical nature* of the dilated function $f(hx)$ is the same as that of the original function $f(x)$. We will see that this is not the case for the general L-spline wavelets.

---

[5] In the setting of some authors these are *prewavelets* (de Boor et al. [9]), in the setting of Chui these are *semi-orthogonal wavelets*.

16.3 The mother wavelet ψ

There is an **explicit formula** for the function ψ in *cardinal spline wavelet analysis*. Since $W_0 \subset V_1$, we have the representation

$$\psi(x) = \sum_{j=-\infty}^{\infty} r_j \phi(2x - j), \qquad (16.19)$$

where r_j are the coefficients of a Laurent polynomial $R(z)$. A Laurent polynomial also has negative exponents. We have

$$R(z) = \frac{1}{2} \cdot \sum_{j=-\infty}^{\infty} r_j z^j.$$

An important property is that only a finite number of the coefficients r_j are non-zero, hence ψ has a **compact support**. Indeed, if we pass to the Fourier images (in the frequency variable ξ), we obtain

$$\widehat{\psi}(\xi) = R\left(e^{-i\xi/2}\right) \widehat{\phi}\left(\frac{\xi}{2}\right) \quad \text{for all } \xi \text{ in } \mathbb{R}. \qquad (16.20)$$

On the other hand, $\psi \perp V_0$. These two conditions and the fact that ψ has shifts $\{\psi(x - \ell) : \text{for all } \ell \in \mathbb{Z}\}$ which are the basis of W_0 determine ψ uniquely up to a constant factor

$$R(z) = -z^{2m-1} P\left(-\frac{1}{z}\right) E_\phi\left(-\frac{1}{z}\right).$$

Here we have denoted by $E_\phi(z)$ the so-called **Euler–Frobenius polynomial** (which is a Laurent polynomial) which was introduced by Schoenberg. It is given by

$$E_\phi(z) = \sum_{\ell=-m+1}^{m-1} \phi_{2m}(m + \ell) z^\ell; \qquad (16.21)$$

where again

$$\phi_{2m}(x) := N_{2m}(x).$$

We see that $R(z)$ is a classical polynomial (it does not contain negative exponents of z) and its degree is evidently $3m - 2$. Hence, we easily prove by equality (16.19), p. 317, that the wavelet function $\psi(x)$ has a compact support coinciding with the interval $[0, 2m - 1]$.

The polynomial $E_\phi(z)$ is the *key function* to the whole approach. Let us note the important equality

$$\lfloor \phi \rfloor = \sum_{\ell=-\infty}^{\infty} |\widehat{\phi}(\xi + 2\pi \ell)|^2 = E_\phi(e^{-i\xi}), \qquad (16.22)$$

where we have used the notation for this norm introduced above in (16.12), p. 315. Since the function $E_\phi(z)$ has no zeros on the unit circle $|z| = 1$, owing to the Riesz inequalities

(16.11) and the norm in (16.12), p. 315, this might be considered as a proof that the shifts of the function $\phi(x)$ provide a Riesz basis in V_0. Due to the symmetry of the cardinal spline $N_{2m}(x)$ around the point m we see that

$$E_\phi(z) = E_\phi\left(\frac{1}{z}\right).$$

16.4 The dual mother wavelet $\tilde{\psi}$

Another important property of the function $\psi(x)$ which *makes it a wavelet* is the existence of the **dual wavelet** $\tilde{\psi}(x)$. The dual wavelet function gives rise to the functions

$$\tilde{\psi}_{j,\ell}(x) := 2^{j/2}\tilde{\psi}(2^j x - \ell), \tag{16.23}$$

which satisfy the basic **bi-orthogonality** property, for all indexes $j, j_1, \ell, \ell_1 \in \mathbb{Z}$, namely[6]

$$\langle \tilde{\psi}_{j,\ell}, \psi_{j_1,\ell_1} \rangle := \int_{-\infty}^{\infty} \tilde{\psi}_{j,\ell}(x)\overline{\psi_{j_1,\ell_1}(x)}\,dx = \delta_{j,j_1} \cdot \delta_{\ell,\ell_1}.$$

Now if the generalized Fourier series representation (16.16), p. 316, is written as follows, see Chui [5, p. 89]:

$$f(x) = \sum_{j,\ell=-\infty}^{\infty} d_{j,\ell} \psi_{j,\ell}(x),$$

we see that after taking the scalar product with $\tilde{\psi}_{j,\ell}(x)$ for every two indexes j and ℓ in $\mathbb{Z}$ we obtain

$$d_{j,\ell} = \langle f, \tilde{\psi}_{j,\ell} \rangle \quad \text{for all } j, \ell \text{ in } \mathbb{Z}.$$

Let us denote

$$\widetilde{W}_j := \operatorname*{clos}_{L_2(\mathbb{R})} \{\tilde{\psi}_{j,\ell}: \text{ for all } j, \ell \text{ in } \mathbb{Z}\}.$$

Then the following orthogonality properties are satisfied:

$$\begin{cases} \widetilde{W}_j \perp V_j & \text{for all } j \in \mathbb{Z}, \\ W_j \perp V_j & \text{for all } j \in \mathbb{Z}. \end{cases} \tag{16.24}$$

There is an explicit expression for the *unique dual wavelet* in terms of its Fourier transform

$$\widehat{\tilde{\psi}}(\xi) = \frac{\widehat{\psi}(\xi)}{\sum_{\ell=-\infty}^{\infty} |\widehat{\psi}(\xi + 2\pi\ell)|^2}.$$

---

[6] Here δ is the Kronecker symbol defined as $\delta_{\alpha\beta} = 1$ for $\alpha = \beta$ and $\delta_{\alpha\alpha} = 0$.

The last expression makes sense since the sum is convergent, and we have the following elegant explicit expression [5, p. 106, formula (5.3.11)]:

$$\lfloor \psi \rfloor = \sum_{\ell=-\infty}^{\infty} |\widehat{\psi}(\xi + 2\pi\ell)|^2 = |R(z)|^2 E_\phi(z) + |R(-z)|^2 E_\phi(-z) \qquad (16.25)$$

$$= E_\phi(z) E_\phi(-z) E_\phi(z^2) \quad \text{for all } z = e^{-i\xi/2}.$$

The last is nonzero since the Euler–Frobenius polynomial $E_\phi(z)$ has no zeros on the unit circle $|z| = 1$. This remark provides a proof that the set of shifts $\{\psi(x-\ell): \text{ for all } \ell \in \mathbb{Z}\}$ is a Riesz basis of W_0 by applying the criterion for a Riesz basis using the Fourier transform of ψ. It is similar to that for the scaling function $\phi(x)$ which we have seen in (16.11) and (16.12), p. 315.

Another important identity is

$$E_\phi(z^2) = |P(z)|^2 E_\phi(z) + |P(-z)|^2 E_\phi(-z). \qquad (16.26)$$

16.5 The dual scaling function $\widetilde{\phi}$

Now we consider the **dual scaling function** $\widetilde{\phi}(x) \in V_0$. It satisfies the orthogonality property

$$\langle \phi(x-j), \widetilde{\phi}(x-\ell) \rangle = \delta_{j,\ell} \quad \text{for all } j, \ell \text{ in } \mathbb{Z}.$$

There is an explicit expression for $\widetilde{\phi}(x)$ in terms of its Fourier transform

$$\widehat{\widetilde{\phi}}(\xi) = \frac{\widehat{\phi}(\xi)}{\sum_{\ell=-\infty}^{\infty} |\widehat{\phi}(\xi+2\pi\ell)|^2} = \frac{\widehat{\phi}(\xi)}{E_\phi(e^{-i\xi})}. \qquad (16.27)$$

If we denote

$$\widetilde{V}_j := \operatorname*{clos}_{L_2(\mathbb{R})} \{\widetilde{\phi}(x-\ell): \text{ for all } \ell \in \mathbb{Z}\},$$

then we obtain

$$\widetilde{V}_j = V_j$$

and also the following orthogonality relations which together with the relations in (16.24) read as follows:

$$W_j \perp \widetilde{V}_j, \quad \widetilde{W}_j \perp V_j, \quad W_j \perp V_j \quad \text{for all } j \text{ in } \mathbb{Z}.$$

16.6 Decomposition relations

Since the **Euler–Frobenius** polynomial $E_\phi(z)$ does not take on zero values on the unit circle $|z| = 1$, there always exists the inverse Laurent polynomial which is convergent

320 Multivariate polysplines

near the unit circle, and we may put

$$E_\phi(z) = \sum_{j=-\infty}^{\infty} \beta_j z^j,$$

$$\frac{1}{E_\phi(z)} = \sum_{j=-\infty}^{\infty} \alpha_j z^j.$$

The coefficients α_j decay exponentially. By taking the inverse Fourier transform in equality (16.27) we see that we may express the basis of V_0 in both directions

$$\widetilde{\phi}(x) = \sum_{j=-\infty}^{\infty} \alpha_j \phi(x-j),$$

$$\phi(x) = \sum_{j=-\infty}^{\infty} \beta_j \widetilde{\phi}(x-j).$$

Since

$$V_1 = V_0 \oplus W_0,$$

it is obvious that the functions $\phi(2x)$ and $\phi(2x-1)$ may be represented as

$$\phi(2x) = \sum_{s=-\infty}^{\infty} \{a_{-2s}\phi(x-s) + b_{-2s}\psi(x-s)\},$$

$$\phi(2x-1) = \sum_{s=-\infty}^{\infty} \{a_{1-2s}\phi(x-s) + b_{1-2s}\psi(x-s)\},$$

by means of the sequences a_{2j}, b_{2j} and a_{2j+1}, b_{2j+1}, respectively. We may combine these two representations in one called the **decomposition relation**, distinguishing only the case of odd and even index ℓ, namely

$$\phi(2x-\ell) = \sum_{s=-\infty}^{\infty} \{a_{\ell-2s}\phi(x-s) + b_{\ell-2s}\psi(x-s)\}. \qquad (16.28)$$

For the sequences a_j and b_j we define the corresponding symbols by putting

$$A(z) = \frac{1}{2} \cdot \sum_{j=-\infty}^{\infty} a_j z^j, \qquad (16.29)$$

$$B(z) = \frac{1}{2} \cdot \sum_{j=-\infty}^{\infty} b_j z^j. \qquad (16.30)$$

The functions $A(z)$ and $B(z)$, which are Laurent polynomials, may be found as a solution of the following algebraic system:

$$P(z)A(\bar{z}) + R(z)B(\bar{z}) = 1 \quad \text{for all } |z| = 1,$$
$$P(-z)A(\bar{z}) + R(-z)B(\bar{z}) = 0 \quad \text{for all } |z| = 1.$$

They may be found explicitly by applying identities (16.25) and (16.26), p. 319.

$$A(z) = \frac{E_\phi(z)}{E_\phi(z^2)} P(z),$$

$$B(z) = -\frac{z^{2m-1}}{E_\phi(z^2)} P\left(-\frac{1}{z}\right).$$

Evidently, $A(z)$ and $B(z)$ are Laurent polynomials.

It is interesting that the symbols $A(z)$ and $B(z)$ also provide the **two-scale relations** for $\tilde{\phi}$ and $\tilde{\psi}$ (which are the "dual" relations to the two-scale relations for ϕ and ψ in (16.8), p. 314, and (16.19), p. 317, respectively, by

$$\begin{cases} \widehat{\tilde{\phi}}(\xi) = A(z)\widehat{\tilde{\phi}}(\xi/2), \\ \widehat{\tilde{\psi}}(\xi) = B(z)\widehat{\tilde{\phi}}(\xi/2). \end{cases} \quad (16.31)$$

16.7 Decomposition and reconstruction algorithms

Let us return to the main point of MRA. Assume that we are given an arbitrary function $f \in L_2(\mathbb{R})$. Then for every $\varepsilon > 0$ we find an approximation $f_N \in V_N$ for a sufficiently large N such that

$$\|f - f_N\| < \varepsilon.$$

We consider the expansion of the function $f_N(x) \in V_N$ given by

$$f_N(x) = \sum_{j=-\infty}^{\infty} c_{N,\ell} \phi(2^N x - \ell).$$

Due to

$$V_N = V_{N-1} \oplus W_{N-1}, \quad (16.32)$$

we have the representation

$$f_N(x) = f_{N-1}(x) + g_{N-1}(x)$$

with $f_{N-1} \in V_{N-1}$ and $g_{N-1} \in W_{N-1}$. Then the coefficients in the representations

$$f_{N-1}(x) = \sum_{j=-\infty}^{\infty} c_{N-1,\ell} \phi(2^{N-1} x - \ell),$$

$$g_{N-1}(x) = \sum_{j=-\infty}^{\infty} d_{N-1,\ell} \psi(2^{N-1} x - \ell),$$

may be computed by applying the above decomposition relations. We obtain the **decomposition algorithm**, holding for all s in $\mathbb{Z}$, namely

$$c_{N-1,s} = \sum_{\ell=-\infty}^{\infty} a_{\ell-2s} c_{N,\ell},$$

$$d_{N-1,s} = \sum_{\ell=-\infty}^{\infty} b_{\ell-2s} d_{N,\ell}.$$

Conversely, we have the representation for the coefficients of f_N by means of the coefficients of f_{N-1} and g_{N-1}, which is known as the **reconstruction algorithm**, and for every $s \in \mathbb{Z}$ given by

$$c_{N,s} = \sum_{\ell=-\infty}^{\infty} \{p_{s-2\ell} c_{N-1,\ell} + r_{s-2\ell} d_{N-1,\ell}\}. \tag{16.33}$$

The "reconstruction algorithm" is practically reasonable since the sequences p_j and r_j are finite. However, from this point of view the "decomposition algorithm" is not very practical since the sequences a_j and b_j are infinite. For that reason, for the decomposition it is better to apply the dual representation of f_N, namely

$$f_N(x) = \sum_{j=-\infty}^{\infty} \tilde{c}_{N,\ell} \tilde{\phi}(2^N x - \ell)$$

$$= \sum_{j=-\infty}^{\infty} \tilde{c}_{N-1,\ell} \tilde{\phi}(2^{N-1} x - \ell) + \sum_{j=-\infty}^{\infty} \tilde{d}_{N-1,\ell} \tilde{\psi}(2^{N-1} x - \ell).$$

Now thanks to the dual two-scale relations (16.31), p. 321, we obtain the *dual decomposition algorithm*, holding for all s in $\mathbb{Z}$, namely

$$\tilde{c}_{N-1,s} = \sum_{\ell=-\infty}^{\infty} p_{2s-\ell} \tilde{c}_{N,\ell},$$

$$\tilde{d}_{N-1,s} = \sum_{\ell=-\infty}^{\infty} r_{2s-\ell} \tilde{d}_{N,\ell}.$$

Note that the order of the indices of the coefficients p_j and r_j has changed.

16.8 Zero moments

An important property of the spline wavelets is that they have **zero moments** up to order m, i.e.

$$\int_{-\infty}^{\infty} x^{\ell} \psi(x) \, dx = 0 \quad \text{for } \ell = 0, \ldots, m, \tag{16.34}$$

the last being equivalent to the fact that $W_0 \perp V_0$ [5, p. 59, p. 61]. This is due to the fact that every compactly supported spline is a finite linear combination of the shifts of the function $\phi(x)$. An equivalent statement is that for every integer $\ell \leq m$ and for every compact interval $[a, b]$ we have the representation

$$x^\ell = \sum_{i=-\infty}^{\infty} a_{\ell,i} \phi(x-i),$$

where $a_{\ell,i}$ is a finite sequence.

16.9 Symmetry and asymmetry

The **symmetry** and the **antisymmetry** properties distinguish the cardinal spline wavelets from other wavelets. Namely, if we put $\psi_1(x) = \psi\left(x + \frac{2m-1}{2}\right)$ then we have

$$\psi_1(x) = \begin{cases} \psi_1(-x) & \text{for even } m, \\ -\psi_1(-x) & \text{for odd } m. \end{cases} \quad (16.35)$$

Chapter 17

Cardinal L-spline wavelet analysis

In the terminology of de Boor *et al.* [9] the L-spline wavelets we are studying here are called prewavelets. Chui uses the name semi-orthogonal wavelets.

It is remarkable that the most essential elements of the *polynomial cardinal spline wavelets* theory developed extensively by Chui may be used for cardinal L-splines.

An important observation which will play a key role throughout Part III of this book is the following: for simplicity assume that in the nonordered vector $\Lambda = [\lambda_1, \lambda_2, \ldots, \lambda_{Z+1}]$ all λ_js are pairwise different. We have the corresponding differential operator

$$\mathcal{L} = \mathcal{L}[\Lambda] = \prod_{j=1}^{Z+1} \left(\frac{d}{dx} - \lambda_j\right).$$

Then the L-splines are piecewise linear combinations of the type

$$f(x) = \sum_{j=1}^{Z+1} c_j e^{\lambda_j x},$$

which are solutions of the equation $\mathcal{L}[\Lambda]f(x) = 0$. We see that for an arbitrary real number h the function

$$f(hx) = \sum_{j=1}^{Z+1} c_j e^{h\lambda_j x}$$

is not a solution of $\mathcal{L}[\Lambda]f(hx) = 0$. But it is a solution of the equation

$$\mathcal{L}[h\Lambda]f(hx) = \prod_{j=1}^{Z+1} \left(\frac{d}{dx} - h\lambda_j\right) f(hx),$$

where by definition the differential operator $\mathcal{L}[h\Lambda]$ is associated with the nonordered vector $h\Lambda = [h\lambda_1, \ldots, h\lambda_{Z+1}]$.

326 *Multivariate polysplines*

We want to warn the reader that although the present chapter is completely self-contained, its digestion will be much eased if the results of [4, 5] are familiar to the reader. We also provide a proper quotation in order to compare the results of those references with ours.

Let us introduce wavelet analysis using cardinal *L*-splines.

17.1 Introduction: the spaces V_j and W_j

Throughout the present chapter we assume that the nonordered vector $\Lambda = [\lambda_1, \lambda_2, \ldots, \lambda_{Z+1}]$.

Let the number $h > 0$ be fixed. We will denote by $\mathcal{S}_{Z+1,h}$ the **space of cardinal *L*-splines** on the mesh $h\mathbb{Z}$. We will be interested in those cardinal splines which are in $L_2(\mathbb{R})$. To define an *MRA* we will introduce the spaces V_j. Denote by V_j the closure in $L_2(\mathbb{R})$ of the **space of cardinal *L*-splines** (which are in $L_2(\mathbb{R})$) with respect to the operator $\mathcal{L}[\Lambda]$ with knots at the points

$$2^{-j}h \cdot \mathbb{Z} = \left\{ \frac{\ell}{2^j} h : \text{ for } \ell \text{ in } \mathbb{Z} \right\},$$

i.e.

$$V_j := \operatorname*{clos}_{L_2(\mathbb{R})} (\mathcal{S}_{Z+1, 2^{-j}h} \cap L_2(\mathbb{R})).$$

By definition these *L*-splines are in $C^{Z-1}(\mathbb{R})$. Evidently, we have the inclusion

$$\cdots \subset V_{-2} \subset V_{-1} \subset V_0 \subset V_1 \subset V_2 \subset \cdots,$$

and the **wavelet spaces** are defined by

$$W_j := V_{j+1} \ominus V_j. \tag{17.1}$$

Recall that in Definition 13.34, p. 239, we denoted by $Q[\Lambda; h](x)$ the compactly supported cardinal *L*-spline with support coinciding with the interval $[0, Zh + h]$, and defined on the grid $h\mathbb{Z}$. On the other hand $Q[\Lambda](x)$ without parameter h denotes the compactly supported *L*-spline defined on the grid $\mathbb{Z}$, see formula (13.24a), p. 242.[1].

Proposition 17.1 *For every $h > 0$ the set of shifts $\{Q[\Lambda; h](x - \ell h) : \text{ for all } \ell \in \mathbb{Z}\}$ is a basis of the set of cardinal L-splines $\mathcal{S}_{Z+1,h}$.*

This is a classical fact which may be proved as in Schoenberg [16, p. 12, Theorem 1], or in Chui [4, p. 84]. It follows that in particular we have:

$$V_0 := \operatorname*{clos}_{L_2(\mathbb{R})} \{Q[\Lambda; h](x - \ell h) : \text{ for all } \ell \text{ in } \mathbb{Z}\}. \tag{17.2}$$

---

[1] Much as we do not like it we will have to use this clumsy notation indicating the dependence of the function Q on both the set Λ and the parameter h since we will have many different Λ's and h's in the present Part.

In Theorem 14.6, p. 272, we proved that this is a Riesz basis of V_0. It is obviously that, by the same theorem – putting $2^{-j}h$ instead of h – we obtain that the set of shifts $\{Q[\Lambda; 2^{-j}h](x - \ell 2^{-j}h):$ for ℓ in $\mathbb{Z}\}$ is a Riesz basis for the space V_j. Thus a basic point of the MRA is settled.

Following tradition we will define the **scaling function**

$$\phi_h(x) := Q_{\mathbb{Z}+1}[\Lambda; h](x). \tag{17.3}$$

We will also use the "scaled version" obtained by the substitution $h \longrightarrow 2^{-j}h$, namely

$$\phi_{2^{-j}h}(x) = Q_{\mathbb{Z}+1}[\Lambda; 2^{-j}h](x).$$

It should be noted that the parameter h in the scaling function ϕ_h will help us achieve flexibility in our results. Thus we may formulate some result only for V_0, V_1, and W_0, but by putting $h \longrightarrow (1/2^{j-1})h$ we will immediately obtain it for V_j and W_j etc. As we have already mentioned, if in equality (17.2) we make the substitution $h \longrightarrow 2^{-j}h$ we obtain

$$V_j = \operatorname*{clos}_{L_2(\mathbb{R})} \{\phi_{2^{-j}h}(x - \ell 2^{-j}h): \text{ for all } \ell \text{ in } \mathbb{Z}\}. \tag{17.4}$$

Now recall that in the classical polynomial spline case

$$f(x) \text{ belongs to } V_j \iff f(2x) \text{ belongs to } V_{j+1}, \tag{17.5}$$

since $f(2x)$ is a piecewise polynomial spline again, and if y is a knot of $f(x)$ then $y/2$ is a knot of $f(2x)$. It is important to note that in the polynomial case one considers the scaling

$$\phi(2^j x - \ell) = \phi\left(2^j\left(x - \frac{\ell}{2^j}\right)\right),$$

and if we put now $h = 1/2^j$ we see that for the "scaled" function

$$\phi_h(y) := \phi(hy)$$

we have to consider the mesh $h^{-1}\mathbb{Z}$.

Things are more complicated in the case of the L-splines. However, the specifics of the L-splines which we study permits us to consider only such L-splines with knots on the mesh $\mathbb{Z}$. Indeed, using the freedom of the parameter h, by formula (13.26), p. 243, if we make the substitution $h \longrightarrow 2^{-j}h$, we obtain the basic representation

$$\boxed{Q[\Lambda; 2^{-j}h](x) = \left(\frac{h}{2^j}\right)^{\mathbb{Z}} Q[2^{-j}h \cdot \Lambda]\left(\frac{2^j}{h}x\right)}$$

which shows that we need to consider L-splines only on the mesh $\mathbb{Z}$, but for other differential operators L, namely $\mathcal{L}[2^{-j}h \cdot \Lambda]$.

It is clear from the above that the function

$$\phi_{2^{-j}h}(x) = \left(\frac{h}{2^j}\right)^{\mathbb{Z}} Q[2^{-j}h \cdot \Lambda]\left(\frac{2^j}{h}x\right),$$

which is (up to a constant factor) the *TB*-spline for the operator $\mathcal{L}[2^{-j}h \cdot \Lambda]$ with knots at $\mathbb{Z}$, is also a *TB*-spline for the operator $\mathcal{L}[\Lambda]$ with knots at $2^{-j}h \cdot \mathbb{Z}$. Consequently, the shifts $Q[2^{-j}h \cdot \Lambda](y - \ell)$ generate the space V_j, i.e.

$$V_j = \operatorname*{clos}_{L_2(\mathbb{R})} \{Q[2^{-j}h \cdot \Lambda](y - \ell) : \text{ for all } \ell \in \mathbb{Z}\}.$$

Thus we have a situation *similar to the classical one* where one considers the function $\phi(x)$ and the shifts $\phi(2^j x - \ell)$. However, we also have to change the Λ-argument by the same factor.

So the analog to the equivalence relation (17.5), p. 327, is the following: if we have the sum

$$f(x) = \sum_{\ell=-\infty}^{\infty} c_\ell Q\left[\Lambda; \frac{h}{2^j}\right]\left(x - \frac{h}{2^j}\ell\right)$$

$$= \sum_{\ell=-\infty}^{\infty} c_\ell \left(\frac{2^j}{h}\right)^Z Q\left[\frac{h}{2^j}\Lambda\right]\left(\frac{2^j}{h}x - \ell\right),$$

and take for simplicity $j = 0$, instead of the scaling $f(2x)$ we consider the **nonstationary scaling operator**[2]

$$f_{\text{sc}}(x) := \sum_{\ell=-\infty}^{\infty} c_\ell \left(\frac{2}{h}\right)^Z Q\left[\frac{h}{2}\Lambda\right]\left(\frac{2}{h}x - \ell\right). \tag{17.6}$$

We replace the usual scaling condition (17.5), p. 327, with

$$f(x) \text{ in } V_j \iff f_{\text{sc}}(x) \text{ in } V_{j+1}.$$

Another property of the MRA which we have defined above is the equivalence for all j in $\mathbb{Z}$, provided by

$$f(x) \text{ belongs to } V_j \iff f\left(x + \frac{h}{2^j}\right) \text{ belongs to } V_j. \tag{17.7}$$

We have defined the **wavelet spaces** W_j by putting

$$W_j := V_{j+1} \ominus V_j.$$

For simplicity we may put $h = 1$. Our main purpose in this chapter will be the construction of a compactly supported wavelet function $\psi_{2^{-j}}(x) = \psi_{2^{-j}}[\Lambda](x)$ which generates W_j through 2^{-j}-shifts, i.e.

$$W_j = \operatorname*{clos}_{L_2(\mathbb{R})} \{\psi_{2^{-j}}(2^j x - \ell) : \text{ for all } \ell \in \mathbb{Z}\}.$$

---
[2] Here we use the terminology of de Boor *et al.* [9, p. 128] who consider "nonstationary wavelets".

In particular, we will have for $j = 0$

$$W_0 = \operatorname*{clos}_{L_2(\mathbb{R})} \{\psi_1(x - \ell) : \text{ for all } \ell \in \mathbb{Z}\}.$$

In Section 17.5 we will provide an explicit construction of this function ψ_h which is analogous to the construction of Chui [4], and was briefly summarized in Section 16.3.

17.2 Multiresolution analysis using L-splines

Here we provide some basic properties of the sets V_j which constitute a kind of MRA.

Proposition 17.2 *The intersection space $\bigcap_{j=-\infty}^{\infty} V_j$ is of dimension ≤ 1. If all $\lambda_j < 0$ then it is one-dimensional, i.e.*

$$\bigcap_{j=-\infty}^{\infty} V_j = \{\phi_Z^+(x)\}_{\mathrm{lin}} = \{C\phi_Z^+(x) : \text{ for all } C \in \mathbb{R}\}.$$

If there exists at least one index j for which $\lambda_j \geq 0$ then we have a zero-dimensional intersection

$$\bigcap_{j=-\infty}^{\infty} V_j = \{0\}.$$

Proof Let $f \in \bigcap_{j=-\infty}^{\infty} V_j$. Now let us consider the function f on an interval $[\varepsilon, N]$, where $\varepsilon > 0$ and $\varepsilon < N$. Then for a sufficiently large j_1 with $2^{j_1} > N$ we have that V_{-j_1} has knots at the set $2^{j_1}\mathbb{Z}$. Hence by the definition of the spaces V_j, on $[\varepsilon, N]$ the function f is an L_2-limit of L-polynomials (for the operator $\mathcal{L}[\Lambda]$). Since the space of L-polynomials is finite-dimensional it follows that f is itself an L-polynomial. Letting $N \longrightarrow \infty$ and $\varepsilon \longrightarrow 0$ we see that on $[0, \infty)$ the function f coincides with an L-polynomial. Similarly we see that it is an L-polynomial on the interval $(-\infty, 0]$. It follows that f is a cardinal L-spline with the only knot at $\{0\}$. According to Corollary 13.16, p. 232 there exists at most one such cardinal spline (up to a constant factor) which is precisely the Green function ϕ_Z^+ if $\phi_Z^+ \in L_2(\mathbb{R})$. By Corollary 13.16 the last holds if and only if the elements of the vector Λ satisfy $\lambda_j < 0$ for all $j = 1, 2, \ldots, Z+1$. This completes the proof. ∎

Thus we see that *not every* nonordered vector Λ generates an MRA.

It is interesting to note that the approximation property of the MRA, namely

$$\bigcup_{j=-\infty}^{\infty} V_j \text{ is dense in } L_2(\mathbb{R})$$

follows from the very general abstract principle about translation invariant subspaces of $L_2(\mathbb{R}^n)$, and results from very simple facts, like $\operatorname{supp}(\widehat{\phi_h}) = \mathbb{R}$ and $V_j \subset V_{j+1}$. We now provide an elegant proof following [9, p. 140].

Theorem 17.3 *The basic approximation condition of MRA holds, namely the L_2-closure of the union of spaces*

$$\operatorname*{clos}_{L_2(\mathbb{R})} \left[\bigcup_{j=-\infty}^{\infty} V_j \right] = L_2(\mathbb{R}).$$

Proof 1. Let us prove that the closure

$$V := \operatorname*{clos}_{L_2(\mathbb{R})} \left[\bigcup_{j=-\infty}^{\infty} V_j \right]$$

is translation invariant, i.e. if $f(x) \in V$ then for every number $t \in \mathbb{R}$ the function $f(x-t)$ belongs to V. Indeed, first let $f(x) \in \bigcup_{j=-\infty}^{\infty} V_j$. Then if $f(x) \in V_N$ for some large N, hence for $N_1 > N$ we also have $f(x) \in V_{N_1}$. Since every space V_ν is $2^{-\nu}$-invariant, it follows that $f(x - \ell 2^{-\nu}) \in \bigcup_{j=-\infty}^{\infty} V_j$ for every couple of integers ℓ and ν. Now recall that for $t_2 \to t_1$ we have

$$\|f(x-t_1) - f(x-t_2)\|_{L_2(\mathbb{R})} \longrightarrow 0.$$

Hence, since every number $t \in \mathbb{R}$ is approximable through rational numbers of the type $\ell 2^{-\nu}$ we see that $f(x-t)$ is approximated in $L_2(\mathbb{R})$ through the translation $f(x-\ell 2^{-\nu})$, hence

$$f(x-t) \in V.$$

Now if $f \in V$ we have a sequence $f_N(x) \xrightarrow{N\to\infty} f(x)$ in $L_2(\mathbb{R})$ with $f_N \in V_N$. For a fixed $t \in \mathbb{R}$ we have seen that $f_N(x-t) \in V$. On the other hand owing to the translation invariance of the Lebesgue measure it follows that $f_N(x-t) \xrightarrow{N\to\infty} f(x-t)$ in $L_2(\mathbb{R})$. This implies that $f(x-t) \in V$.

2. Here we are in the situation to apply [15, pp. 203–206, Theorem 9.17], which says that every closed translation invariant subspace of $L_2(\mathbb{R})$ is a set of Fourier transforms of all functions in $L_2(\Omega)$ for some measurable set Ω in $\mathbb{R}$, i.e. $V = \widehat{L_2(\Omega)}$; the set Ω is determined up to measure zero. Since the Fourier transform is an isomorphism of $L_2(\mathbb{R})$ it follows that $V = L_2(\mathbb{R})$ if and only if $\mathbb{R}\setminus\Omega$ is a set of measure zero. We obtain $\widehat{V} = L_2(\Omega)$. Due to formula (13.23), p. 241, applied to the Fourier transform of $\phi_h(x-\ell h)$, all functions of the type

$$\widehat{\phi_h(x-\ell h)}(\xi) = e^{-i\ell h \xi} \cdot \frac{\prod_{j=1}^{Z+1}(e^{-\lambda_j h} - e^{-i\xi h})}{\prod_{j=1}^{Z+1}(i\xi - \lambda_j)} \quad \text{with}$$

$$h = \nu \cdot 2^{-j} \quad \text{and for all } \nu, j \in \mathbb{Z},$$

belong to $\widehat{V}$. Since their zero sets have measure zero it follows that their supports are equal to $\mathbb{R}$. Hence we see that Ω may be only $\mathbb{R}$. ∎

Thus the spaces V_j satisfy all axioms of (nonstationary) MRA.
We have the following important property: the L-polynomials are reproduced locally.

Proposition 17.4 *Let for some function f and for all real numbers x we have $\mathcal{L}[\Lambda]f(x) = 0$, i.e. f is a global solution (or L-polynomial). Let us denote as above $\phi_h(x) = Q_{Z+1}[\Lambda; h](x)$. Then for every compact interval $[a, b]$ there exist coefficients α_ℓ such that*

$$f(x) = \sum_{\ell=-\infty}^{\infty} \alpha_\ell \phi_h(x - \ell h) \quad \text{for } x \text{ in } [a, b].$$

The proof is immediate due to the fact that f belongs to $\mathcal{S}_{Z+1,h}$ and that the h-shifts of $Q_{Z+1}[\Lambda; h](x)$ are a basis of $\mathcal{S}_{Z+1,h}$.

The above proposition will be generalized for *polyspline wavelet analysis*. It has an important consequence about the "vanishing moments". In order to satisfy the reader's curiosity we will state that the wavelet ψ which we will obtain will, as usual, be orthogonal to V_0, and by the above all "exponential moments" will be zero, i.e.

$$\int_{-\infty}^{\infty} R_j(x) e^{\lambda_j x} \psi(x)\, dx = 0 \quad \text{for } j = 1, \ldots, Z+1;$$

here R_j is a polynomial satisfying $\deg R_j \leq$ (multiplicity of λ_j) -1. This is due to the fact that $f(x) = R_j(x) e^{\lambda_j x}$ is a global solution to $\mathcal{L}[\Lambda]f = 0$.

17.3 The two-scale relation for the *TB*-splines $Q_{Z+1}(x)$

The L-spline wavelet analysis is based on the *two-scale relations* or *refinement equations*. Let us again assume that the nonordered vector Λ be given by

$$\Lambda = [\lambda_1, \lambda_2, \ldots, \lambda_{Z+1}].$$

We have seen in formula (13.50), p. 259, that the *TB*-spline $Q_{Z+1}[\Lambda; h](x)$ which generates V_0 satisfies the following *two-scale relation*:

$$Q_{Z+1}[\Lambda; h](x) = \sum_j p_j Q_{Z+1}\left[\Lambda; \frac{h}{2}\right]\left(x - j\frac{h}{2}\right).$$

Let us recall how this was done by emphasizing the analogy with the case of the classical cardinal splines as in formulas (16.8) and (16.9), p. 314. After taking the Fourier transform on both sides of the above equality, we obtain

$$\widehat{Q_{Z+1}[\Lambda; h]}(\xi) = \frac{s_h(e^{-i\xi h})}{q_{Z+1}(i\xi)} = \frac{s_{h/2}(e^{-i\xi h/2}) s_{h/2}(-e^{-i\xi h/2})}{q_{Z+1}(i\xi)}$$

$$= s_{h/2}(-e^{-i\xi h/2}) \widehat{Q_{Z+1}\left[\Lambda; \frac{h}{2}\right]}(\xi)$$

$$= P_h(e^{-i(h/2)\xi}) \widehat{Q_{Z+1}\left[\Lambda; \frac{h}{2}\right]}(\xi), \qquad (17.8)$$

332 *Multivariate polysplines*

where we have put as usual

$$P_h(z) := \sum_{j=0}^{Z+1} p_{j,h} \cdot z^j = s\left[\frac{h}{2}\Lambda\right](-z) := s_{h/2}(-z) = \prod_{j=1}^{Z+1}(e^{-\lambda_j h/2} + z). \quad (17.9)$$

Remark 17.5 *In the classical one-dimensional case one puts, see Chui [4, p. 91, formula (4.3.2)],*

$$P(z) = \frac{1}{2} \cdot \sum_{j=0}^{Z+1} p_j \cdot z^j$$

as in (16.9), p. 314, in order to swallow the constant $1/2$ *which appears after taking the Fourier transform in formula (16.10), p. 314. From the point of view of the present notation, where we consider different functions* $\phi_{2^{-j}}(x)$ *for the different levels, one also has to consider in the classical case*

$$\phi(2x - j) = \phi\left(2\left(x - \frac{j}{2}\right)\right);$$

the scaling function should be

$$\phi_2(x) = \phi(2x)$$

etc., which we have done in our case.

Now let us pass to splines Q_{Z+1} on the cardinal mesh $\mathbb{Z}$ by applying the Fourier transform which easily reduces the spline $Q_{Z+1}[\Lambda; h]$ to a spline $Q_{Z+1}[\Lambda]$ but on a different mesh. Namely, by equality (13.25), p. 242, and by the above relation (17.8) we obtain for every $\xi \in \mathbb{R}$ the equality

$$h^{Z+1}\widehat{Q_{Z+1}[h\Lambda]}(h\xi) = s\left[\frac{h}{2}\Lambda\right](-e^{-i(h/2)\xi}) \cdot \left(\frac{h}{2}\right)^{Z+1} \widehat{Q_{Z+1}\left[\frac{h}{2}\Lambda\right]}\left(\frac{h}{2}\xi\right),$$

which implies after the change $h\xi \longrightarrow \xi$ the following equality for every $\xi \in \mathbb{R}$:

$$\widehat{Q_{Z+1}[h\Lambda]}(\xi) = s\left[\frac{h}{2}\Lambda\right](-e^{-i\xi/2}) \cdot \left(\frac{1}{2}\right)^{Z+1} \widehat{Q_{Z+1}\left[\frac{h}{2}\Lambda\right]}\left(\frac{\xi}{2}\right)$$

$$= P(e^{-i\xi/2}) \cdot \left(\frac{1}{2}\right)^{Z+1} \widehat{Q_{Z+1}\left[\frac{h}{2}\Lambda\right]}\left(\frac{\xi}{2}\right). \quad (17.10)$$

Remark 17.6 *The classical polynomial cardinal splines are obtained for* $\Lambda = [\lambda_1, \lambda_2, \ldots, \lambda_{Z+1}]$, *with all* $\lambda_j = 0$. *The refinement equation for* ϕ *in (16.8), p. 314, corresponds to the above equality (17.10) up to the constant* 2^{Z+1} *so we have to insert*

$$\frac{P(z)}{2^{Z+1}}$$

into the present notation in order to achieve conformity of the notation with the polynomial case $\Lambda = [0, \ldots, 0]$.

17.4 Construction of the mother wavelet ψ_h

Here we will construct the mother wavelet ψ_h which generates, through shifts, the space W_0. The idea and the techniques are similar to that of Chui [4] and we have already provided some formulas in Section 16.3. The reader has to keep this in mind although the present situation is overburdened with new technical details.

Recall that

$$V_0 = \operatorname*{clos}_{L_2(\mathbb{R})} \{Q_{Z+1}[\Lambda; h](x - \ell h) \text{ for all } \ell \in \mathbb{Z}\}$$

$$V_1 = \operatorname*{clos}_{L_2(\mathbb{R})} \left\{ Q_{Z+1}\left[\Lambda; \frac{h}{2}\right]\left(x - \ell\frac{h}{2}\right) : \text{ for all } \ell \in \mathbb{Z} \right\},$$

evidently satisfying

$$V_0 \subset V_1.$$

The question is how to describe the wavelet space

$$W_0 := V_1 \ominus V_0 \tag{17.11}$$

or the equivalent relation

$$V_1 = V_0 \oplus W_0.$$

The reader may put here for simplicity $h = 1$. So far we will need the constant h for further scaling and the exact result will be important. Actually the result for $V_{j+1} = V_j \oplus W_j$ will be obtained by $h \to h/2^{j-1}$.

As we have discussed at Theorem 14.6, p. 273, the shifts $Q_{Z+1}[\Lambda; h](x - \ell h)$ generate a Riesz basis and it follows that every element in V_0 can be represented by a series

$$\sum_{\ell=-\infty}^{\infty} c_\ell Q_{Z+1}[\Lambda; h](x - \ell h),$$

where

$$\sum_{\ell=-\infty}^{\infty} |c_\ell|^2 < \infty.$$

Further we will be looking for a *semi-orthogonal wavelet* $\psi_h \in W_0$ which has shifts generating W_0. Since $W_0 \subset V_1$, every element ψ in W_0 can be represented in the form

$$\psi(x) = \sum_{j=-\infty}^{\infty} r_j Q_{Z+1}\left[\Lambda; \frac{h}{2}\right]\left(x - j\frac{h}{2}\right) \tag{17.12}$$

for some sequence $\{r_j\}$ satisfying

$$\sum_{j=-\infty}^{\infty} |r_j|^2 < \infty.$$

After taking the Fourier transform on both sides of equality (17.12) we obtain

$$\widehat{\psi}(\xi) = R(e^{-i(h/2)\xi})\widehat{Q_{Z+1}\left[\Lambda; \frac{h}{2}\right]}(\xi), \qquad (17.13)$$

where

$$R(z) = \frac{1}{2} \cdot \sum_j r_j z^j.$$

Due to Parseval's equality (12.1), p. 210, we obtain $R \in L_2(\mathbb{S}^1)$. Here as usual we have denoted the unit circle

$$\mathbb{S}^1 := \{z \in \mathbb{C} : \text{for } |z| = 1\}.$$

Now we change to a spline Q_{Z+1} on the mesh $\mathbb{Z}$ by applying equality (13.25), p. 242. From equality (17.13) we obtain

$$\widehat{\psi}(\xi) = R(e^{-i(h/2)\xi})\left(\frac{h}{2}\right)^{Z+1}\widehat{Q_{Z+1}\left[\frac{h}{2}\Lambda\right]}\left(\frac{h}{2}\xi\right).$$

Hence, after the change $h\xi \longrightarrow \xi$, we obtain the equality

$$\widehat{\psi}\left(\frac{\xi}{h}\right) = \left(\frac{h}{2}\right)^{Z+1} R(e^{-i\xi/2})\widehat{Q_{Z+1}\left[\frac{h}{2}\Lambda\right]}\left(\frac{\xi}{2}\right). \qquad (17.14)$$

We know that

$$\psi \perp V_0$$

and this may be used in order to provide an elegant analytic formula for the coefficients r_j.
Lemma 17.7 provides a description of the elements of the *wavelet* space W_0.

Lemma 17.7 *Every element $\psi \in W_0$ having a **two-scale symbol (refinement mask)** $R(z)$ satisfies*

$$P(z)\overline{R(z)} \cdot \Phi_{2Z+1}\left[\frac{h}{2}\widetilde{\Lambda}\right]\left(Z+1; \frac{1}{z}\right)$$

$$+ P(-z)\overline{R(-z)} \cdot \Phi_{2Z+1}\left[\frac{h}{2}\widetilde{\Lambda}\right]\left(Z+1; -\frac{1}{z}\right) = 0 \qquad (17.15)$$

for all z with $|z| = 1$.

Proof Since $Q_{Z+1}[\Lambda; h](x - jh)$ is a basis for the space V_0 and $\psi \perp V_0$ we obtain for all $j \in \mathbb{Z}$ the equality

$$I := \int_{-\infty}^{\infty} Q_{Z+1}[\Lambda; h](x - jh)\psi(x)\,dx = 0. \qquad (17.16)$$

We apply Parseval's equality of formula (12.5), p. 212, and equality (17.8), p. 331, in order to obtain

$$I = \frac{1}{2\pi}\int_{-\infty}^{\infty} e^{-jh\xi i} \widehat{Q_{Z+1}[\Lambda; h]}(\xi) \overline{\widehat{\psi}(\xi)}\, d\xi$$

$$= \frac{1}{2\pi}\int_{-\infty}^{\infty} e^{-jh\xi i} P(e^{-i(h/2)\xi}) \left|\widehat{Q_{Z+1}\left[\Lambda; \frac{h}{2}\right]}(\xi)\right|^2 \overline{R(e^{-i(h/2)\xi})}\, d\xi.$$

By formula (13.25), p. 242, namely,

$$\widehat{Q_{Z+1}[\Lambda; h]}(\xi) = h^{Z+1}\widehat{Q_{Z+1}[h\Lambda]}(h\xi),$$

we obtain $\widehat{Q_{Z+1}[\Lambda; h/2]}(\xi) = (h/2)^{Z+1}\widehat{Q_{Z+1}[(h/2)\Lambda]}((h/2)\xi)$. This and a change of variable $h\xi \longrightarrow \xi$ imply

$$I = \frac{1}{2\pi}\int_{-\infty}^{\infty} e^{-jh\xi i} P(e^{-i(h/2)\xi}) \left(\frac{h}{2}\right)^{2Z+2} \left|\widehat{Q_{Z+1}\left[\frac{h}{2}\Lambda\right]}\left(\frac{h}{2}\xi\right)\right|^2 \overline{R(e^{-i(h/2)\xi})}\, d\xi$$

$$= \frac{h^{2Z+1}}{2^{2Z+2}} \cdot \frac{1}{2\pi}\int_{-\infty}^{\infty} e^{-j\xi i} P(e^{-i\xi/2}) \left|\widehat{Q_{Z+1}\left[\frac{h}{2}\Lambda\right]}\left(\frac{\xi}{2}\right)\right|^2 \overline{R(e^{-i\xi/2})}\, d\xi.$$

This implies, by using the 4π periodicity of $e^{-i\xi/2}$, the following equalities:

$$I = \frac{h^{2Z+1}}{2^{2Z+2}} \cdot \frac{1}{2\pi} \sum_{\ell=-\infty}^{\infty} \int_{4\pi\ell}^{4\pi(\ell+1)} e^{-j\xi i} P(e^{-i\xi/2})$$

$$\times \left|\widehat{Q_{Z+1}\left[\frac{h}{2}\Lambda\right]}\left(\frac{\xi}{2}\right)\right|^2 \cdot \overline{R(e^{-i\xi/2})}\, d\xi$$

$$= \frac{h^{2Z+1}}{2^{2Z+2}} \cdot \frac{1}{2\pi} \int_0^{4\pi} e^{-j\xi i} P(e^{-i\xi/2})$$

$$\times \sum_{\ell=-\infty}^{\infty} \left|\widehat{Q_{Z+1}\left[\frac{h}{2}\Lambda\right]}\left(\frac{\xi}{2} + 2\pi\ell\right)\right|^2 \cdot \overline{R(e^{-i\xi/2})}\, d\xi.$$

Recalling the notation in Theorem 14.6, p. 271, namely the definition of the function $S(\xi)$ in (14.3), we can put

$$S(\eta) := S\left[\frac{h}{2}\Lambda\right](\eta) := \sum_{\ell=-\infty}^{\infty} \left|\widehat{Q_{Z+1}\left[\frac{h}{2}\Lambda\right]}(\eta + 2\pi\ell)\right|^2. \qquad (17.17)$$

By applying Theorem 14.6, we obtain

$$S\left(\frac{\xi}{2}\right) = e^{(-\lambda_1 - \cdots - \lambda_{Z+1})h/2} \cdot \sum_{j=-\infty}^{\infty} Q_{2Z+2}\left[\frac{h}{2}\tilde{\Lambda}\right](Z+1+j) \cdot e^{-ij\xi/2}$$

336 Multivariate polysplines

where $\widetilde{\Lambda}$ is the symmetrized vector of the vector Λ.[3] By the definition of the Schoenberg's "exponential Euler spline" Φ_{2Z+1} in (13.42), p. 254, we have

$$\Phi_{2Z+1}\left[\frac{h}{2}\widetilde{\Lambda}\right](Z+1;\lambda) = \sum_{j=-Z}^{Z} Q_{2Z+2}\left[\frac{h}{2}\widetilde{\Lambda}\right](Z+1+j)\lambda^{-j},$$

since the support of the *TB*-spline Q_{2Z+2} coincides with the interval $[0, 2Z+2]$, and $Q_{2Z+2}[(h/2)\widetilde{\Lambda}](0) = Q_{2Z+2}[(h/2)\widetilde{\Lambda}](2Z+2) = 0$. Hence,

$$S\left(\frac{\xi}{2}\right) = e^{(-\lambda_1 - \cdots - \lambda_{Z+1})h/2} \cdot \Phi_{2Z+1}\left[\frac{h}{2}\widetilde{\Lambda}\right](Z+1; e^{i\xi/2}).$$

It follows that if we put $z = e^{-i\xi/2}$ and the constant

$$C_S = e^{(\lambda_1 + \cdots + \lambda_{Z+1})h/2},$$

we obtain the equality

$$I = \frac{1}{C_S}\frac{h^{2Z+1}}{2^{2Z+2}} \cdot \frac{1}{2\pi}\int_0^{4\pi} e^{-j\xi i} P(e^{-i\xi/2})\overline{R(e^{-i\xi/2})}$$
$$\times \Phi_{2Z+1}\left[\frac{h}{2}\widetilde{\Lambda}\right](Z+1; e^{i\xi/2})\, d\xi$$
$$= \frac{1}{C_S}\frac{h^{2Z+1}}{2^{2Z+2}} \cdot \frac{1}{2\pi}\int_0^{2\pi} e^{-j\xi i} P(z)\overline{R(z)} \cdot \Phi_{2Z+1}\left[\frac{h}{2}\widetilde{\Lambda}\right]\left(Z+1; \frac{1}{z}\right) d\xi$$
$$+ \frac{1}{C_S}\frac{h^{2Z+1}}{2^{2Z+2}} \cdot \frac{1}{2\pi}\int_0^{2\pi} e^{-j\xi i} P(-z)\overline{R(-z)} \cdot \Phi_{2Z+1}\left[\frac{h}{2}\widetilde{\Lambda}\right]\left(Z+1; -\frac{1}{z}\right) d\xi.$$

Let us put

$$\rho(z) := P(z)\overline{R(z)} \cdot \Phi_{2Z+1}\left[\frac{h}{2}\widetilde{\Lambda}\right]\left(Z+1; \frac{1}{z}\right)$$
$$+ P(-z)\overline{R(-z)} \cdot \Phi_{2Z+1}\left[\frac{h}{2}\widetilde{\Lambda}\right]\left(Z+1; -\frac{1}{z}\right)$$

for the function under the sign of the integral. This function is in $L_2(0, 2\pi)$ for $|z| = 1$ since the "exponential Euler function" Φ_{2Z+1} is continuous, the polynomial $P(z)$ is continuous, and $R(z)$ is in $L_2(0, 2\pi)$. The function $\rho(z)$ is 2π-periodic of $z = e^{-i\xi/2}$ due to $\rho(-z) = \rho(z)$, and since $-z = e^{-i((\xi+2\pi)/2)}$! By equality (17.16) we have $I = 0$ which implies

$$\int_0^{2\pi} e^{-j\xi i} \rho(e^{-i\xi/2})\, d\xi = 0.$$

---
[3] Recall Definition 14.5, p. 271.

for every integer $j \in \mathbb{Z}$. Hence it follows that all Fourier coefficients of the function $\rho(z)$ are zero. By the representation of the functions in $L_2(0, 2\pi)$ through their Fourier series, see Theorem 12.1, p. 209, it follows that for $|z| = 1$ we have

$$P(z)\overline{R(z)} \cdot \Phi_{2Z+1}\left[\frac{h}{2}\widetilde{\Lambda}\right]\left(Z+1; \frac{1}{z}\right)$$
$$+ P(-z)\overline{R(-z)} \cdot \Phi_{2Z+1}\left[\frac{h}{2}\widetilde{\Lambda}\right]\left(Z+1; -\frac{1}{z}\right) = 0.$$

This completes the proof of the lemma. ∎

In particular, for $p = 2$ we obtain $Z = 2p - 1 = 3$, $2Z + 2 = 8$, which gives

$$\Phi_7\left[\frac{h}{2}\widetilde{\Lambda}\right](4; \lambda) = \sum_{j=-3}^{3} Q_8\left[\frac{h}{2}\widetilde{\Lambda}\right](4+j)\lambda^{-j}, \qquad (17.17a)$$

and, due to $Q_8[(h/2)\widetilde{\Lambda}](0) = Q_8[(h/2)\widetilde{\Lambda}](8) = 0$, it follows that:

$$S\left(\frac{\xi}{2}\right) = \left(\frac{1}{2}\right)^8 e^{-\lambda_1 - \cdots - \lambda_4} \cdot \Phi_7\left[\frac{h}{2}\widetilde{\Lambda}\right](4; e^{i\xi/2}).$$

Remark 17.8 *After the substitution $h \longrightarrow h/2^{j-1}$ we obtain the wavelet function ψ of the spaces W_j.*

17.5 Some algebra of Laurent polynomials and the mother wavelet ψ_h

The problem of finding $R(z)$ has been considered by Chui in [5, p. 100], and in greater detail in [4, Chapter 5, pp. 169, 183]. For its solution it is important that $\Phi_{2Z+1}[(h/2)\widetilde{\Lambda}](4; z)$ is a Laurent polynomial and also that $P(z)$ is a polynomial.

Without going deep into the heuristics, which have been exhaustively considered by Chui in the above references, we provide the general solution to problem (17.15), p. 334, by the formula

$$R(z) = zK(z^2)P\left(-\frac{1}{z}\right)\Phi_{2Z+1}\left[\frac{h}{2}\widetilde{\Lambda}\right](Z+1; -z), \qquad (17.18)$$

where K is an arbitrary function in $L_2(0, 2\pi)$. By formula (17.9), p. 332, we have

$$P(z) = \frac{1}{2}\sum_{j=0}^{Z+1} p_{j,h}z^j = s_{h/2}(-z).$$

Naturally, we are looking for a polynomial solution $R(z)$ of the above type (17.18), and with the smallest possible degree. As we will see the last requirement implies a

338 Multivariate polysplines

minimal support of the function ψ, and also that the shifts of the function ψ will be a Riesz basis of W_0. We put
$$K(z) := c \cdot z^Z,$$
and, as is not difficult to see, this is the polynomial of minimal possible degree which guarantees that $R(z)$ is a polynomial.

We obtain, due to $3(Z+1) - 2 = 3Z + 1$ the concrete solution

$$R(z) := R_h(z) := -z^{2Z+1} P\left(-\frac{1}{z}\right) \Phi_{2Z+1}\left[\frac{h}{2}\widetilde{\Lambda}\right](Z+1;-z) \quad (17.19)$$

$$= \frac{1}{2} \sum_{j=0}^{3(Z+1)-2} r_{j,h} \cdot z^j.$$

Although we have yet to prove it, we will now define the function ψ as a "wavelet".

Definition 17.9 We define the **mother wavelet** function $\psi_h(x) = \psi[\Lambda; h](x)$ in the space W_0 by defining its Fourier transform

$$\widehat{\psi_h}(\xi) := R(e^{-i(h/2)\xi}) \cdot \widehat{Q_{Z+1}[\Lambda; h]}(\xi), \quad (17.20)$$

where the polynomial $R(z)$ is given by equality (17.19).

Clearly, by applying the inverse Fourier transform, we obtain

$$\psi_h(x) = \mathcal{F}^{-1}[R(e^{-i(h/2)\xi}) \cdot \widehat{Q_{Z+1}[\Lambda; h]}(\xi)](x).$$

In what follows we will see that ψ_h generates through h-shifts the space W_0 and for that reason is indeed a *mother wavelet* function.

In particular, for $p = 2$ we obtain $P(z) = 1/2 \sum_{j=0}^{4} p_{j,h} z^j$ with

$$K(z) = c \cdot z^3,$$

and since $3(Z+1) - 2 = 3Z + 1 = 3(2p-1) + 1 = 6p - 2 = 10$, we obtain the concrete solution

$$R(z) = -z^7 P\left(-\frac{1}{z}\right) \Phi_7\left[\frac{h}{2}\widetilde{\Lambda}\right](4;-z)$$

$$= -z^7 s_{h/2}\left(\frac{1}{z}\right) \Phi_7\left[\frac{h}{2}\widetilde{\Lambda}\right](4;-z)$$

$$=: \frac{1}{2} \cdot \sum_{j=0}^{10} r_{j,h} \cdot z^j.$$

The function $\Phi_7\left[\frac{h}{2}\widetilde{\Lambda}\right](4;\lambda)$ is found above in formula (17.17a), p. 337. Thus we have the whole link for computing the mother wavelet ψ_h.

This **two-scale symbol (refinement mask)** $R(z)$ implies many useful properties of the wavelet function ψ_h. Some of them follow immediately but others will take longer to prove. In order to give more elegant proofs of the last we will first concentrate on some algebraic properties of the polynomials $P(z)$ and $R(z)$. They will be of great importance later for defining the duals of the *TB*-spline $\phi_h(x) = Q_{Z+1}[\Lambda; h]$ and the wavelet ψ_h, as well as for the *decomposition* and *reconstruction* relations.

17.6 Some algebraic identities

We will prove the following result which is a key for solving algebraic systems with polynomials, which are analogous to [4, p. 135].

Proposition 17.10 *Let the polynomial $P(z)$ be the one given by formula (17.8), p. 331, in the two-scale relation (refinement equation) of $Q[\Lambda; h](x)$. Then for all complex numbers z with $|z| = 1$ we have*

$$|P(z)|^2 \Phi_{2Z+1}\left[\frac{h}{2}\widetilde{\Lambda}\right](Z+1; z) + |P(-z)|^2 \Phi_{2Z+1}\left[\frac{h}{2}\widetilde{\Lambda}\right](Z+1; -z)$$

$$= \frac{2^{2Z+2}}{C_S} \Phi_{2Z+1}[h\widetilde{\Lambda}](Z+1; z^2), \tag{17.21}$$

where the constant C_S is defined by

$$C_S = e^{(\lambda_1 + \cdots + \lambda_{Z+1})h/2},$$

and the nonordered vector $\widetilde{\Lambda}$ denotes the symmetrization of the vector Λ.

Proof We put

$$z = e^{-i\xi/2}$$

and

$$D_1 = |P(z)|^2 \Phi_{2Z+1}\left[\frac{h}{2}\widetilde{\Lambda}\right](Z+1; \overline{z}) + |P(-z)|^2 \Phi_{2Z+1}\left[\frac{h}{2}\widetilde{\Lambda}\right](Z+1; -\overline{z}).$$

Note that taking the complex conjugate on both sides of (17.21) does not essentially change the result. Thus we have to prove the equality

$$D_1 = \frac{2^{2Z+2}}{C_S} \Phi_{2Z+1}[h\widetilde{\Lambda}](Z+1; \overline{z}^2).$$

In equality (17.17), p. 335, we have defined the sum

$$S\left(\frac{\xi}{2}\right) = \sum_{\ell=-\infty}^{\infty} \left|\widehat{Q_{Z+1}\left[\frac{h}{2}\Lambda\right]}\left(\frac{\xi}{2} + 2\pi\ell\right)\right|^2.$$

In Theorem 14.6, p. 273, we saw that

$$S\left(\frac{\xi}{2}\right) = \frac{1}{C_S} \cdot \Phi_{2Z+1}\left[\frac{h}{2}\widetilde{\Lambda}\right](Z+1; e^{i\xi/2}),$$

where the "exponential Euler" function Φ_{2Z+1} was defined by us in (13.42), p. 254. Hence, we obtain

$$D_1/C_S = |P(z)|^2 S\left(\frac{\xi}{2}\right) + |P(-z)|^2 S\left(\frac{\xi+2\pi}{2}\right)$$

$$= |P(z)|^2 \sum_{\ell=-\infty}^{\infty} \left|\widehat{Q_{Z+1}\left[\frac{h}{2}\Lambda\right]}\left(\frac{\xi}{2}+2\pi\ell\right)\right|^2$$

$$+ |P(-z)|^2 \sum_{\ell=-\infty}^{\infty} \left|\widehat{Q_{Z+1}\left[\frac{h}{2}\Lambda\right]}\left(\frac{\xi}{2}+\pi+2\pi\ell\right)\right|^2.$$

By formula (17.10), p. 332, for every $\xi \in \mathbb{R}$ we obtain the equality

$$\widehat{Q_{Z+1}[h\Lambda]}(\xi) = P(e^{-i\xi/2}) \cdot \left(\frac{1}{2}\right)^{Z+1} \widehat{Q_{Z+1}\left[\frac{h}{2}\Lambda\right]}\left(\frac{\xi}{2}\right),$$

which, due to the 4π-periodicity of $P_h(e^{-i\xi/2})$ in ξ, and $P(-z) = P_h(e^{-i\xi/2-i\pi})$, implies

$$D_1/C_S = 2^{2Z+2} \sum_{\ell=-\infty}^{\infty} |\widehat{Q_{Z+1}[\Lambda]}(\xi+4\pi\ell)|^2$$

$$+ 2^{2Z+2} \sum_{\ell=-\infty}^{\infty} |\widehat{Q_{Z+1}[\Lambda]}(\xi+2\pi+4\pi\ell)|^2$$

$$= 2^{2Z+2} \sum_{\ell=-\infty}^{\infty} |\widehat{Q_{Z+1}[\Lambda]}(\xi+2\pi\ell)|^2.$$

The last is equal, again by Theorem 14.6, p. 273, for all $z = e^{-i\xi/2}$ to

$$2^{2Z+2} \cdot e^{(-\lambda_1 - \cdots - \lambda_{Z+1})h} \Phi_{2Z+1}[h\widetilde{\Lambda}](Z+1, e^{i\xi}) = D_1/C_S.$$

After taking the complex conjugate this shows that equality (17.21) is true. This completes the proof. ∎

Remark 17.11 *We see that for $\Lambda = [0, \ldots, 0]$ we obtain the classical result for polynomial cardinal splines, the norming constant 2^{2Z+2} arises through the norming of the polynomial $P(z)$ as noted in Remark 17.6, p. 332.*

As in [4, p. 181], Proposition 17.12 follows from the above theorem, using the identities which we have proved.

Proposition 17.12 *Let the nonordered vector Λ be given and the two-scale symbol (refinement mask) $P(z)$ be given by (17.8), p. 331. Then if the symbol R of the wavelet ψ_h is given by (17.19), p. 338, i.e. $R(z) = -z^{2Z+1} P(-1/z) \Phi_{2Z+1}[(h/2)\widetilde{\Lambda}](Z+1; -z)$, the determinant*

$$D := \det \begin{bmatrix} P(z) & R(z) \\ P(-z) & R(-z) \end{bmatrix}$$

satisfies for all $|z| = 1$ the relations

$$D = \frac{2^{2Z+2}}{C_S} \cdot z^{2Z+1} \Phi_{2Z+1}[h\widetilde{\Lambda}](Z+1; z^2) \neq 0.$$

Proof For $z = e^{-i\xi/2}$, applying $\bar{z} = 1/z$ we obtain the following equalities:

$$D = \det \begin{vmatrix} P(z) & R(z) \\ P(-z) & R(-z) \end{vmatrix}$$

$$= P(z)R(-z) - P(-z)R(z)$$

$$= P(z)z^{2Z+1} P\left(\frac{1}{z}\right) \Phi_{2Z+1}\left[\frac{h}{2}\widetilde{\Lambda}\right](Z+1; z)$$

$$+ P(-z)z^{2Z+1} P\left(-\frac{1}{z}\right) \Phi_{2Z+1}\left[\frac{h}{2}\widetilde{\Lambda}\right](Z+1; -z)$$

$$= z^{2Z+1} \left\{ |P(z)|^2 \Phi_{2Z+1}\left[\frac{h}{2}\widetilde{\Lambda}\right](Z+1; z) + |P(-z)|^2 \Phi_{2Z+1}\left[\frac{h}{2}\widetilde{\Lambda}\right](Z+1; -z) \right\}.$$

We apply Proposition 17.10, p. 339, which gives the equality

$$D = \frac{2^{2Z+2}}{C_S} \cdot z^{2Z+1} \Phi_{2Z+1}[h\widetilde{\Lambda}](Z+1; z^2).$$

By formula (13.43), p. 255, we have $\Phi_{2Z+1}(Z+1, z) = z^{Z+1}\Phi_{2Z+1}(0, z)$, and by Corollary 13.57, p. 256, the function $\Phi_{2Z+1}(0, z)$ never vanishes on the circle $|z| = 1$. This ends the proof. ∎

We now prove the second important identity.

Proposition 17.13 *Let the polynomial $P(z)$ be given by (17.8), p. 331, in the two-scale relation between $Q_{Z+1}[\Lambda; h/2](x)$ and $Q_{Z+1}[\Lambda; h](x)$, and let the wavelet function $\psi_h(x) = \psi[\Lambda; h](x)$ be given by (17.20), p. 338. Then*

$$\sum_{j=-\infty}^{\infty} \left| \widehat{\psi_h}\left(\frac{\xi + 2\pi j}{h}\right) \right|^2 = \left(\frac{h}{2}\right)^{2Z+2} \frac{1}{C_S} \left\{ |R(z)|^2 \Phi_{2Z+1}\left[\frac{h}{2}\widetilde{\Lambda}\right](Z+1; \bar{z}) \right.$$

$$\left. + |R(-z)|^2 \Phi_{2Z+1}\left[\frac{h}{2}\widetilde{\Lambda}\right](Z+1; -\bar{z}) \right\}$$

$$= \frac{h^{2Z+2}}{C_S^2} \Phi_{2Z+1}\left[\frac{h}{2}\widetilde{\Lambda}\right](Z+1; -\bar{z})$$

$$\times \Phi_{2Z+1}\left[\frac{h}{2}\widetilde{\Lambda}\right](Z+1; \bar{z}) \cdot \Phi_{2Z+1}[h\widetilde{\Lambda}](Z+1; z^2),$$

(17.22)

where the nonordered vector $\widetilde{\Lambda}$ is the symmetrization of Λ, the constant C_S is defined by

$$C_S = e^{(\lambda_1 + \cdots + \lambda_{Z+1})h/2},$$

and $z = e^{-i\xi/2}$.

Proof Let us put

$$S_1(\xi) = \sum_{j=-\infty}^{\infty} \left| \widehat{\psi_h}\left(\frac{\xi + 2\pi j}{h}\right) \right|^2.$$

By formula (17.14), p. 334, we have

$$\widehat{\psi_h}\left(\frac{\xi}{h}\right) = \left(\frac{h}{2}\right)^{Z+1} R(e^{-i\xi/2}) Q_{Z+1}\left[\frac{h}{2}\Lambda\right]\left(\frac{\xi}{2}\right).$$

Applying it for $\xi + 2\pi j$ we obtain

$$S_1(\xi) = \left(\frac{h}{2}\right)^{2Z+2} \sum_{j=-\infty}^{\infty} \left| R(e^{-i(\xi+2\pi j)/2}) Q_{Z+1}\left[\frac{h}{2}\Lambda\right]\left(\frac{\xi + 2\pi j}{2}\right) \right|^2,$$

which after splitting for even and odd indices j gives for all $z = e^{-i\xi/2}$ the equality

$$S_1(\xi) = \left(\frac{h}{2}\right)^{2Z+2} \sum_{j=-\infty}^{\infty} \left| R\left(e^{-i((\xi/2)+\pi j)}\right) Q_{Z+1}\left[\frac{h}{2}\Lambda\right]\left(\frac{\xi + 2\pi j}{2}\right) \right|^2$$

$$= |R(z)|^2 \left(\frac{h}{2}\right)^{2Z+2} \cdot \sum_{j=-\infty}^{\infty} \left| Q_{Z+1}\left[\frac{h}{2}\Lambda\right]\left(\frac{\xi}{2} + 2\pi j\right) \right|^2$$

$$+ |R(-z)|^2 \left(\frac{h}{2}\right)^{2Z+2} \cdot \sum_{j=-\infty}^{\infty} \left| Q_{Z+1}\left[\frac{h}{2}\Lambda\right]\left(\frac{\xi}{2} + \pi + 2\pi j\right) \right|^2.$$

Recalling the definition of the function $S(\xi)$ in (17.17), p. 335, we obtain after putting

$$C_S = e^{(\lambda_1 + \cdots + \lambda_{Z+1})h/2},$$

and applying Theorem 14.6, p. 273, the following equality:

$$S_1(\xi) = \left(\frac{h}{2}\right)^{2Z+2} \left\{ |R(z)|^2 S\left(\frac{\xi}{2}\right) + |R(-z)|^2 S\left(\frac{\xi + 2\pi}{2}\right) \right\}$$

$$= \left(\frac{h}{2}\right)^{2Z+2} \frac{1}{C_S} \left\{ |R(z)|^2 \Phi_{2Z+1}\left[\frac{h}{2}\widetilde{\Lambda}\right](Z+1; e^{i\xi/2}) \right.$$

$$\left. + |R(-z)|^2 \Phi_{2Z+1}\left[\frac{h}{2}\widetilde{\Lambda}\right](Z+1; -e^{i\xi/2}) \right\}.$$

This proves the first equality since $z = e^{-i\xi/2}$.

To obtain the second equality we substitute the value of $R(z)$ given by (17.19), p. 338, i.e.

$$R(z) = -z^{2Z+1} P\left(-\frac{1}{z}\right) \Phi_{2Z+1}\left[\frac{h}{2}\widetilde{\Lambda}\right](Z+1; -z),$$

and obtain, having in mind that $|z| = 1$, the equality

$S_1(\xi)$

$$= \left(\frac{h}{2}\right)^{2Z+2} \frac{1}{C_S} \left\{ \left| P\left(-\frac{1}{z}\right) \Phi_{2Z+1}\left[\frac{h}{2}\widetilde{\Lambda}\right](Z+1; -z) \right|^2 \Phi_{2Z+1}\left[\frac{h}{2}\widetilde{\Lambda}\right](Z+1; \bar{z}) \right.$$

$$\left. + \left| P\left(\frac{1}{z}\right) \Phi_{2Z+1}\left[\frac{h}{2}\widetilde{\Lambda}\right](Z+1; z) \right|^2 \Phi_{2Z+1}\left[\frac{h}{2}\widetilde{\Lambda}\right](Z+1; -\bar{z}) \right\}$$

$$= \left(\frac{h}{2}\right)^{2Z+2} \frac{1}{C_S} \Phi_{2Z+1}\left[\frac{h}{2}\widetilde{\Lambda}\right](Z+1; -\bar{z}) \cdot \Phi_{2Z+1}\left[\frac{h}{2}\widetilde{\Lambda}\right](Z+1; \bar{z})$$

$$\times \left\{ \left| P\left(-\frac{1}{z}\right) \right|^2 \cdot \Phi_{2Z+1}\left[\frac{h}{2}\widetilde{\Lambda}\right](Z+1; -z) + \left| P\left(\frac{1}{z}\right) \right|^2 \cdot \Phi_{2Z+1}\left[\frac{h}{2}\widetilde{\Lambda}\right](Z+1; z) \right\}.$$

Let us use $|P(1/z)| = |P(z)|$ and apply Proposition 17.10, p. 339. We obtain the equality

$$S_1(\xi) = \frac{h^{2Z+2}}{C_S^2} \Phi_{2Z+1}\left[\frac{h}{2}\widetilde{\Lambda}\right](Z+1; -\bar{z}) \times \Phi_{2Z+1}\left[\frac{h}{2}\widetilde{\Lambda}\right](Z+1; \bar{z})$$

$$\times \Phi_{2Z+1}[h\widetilde{\Lambda}](Z+1; z^2).$$

This ends the proof. ∎

Remark 17.14 *In particular, when $\Lambda = [0, \ldots, 0]$ then for $h = 1$ we obtain the polynomial case result*

$$\sum_{j=-\infty}^{\infty} \left| \widehat{\psi_h}\left(\frac{\xi + 2\pi j}{h}\right) \right|^2 = \Phi_{2Z+1}[\widetilde{\Lambda}_0](Z+1; -\bar{z}) \times \Phi_{2Z+1}[\widetilde{\Lambda}_0](Z+1; \bar{z})$$

$$\times \Phi_{2Z+1}[\widetilde{\Lambda}_0](Z+1; z^2),$$

which is provided in (16.25), p. 319, or see [4].

17.7 The function ψ_h generates a Riesz basis of W_0

We will now provide a partial justification for giving the name "wavelet function" to ψ_h. The proof that the shifts of ψ_h are a basis of W_0 will be given later using the *decomposition relations* in Theorem 17.23, p. 354.

Proposition 17.15 *1. The functions $\Phi_{2Z+1}[(h/2)\widetilde{\Lambda}](Z+1; z)$, $\Phi_{2Z+1}[\widetilde{\Lambda}](Z+1; z)$, $\Pi_{2Z+1}[(h/2)\widetilde{\Lambda}](0; -z)$, and $\Pi_{2Z+1}[\widetilde{\Lambda}](0; -z)$ never vanish for $|z| = 1$.*

2. The polynomial $P(z)$ does not have symmetric zeros on the circle $|z| = 1$, i.e. number z such that $P(z) = P(-z) = 0$.

Proof 1. The symmetry of the vectors $(h/2)\widetilde{\Lambda}$ and $\widetilde{\Lambda}$ is essential here. Then the result follows from Corollary 13.53, p. 253, and Corollary 13.57, p. 256, using

$\Phi_{2Z+1}[(h/2)\widetilde{\Lambda}](Z+1;z) = z^{Z+1}\Phi_{2Z+1}[(h/2)\widetilde{\Lambda}](0;z)$, and $\Phi_{2Z+1}[\widetilde{\Lambda}](Z+1;z) = z^{Z+1}\Phi_{2Z+1}[\widetilde{\Lambda}](0;z)$.

2. This point follows from identity (17.21), p. 339, and by point (1). ∎

Remark 17.16 Let us remark that point (2) of Proposition 17.15 is used [4, p. 136, Theorem 5.11] to prove that Q_{Z+1} has minimal support.

We can prove now the following fundamental Lemma.

Lemma 17.17 *The shifts of the function ψ_h form a Riesz basis for the space*

$$\operatorname*{clos}_{L_2(\mathbb{R})} \{\psi_h(x-jh): \text{ for all } j \in \mathbb{Z}\}.$$

Proof The "stability" Riesz estimate is proved as in [4, p. 145]. Let us use Theorem 14.1, p. 268, according to which the stability estimate is equivalent to the existence of two constants A, B with $0 < A \leq B < \infty$ such that for all $\xi \in \mathbb{R}$

$$A \leq S_1(\xi) = \sum_{j=-\infty}^{\infty} |\widehat{\psi_h(hx)}(\xi + 2\pi j)|^2 \leq B.$$

After changing the variable $h\xi \longrightarrow x$ we obtain the equality

$$\widehat{\psi_h(hx)}(\xi) = \int_{-\infty}^{\infty} e^{-i\xi x} \psi_h(hx)\, dx = \frac{1}{h} \int_{-\infty}^{\infty} e^{-i(\xi/h)x} \psi_h(x)\, dx$$

$$= \widehat{\psi_h(x)}\left(\frac{\xi}{h}\right),$$

i.e.

$$S_1(\xi) = \sum_{j=-\infty}^{\infty} \left|\widehat{\psi_h}\left(\frac{\xi + 2\pi j}{h}\right)\right|^2.$$

However, in Proposition 17.13, p. 341, we have proved the equality

$$S_1(\xi) = \frac{h^{2Z+2}}{C_S^2} \Phi_{2Z+1}\left[\frac{h}{2}\widetilde{\Lambda}\right](Z+1;-\bar{z}) \cdot \Phi_{2Z+1}\left[\frac{h}{2}\widetilde{\Lambda}\right](Z+1;\bar{z})$$

$$\times \Phi_{2Z+1}[h\widetilde{\Lambda}](Z+1;z^2).$$

The most essential feature is that the nonordered vectors $(1/2)\widetilde{\Lambda}$ and $\widetilde{\Lambda}$ are symmetric and we are within the conditions of Corollary 13.57, p. 256, whereby we use $\Phi_{2Z+1}[(h/2)\widetilde{\Lambda}](Z+1;z) = z^{Z+1}\Phi_{2Z+1}[(h/2)\widetilde{\Lambda}](0;z)$. It gives us the result that all of these Schoenberg "exponential Euler polynomials" $\Phi_{2Z+1}[h\widetilde{\Lambda}](0;z)$ are nonzero on the unit circle $|z| = 1$. However, they are continuous functions on the circle $\{|z| = 1\}$, hence $S_1(\xi)$ is bounded from above and below. ∎

After the substitution $h \to h/2^{j-1}$ we see that the functions $\psi_{2^{-j}}$ satisfy the same Riesz basis properties.

17.8 Riesz basis from all wavelet functions $\psi_{2^{-j}h}(x)$

We will now see how to norm the mother wavelets $\psi_{2^{-j}h}(x)$ for all $j \in \mathbb{Z}$, to generate a Riesz basis of the whole space $L_2(\mathbb{R})$. This question has been left open in de Boor et al. [9, p. 153].

Let us recall that in the *classical wavelet construction* where we have one wavelet function ψ, and we consider the functions

$$\psi_{j,\ell}(x) := 2^{j/2}\psi(2^j x - \ell),$$

then at every level j the functions $\psi(2^j x - \ell)$ and $\psi(2^{j_1} x - \ell_1)$ are orthogonal whenever $j \neq j_1$. Hence, the Riesz basis inequality splits over j, and after we make the change $2^j x \to x$, we obtain

$$I := \int_{-\infty}^{\infty} \left| \sum_{j=-\infty}^{\infty} \sum_{\ell=-\infty}^{\infty} c_{j\ell} 2^{j/2} \psi(2^j x - \ell) \right|^2 dx$$

$$= \sum_{j=-\infty}^{\infty} \int_{-\infty}^{\infty} \left| \sum_{\ell=-\infty}^{\infty} c_{j\ell} 2^{j/2} \psi(2^j x - \ell) \right|^2 dx$$

$$= \sum_{j=-\infty}^{\infty} \int_{-\infty}^{\infty} \left| \sum_{\ell=-\infty}^{\infty} c_{j\ell} \psi(x - \ell) \right|^2 dx.$$

Thus if ψ satifies the Riesz inequality with constants A, B where $0 < A \leq B < \infty$, i.e.

$$A \sum_{\ell=-\infty}^{\infty} |c_\ell|^2 \leq \int_{-\infty}^{\infty} \left| \sum_{\ell=-\infty}^{\infty} c_\ell \psi(x - \ell) \right|^2 dx \leq B \sum_{\ell=-\infty}^{\infty} |c_\ell|^2,$$

we see that

$$A \sum_{j=-\infty}^{\infty} \sum_{\ell=-\infty}^{\infty} |c_{j\ell}|^2 \leq \sum_{j=-\infty}^{\infty} \int_{-\infty}^{\infty} \left| \sum_{\ell=-\infty}^{\infty} c_{j\ell} \psi_{j,\ell}(x) \right|^2 dx \leq B \sum_{j=-\infty}^{\infty} \sum_{\ell=-\infty}^{\infty} |c_{j\ell}|^2,$$

i.e. the whole basis has the same Riesz constants. We see that the norming constant $2^{j/2}$ in front of $\psi(2^j x - \ell)$ has been correctly chosen in order to balance the behavior of the dilates.

Here we assume as usual that we are given the nonordered vector Λ by

$$\Lambda = [\lambda_1, \lambda_2, \ldots, \lambda_{Z+1}],$$

and we denote by

$$\tilde{\Lambda} = [\tilde{\lambda}_1, \ldots, \tilde{\lambda}_{2Z+2}]$$

its symmetrization as introduced in Definition 14.5, p. 271.

346 *Multivariate polysplines*

Now we want to obtain the same Riesz inequality for properly normed functions

$$\psi_{2^{-j}h}\left(t - \ell\frac{h}{2^j}\right) = \psi_{2^{-j}h}\left(2^{-j}h\left(\frac{t}{2^{-j}h} - \ell\right)\right),$$

which will play the role of the one-dimensional wavelets $\psi_{j,\ell}(x)$.

Theorem 17.18 *Let us put $H = 2^{-j}h$ for all $j \in \mathbb{Z}$.*
1. For all $j \in \mathbb{Z}$ let the constants N_j be such that

$$0 < A \leq \frac{N_j^2}{H} \sum_{\ell=-\infty}^{\infty} \left|\widehat{\psi_H}(t)\left(\frac{\xi + 2\pi\ell}{H}\right)\right|^2 \leq B < \infty, \tag{17.23}$$

for some constants A, B. Then the set of shifts of the wavelet functions

$$\left\{N_j \psi_{2^{-j}h}\left(t - \ell\frac{h}{2^j}\right): \text{ for } j, \ell \in \mathbb{Z}\right\}$$

form a Riesz basis of $L_2(\mathbb{R})$ with Riesz bounds A, B.
2. In particular, let us assume that $0 \notin \Lambda$. Then the norming constants given by

$$N_j = H^{2Z+1}\left(\prod_{\lambda_\ell > 0} e^{-\lambda_\ell} \cdot \prod_{\lambda_\ell < 0} e^{-3\lambda_\ell}\right)^{H/2} \quad \text{for all } j \in \mathbb{Z}$$

satisfy the above inequality (17.23).

Remark 17.19 *The interesting point is when $H = 2^{-j}h \longrightarrow \infty$. When $H \longrightarrow 0$ there is no need of norming. In this theorem, for simplicity, one has to put $h = 1$ in order to understand easier the asymptotic order.*

Proof 1. We consider the sum

$$\sum_{j=-\infty}^{\infty} \sum_{\ell=-\infty}^{\infty} c_{j,\ell} N_j \psi_{2^{-j}h}\left(t - \ell\frac{h}{2^j}\right)$$

for an arbitrary double-indexed sequence $c_{j,\ell}$ with

$$\sum_{j,\ell=-\infty}^{\infty} |c_{j,\ell}|^2 < \infty.$$

Due to the orthogonality of the different levels (for different js) we have the norm

$$\left\|\sum_{j=-\infty}^{\infty} \sum_{\ell=-\infty}^{\infty} c_{j,\ell} N_j \psi_{2^{-j}h}\left(t - \ell\frac{h}{2^j}\right)\right\|_{L_2(\mathbb{R})}^2$$

$$= \sum_{j=-\infty}^{\infty} \left\|\sum_{\ell=-\infty}^{\infty} c_{j,\ell} N_j \psi_{2^{-j}h}\left(t - \ell\frac{h}{2^j}\right)\right\|_{L_2(\mathbb{R})}^2.$$

Cardinal L-spline wavelet analysis 347

We will use the notation $H = 2^{-j}h$. Thus for every $j \in \mathbb{Z}$ and $h \neq 0$, we have the integrals

$$I_j := \int_{-\infty}^{\infty} \left| \sum_{\ell=-\infty}^{\infty} c_{j,\ell} N_j \psi_H(t - \ell H) \right|^2 dt$$

$$= \int_{-\infty}^{\infty} \left| \sum_{\ell=-\infty}^{\infty} c_{j,\ell} N_j \psi_H \left(H \left(\frac{t}{H} - \ell \right) \right) \right|^2 dt$$

$$= N_j^2 H \int_{-\infty}^{\infty} \left| \sum_{\ell=-\infty}^{\infty} c_{j,\ell} \psi_H(H(t - \ell)) \right|^2 dt$$

for which we want to have the following inequality:

$$A \sum_{\ell=-\infty}^{\infty} |c_{j,\ell}|^2 \leq \sum_{j=-\infty}^{\infty} I_j \leq B \sum_{\ell=-\infty}^{\infty} |c_{j,\ell}|^2 \quad \text{for all } j \in \mathbb{Z}$$

satisfied with constants A, B, such that $0 < A \leq B < \infty$. Now by Theorem 14.1, p. 268, this holds if and only if

$$A \leq N_j^2 H \sum_{\ell=-\infty}^{\infty} |\widehat{\psi_H(Ht)}(\xi + 2\pi \ell)|^2 \leq B.$$

By a simple property of the Fourier transform we have $\widehat{f(ux)}(\xi) = (1/u)\widehat{f(x)}(\xi/u)$, the last is equivalent to

$$A \leq N_j^2 H \sum_{\ell=-\infty}^{\infty} \frac{1}{H^2} \left| \widehat{\psi_H(t)} \left(\frac{\xi + 2\pi \ell}{H} \right) \right|^2 \leq B.$$

Taking sum over j proves the first statement of the theorem.
2. By Proposition 17.13, p. 341, we have

$$\sum_{\ell=-\infty}^{\infty} \left| \widehat{\psi_H} \left(\frac{\xi + 2\pi \ell}{H} \right) \right|^2 = \frac{H^{2Z+2}}{C_S^2} \Phi_{2Z+1} \left[\frac{H}{2} \widetilde{\Lambda} \right] (Z+1; -\bar{z}) \quad (17.24)$$

$$\times \Phi_{2Z+1} \left[\frac{H}{2} \widetilde{\Lambda} \right] (Z+1; \bar{z}) \cdot \Phi_{2Z+1}[H\widetilde{\Lambda}](Z+1; z^2),$$

where $z = e^{-i\xi/2}$ and $C_S = e^{(\lambda_1 + \cdots + \lambda_{Z+1})H/2}$. Here the nonordered symmetric vector $\widetilde{\Lambda} = [\widetilde{\lambda}_1, \ldots, \widetilde{\lambda}_{2Z+2}]$ is the symmetrization of the nonordered vector Λ. We assume that

$$\widetilde{\lambda}_1 \leq \cdots \leq \widetilde{\lambda}_{2Z+2}.$$

By Lemma 17.20, p. 348 below, for z with $|z| = 1$, the asymptotic order for $H \longrightarrow \infty$ of the function

$$\Phi_{2Z+1}[H\widetilde{\Lambda}](Z+1; z) = z^{Z+1} \Phi_{2Z+1}[H\widetilde{\Lambda}](0; z)$$

is
$$|\Phi_{2Z+1}[H\tilde{\Lambda}](Z+1;z)| \approx C \cdot \prod_{\tilde{\lambda}_\ell > 0} e^{\tilde{\lambda}_\ell H} \cdot \frac{1}{H^{2Z+1}},$$

and after putting $H \longrightarrow H/2$ we obtain

$$\left|\Phi_{2Z+1}\left[\frac{H}{2}\tilde{\Lambda}\right](Z+1;z)\right| \approx C \cdot \prod_{\tilde{\lambda}_\ell > 0} e^{\tilde{\lambda}_\ell (H/2)} \cdot \frac{1}{(H/2)^{2Z+1}}.$$

This implies

$$\frac{N_j^2}{H} \cdot \sum_{\ell=-\infty}^{\infty} \left|\widehat{\psi_H}\left(\frac{\xi + 2\pi\ell}{H}\right)\right|^2 \approx N_j^2 \cdot \frac{H^{2Z+1}}{e^{(\lambda_1 + \cdots + \lambda_{Z+1})H}} C_1 \cdot \prod_{\tilde{\lambda}_\ell > 0} e^{2\tilde{\lambda}_\ell H} \cdot \frac{1}{H^{6Z+3}}$$

$$= N_j^2 \cdot C_1 \cdot \prod_{\lambda_\ell > 0} e^{\lambda_\ell H} \cdot \prod_{\lambda_\ell < 0} e^{3\lambda_\ell H} \cdot \frac{1}{H^{4Z+2}}$$

for some constant $C_1 > 0$. Thus we may choose

$$N_j^2 = H^{4Z+2} \prod_{\lambda_\ell > 0} e^{-\lambda_\ell H} \cdot \prod_{\lambda_\ell < 0} e^{-3\lambda_\ell H},$$

where $H = 2^{-j}h$, which proves the second part of the theorem. ∎

In the proof above we have used the following technical

Lemma 17.20 *Let $\Lambda = [\lambda_1, \lambda_2, \ldots, \lambda_{Z+1}]$ be a nonordered vector with $\lambda_1 \leq \cdots \leq \lambda_{Z+1}$, and for every number H put $H\Lambda := [H\lambda_1, \ldots, H\lambda_{Z+1}]$. Assume that Λ is symmetric, i.e. $\Lambda = -\Lambda$, and $0 \notin \Lambda$. Then for λ with $|\lambda| = 1$ we have the following asymptotic result for $H \longrightarrow \infty$:*

$$\Phi_Z[H\Lambda](0;\lambda) \approx \frac{C_1}{H^Z} \exp\left(H \sum_{\lambda_\ell > 0} \lambda_\ell\right),$$

where the polynomial $q_{Z+1}(\lambda) = \prod_{\ell=1}^{Z+1}(\lambda - \lambda_\ell)$. Here the constant $C_1 \neq 0$[4].

Proof 1. Let us assume for simplicity that all λ_ℓs are pairwise different. Then by formula (13.44), p. 255, we have

$$\Phi_Z[H\Lambda](0;\lambda) = \frac{(-1)^Z}{\lambda^Z} e^{(-\lambda_1 - \cdots - \lambda_{Z+1})H} \cdot r[H\Lambda](\lambda) \cdot A_Z(0;\lambda)$$

$$= \frac{(-1)^Z}{\lambda^Z} \cdot r[H\Lambda](\lambda) \cdot \sum_{\ell=1}^{Z+1} \frac{1}{q'_{Z+1}[H\Lambda](H\lambda_\ell)} \cdot \frac{1}{e^{H\lambda_\ell} - \lambda}.$$

---
[4] The equivalence $\approx$ is meant in the sense explained in the footnote to p. 279.

Since Λ is symmetric there is always some λ_ℓ satisfying $\lambda_\ell < 0$. Then

$$A_Z[H\Lambda](0;\lambda) \approx C \cdot \frac{1}{H^Z} \left\{ \sum_{\lambda_\ell < 0} \frac{1}{q'_{Z+1}(\lambda_\ell)} \right\}.$$

Let us make sure that the quantity

$$S := \sum_{\lambda_\ell < 0} \frac{1}{q'_{Z+1}(\lambda_\ell)} \neq 0$$

so that the above equivalence makes sense. Indeed, let us denote by Γ_R the circle of radius $R > 0$ and the half-circle

$$\Gamma_1 := \Gamma_R \cap \{\operatorname{Im} z < 0\}.$$

By Γ_2 we denote the interval on the imaginary axis,

$$\Gamma_2 := \{z = iy: -R \leq y \leq R\}.$$

We assume that the union $\Gamma_- = \Gamma_1 \cup \Gamma_2$ is clockwise oriented. We easily see that for sufficiently large R we have

$$S = \int_{\Gamma_-} \frac{1}{q(z)} dz = \int_{\Gamma_1} \frac{1}{q(z)} dz + \int_{\Gamma_2} \frac{1}{q(z)} dz.$$

Again it is obvious that the first integral approaches 0 when $R \longrightarrow \infty$. On the other hand due to the symmetry of the set Λ we have

$$q(z) = \prod_{j=1}^{(Z+1)/2} \left(z^2 - \lambda_j^2 \right),$$

hence for $z = iy$ we obtain

$$q(iy) = \prod_{j=1}^{(Z+1)/2} \left(-y^2 - \lambda_j^2 \right).$$

It follows that the integral over Γ_2 is non-zero. The last implies immediately that $S \neq 0$.
  Obviously for $H \longrightarrow \infty$ we have

$$|r[H\Lambda](\lambda)| \approx \exp\left(H \cdot \sum_{\lambda_\ell > 0} \lambda_\ell \right)$$

and for $H \longrightarrow \infty$,

$$|\Phi_Z[H\Lambda](0;\lambda)| \approx \frac{C}{H^Z} \exp\left(H \cdot \sum_{\lambda_\ell > 0} \lambda_\ell \right) \left\{ \sum_{\lambda_\ell < 0} \frac{1}{q'_{Z+1}(\lambda_\ell)} \right\}.$$

2. In order to consider the case of coinciding λ_ℓs we may use the integral representation for $A_Z(x;\lambda)$ in (13.11), p. 235. We have

$$A_Z[H\Lambda](0;\lambda) = \int_{\Gamma_H} \frac{1}{q_{Z+1}[H\Lambda](z)} \frac{1}{e^z - \lambda} dz,$$

where the closed contour Γ_H surrounds the zeros of

$$q_{Z+1}[H\Lambda](z) := \prod_{\ell=1}^{Z+1} (z - H\lambda_\ell),$$

$$q_{Z+1}(z) := q_{Z+1}[\Lambda](z) = \prod_{\ell=1}^{Z+1} (z - \lambda_\ell),$$

i.e. all points of $H\Lambda = [H\lambda_1, \ldots, H\lambda_{Z+1}]$, and excludes the zeros of the function $1/(e^z - \lambda)$.

Since $0 \notin \Lambda$ we may take $\Gamma = \Gamma^{(1)} \cup \Gamma^{(2)}$, where $\Gamma^{(1)}$ is a union of the circles in the half-plane Re $z < 0$ surrounding the elements of Λ which are negative; namely, there exists a sufficiently small $\rho > 0$ such that for $H \geq 1$ the contours (having anticlockwise orientation)

$$\Gamma_\ell = \{z = \lambda_\ell + \rho e^{i\theta} : \text{ for } 0 \leq \theta \leq 2\pi\}$$

do not intersect each other and lie in the half-plane Re $z < 0$. The circles

$$H \cdot \Gamma_\ell := \{z = H\lambda_\ell + H\rho e^{i\theta}\}$$

do not intersect each other.

We put

$$\Gamma^{(2)} = -\Gamma^{(1)},$$

where the equality is considered only as sets, but $\Gamma^{(2)}$ preserves the same (anticlockwise) orientation like $\Gamma^{(1)}$.

For $H > 0$ and $H \longrightarrow \infty$ we have

$$H\lambda_\ell + H\rho \cos\theta \leq H\lambda \longrightarrow +H\rho < 0,$$

since the second value is the distance between the two circles $|\lambda| = 1$ and

$$e^{H\Gamma_\ell} = \{z = e^{z_1} : \text{ for } z_1 \in H\Gamma_\ell\}.$$

Hence, for every $H > 0$, and for λ with $|\lambda| = 1$,

$$\frac{1}{|e^{H\lambda_\ell + H\rho \cdot e^{i\theta}} - \lambda|} \leq \frac{1}{|1 - e^{H\lambda_\ell + H\rho}|}.$$

Thus, we obtain the estimate, for example for $H\Gamma_1$, if we assume that the multiplicity of λ_1 is m:

$$\left| \int_{H\Gamma_1} \frac{1}{q_{Z+1}[H\Lambda](z)} \cdot \frac{1}{e^z - \lambda} dz \right| \tag{17.25}$$

$$\leq \frac{1}{H^{Z+1}} \int_0^{2\pi} \frac{1}{\prod_{\ell=1}^{Z+1} |\lambda_1 + \rho \cdot e^{i\theta} - \lambda_\ell|} \cdot \frac{1}{|e^{H\lambda_1 + H\rho \cdot e^{i\theta}} - \lambda|} (H|\lambda_1| + H\rho) d\theta$$

$$\leq \frac{(|\lambda_1| + \rho)}{\rho^m H^Z |1 - e^{H\lambda_1 + H\rho}|} \cdot \int_0^{2\pi} \frac{1}{\prod_{\ell=m+1}^{Z+1} |\lambda_1 + \rho \cdot e^{i\theta} - \lambda_\ell|} \cdot d\theta$$

$$\leq C_1 \frac{(|\lambda_1| + \rho)}{\rho^m H^Z}$$

$$= \frac{C_2}{H^Z},$$

where the constant $C_2 > 0$ depends only on the set Λ, and also on the minimal distance between the zeros λ_ℓ, counted with multiplicities.

In a similar way we obtain the result for the other circles $H\Gamma_\ell$. We have

$$\int_{H\Gamma^{(1)}} = \sum_{\lambda_\ell \in \Lambda} \int_{H\Gamma_\ell},$$

where we take only one integral for λ_ℓs which coincide.

If we consider the contours $-H\Gamma_\ell$ which surround $H\lambda_\ell \in H\Lambda$ satisfying $\lambda_\ell > 0$, it is also clear from estimates of the type above that we obtain the inequality

$$\left| \int_{-H\Gamma_\ell} \frac{1}{q_{Z+1}[H\Lambda](z)} \cdot \frac{1}{e^z - \lambda} dz \right| \leq \frac{C_1}{e^{H|\lambda_\ell| + H\rho}}.$$

This does not contribute to the asymptotic of the function $A_Z[H\Lambda](0; \lambda)$.

Since

$$A_Z[H\Lambda](0; \lambda) = \left(\int_{H\Gamma^{(1)}} + \int_{H\Gamma^{(2)}} \right) \left(\frac{1}{q_{Z+1}[H\Lambda](z)} \cdot \frac{1}{e^z - \lambda} dz \right),$$

where the curves in $\Gamma^{(1)}$ and $\Gamma^{(2)}$ have the same (anticlockwise) orientation, it follows that

$$|A_Z[H\Lambda](0; \lambda)| \leq \frac{C}{H^Z}.$$

This completes the proof. ∎

Remark 17.21 *1. The polynomial case when $\Lambda = [0, \ldots, 0]$ is exceptional, since then $H\Lambda = \Lambda = [0, \ldots, 0]$ and we may choose a constant curve Γ_H independent of H. Evidently, we have*

$$A_Z[H\Lambda](x; \lambda) = A_Z[\Lambda](x; \lambda)$$

in that case. The result is then $N_j = 2^{j/2}$.

2. The case $0 \in \Lambda$ is more difficult to treat and we will not need it in our further developments. Let some of the elements of Λ be zero, for example $\lambda_1 = 0$, with multiplicity $m \geq 1$. The estimate will be obtained from the above one by letting $\lambda_1 \longrightarrow 0$. We will take

$$\rho = \varepsilon |\lambda_1|,$$

for arbitrary small $\varepsilon > 0$, and from (17.25), p. 351, we obtain the estimate for Φ by considering

$$I := \prod_{j=1}^{Z+1} |e^{H\lambda_j} - \lambda| \cdot \left| \int_{H\Gamma_1} \frac{1}{q_{Z+1}[H\Lambda](z)} \cdot \frac{1}{e^z - \lambda} dz \right|$$

$$\leq C \prod_{j=m+1}^{Z+1} |e^{H\lambda_j} - 1| \cdot |e^{H\lambda_1} - 1|^m \cdot \frac{(|\lambda_1| + \varepsilon|\lambda_1|)}{(\varepsilon|\lambda_1|)^m H^Z |1 - e^{H\lambda_1 + H\varepsilon|\lambda_1|}|}.$$

We apply the inequality

$$D_1 e^x \leq \left| \frac{e^x - 1}{x} \right| \leq D_2 e^x,$$

which is equivalent to

$$D_1 \leq \left| \frac{e^{-x} - 1}{x} \right| \leq D_2.$$

We obtain

$$I \leq \frac{C}{\varepsilon^m H^{Z+1-m}} \prod_{j=m+1}^{Z+1} |e^{H\lambda_j} - 1| \cdot (e^{H\lambda_1})^m \frac{1}{e^{H\lambda_1 + H\varepsilon|\lambda_1|}}.$$

The above theorem shows that we have to norm the function $\psi_{2^{-j}h}(x)$ if we want to obtain a stable asymptotic for the whole basis. Otherwise we will not be able to obtain the result, as in the polynomial case.

17.9 The decomposition relations for the scaling function Q_{Z+1}

We will now prove the *decomposition relations* of the splitting

$$V_1 = V_0 \oplus W_0.$$

They have been summarized in Section 16.6, p. 319.

Let us consider the following algebraic system on the circle $|z| = 1$:

$$P(z)A(\bar{z}) + R(z)B(\bar{z}) = 1,$$
$$P(-z)A(\bar{z}) + R(-z)B(\bar{z}) = 0. \tag{17.26}$$

Theorem 17.22 provides an explicit solution in analytic functions.

Theorem 17.22 *The solution to system (17.26) is given by the two analytic functions*

$$A(z) = \frac{C_S}{2^{2Z+2}} \cdot \frac{P(z)\Phi_{2Z+1}[(h/2)\widetilde{\Lambda}](Z+1;z)}{\Phi_{2Z+1}[h\widetilde{\Lambda}](Z+1;z^2)}, \qquad (17.27)$$

$$B(z) = -\frac{C_S}{2^{2Z+2}} \cdot z^{2Z+1} \frac{P(-1/z)}{\Phi_{2Z+1}[h\widetilde{\Lambda}](Z+1;z^2)}, \qquad (17.28)$$

which may be expanded into convergent Laurent series near the unit circle $|z| = 1$. Here the constant

$$C_S = e^{(\lambda_1 + \cdots + \lambda_{Z+1})\frac{h}{2}}.$$

Proof We consider the numbers z with $|z| = 1$. Recalling Proposition 17.12, p. 340, the solution of the system is given by

$$A(\bar{z}) = R(-z) \Big/ \det\begin{pmatrix} P(z) & R(z) \\ P(-z) & R(-z) \end{pmatrix}$$

$$= \frac{C_S z^{2Z+1} P(1/z)\Phi_{2Z+1}[(h/2)\widetilde{\Lambda}](Z+1;z)}{2^{2Z+2} \cdot z^{2Z+1}\Phi_{2Z+1}[h\widetilde{\Lambda}](Z+1;z^2)}$$

$$= \frac{C_S P(1/z)\Phi_{2Z+1}[(h/2)\widetilde{\Lambda}](Z+1;z)}{2^{2Z+2}\Phi_{2Z+1}[h\widetilde{\Lambda}](Z+1;z^2)},$$

and

$$B(\bar{z}) = -P(-z) \Big/ \det\begin{pmatrix} P(z) & R(z) \\ P(-z) & R(-z) \end{pmatrix}$$

$$= \frac{-C_S P(-z)}{2^{2Z+2} \cdot z^{2Z+1}\Phi_{2Z+1}[h\widetilde{\Lambda}](Z+1;z^2)}.$$

Hence, after taking the complex conjugate, we obtain

$$A(z) = \frac{C_S P(z)\Phi_{2Z+1}[(h/2)\widetilde{\Lambda}](Z+1;1/z)}{2^{2Z+2} \cdot \Phi_{2Z+1}[h\widetilde{\Lambda}](Z+1;1/z^2)}$$

and

$$B(z) = -z^{2Z+1}\frac{C_S P(-1/z)}{2^{2Z+2} \cdot \Phi_{2Z+1}[h\widetilde{\Lambda}](Z+1;1/z^2)}.$$

This ends the proof since the "exponential Euler polynomials" $\Phi_{2Z+1}[h\widetilde{\Lambda}](Z+1;1/z^2)$ and $\Phi_{2Z+1}[(h/2)\widetilde{\Lambda}](Z+1;1/z)$, are finite Laurent series which are not zero on the unit circle $|z| = 1$, hence, the functions $A(z)$ and $B(z)$ may be expanded into a Laurent series which is convergent near the unit circle.

Due to Corollary 13.57, p. 256, $\Phi_{2Z+1}[\Lambda_1](Z+1;z) = \Phi_{2Z+1}[\Lambda_1](Z+1;1/z)$ for symmetric nonordered vectors Λ_1. This finally proves equalities (17.27) and (17.28). ∎

Only now do we prove that the function ψ_h has shifts generating the space W_0, i.e. it is indeed a wavelet function. We will further use the "smart notations" as in Chui [4, p. 19, formula (1.6.6)] to combine the odd and even cases.

Theorem 17.23 *The wavelet function $\psi_h = \psi[\Lambda; h]$ defined by (17.20), p. 338, generates the wavelet space W_0 defined in (17.11), p. 333, i.e.*

$$V_1 = V_0 \oplus W_0$$

holds. More precisely, we have the following **decomposition relations** *for the basis $\{Q[\Lambda, h/2](x - \ell(h/2)): \ell \in \mathbb{Z}\}$ of the space V_1:*

$$Q\left[\Lambda, \frac{h}{2}\right]\left(x - \ell\frac{h}{2}\right) = \sum_{u=-\infty}^{\infty} \{a_{\ell-2u} \cdot Q[\Lambda, h](x - uh) + b_{\ell-2u} \cdot \psi[\Lambda, h](x - uh)\}.$$

(17.29)

This is equivalent to the following two relations obtained for the case of odd and even ℓ, namely:

$$Q\left[\Lambda; \frac{h}{2}\right](x) = \sum_{u=-\infty}^{\infty} a_{-2u} \cdot Q[\Lambda; h](x - uh)$$

$$+ \sum_{u=-\infty}^{\infty} b_{-2u} \cdot \psi[\Lambda, h](x - uh),$$

$$Q\left[\Lambda; \frac{h}{2}\right]\left(x - \frac{h}{2}\right) = \sum_{u=-\infty}^{\infty} a_{1-2u} \cdot Q[\Lambda; h](x - uh)$$

$$+ \sum_{u=-\infty}^{\infty} b_{1-2u} \cdot \psi[\Lambda, h](x - uh).$$

The coefficients a_j and b_j are obtained from the Laurent series expansions

$$A(z) = \frac{1}{2} \sum_{\ell=-\infty}^{\infty} a_\ell z^\ell,$$

$$B(z) = \frac{1}{2} \sum_{\ell=-\infty}^{\infty} b_\ell z^\ell,$$

of the functions $A(z)$, $B(z)$ which are solutions to the system (17.26), p. 352.

Proof 1. By the definition of the space W_0 in (17.11), p. 333, we have $W_0 \subset V_1$ and $V_0 \oplus W_0 = V_1$. By the definition of ψ_h in (17.20), p. 338, and by Lemma 17.7, p. 334, where we have described an arbitrary element ψ of the space V_1, which is orthogonal to V_0, we see that $\psi_h \perp V_0$. Hence $\psi_h \in W_0$. We have the following equality, easily proved by changing variables:

$$\langle \psi_h(x - \ell h), \phi(x - jh)\rangle = \langle \psi_h(x), \phi(x + \ell h - jh)\rangle.$$

It follows that all shifts $\psi[\Lambda, h](x - \ell h)$ for $\ell \in \mathbb{Z}$ are also in W_0.

It remains to show that every element $v_1 \in V_1$ is a unique linear combination of elements of V_0 and shifts of ψ_h, which will imply

$$v_1 = v_0 + w_0.$$

Due to the relation $V_0 \perp W_0$ it follows that this representation is unique. Indeed, if

$$0 = v_0 + w_0,$$

it follows that $0 = \langle v_0, v_0 \rangle$, hence, $v_0 = 0$. It also follows that $w_0 = 0$.

2. Let us prove the decomposition relation (17.29). Taking its Fourier transform we obtain the equivalent one

$$e^{-i\ell(h/2)\xi} \cdot \widehat{Q\left[\Lambda, \frac{h}{2}\right]}(\xi) = \sum_u \{a_{\ell-2u} e^{-iuh\xi} \cdot \widehat{Q[\Lambda, h]}(\xi) + b_{\ell-2u} e^{-iuh\xi} \cdot \widehat{\psi[\Lambda, h]}(\xi)\},$$

or after putting $z = e^{-i(h/2)\xi}$, we obtain

$$\widehat{Q\left[\Lambda, \frac{h}{2}\right]}(\xi) = \sum_u \{a_{\ell-2u} \bar{z}^{\ell-2u} \cdot \widehat{Q[\Lambda, h]}(\xi) + b_{\ell-2u} \bar{z}^{\ell-2u} \cdot \widehat{\psi[\Lambda, h]}(\xi)\}.$$

If we put

$$U(z) = \frac{1}{2} \sum_{u=-\infty}^{\infty} a_u z^u, \qquad V(z) = \frac{1}{2} \sum_{u=-\infty}^{\infty} b_u z^u,$$

we see that we have two cases: of ℓ even,

$$\widehat{Q\left[\Lambda, \frac{h}{2}\right]}(\xi) = (U(\bar{z}) + U(-\bar{z}))\widehat{Q[\Lambda, h]}(\xi) + (V(\bar{z}) + V(-\bar{z}))\widehat{\psi[\Lambda, h]}(\xi),$$

and of ℓ odd

$$\widehat{Q\left[\Lambda, \frac{h}{2}\right]}(\xi) = (U(\bar{z}) - U(-\bar{z}))\widehat{Q[\Lambda, h]}(\xi) + (V(\bar{z}) - V(-\bar{z}))\widehat{\psi[\Lambda, h]}(\xi).$$

Now using the *refinement equations* for ϕ and ψ, in (17.8), p. 331, and (17.13), p. 334, we obtain

$$\widehat{Q[\Lambda, h]}(\xi) = P(z) \cdot \widehat{Q\left[\Lambda, \frac{h}{2}\right]}(\xi),$$

$$\widehat{\psi[\Lambda, h]}(\xi) = R(z) \cdot \widehat{Q\left[\Lambda, \frac{h}{2}\right]}(\xi).$$

We see that the functions $U(z)$ and $V(z)$ are a solution to the system

$$1 = (U(\bar{z}) + U(-\bar{z})) \cdot P(z) + (V(\bar{z}) + V(-\bar{z})) \cdot R(z),$$
$$1 = (U(\bar{z}) - U(-\bar{z})) \cdot P(z) + (V(\bar{z}) - V(-\bar{z})) \cdot R(z),$$

which after summing and subtracting gives the equivalent system

$$1 = U(\bar{z}) \cdot P(z) + V(\bar{z}) \cdot R(z),$$
$$0 = U(-\bar{z}) \cdot P(z) + V(-\bar{z}) \cdot R(z).$$

356 Multivariate polysplines

As we have seen in Theorem 17.22, p. 353, this system has a unique solution
$$U(z) = A(z), \qquad V(z) = B(z).$$

This completes the proof. ∎

17.10 The dual scaling function $\widetilde{\phi}$ and the dual wavelet $\widetilde{\psi}$

Now we can construct the **duals** to the scaling function $\phi(x) = Q_{Z+1}[\Lambda; h](x)$ and to the wavelet function $\psi_h(x) = \psi[\Lambda; h](x)$. In the polynomial case we have provided a summary of the results in Sections 16.4 and 16.5.

Recall that by (13.25) and (13.26), pp. 242, 243, we have $\widehat{Q_{Z+1}[\Lambda; h]}(\xi) = h^{Z+1}\widehat{Q_{Z+1}[h\Lambda]}(h\xi)$ and $Q_{Z+1}[\Lambda; h](x) = h^Z Q_{Z+1}[h\Lambda](x/h)$.

Let us recall that we have defined in (17.3), p. 327, the scaling function by putting
$$\phi_h(x) = Q_{Z+1}[\Lambda; h](x) = h^Z Q_{Z+1}[h\Lambda]\left(\frac{x}{h}\right).$$

Let us introduce the dual scaling function $\widetilde{\phi}$.

Definition 17.24 *We define the **dual scaling function** $\widetilde{\phi}(x)$ through its Fourier transform*
$$\widehat{\widetilde{\phi}_h}(\xi) := \frac{C_S^2}{h^{2Z+1}} \cdot \frac{\widehat{\phi}(\xi)}{\Phi_{2Z+1}[h\widetilde{\Lambda}](Z+1; e^{i\xi h})} \tag{17.30}$$
$$= \frac{C_S^2}{h^{2Z+1}} \cdot \frac{\widehat{Q_{Z+1}[\Lambda; h]}(\xi)}{\Phi_{2Z+1}[h\widetilde{\Lambda}](Z+1; e^{i\xi h})} \quad \text{for } \xi \in \mathbb{R},$$

where the constant C_S is given by
$$C_S := e^{(\lambda_1 + \cdots + \lambda_{Z+1})h/2},$$

and the nonordered vector $\widetilde{\Lambda}$ is the symmetrization of the vector Λ.

The motivation for Definition 17.24 follows:

Theorem 17.25 *The function $\widetilde{\phi}(x)$ is the dual to the scaling function $\phi_h(x) = Q_{Z+1}[\Lambda; h](x)$ in the sense that:*

1. $\widetilde{\phi}_h(x) \in V_0$ and the set of shifts $\{\widetilde{\phi}_h(x - \ell h): \text{ for all } \ell \in \mathbb{Z}\}$ is a Riesz basis of the space V_0. The dual $\widetilde{\phi}_{2^{-j}h}(x) \in V_j$ and its $2^{-j}h$-shifts form a Riesz basis of the space V_j.

2. The orthogonality $\langle \phi_h(x - kh), \widetilde{\phi}_h(x - \ell h) \rangle = \delta_{k\ell}$ holds for all $k, \ell \in \mathbb{Z}$.

Proof We see that after the change $x \to hx$ we obtain $\phi(hx) = h^Z Q_{Z+1}[h\Lambda](x)$ for every $x \in \mathbb{R}$, and $\widehat{\phi}(\xi) = h^{Z+1}\widehat{Q_{Z+1}[h\Lambda]}(h\xi)$, hence
$$\widehat{\phi(hx)}(\xi) = \frac{1}{h}\widehat{\phi}\left(\frac{\xi}{h}\right) = h^Z \widehat{Q_{Z+1}[h\Lambda]}(\xi).$$

Let us put
$$D(\xi) = h^{2Z+1} \Phi_{2Z+1}[h\widetilde{\Lambda}](Z+1; e^{i\xi h})/C_S^2.$$

1. Since the "exponential Euler spline" $\Phi_{2Z+1}[h\widetilde{\Lambda}](Z+1; z)$ is a finite Laurent series, and by Corollary 13.57, p. 256, $\Phi_{2Z+1}[h\widetilde{\Lambda}](Z+1; z)$ has no zeros around the unit circle $|z| = 1$, we obtain its inverse Laurent series

$$\frac{C_S^2}{\Phi_{2Z+1}[h\widetilde{\Lambda}](Z+1; z)} := \sum_{j=-\infty}^{\infty} s_j z^j$$

with coefficients s_j, which is convergent in the neighborhood of the unit circle.

Taking the inverse Fourier transform of $\widetilde{\phi}$ in (17.30), p. 356, we obtain

$$\widetilde{\phi}(x) = \sum_{j=-\infty}^{\infty} s_j \phi(x-j).$$

The last series is convergent since by the stability of the shifts $\phi_h(x - jh)$ proved in Corollary 14.7, p. 276, we have

$$A \sum_{j=-\infty}^{\infty} |s_j|^2 \le \left\| \sum_{j=-\infty}^{\infty} s_j \phi(x - jh) \right\|_{L_2(\mathbb{R})} \le B \sum_{j=-\infty}^{\infty} |s_j|^2,$$

for some constants A, B with $0 < A \le B < \infty$, and by Theorem 12.1, p. 209, for $z = e^{-i\xi}$, the *Fourier series* and its *symbol* have equal norms, i.e.

$$\sum_{j=-\infty}^{\infty} |s_j|^2 = \frac{1}{2\pi} \int_0^{2\pi} \left| \frac{C_S^2}{\Phi_{2Z+1}[h\widetilde{\Lambda}](Z+1; z)} \right|^2 d\xi.$$

Vice versa, we have the finite Laurent series

$$\frac{\Phi_{2Z+1}[h\widetilde{\Lambda}](Z+1; z)}{C_S^2} =: \sum \widetilde{s}_j z^j,$$

which shows that

$$\phi_h(x) = \sum_{j=-\infty}^{\infty} \widetilde{s}_j \widetilde{\phi}_h(x - jh).$$

It follows that the h-shifts of the function $\widetilde{\phi}_h$ generate the space V_0.

2. From the definition of the scalar product (16.1), p. 313, we have

$$I := \langle \phi(x - kh), \widetilde{\phi}(x - \ell h) \rangle := \int_{-\infty}^{\infty} \phi(x - kh) \widetilde{\phi}(x - \ell h) \, dx.$$

358 Multivariate polysplines

What we do now is a standard step in Fourier analysis. Applying the Parseval identity (12.5), p. 212, the Fourier transform of a translation, and the change $h\xi = \eta$ consecutively, gives

$$2\pi I = \int_{-\infty}^{\infty} \widehat{\phi(x-kh)}(\xi) \cdot \overline{\widehat{\phi(x-\ell h)}(\xi)} \, d\xi$$

$$= \int_{-\infty}^{\infty} e^{-ikh\xi + i\ell h\xi} \cdot \widehat{\phi}(\xi) \cdot \overline{\widehat{\phi}(\xi)} \, d\xi$$

$$= \frac{1}{h} \int_{-\infty}^{\infty} e^{i(\ell-k)\eta} \cdot \widehat{\phi}\left(\frac{\eta}{h}\right) \cdot \overline{\widehat{\phi}\left(\frac{\eta}{h}\right)} \, d\eta$$

$$= \frac{1}{h} \sum_{j=-\infty}^{\infty} \int_{2\pi j}^{2\pi(j+1)} e^{i(\ell-k)\eta} \cdot \widehat{\phi}\left(\frac{\eta}{h}\right) \cdot \overline{\widehat{\phi}\left(\frac{\eta}{h}\right)} \, d\eta$$

$$= \frac{1}{h} \sum_{j=-\infty}^{\infty} \int_{2\pi j}^{2\pi(j+1)} e^{i(\ell-k)\eta} \cdot \frac{|\widehat{\phi}(\eta/h)|^2}{D(\eta/h)} \, d\eta$$

$$= h^{2Z+1} \sum_{j=-\infty}^{\infty} \int_{2\pi j}^{2\pi(j+1)} e^{i(\ell-k)\eta} \cdot \frac{1}{D(\eta/h)} |\widehat{Q_{Z+1}[h\Lambda]}(\eta)|^2 \, d\eta.$$

Let us recall that due to Theorem 14.6, p. 272,

$$\sum_{j=-\infty}^{\infty} |\widehat{Q_{Z+1}[h\Lambda]}(\xi + 2\pi j)|^2 = \frac{1}{C_S^2} \Phi_{2Z+1}[h\widetilde{\Lambda}](Z+1; e^{i\xi}).$$

Hence, after changing the variable in every integral and by using the 2π-periodicity of $D(\eta/h)$ we obtain

$$2\pi I = h^{2Z+1} \int_0^{2\pi} e^{i(\ell-k)\eta} \cdot \sum_{j=-\infty}^{\infty} |\widehat{Q_{Z+1}[h\Lambda]}(\eta + 2\pi j)|^2 \frac{1}{D(\eta/h)} \, d\eta$$

$$= \int_0^{2\pi} e^{i(\ell-k)\eta} \, d\eta$$

$$= 2\pi \delta_{k\ell},$$

which proves the theorem. ■

Now we introduce the dual to the wavelet $\psi_h(x) = \psi[\Lambda; h](x)$.

Definition 17.26 *Let us define the* **dual wavelet function** $\widetilde{\psi}_h$ *through its Fourier transform:*

$$\widehat{\widetilde{\psi}_h}(\xi) := \frac{\widehat{\psi_h[\Lambda; h]}(\xi)}{D(\xi)} \quad \text{for } \xi \in \mathbb{R},$$

and the function $D(\xi)$ is given by

$$D\left(\frac{\xi}{h}\right) = \frac{1}{h} \sum_{j=-\infty}^{\infty} \left|\widehat{\widetilde{\psi}_h}\left(\frac{\xi + 2\pi j}{h}\right)\right|^2.$$

Remark 17.27 1. Due to Proposition 17.13, p. 341, for all $z = e^{-i\xi/2}$

$$D\left(\frac{\xi}{h}\right) = \frac{h^{2Z+1}}{C_S^2} \Phi_{2Z+1}\left[\frac{h}{2}\widetilde{\Lambda}\right](Z+1; -\bar{z}) \cdot \Phi_{2Z+1}\left[\frac{h}{2}\widetilde{\Lambda}\right]$$
$$\times (Z+1; \bar{z}) \cdot \Phi_{2Z+1}[h\widetilde{\Lambda}](Z+1; z^2),$$

where the constant C_S is given by

$$C_S := e^{(\lambda_1 + \cdots + \lambda_{Z+1})h/2},$$

and the nonordered vector $\widetilde{\Lambda}$ is the symmetrization of the vector Λ. Evidently, $D(\xi)$ is a real number.

2. Due to Theorem 13.46, p. 245, we see that if Λ_1 is a symmetric nonordered vector then $\Phi_N[\Lambda_1](N/2; z) = \Phi_N[\Lambda_1](N/2; 1/z)$, hence, for $|z| = 1$ we have $\Phi_N[\Lambda_1](N/2; z) = \Phi_N[\Lambda_1](N/2; \bar{z})$. This shows that the expression for $D(\xi)$ remains the same if we replace z with $\bar{z}$.

The following properties justify the above terminology.

Theorem 17.28 *The function $\widetilde{\psi_h}(x)$ is the dual semi-orthogonal to the wavelet function $\psi_h(x) = \psi_h[\Lambda; h](x)$ in the following sense.*
1. The function $\widetilde{\psi_h}(x) \in W_0$ and the shifts $\{\widetilde{\psi_h}(x - \ell h): \text{ for all } \ell \in \mathbb{Z}\}$ are a Riesz basis of the wavelet space W_0. $\widetilde{\psi_{2^{-j}h}}(x) = \psi[\Lambda, 2^{-j}h](x) \in W_j$ and its shifts are a Riesz basis of the wavelet space W_j, i.e.

$$W_j = \operatorname*{clos}_{L_2(\mathbb{R})} \left\{ \psi_{2^{-j}h}\left(x - \frac{\ell}{2^j}h\right): \text{ for all } \ell \in \mathbb{Z} \right\}$$
$$= \operatorname*{clos}_{L_2(\mathbb{R})} \left\{ \widetilde{\psi_{2^{-j}h}}\left(x - \frac{\ell}{2^j}h\right): \text{ for all } \ell \in \mathbb{Z} \right\}.$$

2. We have orthogonality at the level W_0, i.e. $\langle \psi_h(x - kh), \widetilde{\psi_h}(x - \ell h) \rangle = \delta_{k\ell}$ for all $k, \ell \in \mathbb{Z}$. The same considerations apply to the shifts of the functions $\widetilde{\psi_{2^{-j}h}}(x)$ and $\psi_{2^{-j}h}(x)$ in W_j.
3. We have orthogonality between every pair of different levels, i.e. $\widetilde{\psi_h} \perp W_j$ for all $j \neq 0$. Thus $\widetilde{\psi_{2^{-i}h}} \perp W_j$ for all $i, j \in \mathbb{Z}$ with $i \neq j$.

Remark 17.29 *By putting $h \longrightarrow 2^{-\ell+1}h$ we will obtain the result for all spaces W_ℓ. Thus we only need to prove the above results for W_0.*

Proof 1. This follows as above for $\widetilde{\phi}$ in Theorem 17.25, p. 356, since the "exponential Euler splines" $\Phi_{2Z+1}[\Lambda_1](Z+1; z)$ are nonzero on the unit circle $|z| = 1$ when the nonordered vector Λ_1 is symmetric.

2. We see that similarly we have to check the scalar product

$$2\pi I = \int_{-\infty}^{\infty} \widehat{\psi_h(x-kh)}(\xi) \cdot \overline{\widetilde{\psi_h(x-\ell h)}(\xi)} \, d\xi$$

$$= \int_{-\infty}^{\infty} e^{-ikh\xi + i\ell h\xi} \cdot \widehat{\psi_h}(\xi) \cdot \overline{\widetilde{\psi_h}(\xi)} \, d\xi$$

$$= \frac{1}{h} \int_{-\infty}^{\infty} e^{i(\ell-k)\eta} \cdot \widehat{\psi_h}\left(\frac{\eta}{h}\right) \cdot \overline{\widetilde{\psi_h}\left(\frac{\eta}{h}\right)} \, d\eta$$

$$= \frac{1}{h} \sum_{j=-\infty}^{\infty} \int_{2\pi j}^{2\pi(j+1)} e^{i(\ell-k)\eta} \cdot \widehat{\psi_h}\left(\frac{\eta}{h}\right) \cdot \overline{\widetilde{\psi_h}\left(\frac{\eta}{h}\right)} \, d\eta$$

$$= \frac{1}{h} \sum_{j=-\infty}^{\infty} \int_{2\pi j}^{2\pi(j+1)} e^{i(\ell-k)\eta} \cdot \frac{|\widehat{\psi_h}(\eta/h)|^2}{D(\eta/h)} \, d\eta.$$

For $z = e^{-i\eta/2}$, by Proposition 17.13, p. 341, we have

$$\sum_{j=-\infty}^{\infty} \left|\widehat{\psi_h}\left(\frac{\eta + 2\pi j}{h}\right)\right|^2$$

$$= \frac{h^{2Z+2}}{C_S^2} \Phi_{2Z+1}\left[\frac{h}{2}\widetilde{\Lambda}\right](Z+1; -\overline{z}) \cdot \Phi_{2Z+1}\left[\frac{h}{2}\widetilde{\Lambda}\right]$$

$$\times (Z+1; \overline{z}) \cdot \Phi_{2Z+1}[h\widetilde{\Lambda}](Z+1; z^2),$$

which by the 2π-periodicity of $D(\eta/h)$ implies

$$2\pi I = \frac{1}{h} \int_0^{2\pi} e^{i(\ell-k)\eta} \cdot \sum_{j=-\infty}^{\infty} \left|\widehat{\psi_h}\left(\frac{\eta + 2\pi j}{h}\right)\right|^2 \frac{1}{D(\eta/h)} \, d\eta$$

$$= \int_0^{2\pi} e^{i(\ell-k)\eta} \, d\eta$$

$$= 2\pi \delta_{k\ell}.$$

Evidently, from the above we can see that $\overline{D(\eta/h)}$ is a real constant. It follows that $\overline{D(\xi)} = D(\xi)$.

3. This follows immediately from (1). Indeed, we have seen that due to the construction of the wavelet spaces W_j we have

$$V_{j+k} = V_j \oplus W_{j+k-1} \oplus \cdots \oplus W_j,$$

i.e. they are mutually orthogonal. Since $\psi_h \in W_0$ and, in general, $\psi[\Lambda; 2^{-j}h] \in W_j$ we have the result that they are all mutually orthogonal, which also implies that $\psi_h \perp W_j$ for all $j \neq 0$. ∎

Now we can easily write the **refinement equations (two-scale relations)** for the duals $\widetilde{\phi}$ and $\widetilde{\psi_h}$.

Theorem 17.30 Let the functions $A(z)$ and $B(z)$ be the solutions to system (17.26), p. 352. Then for $z = e^{-i\xi/2}$ we have the **refinement equations**

$$\widehat{\widetilde{\phi}_h}(\xi) = \frac{1}{2} A(z) \widehat{\widetilde{\phi}_{h/2}}(\xi),$$

$$\widehat{\widetilde{\psi}_h}(\xi) = \frac{1}{2} B(z) \widehat{\widetilde{\phi}_{h/2}}(\xi).$$

Proof By the definition of the duals $\widetilde{\phi}_h(\xi)$ in Definition 17.24, p. 356, for $z = e^{-i\xi/2}$, we have

$$\widehat{\widetilde{\phi}_h}(\xi) = C_S^2 \frac{\widehat{Q_{Z+1}[\Lambda;h]}(\xi)}{h^{2Z+1} \Phi_{2Z+1}[h\widetilde{\Lambda}](Z+1;e^{i\xi h})},$$

$$\widehat{\widetilde{\phi}_{h/2}}(\xi) = C_S \frac{2^{2Z+1} \widehat{Q_{Z+1}[\Lambda;h/2]}(\xi)}{h^{2Z+1} \Phi_{2Z+1}[(h/2)\widetilde{\Lambda}](Z+1;e^{i\xi h/2})},$$

where the constant

$$C_S = e^{(\lambda_1 + \cdots + \lambda_{Z+1})h/2}.$$

Hence, if we put $h\xi = \eta$, by formula (13.25), p. 242, we obtain

$$\frac{\widehat{\widetilde{\phi}_h}(\xi)}{\widehat{\widetilde{\phi}_{h/2}}(\xi)} = \frac{C_S}{2^{2Z+1}} \cdot \frac{\widehat{Q_{Z+1}[\Lambda;h]}(\xi)}{\Phi_{2Z+1}[h\widetilde{\Lambda}](Z+1;e^{i\xi h})} \div \left\{ \frac{\widehat{Q_{Z+1}[\Lambda;h/2]}(\xi)}{\Phi_{2Z+1}[(h/2)\widetilde{\Lambda}](Z+1;e^{i\xi h/2})} \right\}$$

$$= \frac{C_S}{2^{2Z+1}} \cdot \frac{\widehat{Q_{Z+1}[h\Lambda]}(\eta)}{\Phi_{2Z+1}[h\widetilde{\Lambda}](Z+1;e^{i\eta})} \div \left\{ \frac{\widehat{Q_{Z+1}[\frac{h}{2}\Lambda]}(\eta/2)}{2^{Z+1}\Phi_{2Z+1}[(h/2)\widetilde{\Lambda}](Z+1;e^{i\eta/2})} \right\},$$

which by means of the two-scale relation for $Q_{Z+1}[h\Lambda]$ provided in formula (17.10), p. 332, namely

$$\widehat{Q_{Z+1}[h\Lambda]}(\xi) = P(e^{-i\xi/2}) \cdot \left(\frac{1}{2}\right)^{Z+1} \widehat{Q_{Z+1}\left[\frac{h}{2}\Lambda\right]}\left(\frac{\xi}{2}\right),$$

implies

$$\frac{\widehat{\widetilde{\phi}_h}(\xi)}{\widehat{\widetilde{\phi}_{h/2}}(\xi)} = \frac{C_S}{2^Z} \frac{P(e^{-i\eta/2})}{2^{Z+1}} \cdot \frac{\Phi_{2Z+1}[(h/2)\widetilde{\Lambda}](Z+1;e^{-i\eta/2})}{\Phi_{2Z+1}[h\widetilde{\Lambda}](Z+1;e^{-i\eta})}$$

$$= \frac{C_S}{2^{2Z+1}} P(e^{-i\eta/2}) \cdot \frac{\Phi_{2Z+1}[(h/2)\widetilde{\Lambda}](Z+1;e^{-i\eta/2})}{\Phi_{2Z+1}[h\widetilde{\Lambda}](Z+1;e^{-i\eta})}$$

$$= \frac{1}{2} A(z).$$

In a similar way we prove the second equality. We obtain

$$\widehat{\widetilde{\psi}_h}(\xi) = \left\{ C_S^2 \cdot \widehat{\psi_h[\Lambda;h]}(\xi) \right\} / \left\{ h^{2Z+1} \Phi_{2Z+1}[(h/2)\widetilde{\Lambda}](Z+1;-z) \right.$$

$$\left. \times \Phi_{2Z+1}[(h/2)\widetilde{\Lambda}](Z+1;z) \cdot \Phi_{2Z+1}[h\widetilde{\Lambda}](Z+1;\bar{z}^2) \right\},$$

where the constant C_S is given by

$$C_S := e^{(\lambda_1 + \cdots + \lambda_{Z+1})h/2},$$

and $z = e^{-i\xi h/2}$. By the definition of $\psi_h[\Lambda; h]$ in equality (17.20), p. 338, we have

$$\widehat{\psi_h[\Lambda; h]}(\xi) = R(e^{-i\xi h/2}) \cdot \widehat{Q_{Z+1}\left[\Lambda; \frac{h}{2}\right]}(\xi),$$

where $R(z)$ is given by formula (17.19), p. 338, namely

$$R(z) = -z^{2Z+1} P\left(-\frac{1}{z}\right) \Phi_{2Z+1}\left[\frac{h}{2}\widetilde{\Lambda}\right](Z+1; -z).$$

By the symmetry of Φ_{2Z+1} for symmetric nonordered vectors from (13.46), p. 256, we obtain

$$\Phi_{2Z+1}\left[\frac{h}{2}\widetilde{\Lambda}\right](Z+1; z) = \Phi_{2Z+1}\left[\frac{h}{2}\widetilde{\Lambda}\right](Z+1; \bar{z}) \quad \text{for all } |z|=1.$$

This implies

$$\widehat{\psi_h}(\xi)/\widehat{\phi_{h/2}}(\xi) = (\{C_S^2 R(e^{-i\xi h/2}) \cdot \widehat{Q_{Z+1}[\Lambda; h/2]}(\xi)\}$$

$$/\{h^{2Z+1}\Phi_{2Z+1}[(h/2)\widetilde{\Lambda}](Z+1; -\bar{z}) \cdot \Phi_{2Z+1}[(h/2)\widetilde{\Lambda}](Z+1; \bar{z})$$

$$\times \Phi_{2Z+1}[h\widetilde{\Lambda}](Z+1; z^2)\})$$

$$\div (\{C_S/h^{2Z+1}\} \cdot \{2^{2Z+1}\widehat{Q_{Z+1}[\Lambda; h/2]}(\xi)\}$$

$$/\{\Phi_{2Z+1}[(h/2)\widetilde{\Lambda}](Z+1; e^{i\xi h/2})\})$$

$$= (C_S/2^{2Z+1}) \cdot \{-z^{2Z+1} P(-1/z)\Phi_{2Z+1}[(h/2)\widetilde{\Lambda}](Z+1; -z)$$

$$\times \Phi_{2Z+1}[(h/2)\widetilde{\Lambda}](Z+1; \bar{z})\}$$

$$/\{\Phi_{2Z+1}[(h/2)\widetilde{\Lambda}](Z+1; -\bar{z}) \cdot \Phi_{2Z+1}[(h/2)\widetilde{\Lambda}](Z+1; \bar{z})$$

$$\times \Phi_{2Z+1}[h\widetilde{\Lambda}](Z+1; \bar{z}^2)\}$$

$$= -(C_S/2^{2Z+1}) \times \{z^{2Z+1}P(-1/z)\}/\{\Phi_{2Z+1}[h\widetilde{\Lambda}](Z+1; \bar{z}^2)\}$$

$$= \frac{1}{2}B(z).$$

The last follows from the exact solution (17.28), p. 353. ∎

17.11 Decomposition and reconstruction by L-spline wavelets and MRA

In the previous sections we have prepared all necessary instruments in order to provide decomposition and reconstruction relations for the MRA using the sets V_j. The summary for the polynomial case has been provided in Section 16.7, p. 321.

Cardinal L-spline wavelet analysis

Following the scheme of multiresolution analysis we consider an arbitrary "signal", i.e. a function $f \in L_2(\mathbb{R})$. For every $\varepsilon > 0$ we may find a sufficiently large N such that for some $f^N \in V_N$

$$\|f - f^N\|_{L_2(\mathbb{R})} = \sqrt{\int_{-\infty}^{\infty} |f(x) - f_N(x)|^2 \, dx} < \varepsilon.$$

holds. For that reason it is important to provide the "decomposition" of an arbitrary element of the space V_N.

We consider the *wavelet decomposition* [3, pp.11, 19]

$$f^N(x) = g^{N-1}(x) + f^{N-1}(x),$$

where g^{N-1} belongs to the wavelet space W_{N-1} and $f^{N-1} \in V_{N-1}$. They are orthogonal since for every $j \in \mathbb{Z}$ we have $V_j = V_{j-1} \oplus W_{j-1}$. However, we have some more problems which arise from the decomposition of the series.

Let us put $h = 1/2^{N-1}$, then $h/2 = 1/2^N$. Then the space V_N is by the equality in (17.4), p. 325, equal to the following:

$$V_N = \left\{ \sum_{j=-\infty}^{\infty} c_j^N Q_{Z+1}[\Lambda; 2^{-N}]\left(u - \frac{h}{2}j\right) : \text{ satisfies } \sum_{j=-\infty}^{\infty} |c_j^N|^2 < \infty \right\}.$$

Thus for some sequence $\{c_j^N\}_j \in \ell_2$ we have

$$f^N(u) = \sum_{j=-\infty}^{\infty} c_j^N Q_{Z+1}[\Lambda; 2^{-N}]\left(u - \frac{h}{2}j\right) \qquad (17.31)$$

and for the L_2-norm by formula (13.26), p. 243, we obtain

$$A \sum_{j=-\infty}^{\infty} |c_j^N|^2 \leq \left\| \sum_{j=-\infty}^{\infty} c_j^N Q_{Z+1}[\Lambda; 2^{-N}]\left(u - \frac{h}{2}j\right) \right\|_{L_2}^2 \leq B \sum_{j=-\infty}^{\infty} |c_j^N|^2 < \infty. \qquad (17.32)$$

Further, we will apply the *decomposition relation* in (17.29), p. 354, [3, p. 19; 4, p. 108], which has two main cases – odd shifts and even shifts $j(h/2)$ – and obtain

$$Q_{Z+1}[\Lambda; 2^{-N}](x) = \sum_{u=-\infty}^{\infty} a_{-2u} \cdot Q_{Z+1}[\Lambda; 2^{-N+1}](x - uh)$$

$$+ \sum_{u=-\infty}^{\infty} b_{-2u} \cdot \psi_{Z+1}^{N-1}(x - uh)$$

$$Q_{Z+1}[\Lambda; 2^{-N}]\left(x - \frac{h}{2}\right) = \sum_{u=-\infty}^{\infty} a_{1-2u} \cdot Q_{Z+1}[\Lambda; 2^{-N+1}](x - uh)$$

$$+ \sum_{u=-\infty}^{\infty} b_{1-2u} \cdot \psi_{Z+1}^{N-1}(x - uh)$$

or, if we use the unified notations and write for every shift $(h/2)\ell$ the equality, we obtain

$$Q_{Z+1}[\Lambda; 2^{-N}]\left(x - \frac{h}{2}\ell\right) = \sum_u a_{\ell-2u} \cdot Q_{Z+1}[\Lambda; 2^{-N+1}](x - uh)$$

$$+ \sum_{u=-\infty}^{\infty} b_{\ell-2u} \cdot \psi_{Z+1}^{N-1}(x - uh).$$

This provides us with the representation

$$f^N(x) = \sum_{j=-\infty}^{\infty} c_j^N Q_{Z+1}[\Lambda; 2^{-N}]\left(x - \frac{h}{2}j\right)$$

$$= \sum_{\ell=-\infty}^{\infty} c_\ell^N \left\{ \sum_u a_{\ell-2u} \cdot Q_{Z+1}[\Lambda; 2^{-N+1}](x - uh) \right.$$

$$\left. + \sum_u b_{\ell-2u} \cdot \psi_{Z+1}^{N-1}(x - uh) \right\}$$

$$= \sum_{\ell=-\infty}^{\infty} \sum_u c_\ell^N a_{\ell-2u} \cdot Q_{Z+1}[\Lambda; 2^{-N+1}](x - uh)$$

$$+ \sum_{\ell=-\infty}^{\infty} \sum_u c_\ell^N b_{\ell-2u} \cdot \psi_{Z+1}^{N-1}(x - uh).$$

Hence, if we put

$$f^{N-1}(x) = \sum_{\ell=-\infty}^{\infty} \sum_{u=-\infty}^{\infty} c_\ell^N a_{\ell-2u} \cdot Q_{Z+1}[\Lambda; 2^{-N+1}](x - uh), \tag{17.33}$$

$$g^{N-1}(x) = \sum_{\ell=-\infty}^{\infty} \sum_{u=-\infty}^{\infty} c_\ell^N b_{\ell-2u} \cdot \psi_{Z+1}^{N-1}(x - uh) Y_{k,l}(\theta), \tag{17.34}$$

there arises the question whether or not these series are convergent. In fact, by inequalities (17.32), p. 363, we have the following asymptotic equivalence:

$$\|f^{N-1}(x)\|^2 = \left\| \sum_{u=-\infty}^{\infty} \left(\sum_{\ell=-\infty}^{\infty} c_\ell^N a_{\ell-2u} \right) \cdot Q_{Z+1}[\Lambda; 2^{-N+1}](x - uh) \right\|_{L_2}^2$$

$$\asymp \sum_{u=-\infty}^{\infty} \left| \sum_{\ell=-\infty}^{\infty} c_\ell^N a_{\ell-2u} \right|^2$$

which is a heuristic suggesting the possible convergence of the series for f^{N-1}. For g^{N-1} we obviously have a similar estimate.

In the following theorem we compute the precise rate of convergence of the series in (17.33) and (17.34).

Theorem 17.31 *The series for $f^{N-1}(x)$ and $g^{N-1}(x)$ in (17.33) and (17.34) are convergent, i.e. $f^{N-1} \in V_{N-1}$ and $g^{N-1} \in W_{N-1}$.*

Proof We apply formula (13.26), p. 243, and change variables $(x/h) \longrightarrow x$, to obtain the equality

$$\|f^{N-1}(x)\|^2 = h^{2Z} \cdot \int_{-\infty}^{\infty} \left| \sum_{u=-\infty}^{\infty} \left(\sum_{\ell=-\infty}^{\infty} c_\ell^N a_{\ell-2u} \right) \cdot Q_{Z+1}[h\Lambda]\left(\frac{x}{h} - u\right) \right|^2 dx$$

$$= h^{2Z+1} \cdot \int_{-\infty}^{\infty} \left| \sum_{u=-\infty}^{\infty} \left(\sum_{\ell=-\infty}^{\infty} c_\ell^N a_{\ell-2u} \right) \cdot Q_{Z+1}[h\Lambda](x - u) \right|^2 dx.$$

We now have to prove the convergence of $f^{N-1}(x)$. We use the techniques in [4, p. 108; 3, p. 76]. We put

$$C(\xi) := C^N(\xi) := \sum_{u=-\infty}^{\infty} \left(\sum_{\ell=-\infty}^{\infty} c_\ell^N a_{\ell-2u} \right) e^{-iu\xi},$$

and using the Parseval theorem 12.7, p. 212, we obtain

$$\|f^{N-1}(x)\|^2 = \frac{h^{2Z+1}}{2\pi} \sum_{j=-\infty}^{\infty} \int_0^{2\pi} \left| C(\xi) \widehat{Q_{Z+1}[h\Lambda]}(\xi + 2\pi j) \right|^2 d\xi$$

$$= \frac{h^{2Z+1}}{2\pi} \int_0^{2\pi} |C(\xi)|^2 \cdot S(\xi) \, d\xi.$$

We have studied the sum

$$S(\xi) = \sum_{j=-\infty}^{\infty} |\widehat{Q_{Z+1}[h\Lambda]}(\xi + 2\pi j)|^2$$

many times. By Theorem 14.6, p. 272, it is equal to

$$S(\xi) = e^{-(\lambda_1 + \cdots + \lambda_{Z+1})h} \Phi_{2Z+1}[h\widetilde{\Lambda}](Z+1, e^{i\xi}),$$

where the nonordered vector $\widetilde{\Lambda}$ is the symmetrization of Λ.

Due to the symmetry of $\widetilde{\Lambda}$ the Laurent polynomial $\Phi_{2Z+1}[h\widetilde{\Lambda}](Z+1, z)$ is not zero on the unit circle $|z| = 1$, as studied in Corollary 13.57, p. 256. Hence, $\Phi_{2Z+1}[h\widetilde{\Lambda}](Z+1, z)$ is bounded by two constants A and B independent of ξ, i.e.

$$A \leq S(\xi) \leq B.$$

This implies

$$A \frac{h^{2Z+1}}{2\pi} \int_0^{2\pi} |C(\xi)|^2 \, d\xi \leq \|f^{N-1}(x)\|^2 \leq B \frac{h^{2Z+1}}{2\pi} \int_0^{2\pi} |C(\xi)|^2 \, d\xi.$$

366 Multivariate polysplines

In a similar way we obtain estimates for the norm $\|g^{N-1}(x)\|$. Actually, we are more interested in the estimate from above which would prove the convergence of the series representing f^{N-1} and g^{N-1}, respectively.

Further let us estimate $\int_0^{2\pi} |C(\xi)|^2 \, d\xi$. Here we apply the techniques of [5, p. 108]. For $z = e^{-i\xi/2}$, and recalling the definition of the function $A(z)$ in (17.26), p. 352, we obtain the equalities:

$$C(\xi) = \sum_{\ell=-\infty}^{\infty} c_\ell^N \left(\sum_{u=-\infty}^{\infty} a_{\ell-2u} e^{-iu\xi} \right)$$

$$= \sum_{\ell=-\infty}^{\infty} c_\ell^N \left(\sum_u a_{\ell-2u} z^{2u} \right)$$

$$= \sum_{\ell=-\infty}^{\infty} c_{2\ell}^N z^{2\ell} \left(\overline{A(z)} + \overline{A(-z)} \right) + \sum_{\ell=-\infty}^{\infty} c_{2\ell+1}^N z^{2\ell+1} \left(\overline{A(z)} - \overline{A(-z)} \right)$$

$$= \overline{A(z)} \sum_{\ell=-\infty}^{\infty} c_\ell^N z^\ell + \overline{A(-z)} \sum_{\ell=-\infty}^{\infty} c_\ell^N (-z)^\ell.$$

Let us put

$$\sigma(z) := \sigma^N(z) := \sum_{\ell=-\infty}^{\infty} c_\ell^N z^\ell.$$

This implies

$$\int_0^{2\pi} |C(\xi)|^2 d\xi \leq \int_0^{2\pi} |A(z)|^2 \cdot |\sigma(z)|^2 d\xi + \int_0^{2\pi} |A(-z)|^2 \cdot |\sigma(-z)|^2 d\xi$$

$$= \int_0^{4\pi} |A(z)|^2 \cdot |\sigma(z)|^2 d\xi.$$

We know by the Parseval formula for Fourier series (12.1), p. 210, that the norms of the series and the coefficients are equal, i.e.

$$\frac{1}{2\pi} \int_0^{2\pi} |C(\xi)|^2 \, d\xi = \|c\|_{\ell_2}^2 := \sum_{u=-\infty}^{\infty} \left| \sum_{\ell=-\infty}^{\infty} c_\ell^N a_{\ell-2u} \right|^2,$$

and similarly

$$\frac{1}{4\pi} \int_0^{4\pi} \left| \sigma(e^{-i\frac{\xi}{2}}) \right|^2 d\xi = \sum_{\ell=-\infty}^{\infty} |c_\ell^N|^2.$$

Thus it remains to see that the above estimate is uniform for all z with $|z| = 1$, i.e.

$$|A(z)| \leq C \quad \text{for } |z| = 1.$$

This implies that

$$\frac{1}{2\pi} \int_0^{2\pi} |C(\xi)|^2 \, d\xi \leq C_1 \sum_{\ell=-\infty}^{\infty} |c_\ell^N|^2,$$

for some constant $C_1 > 0$, the last sum being bounded as we saw in (17.31), p. 363, since the constants c_ℓ^N represent the function $f^N \in V_N$.

(Clearly, the estimate of the series for g^{N-1} is parallel. For its estimate we will need to prove that $B(z)$ is bounded.)

From formulas (17.27) and (17.28), p. 353, we have the exact solutions

$$A(z) = \frac{P(z)\Phi_{2Z+1}[(h/2)\widetilde{\Lambda}](Z+1;1/z)}{\Phi_{2Z+1}[h\widetilde{\Lambda}](Z+1;1/z^2)},$$

$$B(z) = -z^{2Z+1}\frac{P(-1/z)}{\Phi_{2Z+1}[h\widetilde{\Lambda}](Z+1;1/z^2)}.$$

Since all functions included are continuous and by Corollary 13.57, p. 256, the function $\Phi_{2Z+1}[h\widetilde{\Lambda}](Z+1;z)$ is not zero on the unit circle $|z|=1$, it follows that for some constant $C > 0$ we have

$$|A(z)| \leq C \quad \text{for all } |z| = 1,$$
$$|B(z)| \leq C \quad \text{for all } |z| = 1.$$

This ends the proof of the boundedness of the norm $\|f^{N-1}\|$. As we have already commented, the estimate of $\|g^{N-1}\|$ is obtained in a similar manner. ∎

The results obtained in Theorem 17.31 will be used later in the synthesis of polyharmonic wavelets where the set Λ will vary.

Let us now summarize all properties of the basic wavelet functions ψ_h.

Corollary 17.32 *Let us denote by $W_0 = V_1 \ominus V_0$. Then*

$$W_0 = \operatorname*{clos}_{L_2(\mathbb{R})}\{\psi_h(x - \ell h): \text{ for all } \ell \in \mathbb{Z}\}.$$

More generally,

$$W_j = \operatorname*{clos}_{L_2(\mathbb{R})}\{\psi_{h2^{-j}}(x - \ell h2^{-j}): \text{ for all } \ell \in \mathbb{Z}\}.$$

The following properties hold:

1. The function ψ_h generates a Riesz basis of the space W_0, and $\psi_{h2^{-j}}$ generates a Riesz basis of the space W_j.

2. The function ψ_h has a compact support which coincides with the interval $[0, 2(Z+1)h - 2h] = [0, 2Zh]$.

3. We have the orthogonal decomposition $V_1 = V_0 \oplus W_0$, hence $V_0 \cap W_0 = \{0\}$.

4. The wavelet ψ_h is orthogonal to V_0, i.e. all "exponential moments" are zero,

$$\int_{-\infty}^{\infty} R_j(x)e^{\lambda_j x}\psi(x)\,dx = 0 \quad \text{for } j = 1,\ldots,Z+1;$$

here R_j is a polynomial satisfying $\deg R_j \leq$ (multiplicity of λ_j) -1.

Point (1) follows from Lemma 17.17, p. 344; (2) follows by the construction; (3) is by definition; and (4) follows since $R_j(x)e^{\lambda_j x}$ is a solution of $\mathcal{L}[\Lambda](R_j(x)e^{\lambda_j x}) = 0$.

17.12 Discussion of the standard scheme of MRA

Let us analyze the basic scaling condition of the *multiresolution analysis (MRA)*. We recall some remarks about formula (17.5), p. 327.

1. It states that if the set of spaces V_j form an *MRA* of $L(\mathbb{R}^n)$ then

$$f(x) \text{ belongs to } V_0 \iff f(2x) \text{ belongs to } V_1. \tag{17.35}$$

If we assume that the functions or signals f to be analyzed represent a certain physical process, we will see that this condition hides two main properties.

The first one is that if $2x_0$ is a singularity of the signal $f(x)$ then x_0 is a singularity of the signal $f(2x)$. This is a more or less *acceptable restriction*, namely that the singularities of the space of functions V_{j+1} are those of V_j divided by 2. So if the singularities of V_0 are $\mathbb{Z}$ then those of V_1 are $\mathbb{Z}/2$.

The second important consequence of the above rule is that the physical sense of the signal $f(2x)$ is the same as the physical sense of the signal $f(x)$. It is quite clear that this is far from being the case for many signals of physical significance. In particular, we have seen in the study of L-splines that if $f(x)$ is a local solution to the ordinary differential equation $Lf = 0$ then, in general, $f(2x)$ is not! The only exception was the case of the usual polynomials when the operator had the form $L(d/dt) = d^p/dt^p$.

2. We also have the following fundamental consequence of the rule (17.35) which is provided in Meyer [14, p. 33, Theorem 4 and p. 38, Corollary]. It states that if the system $\{V_j\}_{j=-\infty}^{\infty}$ is an r-regular *MRA* of $L_2(\mathbb{R}^n)$ then the orthogonal projection

$$E_j : L_2(\mathbb{R}^n) \longrightarrow V_j$$

preserves the polynomials, i.e. $E_j(P) = P$ for every polynomial P of degree $\leq r$. Let us recall the definition of r-regular *MRA*, (see [11, Definition 1, p. 21]).

Definition 17.33 *An MRA* $\{V_j\}_{j=-\infty}^{\infty}$ *is r-regular if and only if the function $g \in V_0$ which generates a Riesz basis of V_0 can be chosen so that for every multi-index α with $|\alpha| \leq r$ and for every integer $N \geq 1$ there exists a constant $C_N > 0$ such that*

$$|D^\alpha g(x)| \leq C_N (1 + |x|)^{-N}$$

holds for all $x \in \mathbb{R}^n$.

By no means could this property be considered as an *acceptable restriction*. This fact shows that the polynomial character of the signals is presupposed in the very germ of the standard scheme of *MRA*. Working for example with L-polynomials (global solutions to the equation $Lf = 0$) shows that there are objects of a different kind from the polynomial which should not be disregarded. We have seen that the L-spline wavelets which have a definite physical significance need a more general framework of *MRA* in order to be incorporated in a proper setting. In particular, we would want conditions which imply that $E_j(P) = P$ for every P which is a solution of $LP = 0$ on the whole line $\mathbb{R}$. In the one-dimensional case this is the setting of the *prewavelets* of de Boor et al. [9]. We have explained the corresponding "non-stationary scaling" condition on p. 328.

We will see in the next Chapter that we may obtain a generalization of the scheme of multiresolution analysis by taking partial differential operators $L = \Delta^p$. Then the projection operators E_j will preserve the solutions to $\Delta^p u = 0$.

Chapter 18

Polyharmonic wavelet analysis: scaling and rotationally invariant spaces

In this chapter we receive the reward for our previous labours. We will fit the mosaic of one-dimensional cardinal L-spline wavelets together and construct multivariate **polyharmonic wavelets**.

We will specify the L-splines of Chapter 17 for the case of the operators $L = \mathcal{L}[\Lambda_k]$ and $L = \mathcal{L}[\Lambda_k + n/2]$, where for every integer $k \geq 0$ the nonordered vector

$$\Lambda_k = [\lambda_1, \ldots, \lambda_{2p}]$$

will be the one defined in (10.28), p. 170, arising from the *spherical operator* $L^p_{(k)}$ and $M_{k,p}$, considered in formula (10.26), p. 169. We put

$$\Lambda_k = \begin{cases} \lambda_1 = -n - k + 2, \\ \lambda_2 = -n - k + 4, \\ \ldots \ldots \\ \lambda_p = -n - k + 2p \\ \lambda_{p+1} = k, \\ \lambda_{p+2} = k + 2, \\ \ldots \ldots \\ \lambda_{2p} = k + 2p - 2. \end{cases} \quad (18.1)$$

We will actually pay more attention to the case of the "frequencies" $\lambda_j + n/2$ which will be symbolically denoted by $\Lambda_k + n/2$, i.e. we put for the nonordered vector

$$\Lambda_k + \frac{n}{2} := \left[\lambda_1 + \frac{n}{2}, \ldots, \lambda_{2p} + \frac{n}{2}\right].$$

Throughout this chapter the letter k will again be reserved for the index of the spherical harmonics $Y_{k,\ell}(\theta)$.

18.1 The refinement equation for the normed TB-spline $\widetilde{Q}_{Z+1}$

Now let us check how the refinement equation (17.8), p. 331, reacts if we consider the normed TB-spline $\widetilde{Q}_{Z+1}$ (defined in formula (14.14), p. 283) instead of the TB-spline Q_{Z+1}. The effect is a kind of asymptotic stability.

Let us recall that by Corollary 14.13 where the Riesz bounds A, B were considered for arbitrary mesh $h\mathbb{Z}$, we obtain the following proposition.

Proposition 18.1 *The Riesz bounds in Theorem 14.6, p. 273, for the set of functions $\{Q_{Z+1}[\Lambda;h](x-jh):$ for all $j \in \mathbb{Z}\}$, satisfy for $k \longrightarrow \infty$ the following asymptotic:*

$$A = O\left(\frac{e^{2pkh}}{k^{4p-1}}\right), \quad B = O\left(\frac{e^{2pkh}}{k^{4p-1}}\right).$$

Let us recall that using the basic asymptotics of the Riesz bounds in Theorem 14.6, p. 271, we have defined in (14.14), p. 283, the *normed TB-spline function* by putting

$$\boxed{\widetilde{Q}_{Z+1}[\Lambda;h](x) := \sqrt{\frac{k^{4p-1}}{e^{2pkh}}} \cdot Q_{Z+1}[\Lambda;h](x)} \qquad (18.2)$$

which will be further used instead of $Q_{Z+1}[\Lambda;h](x)$. From the *refinement equation* (17.8), p. 331, by multiplication with $\sqrt{k^{4p-1}/e^{2pkh}}$ we obtain the **refinement equation** for $\widetilde{Q}_{Z+1}$

$$\widetilde{Q}_{Z+1}[\Lambda;h](x) = \sqrt{\frac{k^{4p-1}}{e^{2pkh}}} \cdot Q_{Z+1}[\Lambda;h](x) \qquad (18.3)$$

$$= \sum_{\sigma=-\infty}^{\infty} \sqrt{\frac{1}{e^{pkh}}} p_{\sigma,h} \cdot \sqrt{\frac{k^7}{e^{pkh}}} \cdot Q_{Z+1}\left[\Lambda;\frac{h}{2}\right]\left(x - \sigma\frac{h}{2}\right)$$

$$= \sum_{\sigma=-\infty}^{\infty} \frac{1}{e^{pk\frac{h}{2}}} p_{\sigma,h} \cdot \widetilde{Q}_{Z+1}\left[\Lambda;\frac{h}{2}\right]\left(x - \sigma\frac{h}{2}\right).$$

We can make the important observation that the asymptotic of the coefficients of the polynomial defined in (17.9), p. 332, as $P_h(z) = \sum_{j=0}^{Z+1} p_{j,h} z^j$ and satisfying the refinement equation (17.8), p. 331, is given by Proposition 18.2, which is easy to prove.

Proposition 18.2 *For $k \longrightarrow \infty$ the coefficients $p_{j,h}$ satisfy the asymptotic*

$$p_{j,h} = O\left(e^{pk\frac{h}{2}}\right) \quad \text{for } k \longrightarrow \infty. \qquad (18.4)$$

In the refinement equation (18.3) we will put

$$\widetilde{p}_\sigma = \widetilde{p}_{\sigma,k} = \frac{p_{\sigma,k,h}}{e^{pk\frac{h}{2}}}, \qquad (18.5)$$

and we see that this corresponds to the asymptotic in (18.4).

18.2 Finding the way: some heuristics

Further, to explain the wavelet analysis using polysplines we specify for simplicity the case of *biharmonic* polysplines, i.e. we put $p = 2$. The essence of our explanation will remain valid for the more general case of an arbitrary integer $p \geq 1$.

We put $\Lambda_k := [\lambda_1, \lambda_2, \lambda_3, \lambda_4]$, with

$$\lambda_1 = -n - k + 2, \quad \lambda_2 = -n - k + 4, \quad \lambda_3 = k, \quad \lambda_4 = k + 2, \quad (18.6)$$

and by the definition of $\Lambda_k + n/2$ we have

$$\Lambda_k + \frac{n}{2} := \left[\lambda_1 + \frac{n}{2}, \lambda_2 + \frac{n}{2}, \lambda_3 + \frac{n}{2}, \lambda_4 + \frac{n}{2}\right].$$

For every $k = 0, 1, 2, \ldots$, we denote the spaces of the cardinal L-spline wavelets studied in Chapter 17 which correspond to the differential operator $\mathcal{L}[\Lambda_k + n/2]$ through

$$V_j^{(k)}, \quad W_j^{(k)} \quad \text{for all } j \in \mathbb{Z}. \quad (18.7)$$

Our construction of polyharmonic wavelets is based on a complete analogy with the one-dimensional case. In this case the MRA is constructed using the spaces V_j, where V_j is the set of all cardinal splines of degree p in $L_2(\mathbb{R})$ having knots at the points $\nu \cdot 2^{-j}$. In a similar way we wish to define the spaces PV_j (where P comes from *polyharmonic*) which will create a *polyharmonic MRA*.

Here we *suggest* a similar way for every $j \in \mathbb{Z}$ to define the space PV_j as the closure in $L_2(\mathbb{R}^n)$ of those polysplines in $L_2(\mathbb{R}^n)$, which have as break-surfaces the spheres of radii $e^{\nu \cdot 2^{-j}}$, i.e.

$$PV_j := \operatorname*{clos}_{L_2(\mathbb{R})} \{\text{all biharmonic polysplines } h \in L_2(\mathbb{R}^n) \cap C^2(\mathbb{R}^n)$$
$$\text{having break-surfaces } S(0; e^{\nu \cdot 2^{-j}}), \text{ for all } \nu \in \mathbb{Z}\}.$$

So far this definition is *not constructive*. We need a *constructive definition* as in the one-dimensional case where we used the shifts of the *scaling function* $\phi(x)$, see (16.7), p. 314, to obtain the sets V_j.[1] Evidently, we will have the inclusion

$$\cdots \subset PV_{-2} \subset PV_{-1} \subset PV_0 \subset PV_1 \subset PV_2 \subset \cdots.$$

The main *hint* to the above definition of the spaces PV_j is that after the change $\upsilon = \log r$ the above radii pass to $\log(e^{\nu \cdot 2^{-j}}) = \nu \cdot 2^{-j}$ which is the usual cardinal sequence when $\nu \in \mathbb{Z}$.

Now we fix some $j \in \mathbb{Z}$. We put $h = 2^{-j}$. We will use different notations for the variables which are transforms to x, namely $r\theta$, also (r, θ) or (υ, θ), where as usual we will assume that $\upsilon = \log r$.

Assume for simplicity that we have a polyspline $f \in PV_j$ which has a compact support. As we have seen in the fundamental Theorem 9.7, p. 124, about the expansion

---

[1] It is clear that we may take polysplines which belong either to a Sobolev smoothness class or to a Hölder smoothness class. One may prove rigorously that the space PV_j is independent of such a choice.

of polysplines on annuli in spherical harmonics, f is expanded in a Fourier–Laplace series

$$f(x) = \sum_{k=0}^{\infty} \sum_{\ell=1}^{d_k} f^{k,\ell}(\log r) Y_{k,\ell}(\theta),$$

where $f^{k,\ell}(v)$ is a cardinal L-spline for the operator $L = \mathcal{L}[\Lambda_k]$ with knots at $2^{-j} \cdot \mathbb{Z}$. For every $k = 0, 1, 2, \ldots$, the spherical harmonics $\{Y_{k,\ell}(\theta)\}_{\ell=1}^{d_k}$ form an orthonormal basis of the set $\mathcal{H}_k$, cf. Section 10.10, p. 149.

Further, we will use as **scaling functions** the shifts of the normed TB-spline $\tilde{Q}_{Z+1}[\Lambda_k + n/2; 2^{-j}](v)$ defined above in equality (18.2), p. 372. For the sake of brevity and in order to keep to the tradition in wavelet analysis we will put

$$\phi^k_{2^{-j}}(v) := \tilde{Q}_{Z+1}\left[\Lambda_k + \frac{n}{2}; 2^{-j}\right](v). \tag{18.8}$$

Thus for $p = 2$ we have $\tilde{Q}_4[\Lambda_k + n/2; 2^{-j}](v)$.

After changing variables $v = \log r$, we obtain the expansion of the norm of f, by

$$\|f\|^2_{L_2(\mathbb{R}^n)} = \sum_{k=0}^{\infty} \sum_{\ell=1}^{d_k} \int_0^{\infty} |f^{k,\ell}(\log r)|^2 r^{n-1} \, dr \tag{18.9}$$

$$= \sum_{k=0}^{\infty} \sum_{\ell=1}^{d_k} \int_{-\infty}^{\infty} |f^{k,\ell}(v)|^2 e^{nv} \, dv$$

$$= \sum_{k=0}^{\infty} \sum_{\ell=1}^{d_k} \int_{-\infty}^{\infty} |f^{k,\ell}(v) e^{\frac{n}{2}v}|^2 \, dv.$$

This shows that $f^{k,\ell}(v)e^{(n/2)v}$ is a cardinal L-spline for the operator $\mathcal{L}[\Lambda_k + n/2]$ with knots at $2^{-j} \cdot \mathbb{Z}$, which is in $L_2(\mathbb{R})$.

Now recall that we have defined the normed TB-spline $\phi^k_{2^{-j}}(v)$ in formula (18.2), p. 372. Since $f^{k,\ell}(v)e^{(n/2)v}$ has a compact support, we have the following representation:

$$f^{k,\ell}(v)e^{(n/2)v} = \sum_{\rho=-\infty}^{\infty} c^{k,\ell}_{\rho} \phi^k_{2^{-j}}(v - \rho 2^{-j}),$$

where only a finite number of the coefficients $\{c^{k,\ell}_{\rho}\}_{\rho}$ are nonzero. If we take into account that $v = \log r$, and $x = r\theta$, this gives the equalities

$$f(x) = e^{-(n/2)v} \sum_{k=0}^{\infty} \sum_{\ell=1}^{d_k} \sum_{\rho=-\infty}^{\infty} c^{k,\ell}_{\rho} \phi^k_{2^{-j}}(v - \rho 2^{-j}) Y_{k,\ell}(\theta). \tag{18.10}$$

Due to the fact that the shifts $\{\phi^k_{2^{-j}}(v - \rho 2^{-j}):$ for all $\rho \in \mathbb{Z}\}$ form a Riesz basis we have the Riesz inequalities

$$A \cdot \sum_{\rho=-\infty}^{\infty} \left|c^{k,\ell}_{\rho}\right|^2 \leq \int_{-\infty}^{\infty} |f^{k,\ell}(v)e^{(n/2)v}|^2 \, dv \leq B \cdot \sum_{\rho=-\infty}^{\infty} |c^{k,\ell}_{\rho}|^2, \tag{18.11}$$

where A, B are here independent of k thanks to the norming in (18.2), p. 372.

After summing in k and ℓ we obtain the inequalities

$$A \cdot \sum_{k=0}^{\infty} \sum_{\ell=1}^{d_k} \sum_{\rho=-\infty}^{\infty} |c_\rho^{k,\ell}|^2 \leq \|f\|_{L_2(\mathbb{R}^n)}^2 \leq B \cdot \sum_{k=0}^{\infty} \sum_{\ell=1}^{d_k} \sum_{\rho=-\infty}^{\infty} |c_\rho^{k,\ell}|^2. \qquad (18.12)$$

Since the space PV_j is complete these inequalities prove that it is isomorphic to the space of sequences $\{c_\rho^{k,\ell}\}$ which are in ℓ_2 with respect to all indexes k, ℓ, ρ. Indeed, let us choose a three-indexed sequence $c_\rho^{k,\ell}$ as above and such that

$$\sum_{k=0}^{\infty} \sum_{\ell=1}^{d_k} \sum_{\rho=-\infty}^{\infty} |c_\rho^{k,\ell}|^2 < \infty.$$

Then reasoning backwards we construct the function $f(x)$ above, and $f \in PV_j$.

18.3 The sets PV_j and isomorphisms

The above reasoning hints at how to define the spaces PV_j of what we will term **polyharmonic (spherical) multiresolution analysis**.

First, we introduce the following space of sequences. We denote by $\ell_2^{(3)}$ the space of three-indexed sequences by putting

$$\ell_2^{(3)} := \left\{ \{c_\rho^{k,\ell}\} : \text{ satisfies } \sum_{k=0}^{\infty} \sum_{\ell=1}^{d_k} \sum_{\rho=-\infty}^{\infty} |c_\rho^{k,\ell}|^2 < \infty \right\}.$$

Having in mind the notations $v = \log r$, and $x = r\theta$, for every integer $j \in \mathbb{Z}$ we give the following definition.

Definition 18.3 *For every integer $j \in \mathbb{Z}$ the spaces PV_j of the polyharmonic (spherical) MRA are defined by*

$$PV_j := \operatorname*{clos}_{L_2(\mathbb{R})} \left\{ \text{all functions } f(x) = \sum_{k=0}^{\infty} \sum_{\ell=1}^{d_k} f^{k,\ell}(\log r) Y_{k,\ell}(\theta) \quad \text{such that} \right.$$

$$\left. f^{k,\ell}(v) e^{(n/2)v} = \sum_{\rho=-\infty}^{\infty} c_\rho^{k,\ell} \phi_{2^{-j}}^k(v - \rho 2^{-j}) \quad \text{with } \{c_\rho^{k,\ell}\} \in \ell_2^{(3)} \right\}.$$

(18.13)

The special case $j = 0$ is of polysplines with knot-surfaces that are the spheres $S(0; e^v)$. Let us remark that, in particular, the set PV_0 contains the cardinal polysplines f of order p having derivatives of order C^{2p-2}. In the case of biharmonic polysplines, when $p = 2$, we have $f \in C^2(\mathbb{R}^n)$, in particular, $f_{|S(0;e^v)} \in C^2(S(0; e^v))$.

Remark 18.4 *1. The regularity of the elements of PV_j is not necessary, especially on the spheres $S(0; e^{v2^{-j}})$. This will not be essential for the study of the wavelet analysis which we carry out in the sequel.*

2. In Definition 18.3 we may evidently assume that the coefficients $c_\rho^{k,\ell}$ are nonzero only for finitely many values of k and ρ. The corresponding function $f(x)$ is obviously a polyspline.

In the case of a sequence $\{c_\rho^{k,\ell}\} \in \ell_2^{(3)}$ we find that due to the above inequalities the functions

$$g^{k,\ell}(v) = \sum_{\rho=-\infty}^{\infty} c_\rho^{k,\ell} \phi_{2-j}^k(v - \rho 2^{-j})$$

satisfy $g^{k,\ell}(v) \in L_2(\mathbb{R})$ and are limits of $\mathcal{L}[\Lambda_k + n/2]$-splines on the mesh $2^{-j}\mathbb{Z}$. We put

$$f^{k,\ell}(v) = e^{-(n/2)v} g^{k,\ell}(v).$$

It is clear that $f^{k,\ell}(v)$ is a limit of $\mathcal{L}[\Lambda_k]$-splines, and the series

$$f(x) = \sum_{k=0}^{\infty} \sum_{\ell=1}^{d_k} f^{k,\ell}(\log r) Y_{k,\ell}(\theta)$$

is convergent in $L_2(\mathbb{R}^n)$ by the above inequalities. Thus we have shown the isomorphism between the space of sequences $\ell_2^{(3)}$ and the space PV_j through the map

$$\ell_2^{(3)} \ni \{c_\rho^{k,\ell}\} \longmapsto f \in PV_j.$$

Recall that in (18.7), p. 373, we have denoted by $\left\{V_j^{(k)}\right\}_j$ the MRA corresponding to the operator $L = \mathcal{L}[\Lambda_k + n/2]$. The above also shows the isomorphism between the space of sequences of functions[2]

$$\left[\bigoplus_{k=0}^{\infty} \bigoplus_{\ell=1}^{d_k}\right]' \{V_j^{(k)}\}$$

$$:= \left\{\{g^{k,\ell}(v)\}: g^{k,\ell}(v) \in V_j^{(k)} \text{ satisfies } \sum_{k=0}^{\infty} \sum_{\ell=1}^{d_k} \int_{-\infty}^{\infty} |g^{k,\ell}(v)|^2 e^{nv}\, dv < \infty\right\},$$

and the space PV_j through the map

$$\{g^{k,\ell}(v)\}_{k,\ell} \longmapsto f(x).$$

Thus we have proved Theorem 18.5, which we now state.

Theorem 18.5 *The following isomorphism holds:*

$$PV_0 \approx \left[\bigoplus_{k=0}^{\infty} \bigoplus_{\ell=1}^{d_k}\right]' V_0^{(k)},$$

where in $\bigoplus_{\ell=1}^{d_k} V_0^{(k)}$ we have taken d_k copies of the space $V_0^{(k)}$.

---

[2] The prime in the notation $[\,]'$ means that only a part of all elements is taken.

The above evidently holds for every $j \in \mathbb{Z}$, i.e.

$$PV_j \approx \left[\bigoplus_{k=0}^{\infty}\right]' \bigoplus_{\ell=1}^{d_k} V_j^{(k)}.$$

Obviously the above isomorphisms are analogs to the classical one where the spaces V_j are isomorphic to ℓ_2.

Now as may be guessed by the reader let us put for the wavelet spaces

$$PW_{j-1} := PV_j \ominus PV_{j-1}.$$

Our main purpose in Section 18.6 will be to prove that in a similar way we have such a decomposition for the **polyharmonic wavelet spaces** PW_j, i.e.

$$PW_j = \left[\bigoplus_{k=0}^{\infty} \bigoplus_{\ell=1}^{d_k}\right]' W_j^{(k)} \quad \text{for all } j \in \mathbb{Z},$$

where $W_j^{(k)}$ denotes the space of L-spline wavelets for the operators $\mathcal{L}[\Lambda_k + (n/2)]$ with knots at $\{v \cdot 2^{-j} : \text{ for all } v \in \mathbb{Z}\}$ which we introduced in (18.7), p. 373.

18.4 Spherical Riesz basis and father wavelet

Now let us make some remarks about the possible interpretations of the above results and their direct analogy with the one-dimensional case, namely we speculate about the notion of Riesz basis.

First, let us rewrite for every $j \in \mathbb{Z}$ the above sum (18.10), p. 374, in the form

$$f(x) = e^{-(n/2)v} \sum_{\rho=-\infty}^{\infty} \left(\sum_{k=0}^{\infty} \sum_{\ell=1}^{d_k} c_\rho^{k,\ell} Y_{k,\ell}(\theta) \right) \phi_{2^{-j}}^{k}(v - \rho 2^{-j}).$$

For every $j \in \mathbb{Z}$ we have the function (the **father wavelet**)

$$\phi_{2^{-j}}^{k,\ell}(v, \theta) := \phi_{2^{-j}}^{k}(v) Y_{k,\ell}(\theta) = \widetilde{Q}\left[\Lambda_k + \frac{n}{2}; 2^{-j}\right](v) Y_{k,\ell}(\theta), \quad (18.14)$$

which depends not only on the variables (v, θ) but also on the discrete parameters k, ℓ. We see that for every sequence $\{c_\rho^{k,\ell}\} \in \ell_2^{(3)}$ if we define the function

$$g(x) = \sum_{k=-\infty}^{\infty} \sum_{\ell=1}^{d_k} \sum_{\rho=-\infty}^{\infty} c_\rho^{k,\ell} \phi_{2^{-j}}^{k,\ell}(v - \rho 2^{-j}, \theta)$$

then

$$g \in PV_j,$$

and the Riesz-type inequalities (18.12), p. 375, hold.

Remark 18.6 *Thus in a certain sense the function*

$$\phi_{2^{-j}}^{k,\ell}(v,\theta)$$

is a unique generator or "father wavelet" for PV_j for all $j \in \mathbb{Z}$. So far the relation among these "father wavelets" for different j's is non-linear. In order to be more precise about this non-linearity, let us recall that by formula (13.26), p. 243, and the normalization in (18.2), p. 372, we obtain the representation

$$\phi_{2^{-j}}^{k,\ell}(v,\theta) = \tilde{Q}\left[\Lambda_k + \frac{n}{2}; 2^{-j}\right](v) Y_{k,\ell}(\theta)$$

$$= \sqrt{\frac{k^{4p-1}}{e^{2pk/2^j}}} Q\left[\Lambda_k + \frac{n}{2}; 2^{-j}\right](v) Y_{k,\ell}(\theta)$$

$$= \sqrt{\frac{k^{4p-1}}{e^{2pk/2^j}}} \left(\frac{1}{2^j}\right)^Z Q\left[2^{-j}\left(\Lambda_k + \frac{n}{2}\right)\right](2^j v) Y_{k,\ell}(\theta).$$

The last shows the nature of the interdependence between the parameters k and j, especially in the T B-spline $Q[\cdot]$.

By the following Definition we will put the above basis property of the functions $\phi_{2^{-j}}^{k,\ell}(v,\theta)$ in a suitable framework which resembles the classical Riesz basis property.

Definition 18.7 *We say that the function $g(k,\ell;v,\theta)$ generates a **spherical Riesz basis** for the scale H if and only if there exist two constants A, B with $0 < A \leq B < \infty$, and such that for every sequence $\left\{c_\rho^{k,\ell}\right\} \in \ell_2^{(3)}$, i.e. satisfying*

$$\|c\|_{\ell_2^{(3)}}^2 = \sum_{k=-\infty}^{\infty} \sum_{\ell=1}^{d_k} \sum_{\rho=-\infty}^{\infty} \left|c_\rho^{k,\ell}\right|^2 < \infty$$

we have the inequalities

$$A \|c\|_{\ell_2^{(3)}}^2 \leq \left\|e^{-\frac{n}{2}v} G(x)\right\|_{L_2(\mathbb{R}^n)}^2 \leq B \|c\|_{\ell_2^{(3)}}^2,$$

where

$$G(x) = \sum_{k=-\infty}^{\infty} \sum_{\ell=1}^{d_k} \sum_{\rho=-\infty}^{\infty} c_\rho^{k,\ell} g(k,\ell; v - \rho H, \theta).$$

Thus the function $g(k,\ell;v,\theta) = \phi_{2^{-j}}^{k,\ell}(v,\theta)$ is a spherical Riesz basis for the scale $H = 2^{-j}$.

We might consider the analogy from the point of view of *spherical transform*, i.e.

$$\widehat{f}(k,\ell,v) = \sum_{\nu=-\infty}^{\infty} f_\nu^{k,\ell} \phi_{2^{-j}}^k (v - \nu 2^{-j}),$$

hence,
$$f(x) = \sum_{k=0}^{\infty} \sum_{\ell=1}^{d_k} \widehat{f}(k, \ell, v) Y_{k,\ell}(\theta).$$

We see that this is the analogy with the one-dimensional **Riesz basis property**.

18.5 Polyharmonic MRA

We will follow the analogy with the classical *multiresolution analysis* scheme, presented by Meyer [14, p. 21].

Let us fix some integer $p \geq 1$. For simplicity we can take the biharmonic case $p = 2$. We have introduced the sets PV_j of the *spherical MRA* and have the following basic

Theorem 18.8 *The approximation property holds*
$$\operatorname*{clos}_{L_2(\mathbb{R})} \left\{ \bigcup_{j=-\infty}^{\infty} PV_j \right\} = L_2(\mathbb{R}^n).$$

Proof (1) A classical fact is that the space of functions with compact support which are infinitely differentiable, usually denoted by $C_0^{\infty}(\mathbb{R}^n)$, is a dense subset of $L_2(\mathbb{R}^n)$ in the metric of the last, i.e.
$$\operatorname*{clos}_{L_2(\mathbb{R}^n)} (C_0^{\infty}(\mathbb{R}^n)) = L_2(\mathbb{R}^n),$$

see Adams [1, Theorem 2.13]. Hence, it suffices to prove that every function $f \in C_0^{\infty}(\mathbb{R}^n)$ is approximable through elements of $\bigcup_{j=-\infty}^{\infty} PV_j$.

(2) Assume that $\operatorname{supp}(f) \subset B(0; R)$ for some $R > 0$. Let some small number $\varepsilon > 0$ be given. Here we will cut the singularity around the origin.

Evidently, there exists some sufficiently small $\delta > 0$ and a function $f_1 \in C_0^{\infty}(\mathbb{R}^n)$ which is defined in the ball $B(0; 2\delta)$ such that it satisfies the conditions $\operatorname{supp}(f_1) \subset B(0; 2\delta)$, and
$$\|f_1\|_{L_2(\mathbb{R}^n)} < \frac{\varepsilon}{3} \quad \text{and} \quad f_1(x) = f(x) \quad \text{for } x \in B(0; \delta).$$

We put
$$F = f - f_1.$$

It follows that $F \in C_0^{\infty}(\mathbb{R}^n)$, and consequently $F \in L_2(\mathbb{R}^n)$.

(3) For the function F we have the expansion in spherical harmonics [17, Chapter 4, Lemma 2.18][3]
$$F(x) = \sum_{k=0}^{\infty} \sum_{\ell=1}^{d_k} F^{k,\ell}(\log r) Y_{k,\ell}(\theta).$$

---

[3] In fact we need to add more to the result in [17]. It is important that the function F has a compact support! Then for every $R > 0$ the system of functions $f(r)Y_{k,\ell}(\theta)$ generates an orthonormal basis for the space $L_2(B(0; R))$. Indeed, by the Gauss representation in Theorem 10.2, p. 141, we know that every polynomial is decomposed into a sum of such functions, and the polynomials are dense in $L_2(B(0; R))$.

Due to the choice of the function f_1 we see that $F^{k,\ell}(v)$ are functions in $C_0^\infty(\mathbb{R})$. As we have seen above in (18.9), p. 374,

$$\|F\|_{L_2(\mathbb{R}^n)}^2 = \sum_{k=0}^\infty \sum_{\ell=1}^{d_k} \int_{-\infty}^\infty |F^{k,\ell}(v)|^2 e^{nv} \, dv < \infty.$$

(4) However, it is evident that there exists a sufficiently large integer k_1 such that

$$I_1^2 := \sum_{k=k_1+1}^\infty \sum_{\ell=1}^{d_k} \int_{-\infty}^\infty |F^{k,\ell}(v)|^2 e^{nv} \, dv < \left(\frac{\varepsilon}{3}\right)^2.$$

So we only have to work with a finite number of functions $F^{k,\ell}(v)$ for $k = 0, \ldots, k_1$ and $\ell = 1, \ldots, d_k$. Due to Theorem 17.3, p. 329, (the approximation property of all splines with refining knot-sets) every such function $F^{k,\ell}(v)$ may be approximated by a compactly supported L-spline $h^{k,\ell}(v)$ with operator $L = M_{k,p}$ with knots on the grid $2^{-j}\mathbb{Z} = \{v2^{-j} : \text{for all } v \in \mathbb{Z}\}$ for sufficiently large $j \in \mathbb{Z}$. We may choose j the same for all $k = 0, \ldots, k_1$ and $\ell = 1, \ldots, d_k$. (Recall that $M_{k,p}$ is the spherical operator defined in (10.26), p. 169.) Thus we take all these $h^{k,\ell}(v)$, for $k = 0, \ldots, k_1$ and $\ell = 1, \ldots, d_k$, in such a way that they satisfy the inequality

$$I_2^2 := \sum_{k=0}^{k_1} \sum_{\ell=1}^{d_k} \int_{-\infty}^\infty |F^{k,\ell}(v) - h^{k,\ell}(v)|^2 e^{nv} \, dv < \left(\frac{\varepsilon}{3}\right)^2.$$

(Note that we can also choose all functions $h^{k,\ell}$ compactly supported due to the basic property of the spaces $V_j^{(k)}$ in (17.4), p. 327.) We put

$$h_\varepsilon(x) = \sum_{k=0}^{k_1} \sum_{\ell=1}^{d_k} h^{k,\ell}(v) Y_{k,\ell}(\theta).$$

Obviously,

$$I_2^2 = \left\| \sum_{k=0}^{k_1} \sum_{\ell=1}^{d_k} F^{k,\ell}(v) Y_{k,\ell}(\theta) - h_\varepsilon(x) \right\|_{L_2(\mathbb{R}^n)}^2.$$

By the above we see that

$$\|f - h_\varepsilon\|_{L_2(\mathbb{R}^n)} \leq \|F - h_\varepsilon\|_{L_2(\mathbb{R}^n)} + \|f_1\|_{L_2(\mathbb{R}^n)}$$
$$< I_1 + I_2 + \frac{\varepsilon}{3} < \varepsilon.$$

This ends the proof. ■

We now summarize the results proved above. We see that in a certain sense the sets PV_j satisfy the axioms of an MRA which we have prematurely called an *polyharmonic (spherical) multiresolution analysis*.

Theorem 18.9 *The sequence of spaces* PV_j, $j \in \mathbb{Z}$, *has the following properties:*
It is a strictly increasing sequence, i.e.

$$\ldots \subset PV_{-1} \subset PV_0 \subset PV_1 \subset PV_2 \subset \ldots,$$

and it approximates $L_2(\mathbb{R}^n)$, *i.e. hold*

1a.
$$\bigcup_{j=-\infty}^{\infty} PV_j \quad \text{is dense in } L_2(\mathbb{R}^n)$$

and

1b.
$$\bigcap_{j=-\infty}^{\infty} PV_j = \{0\}.$$

2. **Scaling invariance**: *For every* $f \in L_2(\mathbb{R}^n)$ *and every* $\ell \in \mathbb{Z}$ *holds*

$$f(x) \text{ belongs to } PV_0 \iff f\left(x \cdot e^\ell\right) \text{ belongs to } PV_0;$$

3. **Rotational invariance**[4]: *Let U be an orthogonal transform of* $\mathbb{R}^n$. *Then*

$$f(x) \text{ belongs to } PV_0 \iff f(Ux) \text{ belongs to } PV_0;$$

4. **Father wavelet**: *There is an element* $\phi_1^{k,\ell}(v,\theta)$ *such that for every sequence* $\left\{c_\rho^{k,\ell}\right\} \in \ell_2^{(3)}$ *the functions of the form*

$$f(x) = \sum_{\rho=-\infty}^{\infty} \sum_{k=0}^{\infty} \sum_{\ell=1}^{d_k} c_\rho^{k,\ell} \phi_1^{k,\ell}(v-\rho, \theta)$$

are in PV_0 *and exhaust* PV_0.

5. **Refinement operator – Polyharmonic Scaling Operator**: *There is a canonical linear "polyharmonic scaling" operator which provides a way to express every* $f \in PV_j$ *through elements of the space* PV_{j+1}, *i.e.*

$$f(x) = \sum_{\nu=-\infty}^{\infty} M_\nu g\left(re^{-\frac{\nu}{2}h}\theta\right).$$

Here $g \in PV_{j+1}$ and the operators M_ν are defined in a canonical way below in formula (18.14a), p. 383, while the sum on ν is finite. The function g is given by the "polyharmonic scaling" of f, i.e.

$$g(x) = f_{\text{sc}}(x),$$

which is defined in formula (18.14b), p. 383, below.

---

[4] More correct is to say "invariance with respect to orthogonal transforms" as is proved.

382   *Multivariate polysplines*

6. **Riesz inequality**: There is a Riesz type inequality for the elements of PV_j, namely (18.12), p. 375.

Proof (1a) This has been proved in Theorem 18.8, p. 379.
(1b) According to Theorem 18.5, p. 376, we have the isomorphisms

$$PV_j \approx \left[\bigoplus_{k=0}^{\infty}\right]' \bigoplus_{\ell=1}^{d_k} V_j^{(k)}.$$

Taking into account the Riesz type estimates (18.12), p. 375, which were ground for the proof of these isomorphisms we obtain

$$\bigcap_{j=-\infty}^{\infty} PV_j \approx \left[\bigoplus_{k=0}^{\infty}\right]' \bigoplus_{\ell=1}^{d_k} \bigcap_{j=-\infty}^{\infty} V_j^{(k)}.$$

Now we apply the one-dimensional version of our result, which is Proposition 17.2, p. 329. Recall that the space $V_j^{(k)}$ is defined with respect to the operator $\mathcal{L}[\Lambda_k + n/2]$. It follows by Proposition 17.2 that $\bigoplus_{\ell=1}^{d_k} \bigcap_{j=-\infty}^{\infty} V_j^{(k)} = \{0\}$ since among the elements of the vector $\Lambda_k + n/2$ there is always at least one nonnegative element.

(2) This point is evident since the function $f(x)$ is polyharmonic of order p if and only if $f(Cx)$ is polyharmonic of order p for all $C \neq 0$.

(3) This is evident since the function $Y_{k,\ell}(U\theta)$ is again a spherical harmonic of degree k, which follows from Proposition 10.17, p. 154.

(4) We have seen above in (18.14), p. 377, that this is the element

$$\phi_1^{k,\ell}(v,\theta) = \widetilde{Q}\left[\Lambda_k + \frac{n}{2}; 1\right](v) Y_{k,\ell}(\theta)$$

$$= \phi_1^k(v) Y_{k,\ell}(\theta),$$

and for PV_j we thus have the element

$$\phi_1^{k,\ell}(v,\theta) := \widetilde{Q}\left[\Lambda_k + \frac{n}{2}; 2^{-j}\right](v) Y_{k,\ell}(\theta) = \phi_{2-j}^k(v) Y_{k,\ell}(\theta).$$

(5) Here we have to see the nonstationary scaling as in the L-spline wavelets, see formula (17.6), p. 328. Indeed, by the refinement equation (17.8), p. 331, and by the refinement equation for $\widetilde{Q}$, (18.3), p. 372, we have

$$\phi_h^k(v) = \widetilde{Q}\left[\Lambda_k + \frac{n}{2}; h\right](v) = \sum_{\nu=-\infty}^{\infty} \widetilde{p}_{\nu,k} \widetilde{Q}\left[\Lambda_k + \frac{n}{2}; \frac{h}{2}\right]\left(v - \nu\frac{h}{2}\right)$$

$$= \sum_{\nu=-\infty}^{\infty} \widetilde{p}_\nu \phi_{h/2}^k\left(v - \nu\frac{h}{2}\right),$$

where the coefficients $\widetilde{p}_\nu$ have been defined by us in (18.5), p. 372. Note that the coefficients $\widetilde{p}_\nu$ also depend on k and h, as defined in formula (17.9), p. 332,

$$P_h(z) = s\left[\frac{h}{2}\left(\Lambda_k + \frac{n}{2}\right)\right](-z) = \prod_{j=1}^{Z+1}\left(e^{-(\lambda_j+(n/2))h/2} + z\right) = \sum_{j=0}^{Z+1} p_j z^j.$$

Let us put $h = 2^{-j}$ and introduce the notation $f(x) = f(r\theta) = f(v, \theta)$. Now by the definition of PV_j in (18.13), p. 375, for arbitrary $f \in PV_j$ we obtain the equalities

$$f(x) = e^{-(n/2)v} \sum_{k=0}^{\infty} \sum_{\ell=1}^{d_k} \sum_{\rho=-\infty}^{\infty} c_\rho^{k,\ell} \phi_h^k(v - \rho h) Y_{k,\ell}(\theta)$$

$$= e^{-(n/2)v} \sum_{k=0}^{\infty} \sum_{\ell=1}^{d_k} \sum_{\rho=-\infty}^{\infty} c_\rho^{k,\ell} \left(\sum_{v=-\infty}^{\infty} \widetilde{p}_{v,k} \phi_{h/2}^k \left(v - v\frac{h}{2} - \rho h \right) \right) Y_{k,\ell}(\theta)$$

$$= e^{-(n/2)v} \sum_{v=-\infty}^{\infty} \sum_{k=0}^{\infty} \widetilde{p}_{v,k} \sum_{\ell=1}^{d_k} \sum_{\rho=-\infty}^{\infty} c_\rho^{k,\ell} \phi_{h/2}^k \left(v - v\frac{h}{2} - \rho h \right) Y_{k,\ell}(\theta)$$

$$= \sum_{v=-\infty}^{\infty} M_v g \left(v - v\frac{h}{2}, \theta \right)$$

$$= \sum_{v=-\infty}^{\infty} M_v g(r e^{-(v/2)h} \theta),$$

where we have put

$$g(x) = e^{-(n/2)v} \sum_{k=0}^{\infty} \sum_{\ell=1}^{d_k} \sum_{\rho=-\infty}^{\infty} c_\rho^{k,\ell} \phi_{h/2}^k(v)(v - \rho h) Y_{k,\ell}(\theta),$$

and M_v is the operator which acts on an arbitrary function G as follows: if the function G has the expansion in spherical harmonics

$$G(x) = \sum_{k=0}^{\infty} \sum_{\ell=1}^{d_k} G^{k,\ell}(v) Y_{k,\ell}(\theta)$$

then

$$M_v [G] (x) = \sum_{k=0}^{\infty} \widetilde{p}_{v,k} \sum_{\ell=1}^{d_k} G^{k,\ell}(v) Y_{k,\ell}(\theta). \tag{18.14a}$$

Obviously, by the definition of the space PV_{j+1} in (18.13), p. 375 we obtain $g \in PV_{j+1}$. ∎

Remark 18.10 *Clearly, we may define now the **polyharmonic scaling operator** by putting*

$$F_{\mathrm{sc}}(x) := e^{-\frac{n}{2}v} \sum_{k=0}^{\infty} \sum_{\ell=1}^{d_k} \sum_{\rho=-\infty}^{\infty} c_\rho^{k,\ell} \phi_{\frac{h}{2}}^k (v - \rho h) Y_{k,\ell}(\theta), \tag{18.14b}$$

where the function $F(x)$ is given by

$$F(x) := e^{-\frac{n}{2}v} \sum_{k=0}^{\infty} \sum_{\ell=1}^{d_k} \sum_{\rho=-\infty}^{\infty} c_\rho^{k,\ell} \phi_h^k (v - \rho h) Y_{k,\ell}(\theta) = \sum_{k=0}^{\infty} \sum_{\ell=1}^{d_k} F^{k,\ell}(v) Y_{k,\ell}(\theta).$$

Thus we have
$$g(x) = f_{sc}(x).$$

Recall now the one-dimensional scaling operators for the L-spline wavelet analysis, which we have defined by equality (17.6), p. 328. We see that every component $F^{k,\ell}(v)$ is subject to the non-stationary scaling from formula (17.6), p. 328.

Remark 18.11 *Let us note that property 2), the "scaling invariance", corresponds in fact to the shift invariance of the standard wavelet analysis. Thus we have a group of symmetries for the spaces PV_j which is isomorphic (by the exponential map) to the additive group $\mathbb{Z}$. The rotational invariance in property 3) must be understood as the genuine contribution of the multivariate setting. Thus the group of all symmetries of the MRA formed by PV_j is isomorphic to*

$$O(n) \times \mathbb{Z}.$$

In a certain sense this group is much bigger than the symmetry group $\mathbb{Z}^n$ of the usual scheme of Multiresolution Analysis, cf. Meyer [14].

18.6 Decomposition and reconstruction for polyharmonic wavelets and the mother wavelet

Now we will prove the decomposition of the wavelet space PW_j for every $j \in \mathbb{Z}$, namely

$$PW_j = \left[\bigoplus_{k=0}^{\infty} \bigoplus_{\ell=1}^{d_k} \right]' W_j^{(k)}$$

which we have mentioned at the end of Section 18.3, p. 375. We will do this in an elegant way by assembling the one-dimensional decomposition relations for L-spline wavelets in (17.29), p. 354, for the spaces $V_j^{(k)}$ and $W_j^{(k)}$.

In order to conform with the standard notations in wavelet analysis we will use the index j for the mesh scaling. The indexes k and ℓ will be used again for the spherical harmonics. We put $h = 1/2^{j-1}$.

Recall that in equality (18.13), p. 375, we defined the space PV_j as a closure of set of polysplines of order p (or better to say L_2-limits of sequences of such) with breaksurfaces that are the spheres $S(0; e^{v2^{-j}})$. By (18.13), every element f^j of the space PV_j can be represented through the expansion in spherical harmonics

$$f^j(x) = \sum_{k,\ell} f^{j,k,\ell}(\log r) Y_{k,\ell}(\theta),$$

hence

$$e^{(n/2)v} f^{j,k,\ell}(u) = \sum_{v=-\infty}^{\infty} c_v^{j,k,\ell} \widetilde{Q}_{Z+1}\left[\Lambda_k + \frac{n}{2}; \frac{h}{2}\right]\left(u - \frac{h}{2}v\right)$$

$$= \sum_{v=-\infty}^{\infty} c_v^{j,k,\ell} \phi_{2-j}^k \left(u - \frac{h}{2}v\right).$$

Recall that we have defined the wavelet space PW_j by putting

$$PW_{j-1} := PV_j \ominus PV_{j-1}. \tag{18.15}$$

The next step is to consider the *wavelet decomposition*

$$f^j(x) = f^{j-1}(x) + g^{j-1}(x),$$

where g^{j-1} belongs to the wavelet space PW_{j-1} and $f^{j-1} \in PV_{j-1}$. They are orthogonal by (18.15).

We will denote the **polyharmonic mother wavelet** in $W_j^{(k)}$ by putting

$$\psi^{j,k}(v) = \psi\left[\Lambda_k + \frac{n}{2}; 2^{-j}\right](v),$$

where under $\psi[\Lambda_k + n/2; 2^{-j}](v)$ we mean the mother L-spline wavelet $\psi_{2^{-j}}$ of Definition 17.9, p. 338, with respect to the operator $L = \mathcal{L}[\Lambda_k + n/2]$ for the mesh $2^{-j}\mathbb{Z} = \{v2^{-j} : \text{for all } v \in \mathbb{Z}\}$.

Remark 18.12 *We see that the dependence of the **polyharmonic mother wavelet** $\psi^{j,k}$ on the parameter k is of very non-linear character. The dependence on j is somewhat simpler. In fact the mother wavelet is the function*

$$\psi^{j,k}(v) Y_{k,\ell}(\theta)$$

as we will see in the isomorphism Theorem below. From computational point of view we need only to tabulate the function

$$\widehat{\psi}\left[\Lambda_k + \frac{n}{2}; 2^{-j}\right](\xi)$$

as a function of k, j, ξ, and to generate the whole $L_2(\mathbb{R}^n)$. For that reason we may use the formulas of Section 17.5, p. 337, which are completely constructive.

We will prove the following Theorem justifying the term "mother wavelet".

Theorem 18.13 *For every $j \in \mathbb{Z}$ if PW_j is the wavelet space defined by*

$$PW_j = PV_{j+1} \ominus PV_j$$

then we have the decomposition isomorphism

$$PW_j \approx \left[\bigoplus_{k=0}^{\infty} \bigoplus_{\ell=1}^{d_k} W_j^{(k)}\right]',$$

where we have denoted by $W_j^{(k)}$ the space of all polyharmonic wavelets $\psi = \psi^{j,k}$ for the operator $\mathcal{L}[\Lambda_k + n/2]$ on the mesh $2^{-j}\mathbb{Z}$, defined in formula (17.20), p. 338.[5]

---

[5] The sum $\bigoplus_{\ell=1}^{d_k} W_j^{(k)}$ is understood as taking d_k identical copies of the space $W_j^{(k)}$.

Proof (1) Let us make the decomposition of an arbitrary element $f^j \in PV_j$. For convenience let $h = 1/(2^{j-1})$. By the definition of the space PV_j in (18.13), p. 375, we have the representation

$$f^j(x) = \sum_{k=0}^{\infty} \sum_{\ell=1}^{d_k} f^{j,k,\ell}(\log r) Y_{k,\ell}(\theta),$$

$$f^{j,k,\ell}(v) e^{(n/2)v} = \sum_{\nu=-\infty}^{\infty} c_\nu^{j,k,\ell} \phi_{2-j}^k \left(v - \frac{h}{2} \nu \right).$$

Recall that by means of formula (18.9), p. 374, we have proved Theorem 18.5, p. 376, which implies that $\left\{ c_\nu^{j,k,\ell} \right\} \in \ell_2^{(3)}$.

(2) Further, we will apply the L-spline wavelet decomposition relations of Theorem 17.31, p. 365, which have two main cases – odd shifts and even shifts $jh/2$–namely

$$\phi_{2-j}^k(v) = \sum_{\sigma=-\infty}^{\infty} a_{-2\sigma} \cdot \phi_{2-j+1}^k(v - \sigma h) + \sum_{\sigma=-\infty}^{\infty} b_{-2\sigma} \cdot \psi^{j-1,k}(v - \sigma h),$$

$$\phi_{2-j}^k \left(v - \frac{h}{2} \right) = \sum_{\sigma=-\infty}^{\infty} a_{1-2\sigma} \cdot \phi_{2-j+1}^k(v - \sigma h) + \sum_{\sigma=-\infty}^{\infty} b_{1-2\sigma} \cdot \psi^{j-1,k}(v - \sigma h),$$

or we may unify the notations, and using Chui's *smart notations* we write for every shift $(h/2)\ell$ the relations as follows:

$$\phi_{2-j}^k \left(v - \frac{h}{2} \ell \right) = \sum_{\sigma=-\infty}^{\infty} a_{\ell-2\sigma} \cdot \phi_{2-j+1}^k(v - \sigma h) + \sum_{\sigma} b_{\ell-2\sigma} \cdot \psi^{j-1,k}(v - \sigma h).$$

Thus we obtain, at first formally, the representation

$$f^{j,k,\ell}(v) = \sum_{\nu=-\infty}^{\infty} c_\nu^{j,k,\ell} \phi_{2-j}^k \left(v - \frac{h}{2} \nu \right)$$

$$= \sum_{\nu=-\infty}^{\infty} c_\nu^{j,k,\ell} \left\{ \sum_{\sigma=-\infty}^{\infty} a_{\nu-2\sigma} \cdot \phi_{2-j+1}^k(v - \sigma h) \right.$$

$$\left. + \sum_{\sigma} b_{\nu-2\sigma} \cdot \psi^{j-1,k}(v - \sigma h) \right\}$$

$$= \sum_{\nu=-\infty}^{\infty} \sum_{\sigma=-\infty}^{\infty} c_\nu^{j,k,\ell} a_{\nu-2\sigma} \cdot \phi_{2-j+1}^k(v - \sigma h)$$

$$+ \sum_{\nu=-\infty}^{\infty} \sum_{\sigma=-\infty}^{\infty} c_\nu^{j,k,\ell} b_{\nu-2\sigma} \cdot \psi^{j-1,k}(v - \sigma h).$$

(This formula is the basic representation showing that the function $\psi^{j,k}$ is indeed a "mother wavelet".)

Hence, if we put

$$f^{j-1,k,\ell}(v) := \sum_{v=-\infty}^{\infty}\sum_{\sigma=-\infty}^{\infty} c_v^{j,k,\ell} a_{v-2\sigma} \cdot \phi^k_{2^{-j+1}}(v - \sigma h)$$

$$g^{j-1,k,\ell}(v) := \sum_{v=-\infty}^{\infty}\sum_{\sigma=-\infty}^{\infty} c_v^{j,k,\ell} b_{v-2\sigma} \cdot \psi^{j-1,k}(v - \sigma h),$$

we have the question of L_2-convergence of the series

$$f^{j-1}(x) = \sum_{k,\ell}\sum_{v=-\infty}^{\infty}\sum_{\sigma=-\infty}^{\infty} c_v^{j,k,\ell} a_{v-2\sigma} \cdot \phi^k_{2^{-j+1}}(v - \sigma h) Y_{k,\ell}(\theta),$$

$$g^{j-1}(x) = \sum_{k,\ell}\sum_{v=-\infty}^{\infty}\sum_{\sigma=-\infty}^{\infty} c_v^{j,k,\ell} b_{v-2\sigma} \cdot \psi^{j-1,k}(v - \sigma h) Y_{k,\ell}(\theta).$$

(This formula shows the role of the "mother wavelet" $\psi^{j,k}(v) Y_{k,\ell}(\theta)$.)
(3) For the L_2 norm of f^{j-1} we have

$$\|f^{j-1}(x)\|^2 = \sum_{k=0}^{\infty}\sum_{\ell=1}^{d_k} \left\|\sum_{\sigma=-\infty}^{\infty}\left(\sum_{v=-\infty}^{\infty} c_v^{j,k,\ell} a_{v-2\sigma}\right) \cdot \phi^k_{2^{-j+1}}(v - \sigma h)\right\|^2_{L_2(\mathbb{R})},$$
(18.16)

where we have $h = 2^{-j+1}$. Due to the Riesz inequalities (14.15), p. 283, for the normed TB-spline $\phi^k_{2^{-j+1}}(v) = \tilde{Q}[\Lambda_k + n/2; 2^{-j+1}](v)$, we obtain the following inequalities:

$$A_{j-1}\sum_{k=0}^{\infty}\sum_{\ell=1}^{d_k}\sum_{\sigma=-\infty}^{\infty}\left|\sum_{v=-\infty}^{\infty} c_v^{j,k,\ell} a_{v-2\sigma}\right|^2 \le \|f^{j-1}(x)\|^2$$

$$\le B_{j-1}\sum_{k=0}^{\infty}\sum_{\ell=1}^{d_k}\sum_{\sigma=-\infty}^{\infty}\left|\sum_{v=-\infty}^{\infty} c_v^{j,k,\ell} a_{v-2\sigma}\right|^2$$

with the constants A_{j-1} and B_{j-1} satisfying $0 < A_{j-1}, B_{j-1} < \infty$. In a similar way we obtain a similar estimate, for g^{j-1}, which contains the coefficients b_v.

(4) However, we will not refer to that inequality and we will make a proof independent of point (3) owing to the importance of the present theorem.
Let us prove the convergence of $f^{j-1}(x)$ and $g^{j-1}(x)$.
We put

$$C(\xi) := C^{j,k,\ell}(\xi) := \sum_{\sigma}\left(\sum_{v=-\infty}^{\infty} c_v^{j,k,\ell} a_{v-2\sigma}\right) e^{-i\sigma\xi}.$$

Recall that $h = 2^{-j+1}$. In expression (18.16) for $\|f^{j-1}(x)\|^2$ we used the Parseval formula, see Theorem 12.7, p. 212; we now apply formula (13.25), p. 242, and the

2π-periodicity of the function $C(\xi)$, in order to obtain

$$\|f^{j-1}(x)\|^2 = \sum_{k=0}^{\infty}\sum_{\ell=1}^{d_k} \frac{1}{2\pi}\int_{-\infty}^{\infty} \left|\widehat{C(\xi h)\phi_h^k(\xi)}\right|^2 d\xi$$

$$= \sum_{k=0}^{\infty}\sum_{\ell=1}^{d_k} \frac{1}{2\pi h}\int_{-\infty}^{\infty} \left|\widehat{C(\xi)\phi_h^k}\left(\frac{\xi}{h}\right)\right|^2 d\xi$$

$$= \sum_{k=0}^{\infty}\sum_{\ell=1}^{d_k} \frac{h^{2Z+1}}{2\pi}\int_{-\infty}^{\infty} \left|C(\xi)\widetilde{Q}_{Z+1}\left[h\left(\Lambda_k + \frac{n}{2}\right)\right](\xi)\right|^2 d\xi$$

$$= \sum_{k=0}^{\infty}\sum_{\ell=1}^{d_k} \frac{h^{2Z+1}}{2\pi}\sum_{j=-\infty}^{\infty}\int_0^{2\pi} \left|C(\xi)\widetilde{Q}_{Z+1}\left[h\left(\Lambda_k + \frac{n}{2}\right)\right](\xi + 2\pi j)\right|^2 d\xi$$

$$= \sum_{k=0}^{\infty}\sum_{\ell=1}^{d_k} \frac{h^{2Z+1}}{2\pi}\int_0^{2\pi} |C(\xi)|^2 \cdot \sum_{j=-\infty}^{\infty} \left|\widetilde{Q}_{Z+1}\left[h\left(\Lambda_k + \frac{n}{2}\right)\right](\xi + 2\pi j)\right|^2 d\xi.$$

The sum

$$S(\xi) = \sum_{j=-\infty}^{\infty} \left|Q_{Z+1}\left[h\left(\Lambda_k + \frac{n}{2}\right)\right](\xi + 2\pi j)\right|^2$$

has been studied in Theorem 14.6, p. 272. The sum using $\widetilde{Q}_{Z+1}$ instead, as normed in (14.14), will be obtained after the norming.

(5) Let us recall that in Theorem 14.6, we proved for $p = 2$ the biharmonic case, for $k = 1, 2, \ldots$, the estimate

$$A\frac{e^{4kh}}{k^3} \leq S(\xi) \leq B\frac{e^{4kh}}{k^3}$$

where the constants A, B satisfy $0 < A \leq B < \infty$. The general case has been considered in Theorem 14.6, p. 272 and Corollary 14.7, p. 276. Accordingly, the norming in (14.14), p. 283, is such that the sum

$$\widetilde{S}(\xi) = \sum_{j=-\infty}^{\infty} \left|\widetilde{Q}_{Z+1}\left[h\left(\Lambda_k + \frac{n}{2}\right)\right](\xi + 2\pi j)\right|^2$$

satisfies

$$A \leq \widetilde{S}(\xi) \leq B\psi \text{ for all } k = 0, 1, 2, \ldots,$$

where $0 < A \leq B < \infty$. This implies the inequalities

$$A\frac{h^{2Z+1}}{2\pi}\sum_{k,\ell}\int_0^{2\pi} |C(\xi)|^2 d\xi \leq \|f^{j-1}(x)\|^2 \leq B\frac{h^{2Z+1}}{2\pi}\sum_{k,\ell}\int_0^{2\pi} |C(\xi)|^2 d\xi.$$

(6) We have to estimate $\sum_{k=0}^{\infty} \sum_{\ell=1}^{d_k} \int_0^{2\pi} |C(\xi)|^2 \, d\xi$. We proceed as in the one-dimensional case in Theorem 17.31, p. 365. We put $z = e^{-i\xi/2}$, and recalling the definition of the function $A(z)$ in Theorem 17.23, p. 354, we obtain

$$C(\xi) = \sum_{\nu=-\infty}^{\infty} c_\nu^{j,k,\ell} \left(\sum_{\sigma=-\infty}^{\infty} a_{\nu-2\sigma} e^{-i\sigma\xi} \right)$$

$$= \sum_{\nu=-\infty}^{\infty} c_\nu^{j,k,\ell} \left(\sum_{\sigma=-\infty}^{\infty} a_{\nu-2\sigma} z^{2\sigma} \right)$$

$$= \sum_{\nu=-\infty}^{\infty} c_{2\nu}^{j,k,\ell} z^{2\nu} \left(\overline{A(z)} + \overline{A(-z)} \right) + \sum_{\nu=-\infty}^{\infty} c_{2\nu+1}^{j,k,\ell} z^{2\nu+1} \left(\overline{A(z)} - \overline{A(-z)} \right)$$

$$= \overline{A(z)} \sum_{\nu=-\infty}^{\infty} c_\nu^{j,k,\ell} z^\nu + \overline{A(-z)} \sum_{\nu=-\infty}^{\infty} c_\nu^{j,k,\ell} (-z)^\nu.$$

Let us put

$$\sigma(z) := \sigma^{j,k,\ell}(z) := \sum_{\nu=-\infty}^{\infty} c_\nu^{j,k,\ell} z^\nu.$$

This implies

$$\int_0^{2\pi} |C(\xi)|^2 \, d\xi \le \int_0^{2\pi} |A(z)|^2 \cdot |\sigma(z)|^2 \, d\xi + \int_0^{2\pi} |A(-z)|^2 \cdot |\sigma(-z)|^2 \, d\xi$$

$$= \int_0^{4\pi} |A(z)|^2 \cdot |\sigma(z)|^2 \, d\xi.$$

By the Parseval identity for the Fourier series in Theorem 12.2, p. 210, we obtain the equality

$$\frac{1}{2\pi} \int_0^{2\pi} |C(\xi)|^2 \, d\xi = \|c\|_{\ell_2}^2 := \sum_{\sigma=-\infty}^{\infty} \left| \sum_{\nu=-\infty}^{\infty} c_\nu^{j,k,\ell} a_{\nu-2\sigma} \right|^2,$$

and in a similar way

$$\frac{1}{4\pi} \int_0^{4\pi} \left| \sigma(e^{-i\xi/2}) \right|^2 \, d\xi = \sum_{\nu=-\infty}^{\infty} \left| c_\nu^{j,k,\ell} \right|^2.$$

(7) Thus we have to prove see that we have for all $k = 0, 1, 2, \ldots$ a uniform estimate

$$\left. \begin{array}{l} |A(z)| \le C, \\ |B(z)| \le C \end{array} \right\} \text{ for all } (|z| = 1)$$

by a constant $C > 0$ independent of k. This would imply that

$$\frac{1}{2\pi} \int_0^{2\pi} |C(\xi)|^2 \, d\xi \le C_1 \sum_{k,\ell} \sum_{\nu=-\infty}^{\infty} \left| c_\nu^{j,k,\ell} \right|^2,$$

for some constant $C_1 > 0$, the last sum being bounded as we saw above.

390 Multivariate polysplines

For the proof of the convergence of g^{j-1} let us note that we obtain a similar expression containing the function $B(z)$ instead of $A(z)$. For that reason we also consider the boundedness of the function $B(z)$.

As is usual, for a nonordered vector Λ we denote by $\widetilde{\Lambda}$ its symmetrization according to Definition 14.5, p. 271. From formulas (17.27) and (17.28), p. 353, we have the exact solutions given by[6]

$$A(z) = \frac{P(z)\Phi_{2Z+1}[h/2 \cdot (\widetilde{\Lambda_k + n/2})](Z+1; z)}{\Phi_{2Z+1}[h \cdot (\widetilde{\Lambda_k + n/2})](Z+1; z^2)},$$

$$B(z) = -z^{2Z+1} \frac{P(-1/z)}{\Phi_{2Z+1}[h \cdot (\widetilde{\Lambda_k + n/2})](Z+1; z^2)},$$

and we see that for them the maximum for $|z| = 1$ is bounded.

Owing to:

- representation (13.44), p. 255, for arbitrary $Z \geq 1$, namely, $\Phi_Z(x; \lambda) = (-1)^Z \lambda^{-Z} e^{-\lambda_1 - \cdots - \lambda_{Z+1}} r(\lambda) A_Z(x; \lambda)$, and
- the exponential property of Φ_{2Z+1} given through equality (13.43), p. 255, and
- formula (17.9), p. 332, for the polynomial $P(z)$,

we see that

$$A(z) = \frac{-P(z)z^{-(2Z+1)} \cdot z^{Z+1} \cdot r[h/2(\widetilde{\Lambda_k + n/2})](z) \cdot A_{2Z+1}[h/2(\widetilde{\Lambda_k + n/2})](0; z)}{-(z^2)^{-(2Z+1)} \cdot (z^2)^{Z+1} r[h(\widetilde{\Lambda_k + n/2})](z^2) A_{2Z+1}[h(\widetilde{\Lambda_k + n/2})](0; z^2)}$$

$$= z^Z \frac{s[h/2(\widetilde{\Lambda_k + n/2})](e^{-i(h/2)\xi}) \cdot A_{2Z+1}[(h/2)(\widetilde{\Lambda_k + n/2})](0; z)}{r[(h/2)(\widetilde{\Lambda_k + n/2})](-z) \cdot A_{2Z+1}[h(\widetilde{\Lambda_k + n/2})](0; z^2)}.$$

Since the nonordered vectors $h/2(\widetilde{\Lambda_k + n/2})$ and $h(\widetilde{\Lambda_k + n/2})$ are symmetric, we can apply formula (14.13), p. 279, and for all $|\lambda| = 1$ we obtain the asymptotic orders

$$A_{2Z+1}\left[\left(\frac{h}{2}\widetilde{\Lambda_k + \frac{n}{2}}\right)\right](0; \lambda) \approx \frac{C}{k^{4p-1}},$$

$$A_{2Z+1}\left[h\left(\widetilde{\Lambda_k + \frac{n}{2}}\right)\right](0; \lambda) \approx \frac{C_1}{k^{4p-1}},$$

for the constants $C, C_1 > 0$.

However, for $k \longrightarrow \infty$ we evidently have the limit

$$\frac{s[(h/2)(\widetilde{\Lambda_k + n/2})](e^{-i(h/2)\xi})}{r[(h/2)(\widetilde{\Lambda_k + n/2})](-e^{-i(h/2)\xi})} \xrightarrow{k \to \infty} 0,$$

---

[6] We recall that for z with $|z| = 1$ and symmetric nonordered vectors Λ, the functions $\Phi_{2P+1}[\Lambda](P+1; z)$ take on real values, i.e. $\Phi_{2P+1}[\Lambda](P+1; z) = \Phi_{2P+1}[\Lambda](P+1; \bar{z})$, see Theorem 13.56, p. 256.

since roughly speaking all exponents of $\Lambda_k + n/2$ are also available in $\widetilde{\Lambda_k + n/2}$ and we are only interested in those which tend to ∞. Thus we see that $A(z)$ is bounded. In the same way it follows that $B(z)$ is bounded. Thus we have finished the proof of the convergence of the series representing the functions $f^{j-1}(x)$ and $g^{j-1}(x)$.

(8) It remains to see that the function g^{j-1} is orthogonal to every element of the space PV_{j-1}. Indeed we have the representation

$$g^{j-1}(x) = \sum_{k=0}^{\infty} \sum_{\ell=1}^{d_k} g^{j-1,k,\ell}(\log r) Y_{k,\ell}(\theta),$$

where

$$e^{(n/2)v} g^{j-1,k,\ell}(v) \in W_{j-1}^{(k)},$$

and we have denoted by $W_{j-1}^{(k)}$ the L-spline wavelets for the operator $\mathcal{L}[\Lambda_k + n/2]$ on the mesh $2^{-j+1}\mathbb{Z}$. For arbitrary $F^{j-1} \in PV_{j-1}$ we have the representation

$$F^{j-1}(x) = \sum_{k=0}^{\infty} \sum_{\ell=1}^{d_k} F^{j-1,k,\ell}(\log r) Y_{k,\ell}(\theta).$$

It follows that:

$$\int_{\mathbb{R}^n} F^{j-1}(x) g^{j-1}(x) \, dx = \sum_{k=0}^{\infty} \sum_{\ell=1}^{d_k} \int_{-\infty}^{\infty} F^{j-1,k,\ell}(v) g^{j-1,k,\ell}(v) e^{nv} \, dv.$$

We now apply the one-dimensional orthogonality of the spaces $V_{j-1}^{(k)}$ and $W_{j-1}^{(k)}$, and obtain $\langle F^{j-1}, g^{j-1} \rangle = 0$. This completes the proof of the theorem. ∎

18.7 Zero moments of polyharmonic wavelets

In the one-dimensional case the wavelets are orthogonal to the polynomials of order p.

We have the analogous theorem that the *polyharmonic wavelet* is orthogonal to the functions polyharmonic of order p.

Theorem 18.14 *Let f be a polynomial polyharmonic of order p, i.e. $\Delta^p f(x) = 0$. For every $j \in \mathbb{Z}$ if some element $\psi \in PW_j$ has a compact support then the integral*

$$\int_{\mathbb{R}^n} f(x) \psi(x) \, dx = 0.$$

Proof For some points of the proof we could refer to Theorem 18.13, p. 385, but we would like to provide a proof that is completely independent. For simplicity, we take the case $j = 0$. In particular, let us take the function $\psi(x) \in PW_0$ with support the set defined by

$$0 \leq v = \log r \leq 3p - 2,$$

and such that

$$\psi(x) = \sum_{k=0}^{\infty} \sum_{\ell=1}^{d_k} \gamma_{k,\ell} \psi^k(\log r) Y_{k,\ell}(\theta),$$

where the constants $\gamma_{k,\ell}$ are such that $\psi(x) \in L_2(\mathbb{R}^n)$. For $k = 0, 1, 2, \ldots$ the functions $\psi^k(v)$ are L-spline mother wavelets provided in Definition 17.9, p. 338, for the operators $L = \mathcal{L}[\Lambda_k + n/2]$.

However, from Corollary 10.38, p. 173, we know that the polyharmonic function $f(x)$ is expanded in every annulus and in particular in the annulus $\{1 \leq r = e^v \leq e^{3p-2}\}$ in spherical harmonics, namely

$$f(x) = \sum_{k=0}^{\infty} \sum_{\ell=1}^{d_k} f^{k,\ell}(\log r) Y_{k,\ell}(\theta),$$

where the functions $e^{(n/2)v} f^{k,\ell}(v)$ satisfy the following equation, see (10.26),

$$\mathcal{L}\left[\Lambda_k + \frac{n}{2}\right]\left(\frac{d}{dv}\right) e^{(n/2)v} f^{k,\ell}(v) = 0.$$

For every $k = 0, 1, 2, \ldots$, the spherical harmonics $\{Y_{k,\ell}(\theta)\}_{\ell=1}^{d_k}$ form a basis for the set $\mathcal{H}_k$. By Proposition 17.1, p. 326, we know that the set

$$\{\phi_1^k(v - \rho) : \text{ for all } \rho \in \mathbb{Z}\}$$

is a basis for the L-splines in the interval $[0, 3p - 2]$, where $L = \mathcal{L}[\Lambda_k + n/2]$. In particular the functions $e^{(n/2)v} f^{k,\ell}(v)$ are L-polynomials and are hence L-splines on every interval. The last implies the following representation:

$$e^{(n/2)v} f^{k,\ell}(v) = \sum_{\rho=-\infty}^{\infty} c_\rho^{k,\ell} \phi_1^k(v - \rho),$$

where only a finite number of c_ρs are nonzero. On the other hand, due to the orthogonality $W_0^{(k)} \perp V_0^{(k)}$ we obtain

$$\int_{-\infty}^{\infty} \psi^k(v) e^{(n/2)v} \phi_1^k(v - \rho) \, dv = 0$$

for every $\rho \in \mathbb{Z}$. This implies

$$\int_{-\infty}^{\infty} \psi^k(v) e^{(n/2)v} e^{n/2v} f^{k,\ell}(v) \, dv = 0.$$

By the properties of the spherical harmonics we obtain

$$\int_{\mathbb{R}^n} f(x) \psi(x) \, dx = \sum_{k=0}^{\infty} \sum_{\ell=1}^{d_k} \int_{-\infty}^{\infty} \psi^k(v) f^{k,\ell}(v) e^{nv} \, dv,$$

hence we see that
$$\langle f, \psi \rangle = \int_{\mathbb{R}^n} f(x)\psi(x)\, dx = 0.$$
This completes the proof. ∎

It should be noted that the space of the polynomials polyharmonic of order p is infinite-dimensional. Due to the compactness of the support of ψ we see that it is not necessary to consider global solutions of $\Delta^p f = 0$ but only some of local character, e.g. in an annulus.

The above result means that the wavelets $\psi \in PW_j$ have enormous "cancellation" properties. This should be compared with other approaches to multidimensional wavelets where the wavelet functions are only orthogonal to finite dimensional subspaces.

18.8 Bibliographical notes

We have added in the Bibliography below some more references on wavelet analysis which might be of interest for the reader. We comment on some references which might be of use to a reader who is a newcomer in wavelet analysis.

1. The most concentrated and informative exposition for a mathematician is that of Y. Meyer [14], where the basic mathematical principles are available. The fact that the operators E_j are pseudodifferential for example, also the fact that r-smooth MRA retains the polynomials of degree $\leq r$ are to be found in [14] in a very transparent exposition. Also the Radial Basis Functions Wavelets studied by Lemarié are available there.

2. Daubechies [8] is rather encyclopedic, has a lot of material which is interesting with the physical interpretation as well as more concrete results. Also the more recent book of S. Mallat [13] covers a wide spectrum of topics. The exposition in these two books is oriented towards a reader with interests in Signal Processing.

3. As we already said, in [4], [5], Chui has scrupulously carried out the program of computing the case of polynomial cardinal spline Wavelets.. He provides also "signal analysis-oriented" interpretation of many formulas. So far in [4] the polynomial case is mixed with generalized scale functions and wavelet analysis, while in [5] only the polynomial case is considered, which is sometimes more transparent. Let us mention that one of the first constructions of wavelets were based on cardinal splines, about the so-called *Battle-Lemarié* wavelets see e.g. in [8, Section 5.4, p. 146].

4. One has to mention the book of Hernández and Weiss [10] which is perfectly written for a beginner-mathematician.

Bibliography to Part III

[1] Adams R. *Sobolev Spaces*, Academic Press, New York – San Francisco – London, 1975.
[2] Candes, E. and Donoho, D., Curvelets – a surprisingly effective nonadaptive representation for objects with edges, Proceedings Confer. Saint–Malo, Eds. L.L. Schumaker et al., Vanderbilt University Press, 2000.
[3] Chui, Ch. *Multivariate Splines*, SIAM, PA, 1988.
[4] Chui, Ch. *An Introduction to Wavelets*, Academic Press, Boston, 1992.
[5] Chui, Ch. *Wavelets: A Mathematical Tool for Signal Processing*, SIAM, Philadelphia, PA, 1997.
[6] Chui, Ch. (Ed.) *Wavelets: A Tutorial in Theory and Applications*, Academic Press, Boston, 1992.
[7] Chui, Ch., Montefusco, L. and Puccio, L. (Eds) *Wavelets: Theory, Algorithms, and Applications,* Academic Press, Boston, 1994.
[8] Daubechies, I. *Ten Lectures on Wavelets*, SIAM, Philadelphia, Pennsylvania, 1992.
[9] de Boor, C., DeVore, R. and Ron, A. On the construction of multivariate (pre)wavelets. *Constructive Approximation*, 9 (1993), pp. 123–166.
[10] Hernández, E. and Weiss, G. A First Course on Wavelets, CRC Press, Boca Raton, FL, 1996.
[11] Kounchev, O. and Render, H. On scaling and rotationally invariant spaces and polyharmonic wavelets. Manuscript.
[12] Lyche, T. and Schumaker, L.L. L-spline wavelets. In: *Wavelets: Theory, Algorithms, and Applications (Taormina, 1993)*, Academic Press, San Diego, CA 1994, pp. 197–212.
[13] Mallat, S. *A Wavelet Tour of Signal Processing.* Academic Press, Inc., San Diego, CA, 1998.
[14] Meyer, Y. *Wavelets and Operators*, Cambridge University Press, 1992.
[15] Rudin, W. *Real and Complex Analysis*, McGraw–Hill Publ. Co., New York, 1976.
[16] Schoenberg, I.J., *Cardinal Spline Interpolation*, SIAM, Philadelphia, Pennsylvania, 1973.
[17] Stein, E. and Weiss, G. *Introduction to Fourier Analysis on Euclidean Spaces*, Princeton University Press, Princeton, 1971.

Part IV

Polysplines for general interfaces

Chapter 19

Heuristic arguments

19.1 Introduction

In the previous Parts of the present book we have considered polysplines only in the case of special domains where the interfaces have a special geometry – they were spheres or hyperplanes. The results, especially on cardinal polysplines on strips and annuli, including the wavelet analysis, were deep and beautiful and were mainly analytical. Thus one might be left with the impression that the polysplines are objects of *mathematical analysis in the narrow sense* of this notion. However, we will see that they are objects which belong to *mathematical analysis in the wide sense* of this notion, so far as *partial differential equations* are a part of this area. What we mean will become completely clear from the results that follow in the present Part. In brief, we will show that the polysplines exist in the case of rather general interfaces, but the price we pay is the lack of the beautiful theory of cardinal polysplines which we developed in Part II.

As we have seen in Chapter 2, p. 19, there are two ways to establish the one-dimensional spline theory for odd-degree splines: the first way is to consider the splines as sufficiently smooth objects. The second way is to consider the splines as solutions to variational problems (Holladay property). We saw that both ways are equivalent. In the polyspline theory (of even-order polysplines) we have the same two ways. In Part I of the present book we have based the notion of polysplines (of arbitrary order) on strips and annuli by appealing to the *smoothness*, *data* and *objects concepts*. In the present Part we will lay particular emphasis on the fact that the even-order polysplines are solutions to variational problems. The smoothness properties will be derived from the last and we will also define polysplines of odd order. So far we will see that "stability with respect to the interpolation data" have in the general case only the polysplines of even order, and this is due to their uniqueness property (or which is the same, to their variational property).

We will consider interfaces having arbitrary geometry but sufficiently smooth. In order to keep the presentation here technically simple we only discuss domains in $\mathbb{R}^n$ but the final results will be formulated for smooth manifolds with boundaries.

Uniqueness of interpolation polysplines

Right at the start, we note that the *most remarkable* results of Part IV are those about the uniqueness of even-order interpolation polysplines in general interface configurations and for rather general elliptic operators L of the type $L = L_1^2$, where L_1 is a formally self-adjoint elliptic operator of order $2q$. In Part I we proved uniqueness for polysplines arising from the operator $L = \Delta^p$ only for special interface configurations – strips and annuli. Let us recall that the uniqueness theorems are very *rare* in the theory of elliptic boundary value problems; they are available for almost all elliptic operators only in the case of second order thanks to the maximum principle, which is not the case for higher-order elliptic operators. In the case of higher-order elliptic operators we have uniqueness only for special combinations of the operator L and the boundary operators $\{B_k\}$, in fact only in those cases which permit reduction to the second-order case, see Garabedian [23, Chapter 7.1, also p. 239 for the biharmonic equation], Mikhlin [51, p. 256 and p. 270], Agmon et al. [2, Section 12.5, Theorems 12.8 and 12.10], Agranovich [4, Section 3.2]. On the other hand the uniqueness theorems are important since they immediately imply the existence of a solution of elliptic boundary value problems (BVPs).[1]

We will explain the ideas underlying the notion of *polysplines* by providing a direct analogy with the variational approach in the *one-dimensional spline theory* which we considered in Chapter 2. More precisely, in Section 2.3, p. 23, in the one-dimensional case the *interpolation splines* of *odd polynomial degree* were considered as solutions to the following extremal problem:

$$\inf \int_a^b \left(\frac{d^q f}{dt^q}\right)^2 dt, \tag{19.1}$$

where the infimum is over the set of functions f which satisfy the *interpolation* conditions

$$f(t_j) = c_j \qquad \text{for } j = 1, 2, \ldots, N; \tag{19.2}$$

$$\left[\frac{d^k f}{dt^k}\right]_{|t=a} = \left[\frac{d^k f}{dt^k}\right]_{|t=b} = 0, \quad \text{for } k = 1, \ldots, q-1, \tag{19.3}$$

where the knots $t_1, t_2, \ldots, t_N$ satisfy the following condition:

$$a = t_1 < \cdots < t_N = b.$$

Figure 2.3, p. 24, is an appropriate illustration.

The solution to problem (19.1)–(19.3) is a function s in the interval $[a, b]$ which satisfies the following conditions.

(i) The function s is a polynomial of degree $\leq 2q - 1$ in every open interval (t_i, t_{i+1}), $i = 1, 2, \ldots, N - 1$, i.e. it satisfies equation

$$\frac{d^{2q} s}{dt^{2q}} = 0 \quad \text{in } [t_i, t_{i+1}].$$

---

[1] See in particular Theorem 23.19, p. 480, in the general theory of elliptic BVPs where the uniqueness for the adjoint problem means $N^* = \Lambda_1 = \{0\}$, hence existence would follow for every set of data.

(ii) The function s satisfies the following *boundary* conditions:

$$\left[\frac{d^k s}{dt^k}\right]_{|t=a} = \left[\frac{d^k s}{dt^k}\right]_{|t=b} = 0 \quad \text{for } k = 1, \ldots, q-1.$$

(iii) The derivative $d^{2q-2}s/dt^{2q-2}$ is *continuous* everywhere in $[a, b]$, i.e.

$$\frac{d^k s_i(t_{i+1})}{dt^k} = \frac{d^k s_{i+1}(t_{i+1})}{dt^k} \quad \text{for } k = 0, 1, \ldots, 2q-2,$$

$$\text{and } i = 1, 2, \ldots, N-2,$$

where we have denoted by $s_i(t)$ the restriction of $s(t)$ to the interval (t_i, t_{i+1}).

Conditions (i–iii) together with the interpolation property (19.2) may be considered as an alternative definition of splines which, chronologically, is the original, cf. Ahlberg et al. [5, p. 76], or Laurent [45, p. 162].

According to the *operator concept* of the *polyharmonic paradigm* implemented through the replacement scheme (1.1), on p. 10, the operator d/dt has to be replaced with the operator Δ in the multivariate case. Then the natural analog to the extremal problem (19.1) will be

$$\int_D (\Delta^q f)^2 \, dx \longrightarrow \inf. \quad (19.4)$$

Here the minimization is over the functions f defined in the bounded domain $D \subseteq \mathbb{R}^n$ and satisfying some interpolation conditions inside D and boundary conditions on ∂D. Here we recall that Δ^q is the q times iterated Laplace operator Δ, i.e. we put $\Delta^1 = \Delta$ and $\Delta^q = \Delta\Delta^{q-1}$ with $\Delta^0 = id$, the identity operator.

The purpose of the present chapter is to provide a proper setting for the solution of the extremal problem (19.4). In other words, we will find an analog to the interpolation conditions (19.2) for which problem (19.4) has a solution. As a result we obtain a constructive solution to the extremal problem (19.4) and an analog to conditions (i–iii).

19.2 The setting of the variational problem

We have already explained (Chapter 3, p. 29) the *data concept* of the *polyharmonic paradigm*. According to the last, the set of N points in $[a, b]$ which are the knot-points (or break-points) of the spline will be replaced by a set of N closed smooth simple surfaces in $\overline{D}$, namely $\Gamma_1, \Gamma_2, \ldots, \Gamma_N$, such that Γ_j surrounds Γ_{j-1} for $j = 1, 2, \ldots, N$, and

$$\Gamma_N = \partial D.$$

The first $N-1$ surfaces $\Gamma_1, \Gamma_2, \ldots, \Gamma_{N-1}$ will be the surfaces where the pieces of the polysplines will match and will be called, depending on the context,[2]

interfaces = break-surfaces = knot-surfaces.

---

[2] The term "interface" has been used for the so-called transmission problems [9, 49], and this is understood as a "break-surface".

402 Multivariate polysplines

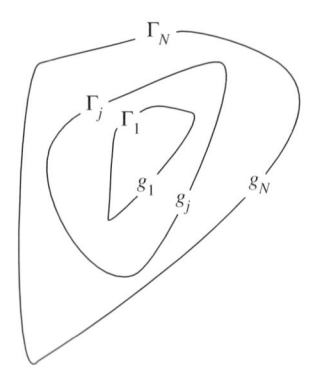

Figure 19.1. The data surfaces Γ_j and the data functions g_j on them.

It is convenient to put $\Gamma_0 = \emptyset$.[3] For the *data set* we put

$$ST = \bigcup_{j=1}^{N} \Gamma_j.$$

(See Figure 19.1 which is very similar to Figure 3.3, p. 31.)

The one-dimensional *interpolation* condition (19.2) is replaced by the *interpolation* condition

$$f = g_j \quad \text{on } \Gamma_j \text{ for } j = 1, 2, \ldots, N, \tag{19.5}$$

where $g = \{g_j\}_{j=1}^{N}$ is a function defined on all surfaces Γ_j for $j = 1, 2, \ldots, N$.

The one-dimensional *boundary* conditions (19.3) are replaced by the following *boundary* conditions:

$$\left.\begin{array}{l} \Delta^k f = 0, \quad \text{on } \Gamma_N, \text{ for } k = 1, \ldots, q-1; \\ \frac{\partial}{\partial n} \Delta^k f = 0, \quad \text{on } \Gamma_N, \text{ for } k = 0, \ldots, q-1; \end{array}\right\} \tag{19.6}$$

here $\vec{n} = \vec{n}_N(x)$ denotes the inner unit normal vector to Γ_N at the point $x \in \Gamma_N$.[4]

The solution s to the extremal problem (19.4)–(19.6) will be called *polysplines of order $2q$*. We will prove in Theorem 20.14, p. 425, that if it exists it satisfies the following analogs to conditions (i–iii):

(i′) The function s is $2q$-harmonic in the complement of all interfaces Γ_j, i.e.

$$\Delta^{2q} s = 0 \quad \text{in } D \setminus ST;$$

---

[3] As is clear, the choice of the surfaces T_j; is not completely arbitrary, in particular they are not intersecting. We also require that T_j be compact, a condition which essentially may be relaxed.

[4] One may choose another set of boundary conditions as will become clear in the rigorous study later.

(ii′) The function s satisfies the following *boundary* conditions on ∂D:
$$\Delta^k s = 0 \quad \text{on } \Gamma_N, \text{ for } k = 1, \ldots, q - 1;$$
$$\frac{\partial}{\partial n}\Delta^k s = 0, \quad \text{on } \Gamma_N, \text{ for } k = 0, \ldots, q - 1;$$

(iii′) If we denote by s_j the restriction of s to the layer lying between Γ_j and Γ_{j-1}, then the following *interface* conditions hold:
$$\frac{\partial^k}{\partial n^k} s_{j+1} = \frac{\partial^k}{\partial n^k} s_j \quad \text{on } \Gamma_j \quad \text{for } k = 0, 1, \ldots, 4q - 2.$$

Here $\vec{n} = \vec{n_j}(x)$ denotes the inner normal vector to the surface Γ_j at the point $x \in \Gamma_j$.

Note that the boundary conditions (ii′) coincide with the boundary conditions (19.6) of the extremal problem (19.4)–(19.6).

In the simplest case, $q = 1$, we have pieces of *biharmonic* functions which match on the interfaces Γ_j in the sense that for all $j = 1, 2, \ldots, N - 1$ the following equalities hold on Γ_j:
$$s_{j+1} = s_j,$$
$$\frac{\partial}{\partial n} s_{j+1} = \frac{\partial}{\partial n} s_j,$$
$$\frac{\partial^2}{\partial n^2} s_{j+1} = \frac{\partial^2}{\partial n^2} s_j;$$

on the boundary $\partial D = \Gamma_N$ we have $\partial/(\partial n)s_N = 0$.

Now, conditions (i′–iii′) may be taken as a definition of a *polyspline of order* $p = 2q$.

19.3 Polysplines of arbitrary order p

Thus we see that the *smoothness property* (iii′) may become our *smoothness concept* for defining polysplines of arbitrary integer order p. We introduce a *polyspline of order* p, as a function s which satisfies the following conditions:

(i″) The function s is a *polyharmonic* function of order p in the complement of all interfaces Γ_j, i.e.
$$\Delta^p s = 0 \quad \text{in } D \setminus ST.$$

(ii″) **Optional** condition: the function s satisfies the following *boundary* conditions on ∂D:
$$\Delta^k s = 0 \quad \text{on } \Gamma_N \quad \text{for } k = 1, \ldots, p - 1.$$

(iii″) If we denote by $s_j(x)$ the restriction of $s(x)$ to the layer lying between Γ_j and Γ_{j-1}, then the following *interface* conditions hold:
$$\frac{\partial^k}{\partial n^k} s_{j+1} = \frac{\partial^k}{\partial n^k} s_j \quad \text{on } \Gamma_j \quad \text{for } k = 0, 1, \ldots, 2p - 2.$$

Here $\vec{n} = \vec{n}_j(x)$ denotes the inner normal vector on the surface Γ_j.

- In addition, the polyspline s is called an *interpolation polyspline* if it satisfies the equalities

$$s = g_j \quad \text{on } \Gamma_j \quad \text{for } j = 1, 2, \ldots, N.$$

The boundary conditions (ii″) are *optional*. As will become clear these conditions may be replaced by other sets of conditions. The interface conditions which have the meaning of smoothness of the polysplines are basic.

19.4 Counting the parameters

For the one-dimensional *spline theory* a simple "counting of parameters (sometimes called the degrees of freedom) against restrictions (which are the interpolation + boundary conditions)" is typical [5, p. 109].

In a similar way when we introduced the polysplines on strips and annuli we have provided some motivation based on the "parametrization" of the space of polyharmonic functions on a rectangle (or annulus). We have established an equality between the number of parameters and the number of restrictions. This reasoning is also available in the case of polysplines on general interface configurations and this lies at the base of the notion of a polyspline.

The following argument shows that we may carry out a completely analogous counting in the polyspline case.

Let us check that every interpolation polyspline of order p satisfies the following *balance* condition:

$$\boxed{\begin{array}{l} \text{the number of degrees of freedom} = \\ \text{the number of interface conditions} + \\ \text{the number of interpolation conditions.} \end{array}} \tag{19.7}$$

According to the theory of higher-order elliptic BVPs[5] every function u_j polyharmonic of order p in the domain D_j enclosed between the two surfaces Γ_{j-1} and Γ_j is determined by p boundary conditions on Γ_{j-1} and by p boundary conditions on Γ_j. Thus on every one of the surfaces Γ_j and Γ_{j-1} we have p boundary conditions at our disposal (playing the role of degrees of freedom). In a similar way, u_{j+1} has p degrees of freedom on Γ_j. On the other hand, by the above definition of polysplines the number of interface conditions on Γ_j is $2p - 1$ and we have one interpolation condition on Γ_j. This manages the balance for $j = 1, 2, \ldots, N - 1$. Only the surface Γ_N is not an interface and there the piece u_N has p degrees of freedom which are fixed by the $p - 1$ boundary conditions on Γ_N plus the interpolation condition on Γ_N. This ends the proof of the above *balance* (19.7).

---

[5] See Chapter 23 and more precisely Theorem 23.19, p. 480, also [3], and [43] which is devoted to higher-order elliptic equations.

19.5 Main results and techniques

First, we consider polysplines not only for the operators $L = \Delta^p$ but also for the rather general elliptic operator L. Our most essential results so far are for operators of the type $L = L_1^2$ where L_1 is an elliptic operator, i.e. for the generalizations of the operator $L = \Delta^{2q}$.

Let us summarize briefly the main results of the present chapter.

(a) In the one-dimensional theory of odd-degree polynomial splines uniqueness of interpolation splines implies their existence through the very simple principle of linear algebra, i.e. we have the property

$$\boxed{\textbf{uniqueness} \Longrightarrow \textbf{existence}.}$$

The remarkable thing about the interpolation polysplines of order $p = 2q$ is that this implication is preserved at a higher level. We will prove *uniqueness* for general interfaces Γ_j in the *Sobolev* spaces and in the *Hölder* spaces, see Theorem 20.9, p. 421. The implication **uniqueness** $\Longrightarrow$ **existence** now follows through the *Fredholm* property of an appropriate operator, to be proved in Theorem 21.4, p. 436. This immediately provides the *existence* of *interpolation polysplines* of order $2q$ for *every* set of interpolation data $\{g_j\}$, see Theorem 22.2, p. 446.[6] There is also a typically infinite-dimensional effect: the polyspline depends continuously (in Sobolev or Hölder norms) on the data $\{g_j\}$ which is obvious in the finite-dimensional case.

However, if no uniqueness is available, which is the case for interpolation polysplines of *arbitrary* order $p \geq 1$ in the spaces of *Sobolev* (or in *Hölder* spaces), one has to check that the interpolation data $\{g_j\}$ satisfy a finite number of "compatibility conditions" arising through the *Fredholm* property of an appropriate operator, see Theorem 21.4 cited above. Only for special interfaces such as parallel hyperplanes (polysplines on strips) and concentric spheres (polysplines on annuli) have we proved the uniqueness of interpolation polysplines for arbitrary order $p \geq 1$, in Proposition 5.8, p. 64, and Proposition 8.6, p. 110. And only then may we apply the principle **uniqueness** $\Longrightarrow$ **existence**.

The core of the present variational approach to polysplines of order $p = 2q$ is a basic identity for them (see Section 20.3, p. 415) which is analogous to the fundamental identity for the one-dimensional odd-degree splines proved in Theorem 2.4, p. 24, and in Lemma 2.5, p. 26. By virtue of this identity we obtain uniqueness for interpolation polysplines (see Theorem 20.9, p. 421), and the equivalence of the two definitions. By virtue of the same identity we prove in Theorem 20.14, p. 425, the analog to the famous extremal property of the one-dimensional splines, called *Holladay*'s theorem (recall the one-dimensional result in Section 2.3, p. 23).

---

[6] Recall that one may consider the Fredholm alternative as an infinite-dimensional generalization of linear algebra for operators acting in Banach spaces. In the finite-dimensional case for a quadratic matrix we have $\operatorname{Ker} T = \{0\}$ implies $(\operatorname{Im} T)^\perp = \{0\}$ which is adequately generalized by the Fredholm alternative.

The above uniqueness results are proved for a wide class of polysplines of order $p = 2q$ which are piecewise solutions of the equation

$$Lu = 0$$

for the operator L of the form $L = L_1^2$, where L_1 is a *uniformly strongly elliptic* operator in the domain D which is *formally self-adjoint* in D.

(b) By using the method of *frozen coefficients* we provide the existence theorem in the form of a *Fredholm alternative* for an arbitrary integer $p \geq 1$. The most technical is Proposition 21.1, p. 430, the proof of which is reduced to checking the conditions for a regular elliptic BVP known from Agmon *et al.* [2], and Lions and Magenes [46, Section 2.4.1], for a special system of boundary operators.

Thus, the reader who is not experienced in *a priori* estimates for higher-order elliptic equations, or is interested in the subject mainly from the point of view of spline theory, may ignore Chapter 21. The final existence results in Chapter 22, p. 445, will provide enough detail. Still, we would encourage such a reader to visit Section 21.1 to make sure that, modulo some rather standard techniques of frozen coefficient, things are essentially reduced to one-dimensional, *Chebyshev splines*,[7] see Proposition 21.1, p. 430, and Lemma 21.2, p. 432. This point of view is strongly supported by the two special cases of polysplines considered up to now in the previous Parts of the book.

For a reader familiar with the *a priori* estimates for higher-order elliptic equations there will be nothing new beyond the classical references [2, 46] and Taylor [67] on manifolds. Proposition 21.1 will be sufficient for the reader to obtain the *a priori estimate* of Theorem 21.3, p. 435, and the *Fredholm* Theorem 21.4, p. 436.

As we have already said, we consider and solve a much more general problem than is necessary for interpolation polysplines, see below problems (21.10)–(21.13), p. 433, whose solution is called the *general polyspline*. This more general result may be used to prove the existence of analogs to *smoothing*, *mono-*, *perfect* and *Bernoulli* splines of the one-dimensional case. We will not consider these generalizations in the present book.

19.6 Open problems

After having introduced the notion of *polyspline* by a direct analogy with the univariate case, it is natural to ask how far the analogy extends. Let us indicate some interesting, unsolved problems.

(1) From the point of view of applications it is natural to have a data set ST which is a union of intersecting smooth manifolds, i.e., when ST itself is a manifold with singularities. In $\mathbb{R}^2$ in case of such a data set ST polysplines are studied in [41]. However, in dimensions greater than *two* there is no complete theory of elliptic BVPs in domains with singularities. For that reason, the results of our research cannot, for the moment be extended to general interface sets with intersections.

(2) The theory of polysplines of odd-order $p = 2q + 1$ needs a conceptual development. As we said above we do not have an identity and uniqueness theorem analogous

---

[7] For the notion of Chebyshev splines see Chapter 11, p. 187, and Schumaker [61, Chapter 10].

to the one for polysplines of order $p = 2q$ provided in Theorem 20.11, p. 423. Let us recall that for the corresponding one-dimensional splines, which are piecewise *even-degree polynomials*, in order to obtain a good theory (and numerics) one has to choose interlacing data points and break-points. [14, Chapter 6]. It is an open question how to find a proper analog of that for polysplines of order $p = 2q + 1$. The question is, which will be the "data surfaces" and which the "interface surfaces". In the special case of cardinal polysplines on annuli we may choose the data on concentric spheres other than the "break-spheres" and the results show that the data functions are (as it may be expected) subject to very strong smoothness restrictions, [43].

Chapter 20

Definition of polysplines and uniqueness for general interfaces

20.1 Introduction

In contrast to Part I we now wish to present the basic concepts of polysplines in nontrivial interface configurations. We will make many simplifications in order to avoid useless formalism. For example we will assume that we work only in domains in $\mathbb{R}^n$ having a C^∞ boundary and the interface surfaces will be C^∞. We will formulate our final results, which act on manifolds with boundaries which have sufficient smoothness, in Chapter 22, p. 445. This will be possible due to the flexibility of the method of frozen coefficients and the use of the continuous parameter method.

Let us introduce the necessary notation.

(a) Let D be a bounded domain in $\mathbb{R}^n$ with infinitely smooth boundary Γ which has a connected complement in $\mathbb{R}^n$. We assume that D lies on one side of the boundary Γ.

We will assume that a subdivision of D is defined in the following sense.

Suppose that a family of subdomains $\widetilde{D}_1, \ldots, \widetilde{D}_N$, where $\widetilde{D}_N = D$, be given such that the topological closure $\text{cl}(\widetilde{D}_j)$ lies in the domain $\widetilde{D}_{j+1}$, i.e.

$$\text{cl}(\widetilde{D}_j) \subseteq \widetilde{D}_{j+1} \quad \text{for } j = 1, 2, \ldots, N-1.$$

We put

$$\Gamma_j = \partial \widetilde{D}_j \quad \text{for } j = 1, 2, \ldots, N,$$

and so we have

$$\Gamma_N = \partial D.$$

We also assume that for every $j = 1, 2, \ldots, N$, the domain $\widetilde{D}_j$ has a connected complement in $\mathbb{R}^n$ and lies on one side of its boundary Γ_j. For convenience we put $\Gamma_0 = \emptyset$.

By D_j we shall denote the layer which lies between the surfaces Γ_j and Γ_{j+1}, i.e.

$$D_j := \widetilde{D}_j \setminus \text{cl}(\widetilde{D}_{j-1}) \quad \text{for } j = 2, \ldots, N;$$

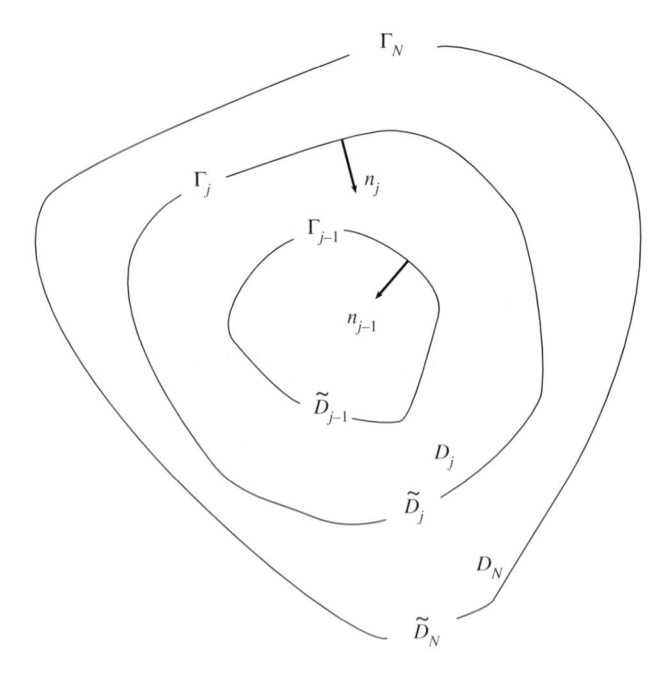

Figure 20.1. A typical geometric configuration of the interfaces.

we put
$$D_1 = \widetilde{D}_1.$$

By $\vec{n} = \vec{n}_j = \vec{n}_j(x)$ we shall denote the *inner* unit normal vector to Γ_j at the point $x \in \Gamma_j$ (see Figure 20.1).

Naturally, for $j = 1, 2, \ldots, N-1$ the interfaces $\Gamma_j = \overline{D_j} \cap \overline{D_{j+1}}$ will be called *interfaces*. The union of the surfaces

$$ST := \bigcup_{j=1}^{N} \Gamma_j$$

is considered to be the *data set* for the polysplines.

Due to the above properties of the interfaces Γ_j it will be possible to apply all results of the theory of elliptic BVPs such as solubility, Green's formulas, etc., cf. Chapter 23, p. 461.

The simplest *examples* of polysplines were considered in Chapters 5 and 8. In Chapter 8 we considered the case of concentric spheres when $D = B(0; r_N)$ is a ball in $\mathbb{R}^n$, and the domains $\widetilde{D}_j = B(0; r_j)$ for $j = 1, 2, \ldots, N$, i.e. they are N concentric balls. In the last case, for $j = 2, \ldots, N$, the domain D_j is the spherical layer between the spheres $\Gamma_{j-1} = \partial \widetilde{D}_{j-1}$ and $\Gamma_j = \partial \widetilde{D}_j$. The domain D_1 is a ball.

(b) In what follows we shall use the standard notation and results concerning spaces of functions on domains and their boundaries which are available in Chapter 23, p. 461, see also [46, 67].

We shall denote by $H^s(D_j)$ the Sobolev space for the domain D_j, and by $H^s(\Gamma_j)$ the Sobolev space for the interface surface $\Gamma_j = \partial \widetilde{D}_j$. For a domain Ω we use $C_0^\infty(\Omega)$ to denote the functions which are C^∞ in Ω and have a compact support in Ω. We use $C^\infty(\overline{\Omega})$ to denote the functions which are C^∞ in the neighborhood of $\overline{\Omega}$.

(c) The space

$$H^s(D \setminus ST) := \bigcup_j H^s(D_j)$$

should be understood as the space of functions f defined on D for which $f \in H^s(D_j)$ for every component D_j of $D \setminus ST$. We use

$$C^{k,\alpha}(D \setminus ST) := \bigcup_j C^{k,\alpha}(\overline{D_j})$$

to denote the set of functions f defined in D and such that $f \in C^{k,\alpha}(\overline{D_j})$ for all $j = 1, 2, \ldots, N$.

20.2 Definition of polysplines

For a function $u(x)$ defined in the domain D we shall denote by $u_j(x)$ its restriction to the subdomain D_j where $j = 1, 2, \ldots, N$. Let L be a *uniformly strongly elliptic operator* of order $2p$ with *real coefficients* in D, and the system of operators $\{B_k\}_{k=0}^{p-1}$ be such that $\{L, B_k, k = 0, 1, \ldots, p-1\}$ is a regular elliptic BVP in the domain D in the sense of Definition 23.11, p. 473.

Now let G be an arbitrary subdomain of D. Owing to the regularity Theorem 23.18, p. 480, any distributional solution to problem

$$Lu = 0 \quad \text{in } G$$

which belongs to $L_2(G)$ satisfies $u \in C^\infty(G)$. The polysplines will be defined as solutions to $Lu = 0$ in every subdomain D_j. As such they will satisfy $u \in C^\infty(D_j)$. We will impose some interface conditions on the common parts of the boundaries ∂D_j.

Definition 20.1 *1. We will say that the function u is a **polyspline** of order p in the Sobolev space $H^{2p}(D \setminus ST)$ with respect to the operator L (for the given subdivision $\{D_j\}$ of the domain D) if and only if the following conditions hold:*[1]

- *All pieces u_j are solutions belonging to $H^{2p}(D_j)$ of the equation*

$$Lu_j = 0 \quad \text{in } D_j, \quad \text{for } j = 1, 2, \ldots, N. \tag{20.1}$$

- *The following interface conditions on Γ_j for $j = 1, 2, \ldots, N-1$, hold:*

$$\frac{\partial^k}{\partial n^k} u_j = \frac{\partial^k}{\partial n^k} u_{j+1} \quad \text{on } \Gamma_j, \quad \text{for } k = 0, 1, \ldots, 2p-2. \tag{20.2}$$

---

[1] Note that although the operator L is of order $2p$ we have called the polyspline u "of order p". Thus we keep close to the terminology in the one-dimensional case.

These equalities should be considered as equalities between traces of functions.

2. We will say that the function u is a **polyspline of class** C^{2p-2} if for all $j = 1, 2, \ldots, N$, the pieces u_j satisfy equation (20.1), u_j belongs to the space $C^{2p-2}(\overline{D_j})$, and the interface conditions (20.2) hold.

3. If the operator L is fixed the space of all such polysplines will be denoted by $PS_p(D)$ or simply by PS. In order to save some notation we will use the same notation for the polysplines in Sobolev spaces or in smoothness classes. The meaning will be clear from the context.

Definition 20.2 *Let the functions f_j be given on the interfaces Γ_j for $j = 1, 2, \ldots, N$. The polyspline u will be called* interpolation with data $\{f_j\}$ *if it satisfies*

$$u = f_j \quad \text{on } \Gamma_j, \quad \text{for } j = 1, 2, \ldots, N. \tag{20.3}$$

In the case of polysplines u belonging to the Sobolev space $H^{2p}(D \setminus ST)$ equality (20.3) is understood as equality between traces on Γ_j.

In addition to the above we may assume that the polyspline u satisfies some boundary conditions on ∂D, e.g. the following *boundary* conditions:

$$B_k u = b_k \quad \text{on } \Gamma_N, \quad \text{for } k = 1, 2, \ldots, p-1, \tag{20.4}$$

where b_k are some prescribed functions. Equalities (20.4) should be considered as equalities between traces of functions. Let us note that just as in the one-dimensional case the boundary conditions (20.4) are optional and they may be chosen in different ways.

Proposition 20.3 is an immediate consequence of the definition of the polyspline.

Proposition 20.3 *Let $u(x)$ be a polyspline in the Sobolev space $H^{2p}(D \setminus ST)$. Then the following properties hold:*

1. For every $j = 1, 2, \ldots, N-1$, and for every multi-index α satisfying $|\alpha| \leq 2p-2$, the following equality holds:

$$D^\alpha u_{j+1} = D^\alpha u_j \quad \text{on } \Gamma_j.$$

2. More generally, let the differential operator A of order $2p - 1$ have the representation

$$A(\widetilde{x}, D) = \sum_{k \leq 2p-2} a_{\beta,k}(\widetilde{x}) D_\tau^\beta \frac{\partial^k}{\partial n^k} \tag{20.5}$$

at every point $\widetilde{x} \in \Gamma_j$, with $k + |\beta| \leq 2p - 1$, where D_τ is the operator of differentiation in the subspace tangential to Γ_j at $\widetilde{x}$ directions. Then

$$A u_{j+1} = A u_j \quad \text{on } \Gamma_j,$$

(see Figure 20.2).

3. Let A be an operator of order $2p - 1$. Then

$$A u_{j+1} - A u_j = c \left(\frac{\partial^{2p-1}}{\partial n^{2p-1}} u_{j+1} - \frac{\partial^{2p-1}}{\partial n^{2p-1}} u_j \right) \quad \text{on } \Gamma_j,$$

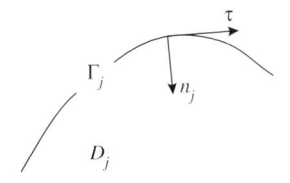

Figure 20.2.

where $c = a_{0,2p-1}(x)$ is the coefficient of the local representation (20.5) of the operator A.

4. The same statement holds if u is a polyspline of class C^{2p-2} and if u belongs to $C^{2p-1}(\overline{D_j})$ for $j = 1, 2, \ldots, N$.

Proof The proofs of (1) and (2) are an immediate consequence of the trace Theorem 23.4, p. 468, since in particular

$$\frac{\partial^k}{\partial n^k} u_j \in H^{2p-k-\frac{1}{2}}(\Gamma_j).$$

For the proof of (3) due to (2) we need to consider only those terms of the operator A, for which $k = 2p - 1$. Since the operator A is of order $2p - 1$ there exists only one such term and it is

$$a_{0,2p-1}(\widetilde{x}) \frac{\partial^{2p-1}}{\partial n^{2p-1}}.$$

This completes the proof of (3).

The proof of (4) follows by the trace properties of the functions of differentiable functions. ∎

We will mainly be interested in the case of the polyharmonic operator,

$$L(x, D) = \Delta^p,$$

and, in particular, in the *even-order polysplines* when $p = 2q$,

$$L(x, D) = \Delta^{2q}.$$

In the last case we will choose for example the boundary operators B_k, for $k = 0, 1, \ldots, 2q - 1$, as the set of the following operators:

$$1, \quad \Delta, \quad \Delta^2, \quad \ldots, \quad \Delta^{q-1},$$

$$\frac{\partial}{\partial n}, \quad \frac{\partial}{\partial n}\Delta, \quad \frac{\partial}{\partial n}\Delta^2, \quad \ldots, \quad \frac{\partial}{\partial n}\Delta^{q-1},$$

i.e. we put $B_k = \Delta^k$ for $k = 0, 1, \ldots, q - 1$, and $B_k = (\partial/\partial n)\Delta^{k-q}$ for $k = q, q+1, \ldots, 2q - 1$.

It is necessary for later purposes to find other equivalent sets of interface conditions in (20.2) which is the subject of Proposition 20.4.

Proposition 20.4 *The set of boundary operators*

$$B_k = \frac{\partial^k}{\partial n^k} \quad \text{for } k = 0, 1, \ldots, 2p - 2$$

is equivalent to the set of boundary operators

$$C_k = \Delta^k \quad \text{for } k = 0, 1, \ldots, p - 1,$$

$$C_k = \frac{\partial}{\partial n} \Delta^{k-p} \quad \text{for } k = p, p+1, \ldots, 2p - 2,$$

on the interface set Γ_j in the following sense. Let the function $u \in H^{2p-1}(D_j)$ and $v \in H^{2p-1}(D_{j+1})$, and satisfy

$$B_k u = B_k v \quad \text{on } \Gamma_j \text{ for } k = 0, 1, \ldots, 2p - 2.$$

Then

$$C_k u = C_k v \quad \text{on } \Gamma_j \text{ for } k = 0, 1, \ldots, 2p - 2$$

and vice versa.

The same statement holds if $u \in C^{2p-2}(\overline{D_j})$ and $v \in C^{2p-2}(\overline{D_{j+1}})$.

The proof follows from Lemma 2.1 in [46, Chapter II.2], where such equivalence is proved for every two Dirichlet systems of the same order. One has to prove that the system of operators $\{C_k\}$ is a Dirichlet system, see Definition 23.12, p. 474.

We will also need a similar equivalence for the arbitrary formally self-adjoint operator L. Assuming that $L = L^*$ let Γ_j, for every interface surface denote by $\{B_k^j\}$ a set of operators such that $\{L, B_k^j, k = 0, 2, \ldots, p-1\}$ is a regular elliptic BVP in every domain D_j in the sense of Definition 23.11, p. 473. By Proposition 23.16, p. 477, we have for every domain D_j the operators $\{S_k^j\}$ related to the operators $\{L, B_k^j, k = 0, 2, \ldots, p-1\}$ through a Green formula, and by Proposition 23.17, p. 478, we have $S_k^j = -S_k^{j+1}$ on the intersection $\Gamma_j = D_j \cap D_{j+1}$ for the neighboring domains D_j and D_{j+1}.

We now have a generalization of the above proposition.

Proposition 20.5 *On the interface Γ_j the set of boundary operators (the so-called Dirichlet boundary conditions)*

$$M_k = \frac{\partial^k}{\partial n^k} \quad \text{for } k = 0, 1, \ldots, 2p - 2$$

is equivalent to the set of boundary operators

$$C_k = B_k^j \quad \text{for } k = 0, 1, \ldots, p - 1,$$

$$C_{p+k} = S_k^j \quad \text{for } k = 1, 2, \ldots, p - 1,$$

in the following sense. Let the function $u \in H^{2p}(D_j)$ and the function $v \in H^{2p}(D_{j+1})$, and

$$M_k u = M_k v \quad \text{on } \Gamma_j \quad \text{for } k = 0, 1, \ldots, 2p-2.$$

Then

$$C_k u = C_k v \quad \text{on } \Gamma_j \quad \text{for } k = 0, 1, \ldots, 2p-2,$$

and vice versa.

For the proof recall that

$$\text{ord } B_0 = 1, \quad \text{hence}$$
$$\text{ord } S_0 = 2p-1.$$

The proof follows further from Lemma 2.1 in [46, Chapter II.2], since the set of operators $\{B_k, S_k, k = 0, 1, \ldots, p-1\}$ forms a Dirichlet system on Γ_j for every $j = 1, 2, \ldots, N$. See Definition 23.12, p. 474.

Remark 20.6 The meaning of Lemma 2.1 in Lions and Magenes [46, Chapter II] may be explained as follows. Let us consider a system of boundary operators $\{B_k\}_{k=0}^{\nu}$ which is a Dirichlet system, see Definition 23.12, p. 474. Let the following boundary conditions hold:

$$B_k u = b_k \quad \text{on } \Gamma_j \quad \text{for } k = 0, 1, \ldots, \nu,$$

as well as boundary conditions containing only the normal derivatives,

$$\frac{\partial^k u}{\partial n^k} = c_k \quad \text{on } \Gamma_j \quad \text{for } k = 0, 1, \ldots, \nu.$$

Then for all functions $u \in H^{\nu+1}(D_j)$ we may express the functions b_k through the functions c_k and vice versa. This is the content of Lemma 2.1 in [46, Chapter II]. We have considered a special case in Proposition 20.5, p. 414.

20.3 Basic identity for polysplines of even order $p = 2q$

Now we come to one of the most remarkable points of the general theory of polysplines. As we have seen in Section 2.4, p. 26, in the one-dimensional theory of *odd-degree polynomial splines* one proves through an ingenious identity (in Theorem 2.4, p. 24) the uniqueness and simultaneously the existence of interpolation splines. In the present section we prove a *similar identity* in the general domain but only for polysplines of even order $p = 2q$. As will be seen in Section 20.4, the uniqueness of interpolation polysplines in the general case holds for such polysplines. Later this will imply *existence* through the Fredholm alternative and stable *a priori* estimates for the dependence of the polysplines on the data. Let us note that uniqueness for the *odd-order* polysplines, for $p = 2q+1$, has been proved by us only in very special configurations of the data set, such as strips and annuli, see Proposition 5.8, p. 64, and Proposition 8.6, p. 110.

Apparently, there is no similar identity and uniqueness in the general interface configuration for odd-order polysplines. The fact that the even-order polysplines behave better than the odd-order polysplines seems to be a demonstration of the one-dimensional phenomenon but at a higher level.[2]

20.3.1 Identity for $L = \Delta^{2q}$

Let us start with the simplest case. The following is a *basic identity* for *polysplines* in the case of the operator $L = \Delta^{2q}$.

Theorem 20.7 *Let the operator* $L = \Delta^{2q}$ *and we have the space of polysplines* $PS_{2q}(D)$ *in the Sobolev space* $H^{4q}(D \setminus ST)$. *Then for every* $h \in H^{2q}(D)$ *and every polyspline* $u \in PS_{2q}(D)$ *the following equality holds true:*

$$\int_D \Delta^q u \Delta^q h \, dx$$

$$= \int_{ST \setminus \partial D} \left\{ \frac{\partial}{\partial n} \Delta^{2q-1} u \right\}_{\text{jump}} h \, d\omega(x)$$

$$- \sum_{l=0}^{q-1} \int_{\Gamma_N} \left\{ \Delta^{l+q} u \frac{\partial}{\partial n} \Delta^{q-1-l} h - \frac{\partial}{\partial n} \Delta^{l+q} u \Delta^{q-1-l} h \right\} d\omega(x), \qquad (20.6)$$

where $\{(\partial^{4q-1}/\partial n^{4q-1})u\}_{\text{jump}}$ *denotes the jump at the point* $x \in \Gamma_j$, *which is equal to*

$$\frac{\partial^{4q-1}}{\partial n^{4q-1}} u_j - \frac{\partial^{4q-1}}{\partial n^{4q-1}} u_{j+1} \quad \text{on } \Gamma_j,$$

for $j = 1, \ldots, N - 1$. *(For* $j = N$ *we have only one term,* $(\partial/\partial n)\Delta^{2q-1} u_N$ *on* Γ_N.) *Here* $\omega(x)$ *denotes the intrinsic measure on the union of the surfaces* Γ_j, $j = 1, \ldots, N$.

The same statement holds if we consider polysplines u *of class* C^{4q-2} *satisfying* $u \in C^{4q-1}(D \setminus ST)$ *(i.e.* $u \in C^{4q-1}(\overline{D_j})$ *for* $j = 1, 2, \ldots, N$) *and take functions* $h \in C^{2q}(\overline{D})$.

Proof By the Green formula in Aronszajn *et al.* [7], see equality (23.6), p. 479, for every j we obtain the following formula:

$$\int_{D_j} \Delta^q u \Delta^q h \, dx + \sum_{l=0}^{q-1} \int_{\partial D_j} \left\{ \Delta^{l+q} u \frac{\partial}{\partial \nu_j} \Delta^{q-1-l} h \right.$$

$$\left. - \frac{\partial}{\partial \nu_j} \Delta^{l+q} u \Delta^{q-1-l} h \right\} d\sigma_j(x) = 0,$$

---

[2] In a certain sense this is expected to also have some consequences for the numerical analysis of the polysplines: the numerical stability of the even-order interpolation polysplines will be greater than that of the odd-order polysplines. Let us note that in the cases of uniqueness for polysplines of order $p = 2q+1$, pointed out above, we will also have the inequality $\|u\| \leq C \sum \|g_j\|$ which provides some stability dependence on the data. However, a small perturbation of the interfaces T_j might spoil the uniqueness and hence that estimate. Thus we might speculate that the numerical stability is not among the virtues of the odd-order polysplines.

Polysplines and uniqueness for general interfaces 417

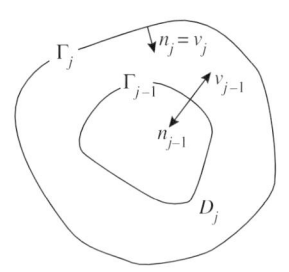

Figure 20.3.

where $d\sigma_j(x)$ denotes the element of the intrinsic measure on ∂D_j, and v_j is the **inner** unit normal vector to the domain D_j.

Taking into account that $v_j = n_j$ on Γ_j and $v_j = -n_{j-1}$ on Γ_{j-1}, after summing up in $j = 1, \ldots, N$, we obtain formula (20.6), see Figure 20.3.

The *jump* term is obtained by the observation that

$$\frac{\partial}{\partial n}\Delta^{2q-1}u_j - \frac{\partial}{\partial n}\Delta^{2q-1}u_{j+1} = \frac{\partial^{4q-1}}{\partial n^{4q-1}}u_j - \frac{\partial^{4q-1}}{\partial n^{4q-1}}u_{j+1} \quad \text{on } \Gamma_j,$$

which follows from Proposition 20.3, (3) on p. 412. ∎

20.3.2 Identity for the operator $L = L_1^2$

Now let us generalize the previous subsection by considering the operator $L = L_1^2$ where L_1 is a *uniformly and strongly elliptic* operator of order $2q$ in the whole domain D and is *formally self-adjoint* there.

On every domain D_j we would like to choose a regular elliptic BVP so that on the intersection $T_j = D_j \cap D_{j+1}$ the boundary operators agree. To achieve this, we assume that on every interface Γ_j, for $j = 1, 2, \ldots, N$, a *normal* system of operators $\{\widetilde{B}_k^j\}_{k=0}^{p-1}$ be given (in the sense of Definition 23.11, (5) on p. 473) such that

$$\widetilde{B}_0^j = 1,$$
$$\text{ord } \widetilde{B}_k^j = k \quad \text{for } k = 0, 1, \ldots, q-1. \tag{20.7}$$

and in the domain $\widetilde{D}_j$ the system of operators

$$\{L_1, \widetilde{B}_k^j, \ k = 0, 1, \ldots, q-1\}$$

forms a *regular elliptic BVP*. For the definition of the domains $\widetilde{D}_j$, see Section 20.1, p. 409, and Figure 20.1, p. 410.

Now let us introduce a regular elliptic BVP on every layer D_j. We note again that all the conditions of a regular elliptic BVP in Definition 23.11, p. 473, carry local character.

Since $\partial D_j = \Gamma_{j-1} \cup \Gamma_j$ (for $j = 2, 3, \ldots, N$) we will define the boundary operator B_k^j as

$$B_k^j := \begin{cases} \widetilde{B}_k^{j-1} & \text{on } \Gamma_{j-1}, \\ \widetilde{B}_k^j & \text{on } \Gamma_j. \end{cases}$$

Thus the system

$$\{L_1, B_k^j, \; k = 0, 1, \ldots, q - 1\}$$

is a regular elliptic BVP on every domain D_j for $j = 1, 2, \ldots, N$.

Now the choice ord $\widetilde{B}_k^j = k$ in (20.7) shows that the system $\{B_k^j\}_{k=0}^{q-1}$ is a *Dirichlet system* of order q (see Definition 23.12, p. 474) and we may apply Proposition 23.16, p. 477, of Chapter 23. According to Proposition 23.16 the adjoint boundary operators for a self-adjoint operator have more special properties, namely for every $j = 1, 2, \ldots, N$, we have only one set of boundary operators $\{S_k^j\}_{k=0}^{q-1}$ in the domain D_j, related through a Green formula to the operators B_k^j. Evidently, every such operator S_k^j is a pair of boundary operators, namely,[3]

$$S_k^j := \begin{cases} \mathcal{L} S_k^{j-1} & \text{on } \Gamma_{j-1}, \\ \mathcal{R} S_k^j & \text{on } \Gamma_j. \end{cases}$$

See Figure 20.4.

According to Proposition 23.17, p. 478, where elliptic BVP in neighboring domains were considered, the following equality is satisfied:

$$\mathcal{L} S_k^{j-1} = -\mathcal{R} S_k^{j-1} \quad \text{for } j = 2, 3, \ldots, N.$$

Due to the choice $\widetilde{B}_0^j = 1$ and by property

$$\text{ord } S_k^j = 2q - 1 - \text{ord } B_k^j,$$

available in Proposition 23.16, p. 477, we see that

$$\text{ord } S_0^j = 2q - 1.$$

By Theorem 23.13, p. 476, about adjoint systems of operators, the set

$$\{\text{ord } S_k^j, \text{ord } B_k^j : \quad \text{for } k = 0, 1, \ldots, q - 1\}$$

---

[3] Here the origin of the notation is: $\mathcal{L}$ from "left" and $\mathcal{R}$ from "right". In the case of the operator B_k^j such notation would not be so appropriate since if we put $\widetilde{B}_k^{j-1} = \mathcal{L} B_k^j$ and $\widetilde{B}_k^j = \mathcal{R} B_k^j$ we would have $\mathcal{L} B_k^{j+1} = \mathcal{L} B_k^j$.

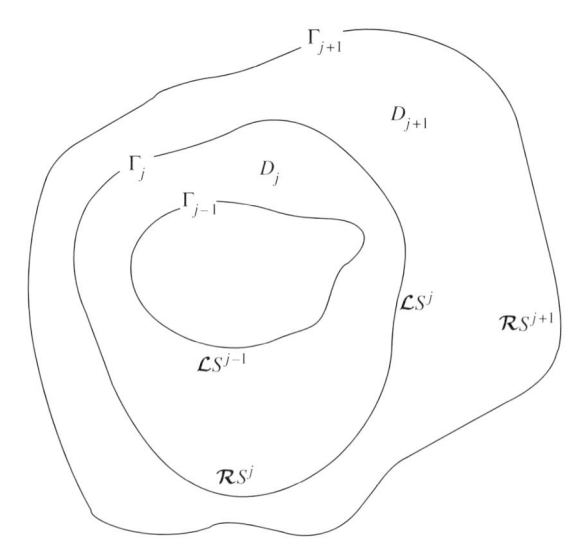

Figure 20.4.

coincides with the set $\{0, 1, \ldots, 2q - 1\}$. An immediate consequence of the last is that for all $k = 1, \ldots, q - 1$ we have

$$\operatorname{ord} S_k^j \leq 2q - 2.$$

Taking into account these properties we may generalize the above basic identity of Theorem 20.7, p. 416, and establish an identity for the polysplines defined with respect to the operator $L = L_1^2$.

Theorem 20.8 *Let $u \in H^{4q}(D \setminus ST)$ be a polyspline of order $2q$ for the operator $L = L_1^2$ in the domain D. Then for every function $h \in H^{2q}(D)$ the following identity holds true:*

$$-\int_D L_1 u \cdot L_1 h \, dx = \sum_{j=1}^{N-1} \int_{\Gamma_j} c_j \left\{ \frac{\partial^{4q-1}}{\partial n^{4q-1}} u \right\}_{\text{jump}} \times h \, d\omega$$

$$+ \sum_{k=0}^{q-1} \int_{\Gamma_N} \left(\mathcal{R} S_k^N L_1 u \cdot B_k^N h \, d\sigma - B_k^N L_1 u \cdot \mathcal{R} S_k^N h \right) d\omega,$$

where $\{(\partial^{4q-1}/\partial n^{4q-1})u\}_{\text{jump}}$ denotes the jump at the point $x \in \Gamma_j$, which is equal to

$$\frac{\partial^{4q-1}}{\partial n^{4q-1}} u_j - \frac{\partial^{4q-1}}{\partial n^{4q-1}} u_{j+1} \quad \text{on } \Gamma_j,$$

for $j = 1, \ldots, N - 1$. Here $\omega = \omega(x)$ denotes the intrinsic measure on the union of the surfaces Γ_j, for $j = 1, \ldots, N$, and $c_j = c_j(x)$ is the coefficient in the representation

$$\mathcal{R} S_0^j L_1 = c_j(x) \frac{\partial^{4q-1}}{\partial n^{4q-1}} + \text{derivatives of lower order in } \frac{\partial}{\partial n}.$$

420 *Multivariate polysplines*

The same statement holds for polysplines u of class C^{4q-2} satisfying $u \in C^{4q-1}(D \setminus ST)$, and functions $h \in C^{2q}(\overline{D})$.

Proof By the Green formula of Proposition 23.16, p. 477, for every j we have the following equality:

$$\int_{D_j} L_1 L_1 u \cdot h \, dx - \int_{D_j} L_1 u \cdot L_1 h \, dx$$

$$= \sum_{k=0}^{q-1} \int_{\Gamma_{j-1}} \left(\mathcal{L} S_k^j L_1 u_j \cdot B_k^j h \, d\sigma - B_k^j L_1 u_j \cdot \mathcal{L} S_k^j h \right) d\omega$$

$$+ \sum_{k=0}^{q-1} \int_{\Gamma_j} \left(\mathcal{R} S_k^j L_1 u_j \cdot B_k^j h \, d\sigma - B_k^j L_1 u_j \cdot \mathcal{R} S_k^j h \right) d\omega.$$

Note that on the left-hand side we have $L_1^2 u = 0$.

To make the proof easier to comprehend, let us consider the case $q = 1$. Then L_1 is an operator of order 2. Taking into account the following two circumstances: (1) the operators $B_0^j = 1$ and (2) the reflection property $\mathcal{R}S_0^j = -\mathcal{L}S_0^{j+1}$, after summing up the above equalities in j we obtain the equality

$$-\int_D L_1 u \cdot L_1 h \, dx = \sum_{j=1}^{N-1} \int_{\Gamma_j} \left(\mathcal{R}S_0^j L_1 u_j \cdot h - L_1 u_j \cdot \mathcal{R}S_0^j h \right) d\omega$$

$$- \sum_{j=1}^{N-1} \int_{\Gamma_j} \left(\mathcal{R}S_0^j L_1 u_{j+1} \cdot h - L_1 u_{j+1} \cdot \mathcal{R}S_0^j h \right) d\omega$$

$$+ \int_{\Gamma_N} \left(\mathcal{R}S_0^N L_1 u_N \cdot h - L_1 u_N \cdot \mathcal{R}S_0^N h \right) d\omega.$$

Now we apply Proposition 20.3, p. 412, which implies that $L_1 u_{j+1} = L_1 u_j$ on Γ_j, hence

$$-\int_D L_1 u \cdot L_1 h \, dx = \sum_{j=1}^{N-1} \int_{\Gamma_j} \{\mathcal{R}S_0^j L_1 u\}_{\text{jump}} \cdot h \, d\omega$$

$$+ \int_{\Gamma_N} \{\mathcal{R}S_0^N L_1 u_N \cdot h - L_1 u_N \cdot \mathcal{R}S_0^N h\} \, d\omega,$$

where for $j = 1, 2, \ldots, N-1$ we have introduced the notation

$$\{\mathcal{R}S_0^j L_1 u\}_{\text{jump}} := \mathcal{R}S_0^j L_1 u_j - \mathcal{R}S_0^j L_1 u_{j+1} \quad \text{on } \Gamma_j.$$

By Proposition 20.3, (3) we obtain (substituting there $p = 1$),

$$\{\mathcal{RS}_0^j L_1 u\}_{\text{jump}} = c_j \left\{ \frac{\partial^3}{\partial n^3} u \right\}_{\text{jump}}$$

$$= c_j \left(\frac{\partial^3}{\partial n^3} u_j - \frac{\partial^3}{\partial n^3} u_{j+1} \right) \quad \text{on } \Gamma_j,$$

where $c_j = c_j(x)$ is the coefficient in the representation

$$\mathcal{RS}_0^j L_1 = c_j(x) \frac{\partial^{4q-1}}{\partial n^{4q-1}} + \text{derivatives of lower order in } \frac{\partial}{\partial n}.$$

This implies the equality

$$-\int_D L_1 u \cdot L_1 h \, dx = \sum_{j=1}^{N-1} \int_{\Gamma_j} c_j \left\{ \frac{\partial^{4q-1}}{\partial n^{4q-1}} u \right\}_{\text{jump}} \cdot h \, d\omega$$

$$+ \int_{\Gamma_N} \left(\mathcal{RS}_0^N L_1 u_N \cdot h - L_1 u_N \cdot \mathcal{RS}_0^N h \right) d\omega.$$

This ends the consideration of the case $q = 1$.

The proof for arbitrary $q \geq 1$ runs along the same lines. The only extra argument which we have to take into account is the inequality ord $B_k^j \leq 2q - 2$, which provides the cancellations of the corresponding terms in the sum. ∎

20.4 Uniqueness of interpolation polysplines and extremal Holladay-type property

The most remarkable thing about the uniqueness theorems below is that we do not impose any special conditions on the interfaces Γ_j and on the interface conditions and yet we still obtain *uniqueness* of the interpolation polysplines.[4] The uniqueness result of the present section is a *complete analog* to the uniqueness of odd-degree polynomial splines in the one-dimensional case, Theorem 2.7, p. 26. The means for proving the uniqueness of interpolation polysplines is again a proper identity which we have proved in Theorem 20.7, p. 416. We start with polysplines for the operator $L = \Delta^{2q}$.

Theorem 20.9 *Let $L = \Delta^{2q}$ and let us suppose that $u(x)$ is a polyspline of order $2q$ in the Sobolev space $H^{4q}(D \setminus ST)$ with respect to the operator $L = \Delta^{2q}$, and $u(x)$ satisfies the following zero boundary conditions (see equalities (20.4), p. 412):*

$$\left. \begin{array}{ll} \Delta^k u_N = 0 & \text{on } \Gamma_N \quad \text{for } k = 1, 2, \ldots, q-1, \\ \frac{\partial}{\partial n} \Delta^k u_N = 0 & \text{on } \Gamma_N \quad \text{for } k = 0, 1, \ldots, q-1. \end{array} \right\} \quad (20.8)$$

If $u = 0$ on the set ST then $u \equiv 0$ in the whole domain D.

---

[4] See the remarks about uniqueness in Chapter 19, p. 400.

The same statement holds for polysplines u of class C^{4q-2} satisfying $u \in C^{4q-1,\alpha}(D \setminus ST)$ for $0 < \alpha < 1$.

Proof Let us put $h(x) = u(x)$ in formula (20.6) of Theorem 20.7, p. 416. The zero interpolation and boundary conditions imply the equality

$$\int_D (\Delta^q u)^2 \, dx = 0.$$

This implies that $\Delta^q u_j = 0$ in $\overline{D}_j$ for $j = 1, \ldots, N$.

From Lemma 20.10 below it follows that $u(x)$ is a q-harmonic function everywhere in D, i.e. $\Delta^q u = 0$. Now the boundary conditions (20.8) and $u = 0$ on Γ_N imply that $u \equiv 0$ in D.

Indeed, since the function $\Delta^{q-1} u$ is harmonic in D, and $\Delta^{q-1} u = 0$ on Γ_N, we may apply the uniqueness of the Dirichlet problem for the Laplace equation, and we obtain $\Delta^{q-1} u \equiv 0$ in D. In the same way we prove $\Delta^{q-2} u \equiv 0$ in D etc.; proceeding inductively we obtain $u \equiv 0$ in D.

Now if u is a polyspline of class C^{4q-2} we will apply Proposition 23.9, p. 471, about smoothness across the interface. We know that the normal derivatives of u_j and u_{j+1} coincide on Γ_j up to order $4q - 2$. Since $\Delta^{2q} u_j = \Delta^{2q} u_{j+1} = 0$ it follows also that the normal derivatives of order $4q - 1$ coincide. Applying Proposition 23.9 implies $u \in C^{4q-1,\alpha}(\overline{D})$. It follows that $u \in C^{2q,\alpha}(\overline{D})$. Since u is a solution to $\Delta^q u = 0$ in D, we may apply the main *a priori* estimate for the elliptic BVP, in Theorem 23.20, p. 482, which implies that $u \in C^\infty(D)$. Further, we proceed as in the Sobolev space case and obtain $u \equiv 0$ in D. Everything is proved. ∎

Lemma 20.10 is a continuation across the boundary result.

Lemma 20.10 *Let the function $u \in H^{2m}(D \setminus ST)$ have the property that $\Delta^m u = 0$ in D_j for every $j = 1, \ldots, N$, and also satisfy the following interface conditions, for $k = 0, 1, \ldots, m-1$, and $j = 1, \ldots, N-1$:*

(i) $\Delta^k u_j = \Delta^k u_{j+1}$ on Γ_j and

(ii) $\partial/\partial n \Delta^k u_j = \partial/\partial n \Delta^k u_{j+1}$ on Γ_j.

Then the function $u \in C^\infty(D)$ and $\Delta^m u = 0$ in D.

Proof Let us apply the second Green formula (see formula (23.6) on p. 479) to every subdomain D_j. We obtain the following formula:

$$u(y) = \Omega_n \sum_{l=0}^{m-1} \int_{\partial D_j} \left(\Delta^l u \cdot \frac{\partial r_{l+1}(x-y)}{\partial \nu_x} - \frac{\partial \Delta^l u}{\partial \nu_x} \cdot r_{l+1}(x-y) \right) d\omega(x)$$

$$- \Omega_n \int_{D_j} \Delta^m u \cdot r_m(x-y) \, dx \qquad (20.9)$$

for all y in D_j, and

$$u(y) = 0 \quad \text{for all } y \notin \overline{D}_j.$$

The constant Ω_n is defined in formula (10.3), p. 130, and $r_l(x)$ is a normalized fundamental solution of the equation $\Delta^l w(x) = 0$ [7, pp. 8–10]. Here $\vec{v} = \vec{v_j}$ means the inner normal vector to D_j at the point $x \in \partial D_j$.

Since $v_j = -v_{j+1}$ on Γ_j, after summing up in j all formulas (20.9), we obtain the second Green formula for $u(y)$ in D, which proves that it is a polyharmonic function of order m in the whole of D. ∎

We generalize Theorem 20.9 for a setting where the operator $L = L_1^2$ and the operator L_1 has been introduced in Section 20.3.2, p. 417.

Theorem 20.11 *Let $L = L_1^2$ where L_1 is a uniformly and strongly elliptic operator of order $2q$ which is formally self-adjoint. Let us assume that the Dirichlet problem for the operator L_1, namely problem*

$$L_1 u = 0 \quad \text{in } D, \tag{20.10}$$

$$\frac{\partial^k}{\partial n^k} u = 0 \quad \text{on } \Gamma_N, \quad \text{for } k = 0, 1, \ldots, q-1, \tag{20.11}$$

has a unique solution for $u \in H^{2q}(D)$. Let $u(x)$ be a polyspline of order $2q$ in the Sobolev space $H^{4q}(D \setminus ST)$ with respect to the operator L and satisfy zero boundary conditions[5] (see equalities (20.4), p. 412)

$$\frac{\partial^k}{\partial n^k} u = 0 \quad \text{on } \Gamma_N, \quad \text{for } k = 1, 2, \ldots, 2q-1. \tag{20.12}$$

If $u = 0$ on the interface set ST then $u \equiv 0$ in the whole domain D.

The statement holds if we consider polysplines u of class C^{4q-2} satisfying $u \in C^{4q-1}(D \setminus ST)$. The uniqueness of the Dirichlet problem (20.10)–(20.11) has to hold for solutions $u \in C^{2q}(\overline{D})$.

Remark 20.12 *Whether we work with the Dirichlet boundary operators $\partial^k/\partial n^k$ for $k = 0, 1, \ldots, q-1$, or with an other system of boundary operators $\{B_k\}_0^{q-1}$ which form a Dirichlet system of order q (in the sense of Definition 23.12, p. 474), makes no difference as is clear from Remark 20.6, p. 415. By Remark 19.6 the uniqueness for the Dirichlet problem (20.12) is equivalent to uniqueness for a boundary problem with conditions*

$$B_k u = 0 \quad \text{on } \Gamma_N \quad \text{for } k = 0, 1, \ldots, q-1.$$

As will become clear from the proof, the uniqueness for the Dirichlet problem (20.10)–(20.11) may be replaced by uniqueness for an arbitrary system of boundary operators $\{B_k\}_{k=0}^{p-1}$ such that $(\partial^k u)/(\partial n^k) = 0$ on Γ_N for $k = 0, 1, \ldots, 2q-1$, implies $B_k u = 0$ on Γ_N for $k = 0, 1, \ldots, p-1$.

Proof (1) We put $h = u$ in Theorem 20.8, p. 419. In order to justify this substitution we have to see that $h \in H^{2q}(D)$. Indeed, all derivatives up to order $2q$ of the function u exist on the interface Γ_j thanks to the inequality $4q - 2 \geq 2q$. By the last and by the

---
[5] As we have said "zero data" or "zero boundary conditions" is used as a synonym for "homogeneous system".

424 Multivariate polysplines

definition of polyspline, all derivatives up to order $4q - 2$ in the normal direction $\vec{n}$ coincide on both sides of Γ_j, i.e.

$$\frac{\partial^{4q-2}}{\partial n^{4q-2}} u_j = \frac{\partial^{4q-2}}{\partial n^{4q-2}} u_{j+1} \quad \text{on } \Gamma_j.$$

In the tangential directions at a point $x \in \Gamma_j$ all derivatives $D_\tau^\alpha u$ exist for $|\alpha| \leq 2q$ since $u_j = u_{j+1}$ on Γ_j implies

$$D_\tau^\alpha u_j = D_\tau^\alpha u_{j+1} \quad \text{on } \Gamma_j.$$

We obtain equality between the mixed derivatives (in τ and n directions) up to order $2q$ by a similar argument.

(2) Since $u = 0$ on all Γ_js, we will apply to Γ_N Proposition 20.3, p. 412, which implies that all differential operators A of order $\leq 2q - 2$ satisfy $Au_N = 0$ on Γ_N. We apply this to the values $B_k^N u_N$ and $B_k^N L_1 u_N$.

Finally, by using equality $u = 0$ on all Γ_js, we obtain the following:

$$-\int_D (L_1 u)^2 \, dx = \sum_{j=1}^{N-1} \int_{\Gamma_j} c_j \left\{ \frac{\partial^{4q-1} u}{\partial n^{4q-1}} \right\}_{\text{jump}} \times u \, d\omega$$

$$+ \sum_{k=0}^{q-1} \int_{\Gamma_N} \left(\mathcal{RS}_k^N L_1 u_N \cdot B_k^N u_N \, d\sigma - B_k^N L_1 u_N \cdot \mathcal{RS}_k^N u_N \right) d\omega$$

$$= 0.$$

It follows that for all $j = 1, 2, \ldots, N$ we have

$$L_1 u = 0 \quad \text{in } D_j.$$

It will suffice to prove that u is a solution of $L_1 u = 0$ in the whole domain D which belongs to $H^{2q}(D)$. The fundamental regularity for the elliptic BVP in Theorem 23.18, p. 480, implies $u \in C^\infty(D)$. Then the boundary conditions (20.12) for $k = 1, 2, \ldots, q - 1$, and the assumption $u = 0$ on Γ_N, represent a full set of Dirichlet boundary conditions on Γ_N. By the assumptions of the Theorem we have uniqueness for the Dirichlet problem for the operator L_1 which implies that $u \equiv 0$ in D.

(3) One may try to look for an analog to Lemma 20.10, which has provided an explicit representation for polyharmonic functions by using the fundamental solution of the operator Δ^{2q}. For that purpose one needs fundamental solutions for the operator L_1 etc. We may surmount the lack of similar representation for arbitrary operators by proving that u is a *weak solution* of the equation $L_1 u = 0$ and then applying results from Schechter [59]. For every function $g \in C_0^\infty(\overline{D})$ we will prove the identity

$$\int_D u \cdot L_1 g \, dx = 0.$$

Indeed, we apply the Green formula for self-adjoint operators of Proposition 23.16, p. 477, to every subdomain D_j and obtain

$$\int_D u \cdot L_1 g \, dx = \sum_{j=1}^{N} \int_{D_j} u \cdot L_1 g \, dx$$

$$= \sum_{j=1}^{N} \int_{D_j} L_1 u \cdot g \, dx + \sum_{k=0}^{q-1} \int_{\partial D_j} (B_k u \cdot S_k g - S_k u \cdot B_k g) \, d\sigma.$$

Since u_j and u_{j+1} have the same derivatives up to order $2q - 2$ on Γ_j, for reasons which we have explained in the proof of Theorem 20.8, p. 419, we can cancel all the terms inside D, and since g has a compact support all terms on the boundary $\partial D = \Gamma_N$ drop and we finally obtain

$$\int_D u \cdot L_1 g \, dx = 0.$$

It follows that u is a weak solution in D, and by Theorem 9–3, 3) in Schechter [59], it is a C^∞ function in D. We apply the uniqueness of the Dirichlet problem, which is among the assumptions of the Theorem and obtain $u = 0$ in D. This ends the proof.

Another proof of point (3) is the following. Since $u_j \in H^{4q}(D_j)$ for $j = 1, 2, \ldots, N$, and u_j and u_{j+1} have coinciding normal derivatives up to order $4q - 2$, and also $L u_j = L u_{j+1} = 0$, it follows that their normal derivatives up to order $4q - 1$ coincide. We may apply to u_j and u_{j+1} Proposition 23.8, p. 470, about smoothness across the interface and obtain $u \in H^{4q}(D)$. By the regularity results we obtain $u \in C^\infty(D)$ etc. ∎

Remark 20.13 *As similar statement holds if we consider polysplines of the smoothness class C^{4q-2}. The proof does not differ essentially.*

20.4.1 Holladay property

Now we can prove that the interpolation polyspline is a solution to the variational problem. In fact, we will prove the complete analog to the one-dimensional extremal property of Holladay (Theorem 2.9, p. 28). For simplicity we have taken zero boundary conditions on $\Gamma_N = \partial D$.

Let the function g be given on the data set ST, i.e. N functions $g = \{g_j\}_{j=1}^N$ be given, where g_j is a function on Γ_j. We assume that $g_j \in H^{4q-1/2}(\Gamma_j)$.

Theorem 20.14 *Let the operator $L = \Delta^{2q}$ and the function $v(x)$ be a polyspline in the Sobolev space $H^{4q}(D \setminus ST)$ with respect to the operator L. Assume that $v(x)$ satisfies the* **boundary** *conditions*

$$\left. \begin{array}{ll} \Delta^k v = 0 & \text{on } \Gamma_N, \quad \text{for } k = 1, \ldots, q-1, \\ \dfrac{\partial}{\partial n} \Delta^k v = 0 & \text{on } \Gamma_N, \quad \text{for } k = 0, \ldots, q-1, \end{array} \right\} \quad (20.13)$$

and the interpolation conditions

$$v = g_j \quad \text{on } \Gamma_j, \quad \text{for } j = 1, 2, \ldots, N.$$

Then the function v is also a solution to the following extremal problem:

$$\inf \int_D (\Delta^q f)^2 \, dx,$$

where the infimum is taken over the set of functions $f \in H^{2q}(D)$ satisfying the same interpolation conditions

$$f = g_j \quad \text{on } \Gamma_j, \quad \text{for } j = 1, 2, \ldots, N,$$

and the same boundary conditions (20.13).

The same statement holds if $f \in C^{2q}(D)$ and $v \in C^{4q-1}(D \setminus ST)$, where v is a polyspline of class C^{4q-2}, while the data g_j satisfy $g_j \in C^{4q-1}(\Gamma_j)$.

Proof For an arbitrary function $f \in H^{2q}(D)$ consider the difference

$$\int_D (\Delta^q f - \Delta^q v)^2 \, dx$$

$$= \int_D (\Delta^q f)^2 \, dx - 2 \int_D (\Delta^q f - \Delta^q v) \cdot \Delta^q v \, dx - \int_D (\Delta^q v)^2 \, dx.$$

We apply Theorem 20.7, p. 416, to the second term and put $h = v - f$ and $u = v$. As a result we obtain the equality

$$\int_D (\Delta^q f - \Delta^q v) \cdot \Delta^q v \, dx$$

$$= \int_{ST \setminus \partial D} (f - v) \cdot \left\{ \frac{\partial}{\partial n} \Delta^{2q-1} v \right\}_{\text{jump}} d\omega(x)$$

$$- \sum_{l=0}^{q-1} \int_{\Gamma_N} \left(\Delta^{l+q} v \cdot \frac{\partial}{\partial n} \Delta^{q-1-l}(f - v) \right.$$

$$\left. - \frac{\partial}{\partial n} \Delta^{l+q} v \cdot \Delta^{q-1-l}(f - v) \right) d\omega(x).$$

All terms on the right-hand side are zero since v and f satisfy the same interpolation conditions and zero boundary conditions (20.13). Thus we obtain

$$\int_D (\Delta^q v)^2 \, dx = \int_D (\Delta^q f)^2 \, dx - \int_D (\Delta^q f - \Delta^q v)^2 \, dx \leq \int_D (\Delta^q f)^2 \, dx$$

which proves our statement. ∎

The proof for an arbitrary operator of the form $L = L_1^2$ is similar. The proof relies upon the identity proved in Theorem 20.8, p. 419.

Theorem 20.15 *We adopt the notation and assumptions of Theorem 20.8. Let the uniformly and strongly elliptic operator L_1 of order $2q$ be given in D which is formally self-adjoint there. We assume that a Dirichlet system of operators $\{B_k\}$ of order q be given with ord $B_k = k$, $B_0 = 1$, and such that the system $\{L, B_k,\ k = 0, 1, \ldots, q-1\}$ forms a regular elliptic BVP on D. We denote by $\{S_k\}$ the adjoint boundary operators. Let the function $v(x)$ be a polyspline in the Sobolev space $H^{4q}(D\backslash ST)$ with respect to the operator $L = L_1^2$. Assume that v satisfies the boundary conditions*

$$\left.\begin{array}{ll} B_k v = 0 & \text{on } \Gamma_N,\ \text{for } k = 1, \ldots, q-1, \\ S_k v = 0 & \text{on } \Gamma_N,\ \text{for } k = 0, \ldots, q-1, \end{array}\right\} \quad (20.14)$$

and the interpolation conditions

$$v = g_j \quad \text{on } \Gamma_j,\ \text{for } j = 1, 2, \ldots, N.$$

Then the function v is also a solution to the following extremal problem:

$$\inf \int_D (L_1 f)^2\, dx$$

where the infimum is taken over the set of functions $f \in H^{2q}(D)$ satisfying the same interpolation conditions

$$f = g_j \quad \text{on } \Gamma_j,\ \text{for } j = 1, 2, \ldots, N,$$

and the same boundary conditions (20.14).

The same statement holds if $f \in C^{2q}(D)$ and $v \in C^{4q-1}(D \setminus ST)$, where v is a polyspline of class C^{4q-2} with respect to the operator L, while the data g_j satisfy $g_j \in C^{4q-1}(\Gamma_j)$.

In order to apply the identity of Theorem 20.8, p. 419, one has to recall that by virtue of Lemma 2.1 in [46, Chapter II] the set of operators

$$\{B_k, S_k,\ k = 0, 1, \ldots, q-1\}$$

is equivalent to the set of operators

$$\left\{1, \frac{\partial}{\partial n}, \ldots, \frac{\partial^{2q-1}}{\partial n^{2q-1}}\right\}.$$

Chapter 21

A priori estimates and Fredholm operators

Following the scheme of *frozen coefficients* one first considers the one-dimensional interface problem obtained by localization about the points of the interface set $\bigcup_{j=1}^{N-1} \Gamma_j$. Next, one obtains *a priori* estimates for two complementing half-spaces – $\mathbb{R}^n_+$ and $\mathbb{R}^n_-$. We adopt the conditions on the domain D, the interfaces Γ_j, and the operator L from Section 20.1, p. 409, and Section 20.2, p. 411.

21.1 Basic proposition for interface on the real line

Following the standard techniques of *frozen* coefficients [46, Chapter II], we assume that we have localized the coordinates about a point $y \in \Gamma_j$ for some interface surface Γ_j. Thus we will work with constant coefficient homogeneous elliptic operators L_0 and $B_{k,0}$ with real coefficients. We refer to Proposition 23.10, p. 472, which means that the operator L_0 is *properly elliptic*.

We will first prove the *main technical result*, Proposition 21.1, p. 430, which is necessary for studying the *a priori* estimates for operators with constant coefficients in the half-space.

We will use the notation $x = (y, t)$ for a point $x \in \mathbb{R}^n$, where $y = (y_1, \ldots, y_{n-1}) \in \mathbb{R}^{n-1}$, $t \in \mathbb{R}$. The variable, dual to x, will be $\xi = (\eta, t')$, where $\eta = (\eta, \ldots, \eta_{n-1}) \in \mathbb{R}^{n-1}$ and $t' \in \mathbb{R}$.

We denote by $\mathbb{R}^n_+$ the half-space of the points $x \in \mathbb{R}^n$ with $t > 0$; by $\mathbb{R}^n_- \mathbb{R}^n_-$ we denote the other half-space, where $t < 0$.

Let us introduce the operators

$$A(D)u = L_0(D_y, D_t)u \tag{21.1}$$

$$S_j(D)u = S_j(D_y, D_t)u := D_t^j u, \quad j = 0, \ldots, 2p-1, \tag{21.2}$$

429

where D_y denotes the multivariate differentiation in y and
$$D_t = \frac{\partial}{\partial t}.$$

We will put $m_j = j = \operatorname{ord} S_j$.

Let us denote by $\tau_i^+ = \tau_i^+(\eta)$ (or $\tau_i^- = \tau_i^-(\eta)$) the roots of $A(\eta, \tau)$ with positive (or negative) imaginary part. Since $A(D)$ has real coefficients, we may write $\tau_j^-(\eta) = \overline{\tau_j^+(\eta)}$, $j = 1, \ldots, m$, where the bar means a complex conjugate.

As usual (cf. [62]), we denote by $S(\mathbb{R}_+)$ the space of functions which are infinitely differentiable for $t \geq 0$ and are rapidly decreasing for $t \to \infty$ (i.e. $t^k \phi^{(j)}(t) \to 0$ for $t \to \infty$, for every k and every j).

We denote by $S(\mathbb{R}_-)$ the similar space for $t \leq 0$, i.e. $\phi(t) \in S(\mathbb{R}_-)$ if and only if $\phi(-t) \in S(\mathbb{R}_+)$.

For every $\eta \in \mathbb{R}^{n-1}$, $\eta \neq 0$, let us consider the system of ordinary differential equations

$$A\left(\eta, \frac{1}{i}\frac{d}{dt}\right) \phi(\eta, t) = 0 \quad \text{for } t \geq 0, \tag{21.3}$$

$$A\left(\eta, \frac{1}{i}\frac{d}{dt}\right) \psi(\eta, t) = 0 \quad \text{for } t \leq 0, \tag{21.4}$$

with *boundary* conditions

$$S_j\left(\eta, \frac{1}{i}\frac{d}{dt}\right) \phi(\eta, t)|_{t=0} = S_j\left(\eta, \frac{1}{i}\frac{d}{dt}\right) \psi(\eta, t)|_{t=0} + c_j, \tag{21.5}$$

for $j = 0, \ldots, 2p - 2$, and with *interpolation* conditions

$$S_0\left(\eta, \frac{1}{i}\frac{d}{dt}\right) \phi(\eta, t)|_{t=0} = \phi(\eta, 0) = \gamma, \tag{21.6}$$

where $\phi(\eta, t) \in S(\mathbb{R}_+)$, $\psi(\eta, t) \in S(\mathbb{R}_-)$, and c_j, γ are given complex numbers.

The following statement is an analog to Proposition 4.2, Chapter II, in [46].

Proposition 21.1 *Let the operators A and S_j, $j = 0, \ldots, 2p-1$, be defined by equalities (21.1)–(21.2). We assume that $\eta \neq 0$ is fixed. Then problem (21.3)–(21.6) has unique solution for every set of constants c_j, $j = 0, \ldots, 2p-1$, and γ.*

Proof By the proper ellipticity of the operator L the polynomial $A(\eta, \tau) = L_0(\eta, \tau)$ has the following representation:

$$A(\eta, \tau) = \prod_{i=1}^{p}(\tau - \tau_i^+) \prod_{i=1}^{p}(\tau - \tau_i^-) \tag{21.7}$$

where $\tau_i^+(\eta) = \overline{\tau_i^-(\eta)}$. Assuming for simplicity that all roots τ_i^+ have multiplicity one, every solution of the ordinary differential equation

$$A\left(\eta, \frac{1}{i}\frac{d}{dt}\right) f(t) = 0$$

will be of the form
$$f(t) = \sum_{i=1}^{p} \left(\alpha_i e^{it\tau_i^+} + \beta_i e^{it\tau_i^-} \right).$$

Due to $\operatorname{Im}(\tau_i^+) > 0$ it follows that $f \in S(\mathbb{R}_+)$ if and only if
$$f(t) = \sum_{i=1}^{p} \alpha_i e^{it\tau_i^+}.$$

Likewise, $f \in S(\mathbb{R}_-)$ if and only if
$$f(t) = \sum_{i=1}^{p} \beta_i e^{it\tau_i^-}.$$

Thus we see that the space of solutions to the system (21.3)–(21.4) has dimension exactly $2p$. It makes no essential difference if we consider multiple roots τ_i^+.

Thus we have to prove that there exist constants α_i and β_i satisfying conditions (21.5) and (21.6). Obviously, this is a linear system for α_i and β_i and by linear algebra we know that uniqueness of the solution of this system will imply the existence of its solution.[1]

To prove uniqueness let us suppose that
$$c_j = 0 \quad \text{for } j = 0, 1, \ldots, 2p-2,$$
$$\gamma = 0.$$

We will use an idea of spline theory where the uniqueness (and existence) of the splines is proved through a suitable identity [5, p. 154] which we have already exploited in Section 2.4, p. 26.

Let us define the function $w = w(\eta, t)$ as coinciding with ϕ on $\mathbb{R}_+$ and with ψ on $\mathbb{R}_-$, and the functions ϕ, ψ satisfy equality (21.5), p. 430. The following identity holds, see Lemma 21.2:

$$\int_{-\infty}^{\infty} \prod_{i=1}^{p} \left(\frac{1}{i} \frac{d}{dt} - \tau_i^+(\eta) \right) f \times \overline{\prod_{i=1}^{p} \left(\frac{1}{i} \frac{d}{dt} - \tau_i^+(\eta) \right) w} \, dt$$
$$= (-1)^{p+1} \{\overline{w}^{(2p-1)}(0)\}_{\text{jump}} \cdot f(0) \tag{21.8}$$

for every $f \in H^p(\mathbb{R})$; here we have used the notation for the jump:
$$\{w^{(2p-1)}(0)\}_{\text{jump}} := \phi^{(2p-1)}(\eta, 0) - \psi^{(2p-1)}(\eta, 0).$$

Let us put $f = w$ in (21.8). Since $\phi(\eta, 0) = \psi(\eta, 0) = w(0) = 0$ we obtain
$$\int_{-\infty}^{\infty} \left| \prod_{i=1}^{p} \left(\frac{1}{i} \frac{d}{dt} - \tau_i^+(\eta) \right) w \right|^2 dt = 0,$$

---

[1] Let us note that the emerging problem is quite close to the so-called *natural splines* for Chebyshev systems and is in fact a problem of that theory. For Chebyshev systems see Chapter 11, and for natural Chebyshev splines see Section 11.5, p. 204, or [61, Chapter 9.8, p. 396].

and, consequently, the following equation holds:

$$\prod_{i=1}^{p}\left(\frac{1}{i}\frac{d}{dt}-\tau_i^+\right)w=0 \quad \text{on } \mathbb{R}. \tag{21.9}$$

But it is evident that $\psi(\eta, t)$ satisfies the last equation only if $\psi \equiv 0$. Conditions (21.5) also imply $\phi \equiv 0$. This proves the uniqueness and, as said above, the existence of the solution to problem (21.3)–(21.6). The proof of the proposition is finished. ∎

For completeness we provide the following lemma.

Lemma 21.2 *Let the function w be a solution to $A(\eta, (1/i)(d/dt))w = 0$ in $\mathbb{R}_+$ and the same equation in $\mathbb{R}_-$. Let the restriction of w to $\mathbb{R}_+$ satisfy $w \in S(\mathbb{R}_+)$, and its restriction to $\mathbb{R}_-$ satisfy $w \in S(\mathbb{R}_-)$, and let, on the entire axis, $\mathbb{R}$ have $w \in C^{2p-2}(\mathbb{R})$. Then for every $f \in H^p(\mathbb{R})$ the following identity holds:*

$$\int_{-\infty}^{\infty} \prod_{i=1}^{p}\left(\frac{1}{i}\frac{d}{dt}-\tau_i^+\right) f \times \overline{\prod_{i=1}^{p}\left(\frac{1}{i}\frac{d}{dt}-\tau_i^+\right) w}\, dt$$
$$= (-1)^{p+1}\{\overline{w}^{(2p-1)}(0)\}_{\text{jump}} \cdot f(0),$$

where we have used the following notation for the jump:

$$\{w^{(2p-1)}(0)\}_{\text{jump}} := \phi^{(2p-1)}(\eta, 0) - \psi^{(2p-1)}(\eta, 0).$$

The **proof** is the same as that of the basic identity for one-dimensional odd-degree splines in Lemma 2.5, p. 26, and consists of simple integration by parts. It uses the fact that $\tau_i^+ = \tau_i^-$, and the representation of $A(\eta, \tau)$ in (21.7), p. 430.

21.2 *A priori* estimates in a bounded domain with interfaces

We drop the *a priori* estimate for the interface problem defined in two half-spaces and provide below only the final *a priori* estimate for an arbitrary bounded domain with interfaces.

Let us return to the *interface BVP* introduced in (20.1) and (20.2), p. 411. We will consider a more general problem than (20.1) and (20.2) and the objects will be called *general polysplines*. The proofs do not become more complicated but these general polysplines cover many of the possible generalizations of polysplines analogous to the one-dimensional *smoothing, mono-, perfect-,* and *Bernoulli* splines.

We will adopt the notation of Section 20.2, p. 411. Accordingly, we will consider polysplines defined for a *uniformly and strongly elliptic* operator L of order $2p$. On the boundary $\partial D = \Gamma_N$ a set of boundary operators $\{B_k\}$ will be given such that $B_0 = 1$, and the system of operators

$$\{L, B_k, \ k = 0, 1, \ldots, p-1\} \quad \text{in } D$$

forms a *regular elliptic BVP* in the sense of Definition 23.11, p. 473. We put

$$m_k := \operatorname{ord} B_k \quad \text{for } k = 0, 1, \ldots, p-1.$$

In order to simplify the notations we introduce a special notation for the "*interface surfaces*", namely

$$ST_1 := \bigcup_{j=1}^{N-1} \Gamma_j,$$

which should be distinguished from the "*data set*" $ST = ST_1 \cup \Gamma_N$.

On every surface Γ_j we define the operators of differentiation with respect to the interior unit normal vector $\vec{n} = \vec{n_x}$ to Γ_j at the point $x \in \Gamma_j$, by putting

$$S_i := \frac{\partial^i}{\partial n^i} \quad \text{for } i = 0, 1, 2, \ldots,$$

for which, obviously,

$$S_0 = 1,$$
$$\operatorname{ord} S_i = i.$$

The *interface* functions $\{G_i\}_{i=0}^{2p-2}$ are defined on ST_1, and every such function has $N-1$ components on all of the surfaces Γ_j. In order to avoid overburdened indices we will not index all components. We assume that the *boundary* data functions $\{h_i\}_{i=1}^{p-1}$ are defined on the boundary $\partial D = \Gamma_N$. For $j = 1, 2, \ldots, N$, the *interpolation* data function g_j is defined on the surface Γ_j. We will consider functions u defined in the whole domain $D \setminus ST$. Recall that the restriction of the function u to the subdomain D_j is denoted by u_j.

We will consider the following *interface BVP* for finding a function u which satisfies:

- the *equation* in every subdomain D_j, i.e.

$$Lu = f \quad \text{in } D \setminus ST, \tag{21.10}$$

- the *boundary* conditions

$$B_k u = h_k \quad \text{on } \Gamma_N, \quad \text{for } k = 1, 2, \ldots, p-1, \tag{21.11}$$

- the *interface* conditions

$$S_i u_j - S_i u_{j+1} = G_i \quad \text{on } \Gamma_j, \quad \text{for } i = 0, 1, \ldots, 2p-2, \tag{21.12}$$

- and the *interpolation* conditions

$$S_0 u_j = g_j \quad \text{on } \Gamma_j, \quad \text{for } j = 1, 2, \ldots, N-1. \tag{21.13}$$

See Figure 21.1.

The solution to problem (21.10)–(21.13) will be called the **general polyspline**.

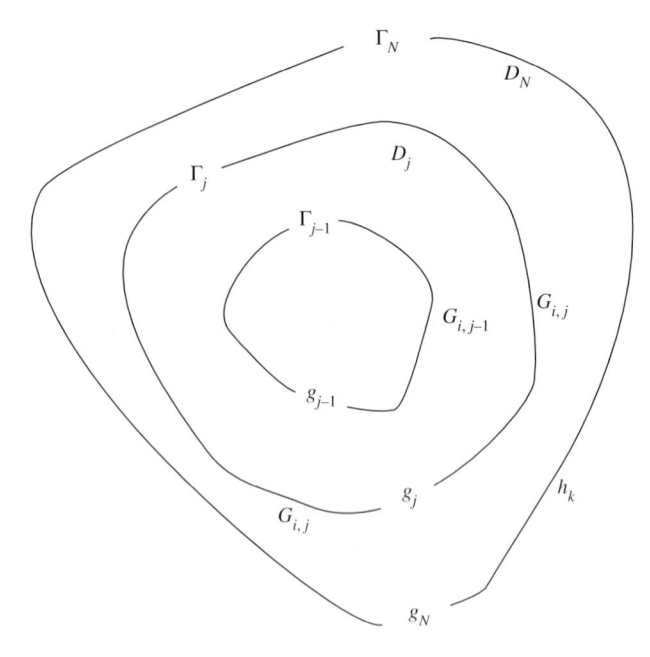

Figure 21.1.

Our aim is to obtain *a priori* estimates for the generalized polysplines for all real numbers $s \geq 0$ which are of the type

$$\|u\|_{H^{s+2p}(D\setminus ST)} \leq C \cdot \left\{ \|f\| + \|g\| + \sum_i \|G_i\| + \sum_k \|h_k\| + \|u\|_{H^{s+2p-1}(D\setminus ST)} \right\},$$

where for simplicity we have not indicated the norms of the functions G_i, h_k and g_j. To simplify the formulation of the results below for every real number $s \geq 0$ we put

$$U_{1,s} := \prod_{i=1}^{p-1} H^{2p+s-m_i-1/2}(\partial D) \times \prod_{i=0}^{2p-2} H^{2p+s-i-1/2}(ST_1) \times H^{2p+s-1/2}(ST),$$

(21.14)

$$U_{2,s} := \prod_{i=1}^{p-1} H^{-2p+s+m_i+1/2}(\partial D) \times \prod_{i=0}^{2p-2} H^{-2p+s+i+1/2}(ST_1) \times H^{-2p+s+1/2}(ST).$$

(21.15)

We consider the *data vector* of problem (21.10)–(21.13) given by

$$F = \{f; h; G_0, G_1, \ldots, G_{2p-2}; g\} \in L^2(D\setminus ST) \times U_{2,0},$$

where

$$f \in L^2(D\setminus ST) \quad \text{and} \quad \{h; G_0, G_1, \ldots, G_{2p-2}; g\} \in U_{2,0}.$$

A priori *estimates and Fredholm operators* 435

The corresponding operator defined by relations (21.10)–(21.13) may be written as

$$\mathcal{P}: u \longmapsto F, \tag{21.16}$$

and is linear and continuous from $H^{2m}(D\backslash ST)$ into $L^2(D\backslash ST) \times U_{1,0}$.

Let us denote by $\mathcal{P}^*$ the operator adjoint to $\mathcal{P}$ mapping $L^2(D\backslash ST) \times U_{2,0}$ into $[H^{2m}(D\backslash ST)]'$, which is defined by the equality

$$\langle u, \mathcal{P}^*\Phi\rangle = \langle \mathcal{P}u, \Phi\rangle, \tag{21.17}$$

which is true for every $\Phi \in L^2(D\backslash ST) \times U_{2,0}$, and every $u \in H^{2m}(D\backslash ST)$. The brackets $\langle\cdot,\cdot\rangle$ denote the pairing between the corresponding dual spaces, which for F as above and for Φ given by

$$\Phi = \{v; \chi; \gamma_0^1, \gamma_1^1, \ldots, \gamma_{2p-2}^1; \gamma\},$$

is defined as

$$\langle F, \Phi\rangle := \int_D f\bar{v}\,dx + \sum_{i=1}^{p-1}\int_{\Gamma_N} h_i\overline{\chi_i}\,d\sigma + \sum_{j=1}^{N-1}\sum_{i=0}^{2p-2}\int_{\Gamma_j} G_i\overline{\gamma_i^1}\,d\sigma + \sum_{j=1}^{N}\int_{\Gamma_j} g\bar{\gamma}\,d\sigma.$$

Following the standard techniques of frozen coefficients we may prove the main *a priori* estimate.

Theorem 21.3 *Within the above assumptions for every given real number $r \geq 0$ the following statements are true:*

- *If $u \in H^{2p}(D\backslash ST)$ and $\mathcal{P}u \in H^r(D\backslash ST) \times U_{1,r}$, then $u \in H^{2p+r}(D\backslash ST)$ and the following inequality holds:*

$$\|u\|_{H^{2p+r}(D\backslash ST)} \leq C_r \cdot \left\{\|\mathcal{P}u\|_{H^r(D\backslash ST)\times U_{1,r}} + \|u\|_{H^{2p+r-1}(D\backslash ST)}\right\},$$

where the constant $C_r > 0$ depends on r but not on the data and $u(x)$. If the homogeneous BVP[2] (21.10)–(21.13) has a unique solution in the class $H^{2p}(D\backslash ST)$ then the term $\|u\|_{H^{2p+r-1}(D\backslash ST)}$ may be dropped (but the constant C_r must be replaced by another one).

- *If $F = \{f; h; G_0, G_1, \ldots, G_{2p-2}; g\} \in L^2(D\backslash ST) \times U_{2,0}$ and $\mathcal{P}^*F \in [H^{2p-r}(D\backslash ST)]'$, then $F \in H^r(D\backslash ST) \times U_{2,r}$.*

- *For real numbers $r > 0$ the following a priori estimate holds:*

$$\|F\|_{H^r(D\backslash ST)\times U_{2,r}} \leq C_r \cdot \left\{\|\mathcal{P}^*F\|_{[H^{2p-r}(D\backslash ST)]'} + \|F\|_{H^{r-1}(D\backslash ST)\times U_{2,r}}\right\},$$

and for $r = 0$, the following:

$$\|F\|_{L^2(D\backslash ST)\times U_{2,0}} \leq C_0 \cdot \left\{\|\mathcal{P}^*F\|_{[H^{2p}(D\backslash ST)]'} + \|\tilde{f}\|_{[H^{-1}(\mathbb{R}^n)]^N} \right.$$

$$\left. + \sum_{i=0}^{2p-2}\|G_i\|_{H^{-2p+i-1/2}(ST_1)} + \|g\|_{H^{-2p-1/2}(ST)}\right\}.$$

---

[2] "Homogeneous" means that all data are zero, i.e. (as vectors) $f = 0$, $h_k = 0$, $G_j = 0$, $\gamma_k = 0$.

where by $\tilde{f}$ we denote the set of N functions $\tilde{f}_j$, $j = 1, \ldots, N$, such that $\tilde{f}_j$ is a continuation like zero of the function $f_j = f_{|D_j}$ outside the domain D_j. Here the constant $C_0 > 0$ is independent of the data.

The proof is similar to that of Theorem 5.1 in Lions and Magenes [46, Chapter II]. For the manifolds case see Taylor [67, Proposition 11.2, p. 382, and Proposition 11.16, p. 394].

21.3 Fredholm operator in the space $H^{2p+r}(D\backslash ST)$ for $r \geq 0$

Here we will prove that the operator $\mathcal{P}$ together with its dual operator $\mathcal{P}^*$ is Fredholm, which is a substitute for the basic duality principle in linear algebra.

We may study the boundary value problem (21.10)–(21.13) in the spaces $H^{2p+r}(D\backslash ST)$ for $r \geq 0$. We have to prove that the operator given by (21.16) is a Fredholm operator from $H^{2p+r}(D\backslash ST)$ into $H^r(D\backslash ST) \times U_{1,r}$ assuming the notation of the previous section.

The proof of the Fredholm theorems for elliptic BVP (see Theorem 23.19, p. 480) is based on a standard argumentation using a lemma by Peetre [46, Theorem 5.2, Chapter II], and in the manifolds case [67, Proposition 11.16, p. 394]. Reasoning in a similar way we obtain the following **Fredholm** operator (the Fredholm Theorem).

Theorem 21.4 *Let the assumptions of Section 20.2, p. 411, hold. Then for every $r \geq 0$ the operator $\mathcal{P}$ defined by (21.16), considered as an operator from $H^{2p+r}(D\backslash ST)$ into $H^r(D\backslash ST) \times U_{1,r}$, is a Fredholm operator.*

- *The kernel of the operator $\mathcal{P}$ defined as the space*

$$\Lambda := \{u : u_{|D_j} = u_j \in C^\infty(\overline{D}_j), \mathcal{P}u = 0\} \qquad (21.18)$$

satisfies

$$\dim \Lambda < \infty.$$

It coincides with the space of elements

$$\mathbf{N} = \begin{cases} u : u_j \in C^\infty(\overline{D}_j) & \text{for all } j = 1, 2, \ldots, N, \text{ and such that:} \\ Lu_j = 0 & \text{in } D_j, \\ S_i u_j = S_i u_{j+1} & \text{on } \Gamma_j, \quad \text{for } j = 1, 2, \ldots, N; \quad i = 0, 1, \ldots, 2p-2, \\ S_0 u_j = 0 & \text{on } \Gamma_j, \quad \text{for } j = 1, 2, \ldots, N, \\ B_i u = 0 & \text{on } \Gamma_N, \quad \text{for } i = 1, 2, \ldots, p-1. \end{cases}$$

- *The image of the operator $\mathcal{P}$ is the set of elements $F = \{f; h; G_0, G_1, \ldots, G_{2p-2}; g\}$ of the space $H^r(D \backslash ST) \times U_{1,r}$ such that the equality*

$$\langle F, \Phi \rangle = 0 \qquad (21.19)$$

holds for every element $\Phi = \{v; \chi; \gamma_0^1, \gamma_1^1, \ldots, \gamma_{2p-2}^1; \gamma\}$ of the space

$$\Lambda_1 = \{\Phi \in \mathcal{L}: \quad \text{such that } \mathcal{P}^*\Phi = 0\}. \tag{21.20}$$

We have put

$$\mathcal{L} = \prod_{j=1}^{N} C^\infty(\overline{D_j}) \times [C^\infty(\partial D)]^{p-1} \times [C^\infty(ST_1)]^{2p-1} \times [C^\infty(ST)]^p, \tag{21.21}$$

and $\mathcal{P}^*$ is the operator adjoint to $\mathcal{P}$, which is defined through the duality relation (21.17) (where $r = 0$).

In order to provide an explicit description of the space Λ_1 we will need the adjoint boundary operators to the basic interface BVP (21.10)–(21.13), p. 433.

Remark 21.5 *Let us note that we have a completely similar result in the Hölder space setting following the elliptic BVP prototype in Theorem 23.20 and Theorem 23.21, p. 483. We will provide a more detailed study of the Hölder space setting in the existence results below.*

21.3.1 The space Λ_1 for $L = \Delta^p$

We will provide a complete description of the set Λ_1 for the operator $L = \Delta^p$. The main reason for the possibility of providing such a complete description is that the Green formulas for the operator Δ^p are relatively simple. Although it is somewhat technical, it is worth undertaking its computation since it provides us with a finished existence result in the case of general domains and interfaces.

We put

$$B_i = \Delta^i \quad \text{for } i = 0, 1, 2, \ldots,$$

$$S_i = \frac{\partial}{\partial n} B_i \quad \text{for } i = 0, 1, 2, \ldots.$$

Recall that $\Delta^0 = 1$ by definition, hence $B_0 = 1$.

According to Proposition 20.4, p. 414, on equivalent systems of operators, we will fix our notation so that the *general polysplines* defined by system (21.10)–(21.13), p. 433, will be defined through the following *interface BVP*: Find a function u satisfying

- the *equation* in every subdomain D_j, i.e.

$$Lu = f \quad \text{in } D \setminus ST; \tag{21.22}$$

- the *boundary* conditions

$$B_i u = h_i \quad \text{on } \Gamma_N, \quad \text{for } i = 1, 2, \ldots, p-1; \tag{21.23}$$

- the *interface* conditions

$$\left.\begin{array}{l} B_i u_j - B_i u_{j+1} = G_i^1 \quad \text{on } \Gamma_j, \quad \text{for } i = 1, \ldots, p-1, \\ S_i u_j - S_i u_{j+1} = G_i^2 \quad \text{on } \Gamma_j, \quad \text{for } i = 0, 1, \ldots, p-1, \end{array}\right\} \tag{21.24}$$

- and the *interpolation* conditions

$$B_0 u_j = g_j \quad \text{on } \Gamma_j, \quad \text{for } j = 1, 2, \ldots, N-1. \tag{21.25}$$

This problem is equivalent to problem (21.10)–(21.13) on p. 433, when considered on functions $H^{2p}(D\setminus ST)$. We have the corresponding operator $\mathcal{P}$ and its dual $\mathcal{P}^*$ and an analog to Theorem 21.4, p. 436, with the corresponding set Λ_1.

The dual space will consist of vectors Φ of the form

$$\Phi = \{v; h; \gamma_0^1, \gamma_1^1, \ldots, \gamma_{p-1}^1; \gamma_0^2, \gamma_1^2, \ldots, \gamma_{p-1}^2; \gamma\}.$$

Here h and γ are vectors of functions: $h = (h_1, h_2, \ldots, h_{p-1})$, where h_k is defined on the boundary Γ_N, and $\gamma = (\gamma_1, \gamma_2, \ldots, \gamma_N)$, where γ_j is defined on Γ_j, and all γ_k^1 and γ_k^2 are functions defined on the interface set ST_1. We will not always write the indexes if it is easily understood by default.

The following proposition describes explicitly the space Λ_1, and the consistency conditions (21.19)–(21.20), p. 437, provide a solution of problem (21.22)–(21.25).

Proposition 21.6 *The space Λ_1 defined by (21.20) coincides with the set run by the element*

$$\Phi = \begin{cases} v & \text{in } D \setminus ST; \\ \chi_k = -S_{p-1-k}v_N & \text{on } \Gamma_N \quad \text{for } k = 1, \ldots, p-1; \\ \gamma_0^1 = -S_{p-1}v_{j+1} & \text{on } \Gamma_j \quad \text{for } j = 1, \ldots, N-1; \\ \gamma_k^1 = -S_{p-1-k}v_j & \text{on } \Gamma_j \quad \text{for } k = 1, \ldots, p-1; \; j = 1, \ldots, N-1; \\ \gamma_k^2 = B_{p-1-k}v_j & \text{on } \Gamma_j \quad \text{for } k = 0, \ldots, p-2; \; j = 1, \ldots, N-1; \\ \gamma_j = -S_{p-1}v_j & \\ \quad + S_{p-1}v_{j+1} & \text{on } \Gamma_j \quad \text{for } j = 1, \ldots, N-1; \\ \gamma_N = -S_{p-1}v_N & \text{on } \Gamma_N; \end{cases}$$

where the function v is such that if we put $v_j = v_{|D_j} \in C^\infty(\overline{D_j})$ the following adjoint interface BVP with zero right sides is satisfied:

$$Lv = 0 \quad \text{in } D \setminus ST;$$
$$B_k v_N = 0 \quad \text{on } \Gamma_N \quad \text{for } k = 1, \ldots, p-1;$$
$$S_k v_j - S_k v_{j+1} = 0 \quad \text{on } \Gamma_j \quad \text{for } k = 0, 1, \ldots, p-2 \text{ and } j = 1, 2, \ldots, N-1;$$
$$B_k v_j - B_k v_{j+1} = 0 \quad \text{on } \Gamma_j \quad \text{for } k = 0, 1, \ldots, p-1 \text{ and } j = 1, 2, \ldots, N-1;$$
$$B_0 v_j = 0 \quad \text{on } \Gamma_j \quad \text{for } j = 1, 2, \ldots, N.$$

Proof The proof is analogous in principle to the proof of Proposition 5.1 in [46, Chapter 2].

We will omit the integration measure $d\sigma$ on the surfaces Γ_j where it is understood by default.

(1) We have introduced above the elements

$$F = \{f; h; G_0^1, G_1^1, \ldots, G_{p-1}^1; G_0^2, G_1^2, \ldots, G_{p-1}^2; g\},$$
$$\Phi = \{v; \chi; \gamma_0^1, \gamma_1^1, \ldots, \gamma_{p-1}^1; \gamma_0^2, \gamma_1^2, \ldots, \gamma_{p-1}^2; \gamma\},$$

where we assume that F arises through the interface BVP (21.22)–(21.25), p. 438. Their pairing is

$$\langle F, \Phi \rangle = \int_D f\bar{v}\, dx + \sum_{k=1}^{p-1} \int_{\Gamma_N} h_k \overline{\chi_k}\, d\sigma + \sum_{j=1}^{N-1} \sum_{k=0}^{p-1} \int_{\Gamma_j} G_k^1 \overline{\chi_k^1}\, d\sigma$$
$$+ \sum_{j=1}^{N-1} \sum_{k=0}^{p-2} \int_{\Gamma_j} G_k^2 \overline{\chi_k^2}\, d\sigma + \sum_{j=1}^{N} \int_{\Gamma_j} g\bar{\gamma}\, d\sigma. \qquad (21.26)$$

(2) Let us recall the Green formula (for real-valued functions) [7, p. 9].

$$0 = \int_{D_j} (u\Delta^p v - \Delta^p u v)\, dx + \sum_{k=0}^{p-1} \int_{\partial D_j} (B_k u \cdot S_{p-1-k} v - S_k u \cdot B_{p-1-k} v).$$

We sum on $j = 1, 2, \ldots, N$, and obtain

$$0 = \int_D (u\Delta^p v - \Delta^p u v)\, dx + \sum_{j=1}^{N-1} \sum_{k=0}^{p-1} \int_{\Gamma_j} \{(B_k u_j S_{p-1-k} v_j - S_k u_j B_{p-1-k} v_j)$$
$$- (B_k u_{j+1} S_{p-1-k} v_{j+1} - S_k u_{j+1} B_{p-1-k} v_{j+1})\}\, d\sigma$$
$$+ \sum_{k=0}^{p-1} \int_{\Gamma_N} (B_k u_N S_{p-1-k} v_N - S_k u_N B_{p-1-k} v_N).$$

Now we will regroup the terms under the double sum in order to shape the conditions of (21.22)–(21.25) with respect to the function u. Evidently we have

$$(B_k u_j S_{p-1-k} v_j - S_k u_j B_{p-1-k} v_j) - (B_k u_{j+1} S_{p-1-k} v_{j+1} - S_k u_{j+1} B_{p-1-k} v_{j+1})$$
$$= (B_k u_j - B_k u_{j+1}) S_{p-1-k} v_j + B_k u_{j+1} (S_{p-1-k} v_j - S_{p-1-k} v_{j+1})$$
$$- (S_k u_j - S_k u_{j+1}) B_{p-1-k} v_j - S_k u_{j+1} (B_{p-1-k} v_j - B_{p-1-k} v_{j+1}).$$

We will shorten the formulas if we substitute the values of the vector F; we also separate the terms where $S_{p-1} u_j$ and $S_{p-1} u_{j+1}$ appear since they are not present as a boundary

440 *Multivariate polysplines*

condition in (21.22)–(21.25). We obtain

$$0 = \int_D (u\Delta^p v - \Delta^p uv)\, dx$$

$$+ \sum_{j=1}^{N-1} \sum_{k=0}^{p-1} \int_{\Gamma_j} \{G_k^1 S_{p-1-k} v_j + B_k u_{j+1}(S_{p-1-k} v_j - S_{p-1-k} v_{j+1})\}$$

$$- \sum_{j=1}^{N-1} \sum_{k=0}^{p-2} \int_{\Gamma_j} \{G_k^2 B_{p-1-k} v_j + S_k u_{j+1}(B_{p-1-k} v_j - B_{p-1-k} v_{j+1})\}$$

$$- \sum_{j=1}^{N-1} \int_{\Gamma_j} \{(S_{p-1} u_j - S_{p-1} u_{j+1}) v_j + S_{p-1} u_{j+1}(v_j - v_{j+1})\}$$

$$+ \sum_{k=0}^{p-1} \int_{\Gamma_N} (B_k u_N S_{p-1-k} v_N - S_k u_N B_{p-1-k} v_N).$$

(3) We are looking for a vector Φ satisfying

$$\langle F, \Phi \rangle = 0 \quad \text{for all } u \in H^{2p}(D \setminus ST).$$

In the expression $\langle F, \Phi \rangle = 0$ (recall the scalar product $\langle \cdot, \cdot \rangle$ given in formula (21.26) in point (1)), we substitute $\int_D \Delta^p uv\, dx$ from the above Green formula of point (2), and obtain the equality

$$0 = \int_D u \Delta^p v\, dx$$

$$+ \sum_{j=1}^{N-1} \sum_{k=1}^{p-1} \int_{\Gamma_j} \{G_k^1(\gamma_k^1 + S_{p-1-k} v_j) + B_k u_{j+1}(S_{p-1-k} v_j - S_{p-1-k} v_{j+1})\}$$

$$+ \sum_{j=1}^{N-1} \sum_{k=0}^{p-2} \int_{\Gamma_j} \{G_k^2(\gamma_k^2 - B_{p-1-k} v_j) - S_k u_{j+1}(B_{p-1-k} v_j - B_{p-1-k} v_{j+1})\}$$

$$+ \sum_{j=1}^{N-1} \int_{\Gamma_j} \{G_0^1(\gamma_0^1 + \gamma + S_{p-1} v_j) + u_{j+1}(S_{p-1} v_j - S_{p-1} v_{j+1} + \gamma)\}$$

$$+ \sum_{k=1}^{p-1} \int_{\Gamma_N} \{B_k u_N(\chi_k + S_{p-1-k} v_N) - S_k u_N B_{p-1-k} v_N\}$$

$$+ \int_{\Gamma_N} \{u_N(\gamma + S_{p-1} v_N) - S_0 u_N B_{p-1} v_N\}.$$

A priori *estimates and Fredholm operators* 441

Now we will extract all the results of Proposition 21.6 from the last identity. First, by taking functions u such that $u_j \in C_0^\infty(D_j)$ for all $j = 1, 2, \ldots, N$, the boundary values on all Γ_js will vanish. Hence, we find that $\Delta^p v = 0$ in $D \setminus ST$. However, since the system of operators

$$\{B_0, B_1, \ldots, B_{p-1}, S_0, S_1, \ldots, S_{p-1}\}$$

forms a Dirichlet system (see Definition 23.12, p. 474) on every boundary Γ_j, we may apply Lemma 2.2 from [46, Chapter 2], which shows that the vector

$$\{B_k u_j|_{\Gamma_j} - B_k u_{j+1}|_{\Gamma_j},\ B_k u_{j+1}|_{\Gamma_j},\ S_k u_j|_{\Gamma_j} - S_k u_{j+1}|_{\Gamma_j},\ S_k u_{j+1}|_{\Gamma_j}\}_{k=0}^{p-1}$$

runs the space

$$\{C^\infty(\Gamma_j) \cup C^\infty(\Gamma_{j-1})\}^{2p}.$$

The last implies two systems of equalities. The first is as follows:

$$\begin{cases}
-S_{p-1-k} v_N = \chi_k & \text{on } \Gamma_N \quad \text{for } k = 1, 2, \ldots, p-1; \\
-S_{p-1} v_N = \gamma & \text{on } \Gamma_N; \\
-S_{p-1-k} v_j = \gamma_k^1 & \text{on } \Gamma_j \quad \text{for } k = 1, 2, \ldots, p-1, \text{ and} \\
& \qquad\quad j = 1, 2, \ldots, N-1; \\
B_{p-1-k} v_j = \gamma_k^2 & \text{on } \Gamma_j \quad \text{for } k = 0, 1, \ldots, p-2, \text{ and} \\
& \qquad\quad j = 1, 2, \ldots, N-1; \\
-S_{p-1} v_j = \gamma_0^1 + \gamma & \text{on } \Gamma_j \quad \text{for } j = 1, 2, \ldots, N-1; \\
-S_{p-1} v_j + S_{p-1} v_{j+1} = \gamma & \text{on } \Gamma_j \quad \text{for } j = 1, 2, \ldots, N-1.
\end{cases}$$

We see that if we know the function v we may generate all other components of the vector Φ. Another system of equalities follows for the function v itself:

$$\begin{aligned}
B_k v_N &= 0 & \text{on } \Gamma_N \quad & \text{for } k = 1, \ldots, p-1; \\
S_k v_j - S_k v_{j+1} &= 0 & \text{on } \Gamma_j \quad & \text{for } k = 0, 1, \ldots, p-2 \text{ and } j = 1, 2, \ldots, N-1; \\
B_k v_j - B_k v_{j+1} &= 0 & \text{on } \Gamma_j \quad & \text{for } k = 0, 1, \ldots, p-1 \text{ and } j = 1, 2, \ldots, N-1; \\
B_0 v_j &= 0 & \text{on } \Gamma_j \quad & \text{for } j = 1, 2, \ldots, N.
\end{aligned}$$

From this sytem we will find a solution v which will generate the rest of the components of the vector Φ. The last system is the *adjoint BVP* to BVP (21.22)–(21.25), with zero data. We see that it coincides with the original BVP for the operator L. This completes the proof. ∎

21.3.2 The case $L = \Delta^2$

It is instructive to consider the simplest nontrivial case $p = 2$. Then the scalar product is given by

$$\langle F, \Phi \rangle = \int_D fv + \int_{\Gamma_N} h_1 \chi_1 + \sum_{j=1}^{N-1} \int_{\Gamma_j} (G_0^1 \gamma_0^1 + G_1^1 \gamma_1^1) + \sum_{j=1}^{N-1} \int_{\Gamma_j} G_0^2 \gamma_0^2 + \sum_{j=1}^{N} \int_{\Gamma_j} g\gamma,$$

and the Green formula is given by

$$\int_D (\Delta^2 uv - u\Delta^2 v) = \sum_{j=1}^{N-1} \int_{\Gamma_j} (u_j S_1 v_j - u_{j+1} S_1 v_{j+1} + B_1 u_j S_0 v_j - B_1 u_{j+1} S_0 v_{j+1})$$

$$- \sum_{j=1}^{N-1} \int_{\Gamma_j} (S_0 u_j B_1 v_j - S_0 u_{j+1} B_1 v_{j+1} + S_1 u_j v_j - S_1 u_{j+1} v_{j+1})$$

$$+ \int_{\Gamma_N} (u_N S_1 v_N - S_0 u_N B_1 v_N + B_1 u_N S_0 v_N - S_1 u_N v_N).$$

21.3.3 The set Λ_1 for general elliptic operator L

We may generalize Proposition 21.6, p. 438, for a *uniformly and strongly elliptic* operator L which is *formally self-adjoint*. We need to define the operators

$$B_i = \frac{\partial^i}{\partial n^i} \quad \text{for } i = 0, 1, \ldots, p - 1,$$

and to apply Proposition 23.16, on p. 477, about the general Green formula. We find that for the system

$$\{L, B_i, \, i = 0, 1, \ldots, p - 1\} \quad \text{in } D_j,$$

there exists a system of boundary operators $\{N_i\}_{i=0}^{p-1}$ on the part of the boundary Γ_j such that

$$\text{ord } B_i + \text{ord } N_i = 2p - 1,$$

for which the Green formula holds:[3]

$$\int_{D_j} (Luv - uLv) = \sum_{i=0}^{p-1} \int_{D_j} (B_i u N_i v - N_i u B_i v). \tag{21.27}$$

In a similar way we again apply Proposition 23.16, p. 477, to the system of operators

$$\{L, B_i, \, i = 0, 1, \ldots, p - 1\}$$

---

[3] In the notation of Theorem 23.13, p. 476, we have $C_i = B_i$, and $S_i = T_i = N_i$.

but in the domain D_{j+1}. By Proposition 23.17, p. 478, about neighboring domains, since $\overline{D_j} \cap \overline{D_{j+1}} = \Gamma_j$ it follows that the system of boundary operators will be $\{-N_i\}_{i=0}^{p-1}$. Thus on $D_j \cup \Gamma_j$ we will have the boundary operators $\{B_i, N_i\}$ and on the set $D_{j+1} \cup \Gamma_j$ we will have the boundary operators $\{B_i, -N_i\}$.

Further we proceed as in Section 21.3.1, p. 437, but we only have to replace the operator S_{p-1-i} with the operator N_i. We use the operators $\{B_i\}_{i=0}^{p-1} \cup \{N_i\}_{i=1}^{p-1}$ to define the smoothness of the polysplines. Just as in (21.22)–(21.25), p. 438, we will define the *general polyspline* as a function u satisfying

- the *equation* in every subdomain D_j, i.e.

$$Lu = f \quad \text{in } D \setminus ST; \tag{21.28}$$

- the *boundary* conditions

$$B_i u = h_i \quad \text{on } \Gamma_N, \quad \text{for } i = 1, 2, \ldots, p-1; \tag{21.29}$$

- the *interface* conditions

$$\left. \begin{array}{ll} B_i u_j - B_i u_{j+1} = G_i^1 & \text{on } \Gamma_j, \quad \text{for } i = 1, \ldots, p-1, \\ N_i u_j - N_i u_{j+1} = G_i^2 & \text{on } \Gamma_j, \quad \text{for } i = 0, 1, \ldots, p-1; \end{array} \right\} \tag{21.30}$$

- and the *interpolation* conditions

$$B_0 u_j = g_j \quad \text{on } \Gamma_j, \quad \text{for } j = 1, 2, \ldots, N-1. \tag{21.31}$$

As usual, we denote by **N** the set of solutions of the system (21.28)–(21.31) when the right sides are zero (the *homogeneous* system). This completes the same description as in Proposition 21.6, p. 438, of the set Λ_1 defined in (21.20), p. 437.

Proposition 21.7 *1. The interface BVP which is adjoint to the interface BVP (21.28)–(21.31) coincides with it, i.e. problem (21.28)–(21.31) is self-adjoint. Then* $\mathbf{N} = \mathbf{N}^*$ *where, as usual,* $\mathbf{N}^*$ *denotes the solutions of the homogeneous adjoint BVP.*

2. The space Λ_1 defined by (21.20) coincides with the set run by the element

$$\Phi = \left\{ v; h; \gamma_0^1, \gamma_1^1, \ldots, \gamma_{p-1}^1; \gamma_0^2, \gamma_1^2, \ldots, \gamma_{p-1}^2; \gamma \right\},$$

where the function $v(x)$ belongs to $\mathbf{N}^*$, i.e. satisfies problem (21.28)–(21.31) with zero right sides, and all components of the vector Φ are expressed through $v(x)$ as follows:

$$\Phi = \left\{ \begin{array}{ll} v & \text{in } D \setminus ST; \\ \chi_k = -N_k v_N & \text{on } \Gamma_N \quad \text{for } k = 1, \ldots, p-1; \\ \gamma_0^1 = -N_0 v_{j+1} & \text{on } \Gamma_j \quad \text{for } j = 1, \ldots, N-1; \\ \gamma_k^1 = -N_k v_j & \text{on } \Gamma_j \quad \text{for } k = 1, \ldots, p-1; \; j = 1, \ldots, N-1; \\ \gamma_k^2 = B_k v_j & \text{on } \Gamma_j \quad \text{for } k = 0, \ldots, p-2; \; j = 1, \ldots, N-1; \\ \gamma_j = -N_0 v_j & \\ \quad + N_0 v_{j+1} & \text{on } \Gamma_j \quad \text{for } j = 1, \ldots, N-1; \\ \gamma_N = -N_0 v_N & \text{on } \Gamma_N. \end{array} \right\}$$

The proof is identical to that of Proposition 21.6. The above Green formula (21.27) plays a decisive role.

Chapter 22

Existence and convergence of polysplines

We will consider all the consequences of the results of the previous chapter for the existence of polysplines including manifolds. As we said in the Introduction to this Part, the *Fredholm operator* theorem (Theorem 21.4, p. 436) is much more general than is necessary and we will use below only data vectors $F = \{f; h; G_0, G_1, \ldots, G_{2p-2}; g\}$ which have components $f = 0$, and $G_i = 0$ for $i = 0, 1, \ldots, 2p-2$; see problem (21.10)–(21.13), p. 433.

22.1 Polysplines of order $2q$ for operator $L = L_1^2$

As in Section 20.3.2, p. 417, we will consider *polysplines* for operators of the type $L = L_1^2$, where L_1 is a *uniformly* and *strongly elliptic* operator of order

$$p = 2q$$

which is formally *self-adjoint*. We assume that the Dirichlet problem for the operator L_1 in the domain D has a unique solution, i.e. problem

$$L_1 u = 0 \quad \text{in } D,$$

$$\frac{\partial^k u}{\partial n^k} = 0 \quad \text{on } \Gamma_N, \quad \text{for } k = 0, 1, \ldots, q-1,$$

has a unique solution for $u \in H^{2p}(D)$.

Remark 22.1 *In the case of the* Hölder *space framework we assume uniqueness for* $u \in C^{2p+s}(D)$ *where $s > 0$ is not an integer (e.g. the classical setting in* Agmon *et al. [2, Theorem 12.7]).*

This implies Theorem 20.11, p. 423, about the *uniqueness* of the interpolation polysplines for the operator L. Thus we may appeal to the description of the set Λ_1

in Proposition 21.7, p. 443. In the terms of Theorem 21.4, p. 436, about the *Fredholm property* of the operator $\mathcal{P}$, it follows that the sets $\mathbf{N}$, Λ and Λ_1 satisfy

$$\mathbf{N} = \Lambda = \{0\},$$
$$\Lambda_1 = \{0\}.$$

In addition, we assume that the polysplines satisfy the Dirichlet boundary conditions

$$\frac{\partial^k u}{\partial n^k} = b_k \quad \text{on } \Gamma_N, \quad \text{for } k = 0, 1, \ldots, p-1, \tag{22.1}$$

for some functions $b_k \in H^{4q-k-(1/2)}(\Gamma_N)$ – or $b_k \in C^{4q-k+s}(\Gamma_N)$ – for $k = 0, 1, \ldots, p-1$, in the case of the *Hölder* space setting.

It is clear that the whole setting may be reformulated for the case of a smooth n-dimensional manifold D with boundary $\partial D = \Gamma_N$. The domains D_j will be diffeomorphic to annuli and the domain D_1 will be diffeomorphic to a ball. Thus only Γ_N is the "exterior" boundary.

By means of the Fredholm operator theorem (Theorem 21.4, p. 436), we obtain the following *uniqueness and existence* result.

Theorem 22.2 *We adopt all the conditions for the manifold D and the interfaces Γ_j given in Section 20.1, p. 409. Let us consider polysplines for the operator $L = L_1^2$ introduced above. We assume that the operator L_1 satisfies the uniqueness for the Dirichlet problem in the domain D. Then for every integer $s \geq 0$, and every set of interpolation data $g_j \in H^{4q-1/2+s}(\Gamma_j)$ and boundary data $b_k \in H^{4q-k-1/2+s}(\Gamma_N)$, there exists a unique interpolation polyspline $u(x)$, i.e. a function $u(x) \in H^{4q+s}(D \setminus ST)$ satisfying*

- *all conditions (20.1)–(20.2), p. 411, of the definition of the polyspline;*
- *the interpolation conditions*

$$u = g_j \quad \text{on } \Gamma_j \quad \text{for } j = 1, 2, \ldots, N;$$

- *the boundary conditions (22.1);*
- *the following inequality holds:*

$$\|u\|_{H^{4q+s}(D\setminus ST)} \leq C_s \cdot \left(\sum_{j=1}^{N} \|g_j\|_{H^{4q-1/2+s}(\Gamma_j)} + \sum_{k=1}^{p-1} \|b_k\|_{H^{4q-k-1/2+s}(\Gamma_N)} \right) \tag{22.2}$$

for an appropriate constant $C_s > 0$.

The last inequality follows from the *a priori* estimate in Theorem 21.3, p. 435, and thanks to the uniqueness of the interpolation polysplines; the norms are obtained by the definition of the set $U_{1,s}$ in (21.14), p. 434.

Remark 22.3 *This inequality means that in the case of uniqueness the polyspline u depends continuously on the data g_j and b_k. Let us recall that this dependence on the data is a trivial fact in the theory of one-dimensional splines.*

We have a similar result and the same accompanying remarks for the *Hölder* space setting.

Theorem 22.4 *Adopting all the conditions of Theorem 22.2 we consider a number $s > 0$ which is not an integer. Let the* interpolation *data satisfy $g_j \in C^{4q+s}(\Gamma_j)$ and the* boundary *data satisfy $b_k \in H^{4q-k+s}(\Gamma_N)$. Then there exists a unique interpolation polyspline $u(x)$, which is of smoothness class C^{4q-2} and such that $u(x) \in C^{4q+s}(\overline{D_j})$ for every $j = 1, 2, \ldots, N$. It satisfies the inequality*

$$\sum_{j=1}^{N} \|u\|_{C^{4q+s}(D_j)} \leq C_s \cdot \left(\sum_{j=1}^{N} \|g_j\|_{C^{4q+s}(\Gamma_j)} + \sum_{k=1}^{p-1} \|b_k\|_{C^{4q-k+s}(\Gamma_N)} \right). \quad (22.3)$$

The simplest special case is the operator $L = \Delta^{2q}$ where $L_1 = \Delta^q$, in particular for $q = 1$ when $L = \Delta^2$. These are the polysplines of Part I of this book. In fact, Theorem 22.2 is the most general result which can be proved in the case of arbitrary interface configuration.

Remark 22.5 *1. In the above theorems we have not treated the case of* Sobolev *spaces of arbitrary exponent. We refer to Lions and Magenes [46, Chapter 2, Section 6 and Section 7] which can be used for a prototype for obtaining polysplines having interpolation data $g_j \in H^s(\Gamma_j)$ for an arbitrary real number s. This is an important point which speaks in favour of the Sobolev space setting, while in the Hölder spaces we have only positive exponents $> 2p$.*

2. Note that we have more smoothness up to the boundary of the domains D_j than the necessary C^{4q-2}! Since the data g_j and b_k are usually given in classical spaces C^l for integer l, we have in such a case, by the inclusion property of the Hölder spaces provided in (23.3), p. 465, that the data belong to $C^{l-1,\alpha}$ for every α with $0 < \alpha < 1$. This is a typical effect for the Hölder spaces approach.

We now consider the operator $L = \Delta^p$ for an arbitrary integer p.

22.2 The case of a general operator L

Let L be a *uniformly strongly elliptic* operator of order $2p$ which is *formally self-adjoint* in the domain D. By Proposition 21.7, p. 443, we know that the adjoint interface BVP coincides with the original, i.e.

$$\dim \mathbf{N} = \dim \Lambda = \dim \Lambda_1.$$

Again if we want to obtain an essential consequence through the *Fredholm operator* theorem (Theorem 21.4, p. 436) we need to *assume uniqueness* of the interpolation polysplines.

Theorem 22.6 *We adopt all conditions about the manifold D and the interfaces Γ_j as described in Section 20.1, p. 409. Let us consider polysplines for the operator*

L introduced above. We assume that there exists a unique interpolation polyspline in $H^{2p}(D \setminus ST)$, i.e. the sets **N**, Λ and Λ_1 satisfy

$$\mathbf{N} = \Lambda = \Lambda_1 = \{0\}.$$

Then for every integer $s \geq 0$ *and every set of interpolation data* $g_j \in H^{2p-1/2+s}(\Gamma_j)$ *and boundary data* $b_k \in H^{2p-k-1/2+s}(\Gamma_N)$ *there exists a unique interpolation polyspline* $u(x)$, *i.e. a function* $u(x) \in H^{2p+s}(D \setminus ST)$ *satisfying*

- *all conditions (20.1)–(20.2), p. 411, of the definition of the polyspline;*
- *the interpolation conditions*

$$u = g_j \quad \text{on } \Gamma_j \quad \text{for } j = 1, 2, \ldots, N;$$

- *the boundary conditions (22.1);*
- *the following inequality holds:*

$$\|u\|_{H^{2p+s}(D\setminus ST)} \leq C_s \cdot \left(\sum_{j=1}^{N} \|g_j\|_{H^{2p-(1/2)+s}(\Gamma_j)} + \sum_{k=1}^{p-1} \|b_k\|_{H^{2p-k-(1/2)+s}(\Gamma_N)} \right)$$

for an appropriate constant $C_s > 0$.

Thus we see that, as was remarked after Theorem 22.2, p. 446, in the case of uniqueness there is a stable dependence of the polyspline on the data.

We have a very similar result for the *Hölder* space setting. In order to formulate it one may take as a prototype Theorem 22.4, p. 447.

Existence for periodic polysplines on strips

Theorem 22.6 provides the *existence* of *periodic polysplines on strips* for arbitrary interpolation and boundary data. As we have already said in Section 4.2, p. 43, the polysplines on strips with periodic data may be naturally considered as polysplines on a cylinder, see Figure 22.1 and Figure 4.2, p. 45.

Indeed, we consider the finite cylinder as a manifold with boundary, and we only have to take into account the fact that the domains D_1 and D_N have pieces which belong

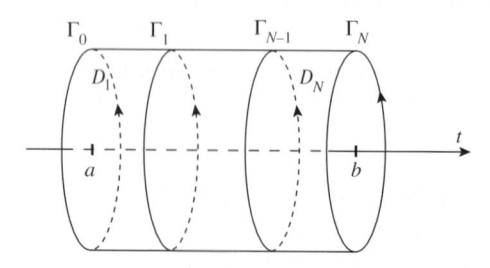

Figure 22.1.

to the boundary of the cylinder. So if we denote

$$\partial D_1 = \Gamma_0 \cup \Gamma_1,$$
$$\partial D_N = \Gamma_{N-1} \cup \Gamma_N,$$

the boundary of the cylinder D is given by

$$\partial D = \Gamma_0 \cup \Gamma_N.$$

Thus we have to impose boundary conditions on both Γ_0 and Γ_N.

Let us recall that in Theorem 9.3, p. 119, we proved the uniqueness of the periodic interpolation polysplines on strips in $\mathbb{R}^n$ and applied Theorem 22.6 to prove their existence.

Existence of polysplines on annuli

We can also apply Theorem 22.6 to the *existence* of *interpolation polysplines* on *annuli*. Indeed, according to Definition 8.1, p. 101, we have two cases.

- If the polyspline $u(x)$ is defined *in a ball* D then by Definition 8.1 it satisfies boundary conditions only on the sphere Γ_N.
- If the polyspline $u(x)$ is *in an annulus* D, then the set D_1 is an annulus, and its boundary consists of two spheres,

$$\partial D_1 = \Gamma_0 \cup \Gamma_1.$$

Hence, the boundary of D consists of two spheres,

$$\partial D = \Gamma_0 \cup \Gamma_N.$$

By Definition 8.1 the polyspline $u(x)$ will also satisfy boundary conditions on the sphere Γ_0.

We proved uniqueness of interpolation polysplines on annuli in Theorem 9.7, p. 124, and then applied Theorem 22.6 to prove the existence.

Remark 22.7 *In the case of an arbitrary elliptic operator L of order $2p$ we cannot say much more than is available in the* Fredholm operator *theorem (Theorem 21.4, p. 436). If the interpolation data $g_j \in H^{4q+r-1/2}(\Gamma_j)$, $j = 1, \ldots, N$, and the boundary data $b_k \in H^{2p-k-1/2}(\Gamma_N)$ are given then they have to satisfy* dim Λ_1 *additional linearly independent conditions of the form (21.19)–(21.20), p. 437, in order to provide existence of interpolation polyspline $u(x)$. In Proposition 21.7, p. 443, we have provided a complete description of the set Λ_1, by which we may choose linearly independent elements of the set $\mathbf{N}^* = \mathbf{N}$ which will generate the linearly independent elements Φ of the set Λ_1. Such a polyspline is* not unique *and we have to impose* dim $\mathbf{N} =$ dim Λ *linearly independent conditions on $u(x)$. Similar remarks hold in the case of polysplines in Hölder spaces.*

22.3 Existence of polysplines on strips with compact data

We recall the setting of the problem in Section 9.1.2, p. 121. It is of practical importance to have interpolation polysplines on strips where the interpolation data $\{g_j\}$ and boundary data $\{b_k\}$ are functions having compact supports or decay sufficiently fast at infinity. Although there is nothing new in principle compared with the case of periodic polysplines on strips it is worth mentioning this result.

We assume that the interfaces Γ_j are defined by

$$\Gamma_j := \{(t, y) \in \mathbb{R}^n : y \in \mathbb{R}^{n-1} \text{ and } t = t_j\},$$

where the numbers t_j satisfy $t_0 < t_1 < \cdots < t_N$. We put $ST := \bigcup_{j=0}^{N} \Gamma_j$ and

$$D := \text{Strip}(t_0, t_N) = \{(t, y) \in \mathbb{R}^n : y \in \mathbb{R}^{n-1} \text{ and } t_0 < t < t_N\},$$

and for $j = 1, 2, \ldots, N$, we consider the domains $D_j := \text{Strip}(t_{j-1}, t_j)$.

We will confine ourselves to the case $L = \Delta^p$ for the operator L. The polysplines $u(x)$ with respect to the operator L are defined by the conditions in Definition 20.1, p. 411. Since the boundary of D has two components, $\partial D = \Gamma_0 \cup \Gamma_N$, we assume that some *boundary* conditions hold on Γ_0 and on Γ_N, e.g.

$$\frac{\partial^k u}{\partial t^k} = c_k \quad \text{for } t = t_0, \text{ and } k = 0, 1, \ldots, p-1,$$

$$\frac{\partial^k u}{\partial t^k} = d_k \quad \text{for } t = t_N, \text{ and } k = 0, 1, \ldots, p-1.$$

The *interpolation* data will be given by functions $\{g_j\}$ so that

$$u = g_j \quad \text{for } t = t_j, \text{ and } j = 0, 1, \ldots, N.$$

Theorem 22.8 *Let us consider the polysplines $u(x)$ with respect to the operator $L = \Delta^p$ in the strip $D = \text{Strip}(t_0, t_N)$, and let $s \geq 0$ be an arbitrary number. Let the boundary functions c_k and d_k have compact supports and belong to $H^{2p-k-1/2+s}(\mathbb{R}^{n-1})$ for $k = 0, 1, \ldots, p-1$; let the interpolation data g_j have compact supports and g_j belong to $H^{2p-1/2+s}(\mathbb{R}^{n-1})$. Then there exists a unique polyspline $u(x)$ with these boundary and interpolation data which satisfies the inequality*

$$\|u\|_{H^{2p+s}(D\setminus ST)}$$

$$\leq C_s \cdot \left(\sum_{j=0}^{N} \|g_j\|_{H^{4q-(1/2)+s}(\Gamma_j)} + \sum_{k=1}^{p-1} \|c_k\|_{H^{4q-k-(1/2)+s}(\Gamma_0)} + \|d_k\|_{H^{4q-k-(1/2)+s}(\Gamma_N)} \right)$$

for an appropriate constant $C_s > 0$ independent of $u(x)$ and the data functions.

The same result holds for the Hölder space setting if the data c_k, d_k belong to $C^{2p-k+s}(\mathbb{R}^{n-1})$, and g_j belongs to $H^{2p+s}(\mathbb{R}^{n-1})$ for some $s > 0$ which is not an integer.

The uniqueness of the interpolation polysplines may be proved as in Theorem 9.3, p. 119, concerning periodic polysplines. Note that taking the Fourier transform in the y-direction will be possible owing to the compactness (or fast decay) in y. The existence is proved using the same techniques as in Lions and Magenes [46, Chapter 2, Theorem 4.3], and Schechter [59, Theorem 6–27, p. 147]. Using [59] we may also consider data g_j, c_k, d_k which decrease sufficiently fast at infinity. In the case of a general operator L without uniqueness of the interpolation polysplines we have to modify slightly the *Fredholm operator* theorem (Theorem 21.4, p. 436), and we have the sets $\mathbf{N} = \Lambda$ and Λ_1 which describe the set of all possible solutions.

22.4 Classical smoothness of the interpolation data g_j

Assuming that the interpolation data g_j are given in the Sobolev spaces then how smooth could they be in the classical spaces C^k?

In this context we want to point out the limits of application of the *Fredholm operator* theorem, which are imposed by the Sobolev imbedding theorem. It is also of practical importance to measure the smoothness of the interpolation data functions $\{g_j\}$ in the classical spaces C^k or in the Hölder spaces $C^{k+\alpha}$.

We will illustrate our observations starting with the simplest case – biharmonic polysplines in the circle in the plane $\mathbb{R}^2$, i.e. when the operator $L = \Delta^2$, i.e. $p = 2$. Assume that we have two concentric circles

$$\Gamma_1 = \{(x_1, x_2) \in \mathbb{R}^2 : r = r_1\},$$
$$\Gamma_2 = \{(x_1, x_2) \in \mathbb{R}^2 : r = r_2\},$$

where $0 < r_1 < r_2$. We assume that the polyspline $u(x)$ is in the circle $D = B(0; r_2)$ with subdomains $D_1 = \{r < r_1\}$ and $D_2 = \{x : r_1 < r < r_2\}$. As usual, we denote by u_j the restriction of the function u to the domain D_j. In order to be able to apply the existence Theorem 22.2, p. 446, we assume that the data functions g_1 and g_2 satisfy $g_j \in H^{3(1/2)}(\Gamma_j)$, and we have

$$u = g_j \quad \text{on } \Gamma_j \text{ for } j = 1, 2.$$

By the Sobolev embedding theorem (Theorem 23.7, p. 470), since $n = 2$ and the dimension of Γ_j (which is obviously a manifold without boundary) is equal to 1, it follows that $H^{3(1/2)}(\Gamma_j) \subset C^2(\Gamma_j)$, hence g_j belongs to $C^2(\Gamma_j)$. Similarly, if we have the boundary condition

$$\frac{\partial u}{\partial r} = b_1 \quad \text{on } \Gamma_2,$$

then by Theorem 22.2 we need to have $b_1 \in H^{2(1/2)}(\Gamma_2)$. By the Sobolev embedding theorem it follows that $H^{2(1/2)}(\Gamma_2) \subset C^1(\Gamma_2)$, i.e. $b_1 \in C^1(\Gamma_2)$.

For the polyspline $u(x)$ itself, the existence theorem (Theorem 22.2, p. 446) says that the two pieces satisfy $u_1 \in H^4(D_1)$ and $u_2 \in H^4(D_2)$. By the Sobolev embedding theorem, since the dimension of D_j is equal to 2 we obtain $H^4(D_j) \subset C^2(\overline{D_j})$, and note that this smoothness holds on the boundary!

Summarizing the above on biharmonic polysplines in $\mathbb{R}^2$ after applying Theorem 23.7 from Chapter 3 we obtain the following corollary.

Corollary 22.9 *If for some number $s \geq 0$ the data satisfy $g_j \in H^{3(1/2)+s}(\Gamma_j)$ and $b_1 \in H^{2(1/2)+s}(\Gamma_2)$ then it will follow that:*

$$g_j \in C^{2+s}(\Gamma_j), \quad b_1 \in C^{1+s}(\Gamma_2).$$

The interpolation polyspline $u(x)$ satisfies

$$u \in C^{2+s}(\overline{D_j}).$$

The $\mathbb{R}^3$ case

Now let us consider a similar example but in $\mathbb{R}^3$, where Γ_1 and Γ_2 are two concentric spheres. The operator will again be $L = \Delta^2$. Since $\dim \Gamma_j = 2$ by the Sobolev embedding we see that $H^{3(1/2)}(\Gamma_j) \subset C^2(\Gamma_j)$, and $H^{2(1/2)}(\Gamma_j) \subset C^1(\Gamma_j)$, hence $g_j \in C^2$ and $b_1 \in C^1$.

But in $\mathbb{R}^4$ we will obtain $g_j \in C^1$ and $b_1 \in C$. This is a typical effect for the so-called L_2 methods where the solutions are in Sobolev spaces, since they rely upon the embedding theorems depending on the dimension.

The general conclusion is that if we wish to obtain polysplines with a reasonable classical smoothness provided by Theorem 22.2 then we need to impose more smoothness on the data g_j.

We leave the reader to draw the conclusions about the case of an arbitrary dimension n and operator L.

22.5 Sobolev embedding in $C^{k,\alpha}$

Assuming that we have obtained a polyspline in a Sobolev space $H^{2p+s}(D)$ by the Sobolev embedding we may see whether or not it belongs to a Hölder space $C^{2p+\alpha}(\overline{\Omega})$ for an appropriate α. By Theorem 23.7, p. 470, we need to have $2p+s-(n/2) > 2p+\alpha$, i.e. $s = (n/2) + \alpha + \varepsilon$ for some $\varepsilon > 0$. Thus applying this result to the polysplines in D we obtain the following corollary.

Corollary 22.10 *For all real numbers $\alpha > 0$ and $\varepsilon > 0$ if $\alpha = s - n/2 - \varepsilon$ then*

$$H^{2p+s}(D_j) \subset C^{2p+\alpha}(\overline{D_j}),$$

i.e. if $u(x)$ is a polyspline in $H^{2p+s}(D_j)$ then it belongs to $C^{2p+\alpha}(\overline{D_j})$.

In particular for the operator $L = \Delta^2$ where $p = 2$ and $D \subset \mathbb{R}^2$ we see that if $u \in H^4(D_j)$ then $u \in C^{2+\alpha}(\overline{D_j})$ for every α with $0 < \alpha < 1$. Similarly, in $\mathbb{R}^3$ we have $H^4(D_j) \subset C^{2+\alpha}(\overline{D_j})$ for all $0 < \alpha < 1/2$.

Smoothness of one-dimensional splines

Let $n = 2$. For the purposes of *CAGD* and smoothing techniques it is important to know to which Sobolev space the data g_j belongs if it is known that g_j is some spline on the curve Γ_j. This question is of interest since if we imagine that the data points are discrete on Γ_j (which is the normal situation) and are not very dense on Γ_j, then we have to join them through a one-dimensional spline of some smoothness which will provide us the data function g_j. The point is how smooth does it have to be to ensure $g_j \in H^{3(1/2)}(\Gamma_j)$. For example, if we choose g_j to be a cubic spline then its third derivative $g_j^{(3)}$ is a piecewise constant function and thus $g_j^{(3)}$ has *Heaviside-like* singularities. It follows that $g_j^{(4)}$ will be a sum of Dirac delta functions. Thus for every $\varepsilon > 0$ we have (see Eskin [17, Example 4.1])

$$g_j^{(3)} \in H^{\frac{1}{2}-\varepsilon}(\Gamma_j),$$

hence

$$g_j \in H^{3\frac{1}{2}-\varepsilon}(\Gamma_j)$$

which fails to conform with the requirement on the data functions. We see that we have to take a spline g_j of degree 4 which will imply

$$g_j \in H^{4\frac{1}{2}-\varepsilon}(\Gamma_j).$$

By a similar argument one immediately proves the following lemma.

Lemma 22.11 *Let the function $s(t)$ be a spline of polynomial degree k (or a Chebyshev spline of order $k + 1$) in the interval $[a, b]$. Then*

$$s \in H^{k-\frac{1}{2}-\varepsilon}((a, b)).$$

Indeed, the derivative $s^{(k-1)}$ is a piecewise constant function.

Remark 22.12 *Note that if the smoothness of the data g_j is not high enough we still obtain a polyspline in the Sobolev spaces with exponents lower than $2p$ but no such result is provided by the Hölder space setting, see Remark 22.5, p. 447, about data in Sobolev spaces with an arbitrary exponent s.*

22.6 Existence for an interface which is not C^∞

By Theorem 23.20, p. 482, concerning the *a priori* estimates and regularity in Hölder space, we know that for solutions of elliptic BVP in the space $C^{2p+k+\alpha}(\overline{D})$ the boundary needs to satisfy the *uniform $C^{2p+k+\alpha}$-regularity*. Similarly, from Theorem 23.18, p. 480, about the *a priori* estimates and regularity in Sobolev spaces, we know that for solutions in the space $H^{2p+k}(D \setminus ST) = \cup H^{2p+k}(D_j)$ we need the *uniform C^{2p+k}-regularity* of the boundary. See Definition 23.3, p. 467, for the notion of *uniform C^m-regularity*.

The basic *Fredholm operator* theorem, p. 436, may be reconsidered and proved in a similar way in domains D and interfaces Γ_j which satisfy the *uniform C^{2p+k}-regularity* in case of solutions in H^{2p+k}. The regularity theorem in Hölder spaces $\cup C^{2p+k+\alpha}(\overline{D_j})$ will follow for domains D and interfaces Γ_j satisfying the *uniform $C^{2p+k+\alpha}$-regularity*.

Let us consider the simplest case of the operator $L = \Delta^2$. Since $p = 2$ and the order of the operator L is 4, the natural space for the polysplines will be the Sobolev space H^4 or H^s for $s \geq 4$. The boundary ∂D and the interfaces Γ_j have to satisfy the *uniform C^4-regularity* at least, or the *uniform C^s-regularity* if s is an integer. This has immediate consequences for the treatment of the data on the interfaces for the purposes of CAGD. Usually the practical data are discrete. We have to join them in order to obtain the contour Γ_j. By the above arguments it follows that in the case of biharmonic polysplines we have to provide smoothness C^4 of Γ_j in order to obtain *uniform C^4-regularity*, i.e. we have to use polynomial splines of degree 5 for determining Γ_j.

The cases we are interested in are the domains $\widetilde{D}$ and interfaces $\widetilde{\Gamma}_j$ with boundary uniformly C^k-regular, but which are piecewise C^∞. Using the continuity method of Schauder as presented in Theorem 23.22, p. 483, we can justify the regularity and existence theorems for such interface configurations. We proceed in several steps.

1. Let $\partial \widetilde{D}$ and $\widetilde{\Gamma}_j$ be of class that is uniformly C^k-regular but piecewise C^∞. It is clear that we may carry out a small perturbation by using a diffeomorphism on a domain D and interface surfaces Γ_j such that ∂D and Γ_j are C^∞ – in fact there is a continuous family of diffeomorphisms ϕ_t defined on a neighborhood of the set $\overline{D}$. It transforms $\partial \widetilde{D}$ and $\widetilde{\Gamma}_j$ to the C^∞ boundary ∂D and Γ_j but the diffeomorphism will be also of class C^k at $t = 0$. We obtain the mapping of the operators

$$\Delta \xrightarrow{\phi_t^*} A_2, \quad \Delta^p \xrightarrow{\phi_t^*} A_2^p.$$

2. We have uniqueness for the elliptic operator A_2^{2q} by Theorem 20.11, p. 423. We have polysplines for the operators A_2^{2q} for every set of interpolation data g_j due to the uniqueness.

3. The boundary conditions will preserve their properties by the diffeomorphism.

22.7 Convergence properties of the polysplines

In one-dimensional spline theory the approximation properties of the splines (the so-called approximation power, see Schumaker [61, Chapter 6, Section 10.5, and references in Section 6.10]) are very important. Let us recall that in order to achieve the precise rate of approximation one needs one of the most subtle devices of spline theory the *B-splines* (splines with compact support). Apparently small, for studying the approximation power of polysplines in the case of general interfaces we will need polysplines with compact support.

In the present book we will pick only the best of the bunch and will prove convergence results for interpolation polysplines which are an immediate consequence of the *Holladay property* of the polysplines. In $\mathbb{R}^2$ and $\mathbb{R}^3$ we will prove the convergence of the interpolation *biharmonic* polysplines in the case of an interface set which becomes denser and denser in the domain D of the definition of the polyspline and the interpolation data will be the values of a given function $f(x)$ defined in $C(D)$. We will provide an elegant proof which is based on the basic *Holladay property* of the interpolation polysplines. Such a proof for cubic splines (and odd-degree polynomial splines) has been given by Ahlberg *et al.* [5, Theorem 3.8.1, Theorem 5.9.1].

Assume that for every integer $\ell = 1, 2, 3, \ldots$, we are given a set of embedded surfaces Γ_j^ℓ which are indexed by $j = 1, 2, \ldots, N_\ell$, and which satisfy

$$\partial D = \Gamma_{N_\ell}.$$

So we have the data set

$$ST_\ell = \bigcup_{j=1}^{N_\ell} \Gamma_j.$$

Now we assume as above in Section 20.2, p. 411, that in the compact domain $D \subset \mathbb{R}^n$ the function $u \in H^4(D \setminus ST)$ is a *biharmonic polyspline*, i.e. it satisfies

$$\Delta^2 u = 0 \quad \text{in } D \setminus ST.$$

As usual, for $j = 2, 3, \ldots, N$, the domain D_j^ℓ is the layer between the surfaces Γ_j^ℓ and Γ_{j-1}^ℓ. The domain D_1^ℓ is diffeomorphic to a ball. We assume that the volume of the set D_j^ℓ is approaching zero with ℓ approaching infinity, i.e.

$$\max_j (\text{vol } D_j^\ell) \overset{\ell \to \infty}{\longrightarrow} 0.$$

Here vol G denotes the n-dimensional Lebesgue measure of the set $G \subset \mathbb{R}^n$. The surfaces Γ_j^ℓ are assumed to be regular enough to provide the existence of the Green function for the Dirichlet problem for the Laplace operator. We assume that the twice continuously differentiable function $f(x)$ be given on the set D, i.e. $f \in C^2(\overline{D})$.

For every $\ell \geq 1$ let us denote by $u^\ell \in H^{4q}(D \setminus ST)$ the *biharmonic interpolation polyspline* with data f as in (20.3), p. 412, i.e. satisfying

$$u^\ell(x) = f(x) \quad \text{on } ST_\ell. \tag{22.4}$$

This exists and is unique according to Theorem 22.2, p. 446.

We have the following approximation result.

Theorem 22.13 *For $n = 2$ and $n = 3$ let us assume that $f \in C^2(\overline{D})$ and within the above introduced notation*

$$\max_j (\text{vol } D_j^\ell) \overset{\ell \to \infty}{\longrightarrow} 0.$$

Then we have

$$u^\ell(x) \overset{\ell \to \infty}{\longrightarrow} f(x)$$

uniformly in $x \in \overline{D}$.

Proof We will omit the index ℓ where it is understood by default.

(1) Let us consider the domain D_j which lies between the surfaces Γ_j and Γ_{j-1}. Denote by $G(x, y)$ the Green function for the Laplace operator with Dirichlet boundary conditions [24, Chapter 2.4], i.e. for every point $y \in D_j$ we have

$$G(x, y) = R_1(x - y) + h_y(x)$$
$$G(x, y) = 0 \quad \text{for all } x \text{ in } \partial D_j.$$

$R_1(\cdot)$ is the fundamental solution of the Laplace operator. Let us note that due to the subharmonicity of the function $R_1(x - y)$ it follows that:

$$|h_y(x)| \leq |R_1(x - y)|,$$

$$G(x, y) \leq 2|R_1(x - y)|. \tag{22.5}$$

Then for every function $v \in C^1(\overline{D_j}) \cap C^2(D_j)$ we have the Green identity [24, formula (2.21)],

$$v(y) = \int_{\partial D_j} v(x) \frac{\partial G(x, y)}{\partial v_x} d\sigma_x + \int_{D_j} G(x, y) \Delta v(x) \, dx \quad \text{for } y \in D_j. \tag{22.6}$$

(2) Now let us substitute in (22.6) the function

$$v(x) = u^\ell(x) - f(x) \quad \text{for } x \text{ in } \overline{D}.$$

Since $u^\ell(x)$ is an interpolation polyspline with data $f(x)$ it follows that

$$v(x) = 0 \quad \text{for } x \text{ in } \partial D_j.$$

There remains only the volume integral from equality (22.6) to which we apply the Cauchy–Schwartz inequality and obtain

$$|u^\ell(y) - f(y)|^2 \leq \int_{D_j} |G(x, y)|^2 dx \cdot \int_{D_j} |\Delta u^\ell(x) - \Delta f(x)|^2 dx.$$

(3) To the second integral on the right-hand side we may apply the extremal *Holladay* property of the *biharmonic polysplines*, Theorem 20.14, p. 425, which implies that

$$|u^\ell(y) - f(y)|^2 \leq \int_{D_j} |G(x, y)|^2 dx \cdot \int_D |\Delta f(x)|^2 dx.$$

We only need to estimate the first integral on the right-hand side of this inequality. From inequality (22.5), p. 456, we obtain

$$\int_{D_j} |G(x, y)|^2 dx \leq 4 \int_{D_j} |R_1(x - y)|^2 dx.$$

Since $R_1(x-y)$ is a multiple of

$$\begin{cases} \log|x-y| & \text{for } n=2, \\ |x-y|^{-1} & \text{for } n=3, \end{cases}$$

it follows that $|R_1(x-y)|^2$ has integrable singularity for $x=y$ and the integral

$$\int_{\mathbb{R}^n} |R_1(x-y)|^2 \, dx$$

is finite. Hence, for $\max_j (\text{vol } D_j^\ell) \xrightarrow{\ell \to \infty} 0$, it follows that

$$\int_{D_j^\ell} |R_1(x-y)|^2 \, dx \xrightarrow{\ell \to \infty} 0,$$

the last being uniform. ∎

Next we prove the convergence of the derivatives.

Theorem 22.14 *Assume that the function $f \in C^4(\bar{D})$ and satisfies*

$$\Delta f = \frac{\partial}{\partial n} f = 0 \quad \text{on } \partial D.$$

Within the conditions of Theorem 22.13, p. 455, we have

$$\max_j (\text{vol } D_j^\ell) \xrightarrow{\ell \to \infty} 0.$$

Assume that the interpolation biharmonic polyspline $u(x)$ with interpolation data f given by (22.4), p. 455, satisfies either the boundary conditions

$$\frac{\partial}{\partial n} u^\ell = \frac{\partial}{\partial n} f \quad \text{on } \partial D, \tag{22.7}$$

or the boundary conditions

$$\Delta u^\ell = \Delta f \quad \text{on } \partial D. \tag{22.8}$$

Then

$$\Delta u^\ell \xrightarrow{\ell \to \infty} \Delta f \quad \text{in } D$$

in the L_2 norm.

Proof By the Green identity [24, formula (2.21)] applied to the functions $f - u$ and $\Delta f - \Delta u$ we obtain

$$\int_{D_j} (\Delta f - \Delta u)^2 \, dx - \int_{D_j} (f - u)(\Delta^2 f - \Delta^2 u) \, dx$$

$$= \int_{\partial D_j} \frac{\partial}{\partial n}(f - u) \cdot (\Delta f - \Delta u) \, d\sigma(x)$$

$$- \int_{\partial D_j} (f - u) \cdot \frac{\partial}{\partial n}(\Delta f - \Delta u) \, d\sigma(x).$$

Here $\vec{n} = \vec{n}_x$ denotes the *exterior unit normal* vector to D_j at the point $x \in \partial D_j$. By the definition of the polysplines we have $\Delta^2 u = 0$ in D_j.

We proceed as in the proof of the basic identity for polysplines in Theorem 20.7, p. 416. We sum up the above equalities on $j = 1, 2, \ldots, N$. By the definition of the biharmonic polyspline the terms u, $\partial u / \partial n$, and Δu, are continuous across the boundary ∂D_j. Since we assume that condition (22.7) or condition (22.8) holds on the exterior boundary ∂D we obtain the following equality:

$$\int_D (\Delta f - \Delta u)^2 \, dx = \int_D (f - u) \cdot \Delta^2 f \, dx.$$

In Theorem 22.13, p. 455, we have proved that, uniformly,

$$u^\ell \xrightarrow{\ell \to \infty} f \quad \text{in } D,$$

which by the above equality implies that in the L_2 metric

$$\Delta u^\ell \xrightarrow{\ell \to \infty} \Delta f.$$

∎

We now prove the convergence of the first derivatives.

Theorem 22.15 *We adopt the assumptions of Theorem 22.13, p. 455. We assume that the domain D satisfies $\partial D \in C^1$. Then the following convergence in the L_2 norm holds:*

$$\frac{\partial}{\partial x_i} u^\ell(x) \xrightarrow{\ell \to \infty} \frac{\partial}{\partial x_i} f(x).$$

Proof For every $\varepsilon > 0$ and for every function $h \in C^2(\overline{D})$ with $h = 0$ on ∂D we have the following interpolation inequality [24, Problem 2.15]:

$$\int_D |\nabla h|^2 \, dx \leq \varepsilon \int_D (\Delta h)^2 \, dx + \frac{1}{4\varepsilon} \int_D h^2 \, dx.$$

We apply this to the function

$$h = f - u^\ell$$

and obtain the L_2-convergence. ∎

22.8 Bibliographical notes and remarks

The main properties that feature in polysplines are the extremal property of Holladay in Theorem 20.8, p. 419, and the uniqueness theorem (Theorem 20.11, p. 423). All these results are new. Here we have to include the uniqueness Proposition 5.8, p. 64, and Proposition 8.6, p. 110. They have been initiated in the papers of the author [35–37,41]. The results on convergence of polysplines in Section 22.7, p. 454, are also new.

The polysplines are related most naturally to the *transmission* or *interface* problems for elliptic equations which have been considered by many authors, especially for second-order equations. For higher-order equations we refer to Chazarain and Piriou [10], where, in the most general framework, interface problems for higher-order equations have been formulated and general conditions providing the Fredholmness theorem are stated. In the same direction see Nicaise and Sändig [53,54], and references therein. Further references on transmission problems have been given by Agranovich [4].

Let us point to the general framework of networks introduced by Lumer [47] and studied further by Nicaise [55]. In [55] interfaces with singularities are considered in the two-dimensional case for fourth-order equations.

In all these references existence and regularity results are the main topic.

In the Bibliography to this Part we have added a number of references to works of Nicolescu, Picone, Bramble and Payne, Vekua, Hayman and Korenblum, and the author, which are related to qualitative properties of the polyharmonic functions. We have added the references to the works of Fisher and Jerome, Freeden, and others, which contain other approaches to multivariate splines by using solutions of polyharmonic equation.

Chapter 23

Appendix on elliptic boundary value problems in Sobolev and Hölder spaces

We will provide here the basic notions and results on elliptic BVPs. This area was almost exhausted in the mid-sixties for differential equations, and some new aspects and solutions were presented by the theory of pseudodifferential operators [4, 67].

For the reader's convenience we recall the fundamental results of existence and regularity in the area of general elliptic BVPs. We refer to Grisvard [25] for a user-friendly introduction to Sobolev and Hölder spaces. We will work in the L_2 norms; see Friedman [22] and Taylor [66, Chapter I.6 and Chapter XI] for the L_p spaces.

23.1 Sobolev and Hölder spaces

23.1.1 Sobolev spaces on manifolds without boundary

Let Ω be a bounded domain in $\mathbb{R}^n$. We will be interested in domains which have a boundary satisfying regularity properties such as the *segment property*, *uniform C^m-regularity* etc., see Adams [1, Sections 4.2–4.4 on p. 67 and Theorem 3.18].

- We define the scalar product

$$\langle u, v \rangle_{H^k(\Omega)} := \sum_{|\alpha| \leq k} \int_\Omega D^\alpha u \, \overline{D^\alpha v} \, dx$$

and the norm

$$\|u\|^2_{H^k(\Omega)} := \sum_{|\alpha| \leq k} \int_\Omega |D^\alpha u|^2 \, dx.$$

We take the closure of the space $C_0^\infty(\Omega)$ in $L_2(\Omega)$ with respect to this norm and obtain the Sobolev space $H^k(\Omega)$.

We will say that a domain Ω satisfies the **segment property** if for every $x \in \partial\Omega$ there exists an open set $U_x \ni x$ and a nonzero vector y_x such that if $z \in \overline{\Omega} \cap U_x$ then $z + ty_x \in \Omega$ for $0 < t < 1$. See [1]. Let us note that a domain satisfying the segment property has by necessity $(n-1)$-dimensional boundary and cannot lie on both sides of its boundary. Let Ω be a **bounded** domain of class C^1 (or at least satisfy the segment property) and let $k \geq 1$. The following are classical results:

- For every $k \geq 0$ the space $H^k(\Omega)$ is **Hilbert** with scalar product

$$\langle u, v \rangle_{H^k(\Omega)} := \sum_{|\alpha| \leq k} \int_\Omega D^\alpha u \, \overline{D^\alpha v} \, dx.$$

- For every $k \geq 0$ the space $C^\infty(\overline{\Omega})$ is dense in $H^k(\Omega)$ [1, Theorem 3.18].

For general exponent $s \in \mathbb{R}$ we introduce

$$H^s(\mathbb{R}^n) = \{u \in \mathcal{S}'(\mathbb{R}^n) : \mathcal{F}^{-1}\big((1+|\xi|^2)^{s/2}\widehat{u}(\xi)\big) \text{ belongs to } L_2(\mathbb{R}^n)\}.$$

Let $\mathbb{M}$ be a compact smooth manifold (without boundary) of dimension n. We assume that it is embedded in a Euclidean space. Let $u \in \mathcal{D}'(\mathbb{M})$. We will say that $u \in H^s(\mathbb{M})$ if and only if on any coordinate patch $U \subset \mathbb{M}$ and for any $\psi \in C_0^\infty(U)$ we have

$$\psi u \in H^s(U),$$

where U is identified with its image in $\mathbb{R}^n$. This definition is independent of the particular coordinate cover of $\mathbb{M}$. We summarize all the classical results in Theorem 23.1.

Theorem 23.1 *Let $\mathbb{M}$ be a compact smooth manifold or $\mathbb{M} = \mathbb{R}^n$. Then the following properties hold:*

- *The space $H^s(\mathbb{R}^n)$ is Hilbert with inner product*

$$\langle u, v \rangle_{H^s} := \int_{\mathbb{R}^n} (1+|\xi|^2)^s \widehat{u}(\xi) \overline{\widehat{v}(\xi)} \, d\xi.$$

- *The dual of $H^s(\mathbb{R}^n)$ is isomorphic to $H^{-s}(\mathbb{R}^n)$, and if $\mathbb{M}$ is also a Riemannian manifold, the dual of $H^s(\mathbb{M})$ is isomorphic to $H^{-s}(\mathbb{M})$.*

- *Let us denote by $C_b(\mathbb{R}^n)$ the set of bounded continuous functions in $\mathbb{R}^n$. Then the following embeddings are continuous for $\alpha > 0$:*

$$H^{(n/2)+\alpha}(\mathbb{M}) \subset \begin{cases} C_b(\mathbb{M}) & \text{if } \mathbb{M} = \mathbb{R}^n, \\ C(\mathbb{M}) & \text{if } \mathbb{M} \text{ is a compact manifold,} \end{cases}$$

$$H^{(n/2)+k+\alpha}(\mathbb{M}) \subset C^k(\mathbb{M}) \quad \text{for all integers } k \geq 0,$$
$$H^{(n/2)+\alpha}(\mathbb{M}) \subset C^\alpha(\mathbb{M}) \quad \text{for all } 0 < \alpha < 1.$$

In the last $C^\alpha(\mathbb{M})$ denotes the space of bounded functions which are Hölder continuous with exponent α (see Section 23.1.4 for the definition of the Hölder space).

- If $\mathbb{M}$ is compact then the embedding

$$H^{s+\alpha}(\mathbb{M}) \subset H^s(\mathbb{M}) \quad \text{for all } \alpha > 0$$

is compact.

- The trace τf of a function $f \in \mathcal{S}'(\mathbb{R}^n)$ on the subspace $\{0\} \times \mathbb{R}^{n-1} \subset \mathbb{R}^1 \times \mathbb{R}^{n-1} = \{(x_1, x')\}$ is defined by

$$[\tau f](x') = f(0, x').$$

For all $s > 1/2$ the map τ extends to a *unique* surjective continuous *map*

$$\tau: H^s(\mathbb{R}^n) \longrightarrow H^{s-1/2}(\mathbb{R}^{n-1}).$$

23.1.2 Sobolev spaces on the torus $\mathbb{T}^n$

Peter Lax observed that in the case of the torus, several inessential technical problems disappear and the theory of the elliptic BVPs is very elegant and concise. One of the reasons is perhaps that the Sobolev spaces have a simple description. Bers *et al.* [8, Part II, Chapter 3] and Warner [73, Chapter 6] give a detailed account of it.

We provide only a few details on the Sobolev spaces on the torus, which is the simplest manifold without a boundary. Let us denote by $\mathcal{P}$ the space $C^\infty(\mathbb{T}^n)$. By $\mathcal{S}$ we denote the space of all sequences $u = (u_\kappa)_{\kappa \in \mathbb{Z}^n}$. For every real number s we denote the space

$$\widetilde{H}^s = \left\{ u \in \mathcal{S} : \sum_{\kappa \in \mathbb{Z}^n} (1 + |\kappa|^2)^s |u_\kappa|^2 < \infty \right\}.$$

The space of all series (in the formal sense) is

$$u(x) \sim \sum_{\kappa \in \mathbb{Z}^n} u_\kappa e^{ix \cdot \kappa}, \tag{23.1}$$

where $u \in \widetilde{H}^s$ will be called the Sobolev space $H^s(\mathbb{T}^n)$. It should be noted that very often one identifies the spaces $\widetilde{H}^s$ and H^s since the space $\widetilde{H}^s$ may be considered to be a space of the Fourier transforms of H^s.

Note that the space $H^s(\mathbb{T}^n)$ is precisely the one which we have defined in Section 23.1.1, p. 461. Indeed, the *Laplace–Beltrami* operator is $\Delta_S = -\sum_{j=1}^n \partial^2/\partial x_j^2$ in the local coordinates of the torus $\mathbb{T}^n$. Obviously, every function $e^{ix \cdot \kappa}$ with $\kappa \in \mathbb{Z}^n$ is an eigenfunction of Δ_S and

$$\Delta_S e^{ix \cdot \kappa} = \lambda_\kappa e^{ix \cdot \kappa} \quad \text{for } \lambda_\kappa = |\kappa|^2.$$

Now we may use the operator $1 + \Delta_S$ to define the Sobolev space [63, Section 7]. Reasoning *formally*, we know that u belongs to $H^s(\mathbb{T}^n)$ if and only if $(1 + \Delta_S)^{\frac{s}{2}} u$ belongs to $L_2(\mathbb{T}^n)$. Applied formally to the expansion (23.1) we obtain

$$(1 + \Delta_S)^{s/2} u(x) = \sum_{\kappa \in \mathbb{Z}^n} (1 + |\kappa|^2)^{s/2} u_\kappa e^{ix \cdot \kappa}.$$

The last expansion is in L_2 if and only if $\sum_{\kappa \in \mathbb{Z}^n} (1 + |\kappa|^2)^s |u_\kappa|^2 < \infty$.

See the detailed explanation of the equivalence of the present definition of H^s and the previous definition in [63, Section 7].

Evidently, for every summable function u, i.e. for $u \in L_1(\mathbb{T}^n)$, the Fourier coefficients

$$u_\kappa = \frac{1}{(2\pi)^n} \int_0^{2\pi} \cdots \int_0^{2\pi} u(x) e^{-ix\cdot\kappa}\, dx$$

make sense and we have convergence of the series in the distributional sense [62, 65]. The main point is whether or not the corresponding Fourier series is convergent in the classical sense. In particular we have the Sobolev-type embedding, see Warner [73, Lemma 6.22 and corollaries] or Stein and Weiss [65, Chapter VII, Corollary 1.9], given in Proposition 23.2.

Proposition 23.2 *Let $s \geq [n/2]+1$ and let $u \in H^s(\mathbb{T}^n)$. Then the series $\sum_{\kappa \in \mathbb{Z}^n} u_\kappa e^{ix\kappa}$ is uniformly convergent to the function $u(x)$. The embedding $H^s(\mathbb{T}^n) \subset C(\mathbb{T}^n)$ then follows. More generally, if $s \geq [n/2] + 1 + k$ for some integer $k \geq 0$ then for $|\alpha| \leq k$ we may differentiate the series, and obtain*

$$D^\alpha u(x) = i^{|\alpha|} \sum_\kappa \kappa^\alpha u_\kappa e^{ix\cdot\kappa}$$

and the series is uniformly convergent.

We refer to [65, Chapter VII] for many subtle convergent questions about multiple Fourier series.

23.1.3 Sobolev spaces on the sphere $\mathbb{S}^{n-1}$

The Laplace series of a function h defined on the sphere provides a simple device for measuring whether or not h belongs to the space $H^s\left(\mathbb{S}^{n-1}\right)$. Let $Y_{k,\ell}$ denote an orthonormal basis of the spherical harmonics on the sphere $\mathbb{S}^{n-1}$ and $Y_{k,\ell} \in \mathcal{H}_k$, recall the definition of the spaces $\mathcal{H}_k$ in formula (10.12), p. 147. Let the Laplace series of h be (in the L_2-metric)

$$h(\theta) = \sum_{k=0}^\infty \sum_{\ell=1}^{d_k} \widehat{h}_{k,\ell} Y_{k,\ell}(\theta) \quad \text{for } \theta \in \mathbb{S}^{n-1}. \tag{23.2}$$

Recall that by Lemma 10.9, p. 147, we know that the eigenvalues of the *Laplace–Beltrami* operator $\Delta_S = -\Delta_\theta$ are $\lambda_k = k(n + k - 2)$ for $k \geq 0$. We can use the operator $1 - \Delta_\theta$ to define the Sobolev space $H^s(\mathbb{S}^{n-1})$ [63, Section 7, Lemma 7.1 and Section 22].

Now let s be a real number. Then the function $h \in L_2(\mathbb{S}^{n-1})$ belongs to the Sobolev space $H^s(\mathbb{S}^{n-1})$ if and only if $(1 + \Delta_S)^{s/2} u(x)$ belongs to $L_2(\mathbb{S}^{n-1})$. If we formally apply the operator $(1 + \Delta_S)^{s/2}$ to the expansion (23.2) we find, as in the previous section (for the torus case), that h belongs to $H^s(\mathbb{S}^{n-1})$ if and only if

$$\sum_{k=0}^\infty \sum_{\ell=1}^{d_k} (1 + \lambda_k)^s |\widehat{h}_{k,\ell}|^2 < \infty.$$

The equivalence with the definition of $H^s(\mathbb{M})$ which we have given above for an arbitrary manifold $\mathbb{M}$ without boundary is explained in [63, Section 7]. See also Lions and Magenes [46, Chapter I, Remark 7.5]. We have to note that in the case of $\mathbb{S}^2$ the Sobolev spaces provide the framework of Freeden *et al.* [20].

23.1.4 Hölder spaces

Let the function u be given in the domain Ω in $\mathbb{R}^n$. We define its *Hölder exponent* $\beta > 0$ by putting
$$[u]^{(\beta)} := \sup_{x \neq y} \frac{|u(x) - u(y)|}{|x - y|^\beta}.$$

If
$$[u]^{(\beta)} < \infty$$
then u is called Hölder continuous with exponent β. The Hölder spaces are now defined for all integers $m \geq 0$ and a number β with $0 < \beta \leq 1$, by letting
$$C^{m,\beta}(\Omega) = \{u \in C^m(\Omega): [D^\alpha u]^{(\beta)} < \infty \quad \text{for all } \alpha \text{ with } |\alpha| = m\}.$$
Some authors use the notation $C^{m+\beta}(\Omega)$ instead of $C^{m,\beta}(\Omega)$.

Likewise we define $C^{m,\beta}(\overline{\Omega})$ but as a subspace of $C^m(\overline{\Omega})$. For relatively compact sets Ω the space $C^{m,\beta}(\overline{\Omega})$ is **Banach** if we introduce the norm
$$\|u\|_{C^{m,\beta}} := \sum_{|\alpha| \leq m} \|D^\alpha u\|_{L_\infty} + \sum_{|\alpha|=m} [D^\alpha u]^{(\beta)}.$$

In a similar way we define the *Hölder* spaces for compact manifolds Ω which, for simplicity, we assume to be embedded in $\mathbb{R}^N$ for a large N.

One has to say that proving that the properties of the Hölder spaces are analogous to those of the Sobolev spaces is much easier.

In the case of a sufficiently smooth boundary of D, which is considered in the present book, if the exponents satisfy
$$k_2 + \alpha_2 < k_1 + \alpha_1$$
then we have the set-theoretic inclusion
$$C^{k_1,\alpha_1}(\overline{D}) \subset C^{k_2,\alpha_2}(\overline{D}). \tag{23.3}$$
Note that if the boundary is not smooth enough such inclusion does not hold.

We will not discuss the extension and trace properties of the spaces $C^{m,\alpha}$ since they are more intuitive and much easier to establish than those for the Sobolev spaces [25, Chapter 6.2]. On the other hand the Sobolev spaces are definitely more flexible. We refer to Gilbarg and Trudinger [24, Chapter 4.1] or Grisvard [25, Chapter 6.2] for a readable account of the properties of the Hölder spaces. Grisvard [25] also considers domains with a non-smooth boundary. Let us note that one of the major drawbacks of the Hölder spaces is the lack of good density results – the smooth functions are not dense in $C^{m,\beta}(\Omega)$ for $0 < \beta \leq 1$. Another is the lack of reasonable spaces with negative exponents etc.

Let us note that the Besov spaces generalize both the Sobolev and the Hölder spaces, see Triebel [69], Meyer [48]. For example $H^s = B_1^{s,2}$ and $C^{[s],s-[s]} = B_\infty^{s,\infty}$.

23.1.5 Sobolev spaces on manifolds with boundary

Further we will consider an n-dimensional smooth *compact* manifold Ω with *boundary* $\partial\Omega$ of smoothness at least C^1. Obviously, $\overline{\Omega} = \Omega \cup \partial\Omega$. The simplest case would be when Ω is a bounded domain in $\mathbb{R}^n$. We assume that Ω is embedded in a manifold $\mathbb{M}$ of dimension n without boundary. In particular, if Ω is a subset of a large cube in $\mathbb{R}^n$ the last may obviously be embedded into $\mathbb{T}^n$ by periodization in each direction. We further assume that $\mathbb{M}$ is C^∞ embedded in $\mathbb{R}^N$ for a sufficiently large N. That means that the coordinate functions of this embedding are C^∞ functions.

We will define the space $H^s(\Omega)$ for an arbitrary real number $s \in \mathbb{R}$. The first step is to define the space $H^s(\mathbb{R}^n_+)$, where the half-space is, as usual, given by

$$\mathbb{R}^n_+ := \{x = (x', x_n) \in \mathbb{R}^n : x_n > 0\}.$$

We define $H^s(\mathbb{R}^n_+)$ as the set of elements $f \in \mathcal{S}'(\mathbb{R}^n)$ which permit an extension lf on $\mathbb{R}^n$ satisfying $lf \in H^s(\mathbb{R}^n)$. The norm in $H^s(\mathbb{R}^n_+)$ is given by

$$\|f\|_{H^s(\mathbb{R}^n_+)} := \inf_{l} \|lf\|_{H^s(\mathbb{R}^n)},$$

where the infimum is taken over all extensions $lf \in H^s(\mathbb{R}^n)$.

The following are the main properties of the Sobolev spaces H^s.

- For integer $s = k \geq 0$ the norm $\|\cdot\|_{H^s(\mathbb{R}^n_+)}$ is equivalent to the following norm [1, 17, Lemma 4.5]:

$$\sum_{|\alpha| \leq k} \int_{\mathbb{R}^n_+} |D^\alpha f(x)|^2 \, dx.$$

- If $s = k + \lambda$, where k is an integer and $0 < \lambda < 1$ then the norm $\|\cdot\|_{H^s(\mathbb{R}^n_+)}$ is equivalent to the following [17, Remark 4.2] and [1, p. 208 and Theorem 7.48]:

$$\sum_{|\alpha|=k} \int_{\mathbb{R}^n_+}\int_{\mathbb{R}^n_+} \frac{|D^\alpha f(x) - D^\alpha f(y)|^2}{|x-y|^{n+2\lambda}} \, dx \, dy + \int_{\mathbb{R}^n_+} |f(x)|^2 \, dx.$$

- Let us denote by $H^s_-(\mathbb{R}^n)$ the subspace of $H^s(\mathbb{R}^n)$ consisting of functions u with support in $\mathbb{R}^n \setminus \mathbb{R}^n_+$. Then we have the isomorphism

$$H^s(\mathbb{R}^n_+) \approx H^s(\mathbb{R}^n)/H^s_-(\mathbb{R}^n).$$

There is an equivalent way to introduce the spaces $H^s(\mathbb{R}^n_+)$ by means of interpolation spaces [46, 66, 67].

Let us return to the compact smooth manifold with boundary Ω embedded into the manifold $\mathbb{M}$ without boundary. Further we have the space $H^s(\mathbb{M})$ and we proceed as in the case of the half-space $\mathbb{R}^n_+$. We define the Sobolev space $H^s(\Omega)$ as the set of elements $f \in \mathcal{S}'(\Omega)$ which permit an extension lf on $\mathbb{R}^n$ satisfying $lf \in H^s(\mathbb{M})$. The norm on $H^s(\Omega)$ is given by

$$\|f\|_{H^s(\Omega)} := \inf_{l} \|lf\|_{H^s(\mathbb{M})},$$

where the infimum is taken over all extensions $lf \in H^s(\mathbb{M})$. Note that since Ω is compact $S'(\Omega) = C^\infty(\Omega)$ [67, Chapter 3.3].

When Ω is a bounded domain in $\mathbb{R}^n$ we may simply imagine a coordinate cover $\bigcup_{j=0}^N U_j$ of $\overline{\Omega}$ and a corresponding partition of the unity $\psi_j \in C_0^\infty(\mathbb{R}^n)$, $\sum_{j=0}^N \psi_j = 1$, where $\mathrm{supp}(\psi_0) \subset \Omega$, through which every element $u \in H^s(\Omega)$ is "localized" in a nonboundary term $\psi_0 u \in H^s(\mathbb{R}^n)$ and boundary terms $\psi_j u \in H^s(U_j) \subset H^s(\mathbb{R}_+^n)$ for $j = 1, 2, 3, \ldots$, the last understood up to isomorphism.

23.1.6 Uniform C^m-regularity of $\partial\Omega$

In the present book the smoothness of the boundary $\Gamma = \partial\Omega$ is important. For that reason we will specify precisely the conditions on Γ which provide natural Sobolev embedding theorems and trace theorems. The main references here are Agmon et al. [2, p. 667 and 704], Adams [1] and Necas [52] who work with Sobolev spaces in domains with C^k boundary. Necas states the majority of the results on elliptic BVP in Sobolev spaces with integer exponent indicating the smoothness of the boundary. The majority of the authors work with C^∞ boundary. Adams provides very precise results for domains in $\mathbb{R}^n$ which in an obvious way hold also for manifolds with boundary.[1]

Since the smoothness properties of the compact manifold $\Omega \subset \mathbb{M}$ are of extreme importance for our applications we will assume that the compact manifold $\mathbb{M}$ is a subset of $\mathbb{R}^N$ with a finite coordinate cover $\{V_k, g_k\}$ where V_k are the coordinate patches and g_k the coordinate functions. As we said above, we assume the functions g_k to be C^∞ together with their inverses. On the other hand we will assume that the manifold Ω satisfies the so-called **uniform C^m-regularity property** for some integer m which will be specified later depending on the problem. This property is important for the formulation of the BVPs since it ensures the applicability of trace theorems and extension theorems in Adams [1, Theorem 4.26, p. 84, Theorem 7.53, p. 216]. We have the following definition [1, p. 67], where compared with the reference we also require some correspondence between the manifolds Ω and $\mathbb{M}$, which makes it somewhat clumsy.

Definition 23.3 *We will say that Ω has the **uniform C^m-regularity property** if the following holds: there exists a finite open cover $\{U_j\}$ of the boundary $\partial\Omega$ which is included in a coordinate cover $\{V_k, g_k\}$ of $\mathbb{M}$ (with mapping coordinate functions $g_k : V_k \to \mathbb{R}^n$) in the sense that for every j there exists a k with $U_j \subset V_k$. In addition, there exists a corresponding sequence $\{\Phi_j\}$ of C^m one-to-one mappings with Φ_j mapping the set $g_k U_j (\subset \mathbb{R}^n)$ into the ball $B(0; 1) = \{x \in \mathbb{R}^n : |x| < 1\}$ and such that it satisfies the following properties.*

1. *Let us denote $\Psi_j = \Phi_j^{-1}$ and $\Omega_\delta := \{x \in \Omega : \mathrm{dist}(x, \partial\Omega) < \delta\}$. Then for some number $\delta > 0$*

$$\bigcup_j g_k^{-1} \Psi_j \left(B\left(0; \frac{1}{2}\right) \right) \supset \Omega_\delta,$$

where for every j we have denoted the map of the set V_k containing U_j by g_k.

---

[1] Let us note that many authors work with the notion "manifold with smooth boundary Γ" meaning by that "Γ belongs to class C^∞ and Ω lies on one side of Γ", due to the tradition in topology where a smooth manifold has an equivalent C^∞ structure.

2. For each j the image of the set $U_j \cap \Omega$ coincides with the half-ball, i.e. $\Phi_j g_k (U_j \cap \Omega) = B(0; 1) \cap \{x_n > 0\}$.
3. If the set U_j is contained in the open set V_k of the coordinate cover of $\mathbb{M}$ with local coordinate function g_k, then the components

$$\Phi_j = (\Phi_{j,1}, \Phi_{j,2}, \ldots, \Phi_{j,n}),$$
$$\Psi_j = (\Psi_{j,1}, \Psi_{j,2}, \ldots, \Psi_{j,n}),$$

satisfy for all $i = 1, 2, \ldots, n$, all α with $|\alpha| \leq m$, and for some constant $M < \infty$, the inequalities

$$|D^\alpha \Phi_{j,i}(x)| \leq M \quad \text{for } x \text{ in } g_k U_j,$$
$$|D^\alpha \Psi_{j,i}(x)| \leq M \quad \text{for } x \text{ in } g_k U_j.$$

We have modified in an obvious way the above definition of uniform C^m-regularity available in Adams [1, p. 67] where only domains $\Omega \subset \mathbb{R}^n$ are considered. The fact that the coordinate functions g_k are smooth enough ensures that all results of that reference hold. We have the *strong extension* theorem [1, p. 84, Theorem 4.26], only for domains having uniform C^m-regularity, and we also have an optimal trace theorem for the spaces $H^m(\Omega) \longrightarrow H^{m-1/2}(\partial\Omega)$ to be defined later, see [1, p. 216, Theorem 7.53].

23.1.7 Trace theorem

In order to formulate the BVPs in Sobolev spaces we need to have traces of the functions in these spaces. Let γ denote the linear mapping provided by the usual **restriction** operator

$$u \mapsto \gamma u = \left(u|_{\partial\Omega}, \partial u/\partial n|_{\partial\Omega}, \ldots, \frac{\partial^{m-1} u}{\partial n^{m-1}}\bigg|_{\partial\Omega} \right)$$

of functions defined in $C_0^\infty(\mathbb{R}^n)$, where $\partial^j u/\partial n^j$ denotes the jth normal derivative and $n = n_x$ is the exterior unit normal vector to $\partial\Omega$ at the point $x \in \partial\Omega$, taken relative to the manifold $\mathbb{M}$.

We have the following **trace** theorem, see Adams [1, Theorem 7.53, p. 216, and Theorem 7.58, p. 218], Lions and Magenes [46, Chapter I, Theorem 8.3, Theorem 9.4], Eskin [17, Theorem 4.2], and Taylor [67, Proposition 4.5, p. 287].

Theorem 23.4 *Let the manifold Ω satisfy the* uniform C^m-*regularity condition of Definition 23.3. Then for every real number s satisfying $s - (m - 1) > 1/2$ the mapping γ defined above extends by continuity to an isomorphism and homeomorphism of the Sobolev space $H^s(\Omega)/\ker \gamma$ onto*

$$\prod_{k=0}^{m-1} H^{s-k-(1/2)}(\partial\Omega).$$

In Eskin [17], the approach to studying the spaces H^s is direct, without the application of interpolation techniques. Let us provide the following result which will be useful for us [17, Theorem 4.2].

Theorem 23.5 *Let $s > 1/2$. Then if the function $u(x', x_n)$ belongs to $H^s(\mathbb{R}^n)$ its trace $u(x', c)$ to every hyperplane $\mathbb{R}^{n-1} \times \{x_n = c\}$ is a continuous function of c in the space $H^{s-(1/2)}(\mathbb{R}^{n-1})$, and the following estimate holds:*

$$\max_{c \in \mathbb{R}} \|u(x', c)\|_{H^{s-(1/2)}(\mathbb{R}^{n-1}_{x'})} \leq C \|u\|_{H^s(\mathbb{R}^n)} \tag{23.4}$$

for some constant $C > 0$ independent of u.

The same result holds if we consider the space $H^s(\mathbb{R}^n_+)$ since it is known that we may extend the space $H^s(\mathbb{R}^n_+)$ through a continuous operator E to the whole space $\mathbb{R}^n$ and $\|Eu\|_{H^s(\mathbb{R}^n)} \leq C_1 \|u\|_{H^s(\mathbb{R}^n_+)}$. Thus the norm $\|u\|_{H^s(\mathbb{R}^n_+)}$ will appear on the right-hand side of inequality (23.4). Consequently we obtain the following corollaries.

Proposition 23.6 *Let $s > 1/2$.*

1. Then every function $u(x', x_n) \in H^s(\mathbb{R}^n_+)$ is a continuous function of the real variable $x_n \geq 0$ with values in $H_{s-(1/2)}(\mathbb{R}^{n-1})$. In particular let the trace u on the boundary $\partial \mathbb{R}^n_+ = \mathbb{R}^{n-1} \times \{0\}$ be equal to a function $g \in H^{s-(1/2)}(\mathbb{R}^{n-1})$. If we denote by $\widehat{u}(\xi, x_n)$ the Fourier transform with respect to the variable $x' \in \mathbb{R}^{n-1}$, then $\widehat{u}(\xi, x_n)$ is a continuous function of the variable $x_n \geq 0$ and $\widehat{u}(\xi, 0) = \widehat{g}(\xi)$ for every $\xi \in \mathbb{R}^{n-1}$.

2. Let $u(x)$ belong to $H^s(B(0; R))$. Then we consider for every constant $c > 0$ with $c \leq R$ the trace on the sphere $S(0; c)$ which we denote naturally by $u(c\theta)$. Then $u(c\theta)$ is a continuous function of c in the space $H^{s-(1/2)}(\mathbb{S}^{n-1})$.

Necas [52] does not consider the fractional order spaces H^s.

It should be noted that the trace results are more sensitive to the smoothness of the boundary than the Sobolev-embedding theorems below. In the last we need only $C^{0,1}$ regularity of the boundary, and we still obtain higher classical smoothness up to the boundary.

23.1.8 The general Sobolev-type embedding theorems

The Sobolev spaces also have suitable **embedding** properties which hold for relatively weak regularity properties of the boundary $\partial \Omega$. The main results belong to Sobolev, Rellich, Kondrachev, and Morrey but are known under the name "Sobolev embedding theorems". We provide the general embedding results as presented in Adams [1, Theorem 5.4, p. 97, Theorem 7.57, Theorem 7.58, p. 218 and Theorem 4.26, p. 84, for conditions on the boundary] and Gilbarg and Trudinger [24, Theorem 7.26] see also Taylor [67, Proposition 4.3, Proposition 4.4, p. 286] for the manifolds case.

Theorem 23.7 *Let the manifold Ω be compact with boundary of class $C^{0,1}$. Then the following statements hold:*

- *The following embeddings are continuous:*

$$H^{\frac{n}{2}+\alpha}(\Omega) \subset C(\overline{\Omega}) \quad \text{for all } \alpha > 0,$$

$$H^{\frac{n}{2}+k+\alpha}(\Omega) \subset C^{k,\alpha}(\overline{\Omega}) \quad \text{for all integers } k \geq 0 \text{ and } 0 < \alpha < 1,$$

$$H^{\frac{n}{2}+\alpha}(\Omega) \subset C^{\alpha}(\overline{\Omega}) \quad \text{for all } 0 < \alpha < 1.$$

In the last $C^{\alpha}(\mathbb{R}^n)$ denotes the space of bounded functions which are Hölder continuous with exponent α. (For the Hölder norms see Section 23.1.4). If the constants ε, ε_1 satisfy $\varepsilon > \varepsilon_1 > 0$ then the embedding $H^{(n/2)+k+\varepsilon}(\Omega) \subset C^{k+\varepsilon_1}(\overline{\Omega})$ is compact.

- *For any real number $s \geq 0$ and $\alpha > 0$ the embedding*

$$H^{s+\alpha}(\Omega) \subset H^s(\Omega) \quad \text{is compact.}$$

The above embedding in the space C needs only the *cone property* of $\partial\Omega$ which is less than $C^{0,1}$. Only the embeddings in the Hölder spaces $C^{k,\alpha}$ proved by Morrey require a Lipschitz boundary.

23.1.9 Smoothness across interfaces

We will need a simple interface result. For an integer $m \geq 1$ we consider the Sobolev spaces on the half-space $H^m(\mathbb{R}^n_+)$ and $H^m(\mathbb{R}^n_-)$, where as usual $\mathbb{R}^n_+ := \{x = (x', x_n) \in \mathbb{R}^n : x_n > 0\}$ and $\mathbb{R}^n_- := \{x = (x', x_n) \in \mathbb{R}^n : x_n < 0\}$. We will make use of the following.

Proposition 23.8 *Let the function $u \in H^m(\mathbb{R}^n_+)$ and the function $v \in H^m(\mathbb{R}^n_-)$. If for $k = 0, 1, \ldots, m-1$, the equalities*

$$\frac{\partial^k}{\partial x_n^k} u = \frac{\partial^k}{\partial x_n^k} v \quad \text{on } \{x_n = 0\}$$

hold, then the function w which is equal to the function u on $\mathbb{R}^n_+$ and to the function v on $\mathbb{R}^n_-$ satisfies

$$w \in H^m(\mathbb{R}^n).$$

Proof We have only to prove that for all multi-indices α with $|\alpha| \leq m$ the derivative $D^{\alpha} w$ exists in the distributional sense. It suffices to consider the case $m = 1$; for $m \geq 1$ one may proceed further by induction.

We will consider functions $\varphi \in C_0^{\infty}(\mathbb{R}^n)$ with support, say in the cylinder

$$Z := \{(x', x_n) \in \mathbb{R}^n : |x'| < 1\}.$$

We put

$$Z_+ := Z \cap \mathbb{R}^n_+, \quad Z_- := Z \cap \mathbb{R}^n_-.$$

Obviously,

$$\partial Z_+ = \partial Z_- = \{x' \in \mathbb{R}^{n-1} : |x'| < 1\}.$$

Since for every α with $|\alpha| \leq m$ the function $D^\alpha u \in L_2$, we have the equality

$$\int D^\alpha w \varphi \, dx = \int_{Z_-} D^\alpha u \varphi \, dx + \int_{Z_+} D^\alpha v \varphi \, dx.$$

Now let us consider the derivative $(\partial/\partial x_n)u$. Integrating by parts gives us the equalities

$$\int_{Z_-} \frac{\partial v}{\partial x_n} \varphi \, dx = -\int_{Z_-} v \frac{\partial \varphi}{\partial x_n} \, dx + \int_{\mathbb{R}^{n-1}} v(x', 0) \varphi(x', 0) \, dx',$$

$$\int_{Z_+} \frac{\partial u}{\partial x_n} \varphi \, dx = -\int_{Z_+} u \frac{\partial \varphi}{\partial x_n} \, dx - \int_{\mathbb{R}^{n-1}} u(x', 0) \varphi(x', 0) \, dx',$$

where $u(x', 0)$ denotes the trace of the function u on the boundary $\partial \mathbb{R}^n_+ = \mathbb{R}^{n-1}$ and likewise $v(x', 0)$ is the trace of v on the same boundary $\partial \mathbb{R}^n_- = \mathbb{R}^{n-1}$. These traces exist since u and v are of class H^1. By the definition of the distributional derivative we obtain

$$\int \frac{\partial w}{\partial x_n} \varphi \, dx = -\int w \frac{\partial \varphi}{\partial x_n} \, dx.$$

The proof that the other derivatives $\partial/\partial x_j w$ exist for $j = 1, 2, \ldots, n-1$ is easier. ∎

We have a similar result for the functions in the Hölder spaces which we formulate only in the case of half-spaces but which obviously holds for a sufficiently smooth boundary which may be "flattened".

Proposition 23.9 *Let the function $u \in C^{m,\alpha}(\mathbb{R}^n_+)$ and the function $v \in C^{m,\alpha}(\mathbb{R}^n_-)$ and let both be bounded. If for $k = 0, 1, \ldots, m$, the equalities*

$$\frac{\partial^k}{\partial x_n^k} u = \frac{\partial^k}{\partial x_n^k} v \quad \text{on } \{x_n = 0\}$$

hold, then the function w, which is equal to the function u on $\mathbb{R}^n_+$ and to the function v on $\mathbb{R}^n_-$, satisfies

$$w \in C^{m,\alpha}(\mathbb{R}^n).$$

Proof Consider the simplest case $m = 0$. We have $u \in C^{0,\alpha}(\mathbb{R}^n_+)$, and $v \in C^{0,\alpha}(\mathbb{R}^n_-)$, and $u = v$ on $\mathbb{R}^{n-1} \times \{0\}$, and we have to prove that $w \in C^{0,\alpha}(\mathbb{R}^n)$. Clearly, it suffices to consider two points, $x \in \mathbb{R}^n_+$ and $y \in \mathbb{R}^n_-$, and to estimate the Hölder fraction. Let the line segment connecting x and y cross the hyperplane $\mathbb{R}^{n-1} \times \{0\}$ at z (see Figure 23.1).

Using the inequality of the triangle we obtain the following inequality:

$$\frac{|u(x) - v(y)|}{|x - y|^\alpha} \leq \frac{|u(x) - u(z)| + |v(z) - v(y)|}{|x - y|^\alpha}$$

$$\leq \frac{|u(x) - u(z)|}{|x - z|^\alpha} + \frac{|v(z) - v(y)|}{|z - y|^\alpha}.$$

The last is bounded by a constant since $u, v \in C^{0,\alpha}$. The case of an arbitrary integer $m \geq 0$ is obtained by induction. ∎

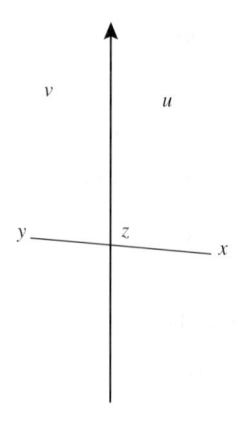

Figure 23.1.

23.2 Regular elliptic boundary value problems

The popular references here are [3, 22, 46, 59] [46](Chapter II) and in the manifolds case [66], [67].

Let Ω be as above an n-dimensional compact manifold with boundary.

We will consider the *elliptic* linear partial differential operator $L(x; D)$ of even order $2p$ in Ω defined formally in local coordinates as the expression

$$L(x; D)u = \sum_{|\alpha|,|\beta|\leq p} (-1)^{|\alpha|} D^\alpha (a_{\alpha,\beta}(x) D^\beta u)$$

where we assume that the coefficients $a_{\alpha,\beta}$ are *real* numbers. The polynomial

$$L_0(x; D) := \sum_{|\alpha+\beta|=2p} a_{\alpha,\beta}(x) D^{\alpha+\beta}$$

is called the *principal part*. The *ellipticity* of L means

$$L_0(x; D) \neq 0 \quad \text{for } \xi \in \mathbb{R}^n \setminus \{0\}.$$

Proposition 23.10 *The operator L is* properly elliptic *in the following sense. Let $\eta_1, \eta_2 \in \mathbb{R}^n$ be linearly independent, and $x \in \overline{\Omega}$. Then the polynomial in τ given by $L_0(x, \eta_1 + \tau \eta_2)$ has exactly p roots with positive imaginary parts.*

This condition is automatically fulfilled for $n \geq 3$, but for $n = 2$ it also follows since the coefficients $a_{\alpha,\beta}$ are real, see [46, Chapter II, Section 1.1, Proposition 1.1] and [59, Theorem 6.31, p. 150].[2] This property, in the case of half-space, is usually written for two orthogonal vectors η_1 and η_2, namely $\eta_1 = \eta \in \mathbb{R}^{n-1}$ and $\eta_2 = (0, \tau)$ with $\tau \in \mathbb{C}$, when we have to consider the roots of the polynomial $L_0(x; \eta, \tau)$.

---

[2] Since the coefficients of L_0 are assumed to be real the roots of the equation $L_0(x, \eta_1 + \tau \eta_2)$ will be in conjugate pairs.

Elliptic BVPs in Sobolev and Hölder spaces 473

We introduce m differential *boundary operators*

$$B_j(x, D) = \sum_{|\alpha| \leq m_j} b_{j,\alpha}(x) D^\alpha \quad \text{for } j = 0, 1, \ldots, p-1,$$

which are defined in the neighborhood of the boundary $\partial \Omega$. Let us denote the *principal part* of B_j by

$$B_{j,0}(x_0; D) := \sum_{|\alpha| = m_j} b_{j,\alpha}(x) D^\alpha.$$

We present the following classical definition.

Definition 23.11 *The system of operators $\{L, B_j, j = 0, 1, \ldots, p-1\}$ form a **regular elliptic boundary value problem** in the manifold Ω if the following conditions hold.*

1. *The operator L is **uniformly strongly elliptic**, i.e. there exists a constant $A > 0$ such that*

$$A^{-1}|\eta|^{2p} \leq |L_0(x; \eta)| \leq A|\eta|^{2p} \quad \text{for all } \eta \text{ in } \mathbb{R}^n \text{ and } x \text{ in } \overline{\Omega}.$$

2. **Smoothness of $\partial \Omega$.** *We assume that $\partial \Omega$ is a surface of the class C^{2p} at least and will be specified later in more detail depending on the Sobolev or the Hölder setting of the BVP.*

3. **Smoothness of L, and B_j.** *We assume that*

$$m_j \leq 2p - 1 \quad \text{for } j = 0, 1, \ldots, p-1,$$

and the coefficients of the boundary operator $B_j(x; D)$ belong to the class C^{2p-m_j} on $\partial \Omega$, while those of L are assumed at least to be continuous.

4. **Complementing (covering) condition.** *For any point $y \in \partial \Omega$ let ν_y denote the outward normal vector to $\partial \Omega$ at the point y. Let $\eta \in \mathbb{R}^n \setminus \{0\}$ be any vector belonging to the hyperplane tangent to $\partial \Omega$ at y, i.e.*

$$(\nu_y, \eta) = 0.$$

Then the polynomials $B_{j,0}(y; \eta + \tau \nu_y)$ in the complex variable $\tau \in \mathbb{C}$ are linearly independent modulo the polynomial

$$M_{2p}^+(\tau) := \prod_{k=1}^{p} (\tau - \tau_k^+(\eta)).$$

Here $\tau_k^+(\eta)$ are the roots of the polynomial $L_0(y, \eta + \tau \nu_y)$.

5. **Normality of the system $\{B_j\}_{j=0}^{p-1}$.**

    (a) *For $i \neq j$ we have $m_i \neq m_j$.*

    (b) *The boundary $\partial \Omega$ is noncharacteristic to B_j for every $j = 0, 1, \ldots, p-1$, i.e. for every $y \in \partial \Omega$ the polynomial $B_j(y; \eta + \tau \nu_y)$ of τ has degree exactly equal to m_j.*

We will also use the notion of the *Dirichlet system* of boundary operators.

Definition 23.12 We will say that the system of boundary operators $\{B_k\}_{k=0}^{\nu}$ is a **Dirichlet system** of order ν if it is normal in the sense of Definition 23.11, (5), and the following coincidence of sets holds:

$$\{\text{ord } B_k, \ k = 0, 1, \ldots, \nu\} = \{0, 1, \ldots, \nu\}.$$

23.2.1 Regular elliptic boundary value problems in $\mathbb{R}_+^n$

The classical method of *frozen coefficients* in the theory of elliptic BVPs is based on proper localization around every point $x \in \overline{\Omega}$. For that reason one only needs the above conditions in two special cases: in the whole space $\Omega = \mathbb{R}^n$ and in the half-space $\Omega = \mathbb{R}_+^n$, and only for operators L and B_j with constant coefficients [46, Chapter II.4].

Some of the conditions of the *regular elliptic BVP* of Definition 23.11 become somewhat simpler for the half-space, and so this is the tool to use. For that reason we write these conditions for the half-space, so let

$$\Omega = \mathbb{R}_+^n = \{x_n > 0\}.$$

First, we assume that the operators L and B_j have constant coefficients and coincide with their principal parts, i.e. $L = L_0$ and $B_j = B_{j,0}$. In other words, $L(\xi)$ and $B_j(\xi)$ are homogeneous polynomials.

At any point $x_0 \in \partial\Omega = \mathbb{R}^{n-1} \times \{0\}$ we have the hyperplane $\mathbb{R}^{n-1} \times \{0\}$ tangential to $\partial\Omega$. For arbitrary $\xi \in \mathbb{R}^{n-1}$ and $\tau \in \mathbb{R}$ we have in the new variables $\eta = (\xi, \tau) \in \mathbb{R}^n$

$$L_0(\xi, \tau) = \sum_{i=0}^{2p} p_i(\xi) \tau^{2p-i},$$

every tangential vector is of the form

$$\eta = (\xi, 0),$$

and for every $x \in \partial\Omega$ the normal vector is

$$\nu = (0, -1).$$

The *root condition* in Proposition 23.10 is equivalent to considering only special vectors. For every $\xi \in \mathbb{R}^{n-1}$ with $\xi \neq 0$ the equation

$$L_0(\xi, \tau) = 0$$

has exactly p zeros with positive imaginary parts, which we denote by

$$\tau_k^+(\xi) = \tau_k^+(x, \xi) \quad \text{for } k = 1, 2, \ldots, p,$$

and the remaining p roots (with negative imaginary parts) by

$$\tau_k^-(\xi) = \tau_k^-(x, \xi) \quad \text{for } k = 1, 2, \ldots, p.$$

We put

$$M_{2p}^+(\xi, \tau) = \prod_{k=1}^{p} (\tau - \tau_k^+(\xi)),$$

$$M_{2p}^-(\xi, \tau) = \prod_{k=1}^{p} (\tau - \tau_k^-(\xi)),$$

where it follows easily that:

$$M_{2p}^+(\xi, \tau) = (-1)^p M_{2p}^-(-\xi, -\tau).$$

See [46, Chapter II.1, Definition 1.2 and Remark 1.2, and Chapter 2.4].

The *complementing condition* in Definition 23.11 may be reformulated as follows: For every $\eta \in R^{n-1}$ with $\eta \neq 0$ the polynomials $B_j(\eta, \tau)$ of τ are linearly independent modulo the polynomial $M_{2p}^+(\eta, \tau)$. If we put

$$B_j(\eta, \tau) = Q_j(\eta, \tau) M_{2p}^+(\eta, \tau) + B'_j(\eta, \tau),$$

where

$$B'_j(\eta, \tau) = \sum_{k=0}^{p-1} b'_{jk}(\eta) \tau^k,$$

then the determinant

$$\det[b'_{jk}] \neq 0 \quad \text{for every } \eta \in \mathbb{R}^{n-1}, \ \eta \neq 0.$$

The *normality* of the system $\{B_j\}$ implies that $B_{j,0}(0, 1) \neq 0$.

23.3 Boundary operators, adjoint problem and Green formula

The present section is of particular importance since the polysplines rely upon properly defined systems of boundary operators on the interfaces and upon their adjoint systems.

We will again be working with compact domains $\Omega \subset \mathbb{R}^n$, or more generally with n-dimensional manifolds Ω with boundary embedded in $\mathbb{M} \subset \mathbb{R}^N$. The boundary $\partial\Omega$ is a manifold of dimension $n-1$ lying in Ω, having smoothness C^m for some m, but is piecewise C^∞, for which there exists a *homotopy of diffeomorphisms* ϕ_t such that

$$\phi_0 = 1 \quad (\text{or } = id),$$
$$\phi_1 D = \widetilde{D},$$
$$\phi_1 T_j = \widetilde{T}_j \quad \text{for } j = 1, 2, \ldots, N,$$

and $\partial \widetilde{D}$ and $\widetilde{T}_j$ are C^∞. It is clear that this diffeomorphism will be at most C^m at the starting point $t = 0$.

476 Multivariate polysplines

We will essentially use the adjoint boundary value problem. We will assume that the operator L is in the convenient "divergent" form:

$$Lu = \sum_{|\alpha|,|\beta| \leq p} (-1)^{|\alpha|} D^\alpha (a_{\alpha,\beta}(x) D^\beta u).$$

Then the formal adjoint is given by

$$L^*u = \sum_{|\alpha|,|\beta| \leq p} (-1)^{|\alpha|} D^\alpha (\overline{a_{\beta,\alpha}(x)} D^\beta u),$$

and satisfies

$$\int_\Omega Lu\bar{v}\, dx - \int_\Omega u\overline{L^*v}\, dx = 0$$

for every $u, v \in C^\infty(\Omega)$ with supports satisfying

$$\mathrm{supp}(u) \cap \partial\Omega = \mathrm{supp}(v) \cap \partial\Omega = \emptyset.$$

Now let the manifold Ω with boundary $\Gamma = \partial\Omega$ be given and a system of operators $\{B_j\}_{j=0}^{p-1}$ normal on the boundary Γ in the sense of Definition 23.11, 5. It generates in a nonunique way an adjoint system of boundary operators $\{C_j\}_{j=1}^{p}$ that are related through a Green formula. Thus, we have Theorem 23.13 [46, Chapter II, Theorem 2.1].

Theorem 23.13 *Let the operator L be elliptic and the system $\{B_j\}_{j=0}^{p-1}$ be normal on Γ with orders $m_j = \mathrm{ord}\, B_j \leq 2p - 1$, and the coefficients $b_{j,\alpha}$ be C^∞ on Γ. Then there exists at least one normal system of boundary operators $\{S_j\}_{j=0}^{p-1}$ on Γ with C^∞ coefficients and orders $\mu_j = \mathrm{ord}\, S_j \leq 2p - 1$, such that all orders m_j and μ_j for $j = 0, 1, \ldots, p - 1$, exhaust the set $\{0, 1, 2, \ldots, 2p - 1\}$. Further, after the choice of the system $\{S_j\}$ there exists a unique system of boundary operators C_j, T_j for $j = 0, 2, \ldots, p - 1$, with C^∞ coefficients and such that:*

- *The orders of the operators satisfy*

$$\mathrm{ord}\, C_j + \mathrm{ord}\, S_j = 2p - 1,$$
$$\mathrm{ord}\, T_j + \mathrm{ord}\, B_j = 2p - 1.$$

- *The system of operators $\{C_j\}$, $\{T_j\}$ are normal in the sense of Definition 23.11, (5), p. 473, and satisfy the Green formula*

$$\int_\Omega Lu\bar{v}\, dx - \int_\Omega u\overline{L^*v}\, dx = \sum_{j=0}^{p-1} \int_\Gamma S_j u \overline{C_j v}\, d\sigma - \sum_{j=0}^{p-1} \int_\Gamma B_j u \overline{T_j v}\, d\sigma \quad (23.5)$$

for all functions $u, v \in C^\infty(\overline{\Omega})$.

We now have the following basic theorem.

Theorem 23.14 *The system $\{L^*, C_j, j = 0, 1, \ldots, p - 1\}$ forms a regular elliptic BVP in the sense of Definition 23.11, p. 473.*

The Green formula plays an important role in our study of polysplines. We may also apply it to the case of interfaces T_j which are not C^∞. As we have stated above, we assume that there exists a family of diffeomorphisms which map $\Omega \subset \mathbb{M}$ onto another manifold $\widetilde{\Omega} \subset \mathbb{M}$ having C^∞ boundary $\partial\widetilde{\Omega}$ and even more we have

$$\partial \phi_t \Omega \in C^\infty \quad \text{for all } 0 < t \leq 1.$$

Thus if the system $\{B_j\}$ with C^∞ coefficients is given we may extend it to a neighborhood of Γ and define the boundary operators $\{C_j\}$, $\{S_j\}$ and $\{T_j\}$ on the boundary of the set $\phi_1 \Omega$. Since the diffeomorphisms represent a regular change in the variables of the integrals, we can also preserve the Green formula.

Recalling the trace Theorem 23.4 for non-C^∞ boundary, we obtain Corollary 22.15, see also [46, Chapter 2, Remark 2.2].

Corollary 23.15 *Let the manifold Ω be* uniform C^{2p}-regular *in the sense of Definition 23.3, p. 467. Then the above Green formula (23.5) holds for functions $u, v \in H^{2p}(\overline{\Omega})$.*

In the case of *formally self-adjoint* operators, i.e. when $L^* = L$, and when the operators $\{B_k\}$ form a *Dirichlet system* of order p (see Definition 23.12, p. 474), we obtain a simpler Green formula [46, Chapter II, Section 2.4], namely.

Proposition 23.16 *If the operator L is formally self-adjoint and the system $\{B_k\}_0^{p-1}$ is Dirichlet then the* normal system *of boundary operators $\{S_k\}_0^{p-1}$ (see Definition 23.11, 5), p. 473 about the normal may be constructed system in such a way that $C_k = B_k$ and $S_k = T_k$, and the following Green formula holds:*

$$\int_\Omega Lu\overline{v}\,dx - \int_\Omega u\overline{Lv}\,dx = \sum_{k=0}^{p-1}\int_\Gamma S_k u \overline{B_k v}\,d\sigma - \sum_{k=0}^{p-1}\int_\Gamma B_k u \overline{S_k v}\,d\sigma.$$

Thus

$$\operatorname{ord} S_k = 2p - 1 - \operatorname{ord} B_k.$$

23.3.1 Boundary operators in neighboring domains

Now let us imagine for simplicity that the domains Ω_1 and Ω_2 in $\mathbb{R}^n$ have a common piece of boundary, call it Γ. The *question* we are interested in may be formulated as follows. Let L be *uniformly strongly elliptic* in $\Omega_1 \cup \Omega_2$ and a *formally self-adjoint* operator on $\Omega_1 \cup \Omega_2$ and let $\{B_j\}$ be a system of boundary operators which are a *Dirichlet system* of order p on Γ (see Definition 23.12, p. 474 for the Dirichlet system). We consider two regular elliptic BVPs, one in Ω_1 and the other in Ω_2, having the same boundary operators $\{B_j\}$ on $\Omega_1 \cap \Omega_2$. Is there a relation between the corresponding boundary operators $\{S_j\}$ arising from Proposition 23.16? (see Figure 23.2 which illustrates the situation).

To illustrate what we mean, we refer to the simplest geometric configuration: let Ω_1 be a larger annulus, $\Omega_1 = \{r_1 < r < r_2\}$, and let Ω_2 be a smaller annulus, $\Omega_2 = \{r_0 < r < r_1\}$, for some constants $0 < r_0 < r_1 < r_2$. Hence,

$$\Gamma = \Omega_1 \cap \Omega_2 = S(0; r_1).$$

478 *Multivariate polysplines*

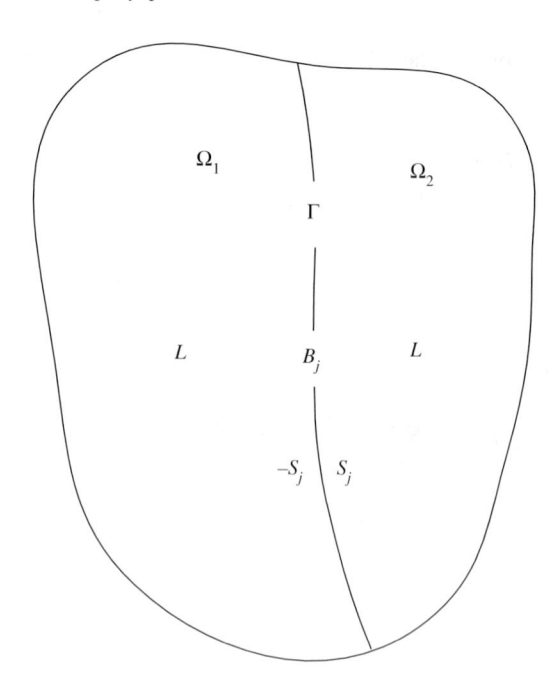

Figure 23.2. An illustration of Proposition 23.17 showing that the adjoint boundary systems $\{S_j\}$ on neighboring domains have opposite signs.

Let $\{L, B_j, j = 0, 1, \ldots, p-1\}$ be a regular elliptic BVP on the domain Ω_1 in the sense of Definition 23.11, p. 473. Since the conditions of this definition have local character, the boundary operators B_j have two components – on the spheres $S(0; r_1)$ and $S(0; r_2)$, respectively. They are independent and do not interfere with each other. However, we assume that a regular elliptic BVP $\{L, \widetilde{B}_j, j = 0, 1, \ldots, p-1\}$ be given on Ω_2 such that B_j and $\widetilde{B}_j$ coincide on Γ (see Figure 23.2).

Proposition 23.17 *We adopt the assumptions of Proposition 23.16. Let the set of boundary operators $\{S_j\}_{j=0}^{p-1}$ be those provided by Proposition 23.16, p. 477, which correspond in the domain Ω_1 to the system $\{L, B_j, j = 0, 1, \ldots, p-1\}$ on the common part of the boundary $\Gamma = \Omega_1 \cap \Omega_2$. Then the boundary operators which, according to Proposition 23.16, correspond to $\{L, \widetilde{B}_j, j = 0, 1, \ldots, p-1\}$ on Γ, but in the domain Ω_2, coincide with $\{-S_j\}_{j=0}^{p-1}$.*

Proof The proof follows the techniques used to prove Lemma 2.2 in Lions and Magenes [46, Chapter II, Sections 2.3 and 2.4]. We localize about a point $x_0 \in \Gamma$ by diffeomorphically mapping a neighborhood of x_0 in Ω_1 onto the half-ball

$$\sigma_+ = \{|y|^2 + t^2 < 1 : t > 0\},$$

and similarly a neighborhood of x_0 in Ω_2 onto the other half-ball

$$\sigma_- = \{|y|^2 + t^2 < 1 : t < 0\}.$$

We note that in Section 2.4 of [46] the following identity (which we have used in some of our notations)

$$\int_{\sigma_+} (Au)\bar{v}\,dy\,dt = \sum_{|\alpha|,|\beta|\le p} \int_{\sigma_+} a_{\alpha,\beta} D^\alpha u \overline{D^\beta v}\,dy\,dt$$

$$+ \sum_{j=0}^{p-1} \int_{\partial_1 \sigma_+} \left[D_t^j u \overline{N_{m-j} v} \right]_{t=0} dy$$

will have an analog on σ_-,

$$\int_{\sigma_-} (Au)\bar{v}\,dy\,dt = \sum_{|\alpha|,|\beta|\le p} \int_{\sigma_-} a_{\alpha,\beta} D^\alpha u \overline{D^\beta v}\,dy\,dt$$

$$+ \sum_{j=0}^{p-1} \int_{\partial_1 \sigma_+} \left[-D_t^j u \overline{N_{m-j} v} \right]_{t=0} dy,$$

where we note that the *minus* sign has appeared in front of D_t^j. This sign change results in the operators $-S_j$ on σ_-. The rest of the proof is standard and is as in Sections 2.3 and 2.4 in [46]. ∎

This result is essential for proving a fundamental identity for polysplines of even order in Section 20.3, p. 415.

23.3.2 The Green formula for the operator $L = \Delta^p$

It is easy to see that the operator $L = \Delta^p$ is formally self-adjoint. We have the following Green identity [7, p. 10, equality (2.9)]:

$$\sum_{j=0}^{p-1} \int_{\partial D} \left\{ \Delta^j f(x) \frac{\partial \Delta^{p-1-j} g(x)}{\partial \nu} - \frac{\partial \Delta^j f(x)}{\partial \nu} \Delta^{p-1-j} g(x) \right\} d\sigma(x)$$

$$= -\int_D (f(x) \Delta^p g(x) - \Delta^p f(x) g(x))\,dx. \qquad (23.6)$$

Here $\vec{\nu} = \vec{\nu}_x$ denotes the normal inner unit vector at every point x of the boundary ∂D. The measure $d\sigma$ is the naturally defined intrinsic measure on ∂D.

23.4 Elliptic boundary value problems

Now we have at our disposal almost all the necessary tools for the correct formulation of the elliptic BVP and its solution.

The majority of the results involve the smoothness property implied by *a priori* estimates in C^∞ domains and the existence results are usually formulated as the *Fredholm property* of a suitable operator. One proceeds further to C^k domains by using Schauder's continuous parameter method.

So we assume that the compact manifold Ω has a smooth boundary Γ, i.e. Γ is an infinitely smooth manifold of dimension $n - 1$, and the set Ω lies on one side of Γ. We assume that the operator

$$\mathcal{P} := \{L, \; B_j, \; j = 0, 1, \ldots, p - 1\} \tag{23.7}$$

forms a regular elliptic BVP in the sense of Definition 23.11, p. 473.

23.4.1 A priori estimates in Sobolev spaces

Let Ω be a compact manifold. We specify the conditions of [2, p. 704].

For the L_2 estimates in *Sobolev spaces* let $l \geq 2p$ be fixed. Now we assume that the manifold Ω satisfies the *uniform C^l-regularity* of Definition 23.3, p. 467.

The coefficients of L belong to C^{l-2p} and those of B_j belong to C^{l-m_j}.

We have the following fundamental theorem [2, Theorem 15.2, p. 704, Remark on p. 706], and for manifolds [67, Proposition 11.2, p. 382].

Theorem 23.18 *Let $u(x)$ belong to $H^{2p}(\Omega)$ while the data satisfy $f \in H^{l-2p}(\Omega)$ and $g_j \in H^{l-m_j-1/2}(\partial\Omega)$. Then u belongs to $H^l(\Omega)$ and the following* a priori *estimate holds for a proper constant $C > 0$ independent of the data:*

$$\|u\|_{H^l(\Omega)} \leq C \left(\|f\|_{H^{l-2p}(\Omega)} + \sum_{j=0}^{p-1} \|g_j\|_{H^{l-m_j-1/2}(\Gamma)} + \|u\|_{L_2(\Omega)} \right). \tag{23.8}$$

In inequality (23.8) the term $\|u\|_{L_2(\Omega)}$ may be replaced by the L_1-norm $\|u\|_{L_1(\Omega)}$ and may be dropped in the case of a unique solution u in $H^{2p}(\Omega)$.

23.4.2 Fredholm operator

The above *a priori* estimates are used for the proof of the Fredholmness of the operator $\mathcal{P}$.

By the results of Theorem 23.13, p. 476, we have the boundary operators $\{S_j\}$, $\{C_j\}$, $\{T_j\}$, related to $\{B_j\}$ through a Green formula. We introduce the *adjoint* elliptic BVP by putting

$$\mathcal{P}^* := \{L^*, \; C_j, \; j = 0, 1, \ldots, p - 1\}.$$

We have the following fundamental *existence* theorem which states that the operator $\mathcal{P}$ is *Fredholm*, see [46, Chapter II, Theorem 5.3], and for the manifolds case [67, Proposition 11.16, p. 394].

Theorem 23.19 *If the operator $\mathcal{P} = \{L, \; B_j, \; j = 0, 1, \ldots, p - 1\}$ defines a regular elliptic BVP on Ω in the sense of Definition 23.11, p. 473, then for every real number*

$s \geq 2p$ the following properties hold:

- The space
$$\mathbf{N} := \{u : u \in C^\infty(\overline{\Omega}),\ Lu = 0,\ B_j u = 0 \quad \text{for } j = 0, 1, \ldots, p-1\}$$
satisfies $\dim(\mathbf{N}) < \infty$.

- The space (zero space of the adjoint operator $\mathcal{P}^*$)
$$\mathbf{N}^* := \{v : v \in C^\infty(\overline{\Omega}),\ L^* v = 0,\ C_j v = 0 \quad \text{for } j = 0, 1, \ldots, p-1\}$$
satisfies $\dim(\mathbf{N}^*) < \infty$.

- The operator $\mathcal{P} = \{L, B_0, B_1, \ldots, B_{p-1}\}$ realizes the mapping
$$H^{2p+r}(\Omega) \longrightarrow H^r(\Omega) \times \prod_{j=0}^{p-1} H^{2p+r-m_j-1/2}(\Gamma)$$

and its kernel and range satisfy
$$\operatorname{Ker}(\mathcal{P}) = \mathbf{N},$$
$$\operatorname{Im}(\mathcal{P}) = (\mathbf{N}^*)^\perp;$$

the second means that $\{f; g_0, g_1, \ldots, g_{p-1}\} \in \operatorname{Im}(\mathcal{P})$ if and only if
$$\{f; g_0, g_1, \ldots, g_{p-1}\} \perp \mathbf{N}^*,$$

i.e.
$$\int_\Omega f \overline{v} + \sum_{j=0}^{p-1} \int_\Gamma g_j \overline{T_j v} = 0 \quad \text{for all } v \text{ in } \mathbf{N}^*.$$

The above is essentially an *existence* result which is established in a domain with C^∞ boundary and operators with C^∞ coefficients. If $d = \dim(\mathbf{N}) \geq 1$ it means that for finding a *unique* solution to problem
$$Lu = f,$$
$$B_j u = g_j \quad \text{on } \Gamma \text{ for } j = 0, 1, \ldots, p-1,$$

we have to impose d linearly independent conditions on u. If, on the other hand, $d_1 = \dim(\mathbf{N}^*) \geq 1$ then we have to impose d_1 linearly independent conditions on the data f, g_j in order to obtain solubility.

In order to obtain a solution for the case of operators L and B_j with nonsmooth coefficients or for a manifold Ω with boundary which is not C^∞ we may apply the Schauder continuous parameter method. It will be formulated below.

First, we consider the *a priori* estimates in domains with a boundary which is not C^∞. If we find a solution in such domains, these estimates provide us with the smoothness of the solution.

482 *Multivariate polysplines*

23.4.3 Elliptic boundary value problems in Hölder spaces

In principle all the above results hold in the framework of Hölder spaces. In fact they hold for the framework of the *Besov* spaces which generalize the Sobolev and the Hölder spaces simultaneously [69, Chapter 5].

We formulate the results in Agmon *et al.* [2, Theorem 7.3, Remark 1, Remark 2, pp. 668–669] and Taylor [66, Theorem 11.5.8].

Let the compact manifold $\Omega \subset \mathbb{M}$ with boundary $\Gamma = \partial \Omega$ be given. We will imagine that $\mathbb{M}$ is imbedded in $\mathbb{R}^N$ for a sufficiently large N. We will impose some regularity conditions on Γ in order to provide solubility of the elliptic BVPs in Sobolev and Hölder spaces. The conditions differ slightly in the two cases.

We assume that the system of operators $\{L, B_j, j = 0, 1, \ldots, p-1\}$ form a regular elliptic BVP in the sense of Definition 23.11, p. 473.

We consider bounded solutions to the BVP

$$\begin{cases} Lu = f & \text{in } \Omega, \\ B_j u = g_j & \text{on the boundary } \Gamma = \partial \Omega, \text{ for } j = 0, 1, \ldots, p-1. \end{cases} \quad (23.9)$$

The differential operator L is of order $2p$ and the differential operators B_j are of orders m_j. We consider

$$l \geq 2p$$

and fix some α satisfying

$$0 < \alpha < 1.$$

Conditions on the coefficients of the operators L, B_j

The coefficients of L have finite $\|\cdot\|_{l-2p+\alpha}$ norms in $\overline{\Omega}$ and those of B_js have finite $\|\cdot\|_{l-m_j+\alpha}$ norms on $\partial \Omega$ [2, p. 667].

Conditions on $\partial \Omega$ for Hölder estimates

We will specify the conditions necessary to obtain *a priori* estimates in the Hölder spaces, as in [2, p. 667].

We assume that the boundary Γ is of class $C^{l+\alpha}$ [2, Section 7, p. 667], in the following sense: all conditions of the uniform C^l-*regularity* in Definition 23.3, p. 467, hold except that in condition (3) the maps $\Phi_{j,i}$ and $\Psi_{j,i}$ have finite $\|\cdot\|_{l+\alpha}$ norms bounded by a constant independent of x.

We have the following *a priori* estimate and *regularity* theorem [2, Theorem 7.3, p. 668, Remark 1 and Remark 2, pp. 668–669]:

Theorem 23.20 *Let $u(x)$ be a bounded solution to problem (23.9) and $u(x)$ belong to $C^{2p+\alpha}(\overline{\Omega})$ while the data satisfy $f \in C^{l-2p+\alpha}(\overline{\Omega})$ and $g_j \in C^{l-m_j+\alpha}(\partial \Omega)$. Then $u(x)$ belongs to $C^{l+\alpha}(\overline{\Omega})$ and the following a priori estimate holds for a proper constant $C > 0$ independent of the data:*

$$\|u\|_{C^{l+\alpha}(\overline{\Omega})} \leq C \left(\|f\|_{C^{l-2p+\alpha}(\overline{\Omega})} + \sum_{j=0}^{p-1} \|g_j\|_{C^{l-m_j+\alpha}(\Gamma)} + \int_\Omega |u| \, dx \right) \quad (23.10)$$

If the solution $u(x)$ of class $C^{2p+\alpha}$ is unique *then we may omit the integral term on the right-hand side of inequality (23.10).*

In the same way as for the Sobolev spaces we find that the operator $\mathcal{P}$ is Fredholm. We only have to show which are the participating spaces [2, 69], and for manifolds [66, Chapter 11, Theorem 5.8].

Theorem 23.21 *We adopt the conditions and notation of Theorem 23.19 about the operator $\mathcal{P}$. Then $\mathcal{P}$ is a Fredholm operator which realizes the mapping*

$$C^{l+\alpha}(\Omega) \longrightarrow C^{l-2p+\alpha}(\overline{\Omega}) \times \prod_{j=1}^{p} C^{l-m_j+\alpha}(\Gamma).$$

All the conclusions of Theorem 23.19 also hold: $\mathrm{Ker}(\mathcal{P}) = \mathbf{N}$ *and* $\mathrm{Im}(\mathcal{P}) = (\mathbf{N}^*)^\perp$.

23.4.4 Schauder's continuous parameter method

As we have already said, the domains and manifolds of interest will be Ω, having a boundary Γ which is *uniformly C^k-regular* and piecewise C^∞. They are such that there exists a family of diffeomorphisms ϕ_t which transforms Γ into a C^∞ boundary. After the transform the coefficients of the operators L and B_j will have smoothness C^k.

Let us formulate the method of continuous parameter [2, Theorem 12.5, p. 689], in the case of the Hölder spaces, the case of the Sobolev spaces is very similar:

Theorem 23.22 *Let L_t and $B_{j,t}$ be a family of systems depending on a parameter $t \in [0, 1]$, satisfying uniformly in t the conditions of Section 23.4.3 and more precisely Theorem 23.20, p. 482. It is assumed that the orders* ord $B_{j,t} = m_j$ *and the coefficients of L_t vary continuously in the norm of C^α and those of $B_{j,t}$ in the norm of $C^{2p-m_j+\alpha}$. Consider the system*

$$\begin{cases} (L_t + \lambda)u = f & \text{in } \Omega, \\ B_{j,t}u = g_j & \text{on } \partial\Omega, \text{ for } j = 0, 1, \ldots, p-1, \end{cases} \quad (23.11)$$

where $f \in C^\alpha$ and $g_j \in C^{2p-m_j+\alpha}$. In addition, we assume that

- *for $t = 0$ and $\lambda = 0$ the system (23.11) is uniquely solvable,*
- *for every t there exists a complex number λ_t such that the system has a unique solution for $\lambda = \lambda_t$.*

Then for every $t \in [0, 1]$ the system (23.11) has a solution in $C^{2p+\alpha}$ which is unique for arbitrary data f, g_j, with the possible exception of a discrete set of values of λ. If for some particular values of t and λ the system (23.11) has the uniqueness property then it also has the existence property.

23.5 Bibliographical notes

Well written books include: Schechter [59], which is an introduction to the topic, although it is not complete. Agmon [3] is also a good approach, with the Green formula proved for the Dirichlet problem. It includes a lot of useful algebra on systems of boundary operators. Many ideas are competently explained in Bers *et al.* [8]. The most complete introduction seems to be Lions and Magenes [46], but it is not for the beginner; Grisvard [25] and Schechter [59] seem to be the best in that regard. Avoiding useless generalizations, Friedman [22] also is a reasonable introduction. It creates the feeling of a closed exposition. Necas [52] is very encyclopaedic and describes the results completely, including the noninfinitely smooth boundary.

Taylor [66, 67] seems to be the most up-to-date and complete reference for elliptic BVPs on manifolds. Hörmander [32]; and Eskin [17] are based on the reduction of the elliptic BVP to pseudodifferential problems. Eskin introduces the Sobolev spaces of fractional order without interpolation. Elliptic operators and Sobolev spaces on manifolds are also considered by Shubin [63]. Warner [73] provides a simple introduction to Sobolev spaces and elliptic operators on manifolds (without BVP). The papers by Agranovich [4] and Vishik and Eskin [71, 72] are very readable, and very useful for understanding the elliptical BVP.

Chapter 24

Afterword

Many things remain beyond the scope of the present book.

- *Polysplines* are a concept that is related to *polynomial splines* and to *RBFs*. They realize a smooth transition and a synthesis between the two notions. As such they inherit virtually all the important applications which are typical for splines or RBF. They also bring in a specific novelty and a concept, creating a genuinely new structure. We have seen this in the wavelets case. We leave to the future the application to neural networks – the so-called *neural networks with structured data.*
- It makes sense to make an *experimental* comparison with some other methods. When the data are on regular grids there are the methods of box-splines. When the data are more scattered, these are RBFs, kriging, etc. The theory of box-splines (polynomial, exponential, etc.) and the theory of simplex splines are completely different since they are based on finite-dimensional spaces of functions while we are using solutions of elliptic equations which are infinite dimensional space. On the other hand, the RBFs are relatives of the polysplines.
- Talking about the sources of the idea of *polysplines* one has to mention the *Peano kernel* of the finite differences in the one-dimensional case. There, the Peano kernel of a finite difference of order n is in fact the famous B-spline of order $n-1$ with a compact support (cf. [14, 61]) which was discovered in the simplest cases by Euler. Our point of view is that the finite differences are *mean-value theorems* for the polynomials of degree $n-1$. In the multivariate case we have some mean-value theorems for polyharmonic functions [9, 56–58]. For these mean-value properties we obtain the result that the Peano kernel is exactly a *polyspline with a compact support*. What a better proof that the polysplines are a genuine multivariate spline concept! We leave further discussion for the future.
- As we saw above, the main point of the *polyharmonic paradigm* is in the way we consider the data. They have to sit on $(n-1)$-dimensional manifolds. That is a very natural concept. So far a very paradoxical situation has occurred in mathematics. If you ask for such data people working in spline theory they will tell you that you are very strange, they will tell you that the data are collected on discrete points. If

you tell such a thing to the people working in mathematical physics, in boundary value problems, for at least hundred and fifty years! You tell them that story with the discrete data and they will tell you that it is not their business while the data are very incomplete: why do you start to solve a problem if you do not have enough data?! One has to appreciate so far the ingenuity of Dirichlet who has realized that what is nowadays called "Dirichlet problem" makes sense and his arguments were based on a good deal of physical intuition.

The present work may be considered as an attempt to start filling the gap between the two points of view.

- The reader who is acquainted with the main results of the present book may be able to answer the question: to which area do the polysplines belong – to mathematical analysis or to PDEs? It seems that they are an example of a very rare synthesis of the two areas.

Bibliography to Part IV

[1] Adams, R. *Sobolev Spaces*, Academic Press, New York–San Francisco–London, 1975.
[2] Agmon, S., Douglis, A. and Nirenberg, L. Estimates near the boundary for solutions of elliptic partial differential equations satisfying general boundary conditions, Part I. *Comm. Pure Appl. Math.*, 12 (1959), pp. 623–727; Part II, *Comm. Pure Appl. Math.*, 17 (1964), pp. 35–92.
[3] Agmon, S. *Lectures on Elliptic Boundary Value Problems*, van Nostrand, 1965.
[4] Agranovich, M.S. Elliptic Boundary Value Problems. In: *Partial Differential Equations IX*, Agranovich, M.S., Egorov, Yu. and Shubin, M. (Eds), Springer-Verlag, 1997.
[5] Ahlberg, J., Nilson, E. and Walsh, J. *The Theory of Splines and their Applications*, Academic Press, New York, 1967. (Russian transl: "Mir", Moscow, 1972)
[6] *Approximation Theory*, Editor C. de Boor, AMS, Providence, RI, 1986.
[7] Aronszajn, N., Creese, T.M. and Lipkin, L.J. *Polyharmonic Functions*, Clarendon Press, Oxford, 1983.
[8] Bers, L., John, F. and Schechter, M. *Partial Differential Equations*, Interscience Publishers, New York, 1964.
[9] Bramble, J.H. and Payne, L.E. Mean value theorems for polyharmonic functions, *Amer. Math. Monthly*, 73 Part II (1966), pp. 124–127.
[10] Chazarain, J. and Piriou A. *Introduction a la theorie des equations aux derivees partielles lineaires*, Gauthier–Villars, Paris 1981.
[11] Chui, Ch. *Multivariate Splines*, SIAM, Philadelphia, 1988.
[12] Dauge, M. *Elliptic Boundary Value Problems on Corner Domains*. Lecture Notes in Mathematics, 1341 Springer-Verlag, Berlin Heidelberg, 1988.
[13] Davis, P. *Interpolation and Approximation*, Dover Publ. Inc., New York, 1975.
[14] de Boor, C. *A Practical Guide to Splines*, Springer-Verlag, New York, 1978.
[15] Dryanov, D. and Kounchev, O. Multivariate formula of Euler–Maclaurin, Bernoulli functions, and Poisson type formula. Preprint, University of Hamburg, 1997.
[16] Dryanov, D. and Kounchev, O. Polyharmonically exact formula of Euler–Maclaurin, multivariate Bernoulli functions, and Poisson type formula. *C. R. Acad. Sci. Paris*, 327, Serie 1 (1998), pp. 515–520.
[17] Eskin, G. *Boundary Value Problems for Elliptic Pseudodifferential Equations*, AMS, Providence, 1981.
[18] Fisher, S. and Jerome, J. *Minimum Norm Extremals in Function Spaces*, Lecture Notes in Mathematics, 479, Springer-Verlag, Berlin, 1975.
[19] Folland, G. *Introduction to Partial Differential Equations*, Princeton University Press, Princeton, 1976.

[20] Freeden, W., Gervens, T. and Schreiner, M. *Constructive Approximation on the Sphere*, Oxford Science Publications, Clarendon Press, Oxford, 1998.
[21] Freeden, W. On spherical spline interpolation and approximation. *Math. Methods Appl. Sci.*, 3 (1981), pp. 551–575.
[22] Friedman, A. *Partial Differential Equations*, Holt, Rinehart and Winston, New York, 1969.
[23] Garabedian, P. *Partial Differential Equations*, John Wiley and Sons, New York, 1964.
[24] Gilbarg, D. and Trudinger, N. *Elliptic Partial Differential Equations of Second Order*, Springer-Verlag, Berlin, 1983.
[25] Grisvard, P. *Elliptic Problems in Nonsmooth Domains*, Pitman, Boston, 1985.
[26] Haussmann, W. and Kounchev, O.I. Peano theorems for linear functionals vanishing on polyharmonic functions. Preprint, University of Duisburg, SM-DU-258, 1994.
[27] Haussmann, W. and Kounchev, O.I. Peano theorem for linear functionals vanishing on polyharmonic functions. In: *Approximation Theory VIII*, Vol. 1, Chui, Ch. and Schumaker, L.L. (Eds), World Scientific, River Edge, NJ, 1995, pp. 233–240.
[28] Haussmann, W. and Kounchev, O.I. Variational property of the Peano kernel for harmonicity differences of order p, In: Clifford Algebras and their Applications in Mathematical Physics (Aachen, 1996), Ed. P. Jank, Kluwer Acad. Publ., Dordrecht, 1998, pp. 185–199.
[29] Haussmann, W. and Kounchev, O.I. Definiteness of the Peano kernel associated with the polyharmonic mean value property, *J. London Math. Soc.* (2) 62 (2000), no. 1, pp. 149–160.
[30] Haussmann W., O. I. Kounchev, Peano kernel associated with the polyharmonic mean value property in the annulus, Numer. Funct. Anal. Optim. 21 (2000), no. 5–6, 683–692.
[31] Hayman, W. K. and Korenblum, B., Representation and uniqueness theorems for polyharmonic functions. *J. Anal. Math.* 60 (1993), pp. 113–133.
[32] Hörmander, L. *The Analysis of Linear Partial Differential Operators III. Pseudo-Differential Operators*, Springer-Verlag, Berlin–Heidelberg–New York–Tokyo, 1985.
[33] Karlin, S. and Studden, W. *Tchebycheff Systems: with Applications in Analysis and Statistics*, Interscience Publishers, New York, 1966.
[34] Karlin, S. *Total Positivity*, Stanford University Press, 1968.
[35] Kounchev, O.I. Definition and basic properties of polysplines, I and II. *C. R. Acad. Bulg. Sci.*, 44 (1991), No. 7, and No. 8, pp. 9–11, pp. 13–16.
[36] Kounchev, O.I. Minimizing the integral of the Laplacian of a function squared with prescribed values on interior boundaries – theory of polysplines, *Trans. Am. Math. Soc.*, 350 (1998), 2105–2128.
[37] Kounchev, O.I. Theory of polysplines – minimizing the integral of the Laplacian of a function squared with prescribed values on interior boundaries with singularities. Preprint University of Duisburg, SM-DU-211 1993.
[38] Kounchev, O. Splines constructed by pieces of polyharmonic functions. In: *Wavelets, Images and Surface Fitting*, Laurent P.-J. et al., (Eds), AK Peters, Mass, 1994, pp. 319–326.

[39] Kounchev, O.I. Optimal recovery of linear functionals of Peano type through data on manifolds. *Computers Math. Applic.*, 30 (1995), No. 3–6, pp. 335–351.

[40] Kounchev, O.I. Sharp estimate of the Laplacian of a polyharmonic function and applications. *Trans. Amer. Math. Soc*, 332 (1992), No. 1, pp. 121–133.

[41] Kounchev, O.I. Theory of polysplines – minimizing the integral of the Laplacian of a function squared with prescribed values on interior boundaries with singularities, II. Preprint SM-DU-212, University of Duisburg, 1993.

[42] Kounchev, O.I. A nonlocal maximum principle for the biharmonic equation and Almansi type formulas for operators which are squares of elliptic operators. In: Jubilee Session Devoted to the Centennial of Acad. L. Chakalov (Samokov, 1986), pp. 88–92.

[43] Kounchev, O.I. and Render, H. The interpolation problem for cardinal polysplines, submitted, 2000.

[44] Krein, M. and Nudel'man, A. *The Markov Moment Problem and Extremal Problems*, AMS Transl., Providence, RI, 1977.

[45] Laurent, P.-J. *Approximation et Optimisation*, Hermann, Paris, 1972.

[46] Lions, J.L. and Magenes, E. *Non-Homogeneous Boundary Value Problems and Applications*, Vol. I, Springer-Verlag, Berlin–Heidelberg–New York, 1972.

[47] Lumer, G. Connecting of local operators and evolution equations on networks. In: *Potential Theory*, Lecture Notes in Mathematics, 787, Springer-Verlag, Berlin, 1980, pp. 219–234.

[48] Meyer, Y. *Wavelets and Operators*, Cambridge University Press, 1992.

[49] Micchelli, Ch. Oscillation matrices and cardinal spline interpolation. In: *Studies in Spline Functions and Approximation Theory*, Karlin, S. et al., (Eds), Academic Press, New York, 1976, pp. 163–202.

[50] Mikhlin, S. *The Problem of the Minimum of a Quadratic Functional*, GITTL, Moscow–Leningrad, 1952 (English transl. Holden-Day, Inc., San Francisco, London, Amsterdam 1965).

[51] Mikhlin, S.G. *Mathematical Physics, an Advanced Course*, North-Holland, Amsterdam, 1970.

[52] Necas, J. *Les Methodes Directes en Theorie des Equations Elliptiques*, Masson, Paris; Academia, 1967.

[53] Nicaise, S. and Saendig, A.-M. General interface problems. I. *Math. Methods Appl. Sci.*, 17 (1994), No. 6, pp. 395–429.

[54] Nicaise, S. and Saendig, A.-M. General interface problems. II. *Math. Methods Appl. Sci.*, 17 (1994), No. 6, pp. 431–450.

[55] Nicaise, S. *Polygonal Interface Problems*, Peter Lang, Frankfurt am Main–Berlin, 1993.

[56] Nicolescu, *Opera Mathematica. Polyharmonic Functions*, Editura Academiei Republicii Socialiste Romania, Bucharest, 1980.

[57] Nicolescu, M. Sur les fonctions de n-variables harmoniques d'ordre p, *Bull. Soc. Math. France*, 60 (1932), pp. 129–151.

[58] Picone, M. Nuovi indirizzi di ricerca teoria e nel calcolo soluzioni di talune equazioni lineari alle derivate parziali della Fizica–Matematica. *Ann. Scuola Norm. Sup. Pisa Cl. Sci.*, (1935), No. 4, pp. 213–288.

[59] Schechter, M. *Modern Methods in Partial Differential Equations*, McGraw-Hill, New York, 1977.
[60] Schulze, B.-W. and Wildenhain, *Methoden der Potentialtheorie in der Partiellen Differentialgleichungen*, Akademie-Verlag, Berlin, 1977.
[61] Schumaker, L.L. *Spline Functions: Basic Theory*, John Wiley and Sons, Chichester–Brisbane–Toronto, 1981.
[62] Schwartz, L. *Theorie des Distributions*, vols. I and II, Hermann, Paris, 1950 and 1951 (2nd edn 1957).
[63] Shubin, M. *Pseudodifferential Operators and Spectral Theory*, Springer-Verlag, Berlin–Heidelberg–New York, 1987.
[64] Sobolev, S.L. *Partial Differential Equations of Mathematical Physics*, Pergamon Press, Oxford, 1964.
[65] Stein, E. and Weiss, G. *Introduction to Fourier Analysis on Euclidean Spaces*, Princeton University Press, Princeton, New Jersey, 1971.
[66] Taylor, M. *Pseudodifferential Operators*, Princeton University Press, Princeton, New Jersey, 1981.
[67] Taylor, M. *Partial Differential Equations I, Basic Theory*, Springer-Verlag, New York, 1996.
[68] Treves, F. *Basic Linear Partial Differential Equations*, Academic Press, New York, 1975.
[69] Triebel, H., *Higher Analysis*, Johann Ambrosius Bath, Leipzig, 1992.
[70] Vekua, I.N. *New Methods for Solving Elliptic Equations*, Wiley, New York, 1967.
[71] Vishik, M. and Eskin, G. Convolution equations in a bounded domain, *Usp. Mat. Nauk*, 20, (1964), No. 3, pp. 89–152. English transl.: *Russ. Math. Surv.* 20 (1964), No. 3, pp. 85–151.
[72] Vishik, M. and Eskin, G. Elliptic convolution equations in a bounded domain and its applications. *Usp. Mat. Nauk*, 22, (1967), No. 1, pp. 15–76. English transl.: *Russ. Math. Surv.*, 22 (1967) No. 1, pp. 13–75.
[73] Warner, F.W. *Foundations of Differentiable Manifolds and Lie Groups*, Springer-Verlag, New York, Berlin, Heidelberg, Tokyo, 1983.
[74] Whittaker, E.T. and Watson, G.N. *A Course of Modern Analysis*, Cambridge University Press, London, 1965.

Index

$A_{a,b}$, 164
$A_{p,k}$, 168
$A(z)$, 320
$A_Z(x;\lambda)$, 231, 233, 235
$A_Z^*(x;\lambda)$, 252
Adams, 379, 461
Agmon, 484
Agmon–Douglis–Nirenberg, 394, 400, 406
Agranovich, 394, 400, 459, 461, 484
Ahlberg–Nilson–Walsh, 21, 23, 200, 398, 401, 404, 447, 455
a priori estimates, 406, 480
Airborne magnetic field data, 67
   Cobb Offset, 67
Algorithm, 46
Almansi formula, 11
Almansi theorem, 179
Almansi representation, 94, 96, 144, 146
   in annulus, 178, 179
   in star–shaped domain, 182, 184
Annulus, 31, 33, 34, 36
   definition, 32, 34
Ansatz, 34
Aronszajn–Creese–Lipkin, 161, 168, 408, 416, 471, 479
Axler, 81
Axler–Bourdon–Ramey, 79, 81, 83, 134

$B(x_0;r)$, 130
$B(z)$, 320
B-splines
   Fourier transform, 222
   polynomial, 221
Baker, A., 291
Ball, 128, 130
   volume, 128, 130
Basic identity
   for one-dimensional odd degree splines, 22, 24
   for polysplines of even order, 407, 415
Bers-John-Schechter, 463, 484
Besov spaces, 465
Biharmonic, 37, 39
Biharmonic functions, 37, 39
   in spherical domains, 84, 86

   in the strip, 44, 46
Bishop–Stone–Weierstrass, 150
Bojanov–Hakopian–Sahakian, 193
Bos–Levenberg–Milnam-Taylor, 159
Boundary
   of class $C^{1+\alpha}$, 482
   not C^∞, 453
   uniformly C^k-regular, 483
Boundary Value Problems,
   see BP, 3
Break-surfaces,
   see kot-surfaces, 1, 219, 401
Briggs algorithm,
   see mnimum curvature, 8
Budak–Samarskii–Tikhonov, 132
Budan–Fourier, 238
BVP, 3
BVP in neighboring domains, 418

$C^{k,\alpha}(D \setminus ST)$, 411
$c_\rho^{k,\ell}$, 374
$\text{clos}_{L_2(\mathbb{R})}$, 272, 314
C_S, 336, 339, 341
$\chi_{[0,h]}(x)$, 240
CAGD, 71, 75, 454
Cardinal L-spline wavelet analysis
   mother wavelet ψ_h, 333
Cardinal interpolation problem, 223, 225
   estimate, 302, 304
   Marsden's identity, 255, 257
   Peano kernel, 256, 257
Cardinal L-splines, 219, 221–4
   compactly supported, 237, 239
   eigenvalues, 223, 225
   interpolation problem, 236, 238
   properties, 268, 270
Cardinal L-spline Wavelet analysis
   decomposition and reconstruction, 360, 362
   decomposition relations, 349, 352
   mother wavelet ψ, 332, 334
      definition, 335, 338
   Multiresolution analysis, 327, 329
   the dual scaling function $\widetilde{\phi}(x)$, 353, 356
   the dual wavelet function $\widetilde{\psi}(x)$, 355, 358
   the scaling function ϕ, 324, 327

Index

Cardinal L-spline Wavelet analysis (*cont.*):
   the sets V_j, 324, 326
   the wavelet spaces W_j, 326, 328
Cardinal Polyspline
   on annuli
      interpolation, 285–8
      problem of interpolation, 291, 293
      synthesis of interpolation, 303, 305
Cardinal spline wavelet analysis, 260
Chazarain–Piriou, 459
Chebyshev splines, 61, 63, 106, 189, 406
   see L-splines
   and one-sided basis, 197, 199
   left-sided natural, 106, 107, 205
   natural, 202, 204, 205
   right-sided natural, 205
Chebyshev systems, 49, 88, 90, 102, 104, 165, 167
   divided difference operators, 192, 194
   Green function, 196, 198
   Lagrange–Hermite interpolation, 194, 196
Chebyshev TB-splines
   as Peano kernel, 199, 201
Chui, 220, 222, 260, 268, 309, 311
Cobb Offset, 67
Compact support, 317
Compendium, 6, 129
Compendium on spherical harmonics and polyharmonic functions, 6, 129
Computer-aided geometric design
   see CGD, 4
Cubic splines, 17, 19

Δ, 130
Δ^p, 130
Δ_S, 463
Δ_θ, 132
Δ_r, 132
D^α, 130
D_j, 409
$\mathcal{D}_j$, 226
$\widetilde{D}_{j+1}$, 409
d_k, 144
Data concept, 29
Data curves
   nonparallel, 74
Data lines
   parallel, 71
Data set, 20
   two-dimensional case, 27, 28, 29, 30
   for polysplines, 402
Data set (sample points)
   for one–dimensional splines, 19, 21
de Boor, 19, 21, 194

de Boor–DeVore–Ron, 5, 221, 223, 325, 329
de Rham, 133
Dgn-Ron, 223
Diffeomorphism, 454, 483
Differential operators
   solution set, 224, 226
Dirichlet problem
   for biharmonic functions in the strip
      periodic, 45, 47
   for biharmonicfunctions in annulus, 90, 92
   for harmonic functions in the strip
      nonperiodic, 50, 52
   for harmonic functions in the annulus, 80, 82
   for harmonic functions in the ball, 80, 82
   for harmonic functions in the strip
      periodic, 41, 43
   for polyharmonic functions in annulus, 90, 92
   for polyharmonic functions in circle, 90, 92
   for polyharmonic functions in the strip
      nonperiodic, 52, 54
      periodic, 44, 46
Dirichlet system, 414, 474
Divided difference, 229
Divided difference operator, 191
   for ECT-systems, 194
   generalized, 234, 236
Donoho, 312
Dual wavelet, 318
Dyn, N., 5
Dyn-Ron, 64, 221

$E_\phi(z)$, 317
$E_\phi(z^2)$, 319
ECT-systems, 187
   see Etented Complete Chebyshev systems
   see Cebyshev systems
   canonical, 189
   dual operator for, 197
   Lagrange–Hermite interpolation, 196
Elliptic B.V.P., 398, 404
   regular, 472
Eskin, 468, 484
Euler polynomial $A_Z(x; \lambda)$, 220, 222, 248, 250
   asymptotic in Λ_k, 276, 278
   estimates, 262, 264
   generalized, 230–3
      residuum representation, 233, 235
Euler–Frobenius polynomial $\Pi_Z(\lambda)$, 220, 222, 319
   asymptotic of the zeros in Λ_k, 296, 298
   generalized, 233, 235
   zeros, 235, 237
   leading coefficient, 251, 253

symmetric representation, 248, 250
symmetry, 261, 263
symmetry of zeros, 259, 261
classical, 294
Exponential Euler, 224
Exponential Euler spline, 336
"Exponential" Euler L-spline, 254
"Exponential" Euler spline of Schoenberg, 224
Extented Complete Chebyshev systems, 91
   see ET-system
   see Cebyshev systems

$f^{k,\ell}(\log r)$, 374
Father wavelet, 377
Fourier series, 209
   sine trigonometric, 211
Fourier transform, 209
   inverse, 212
Fourier–Laplace series, 127, 129
Fredholm alternative, 406
Fredholm operator, 436, 480
Freeden–Gervens–Schreiner, 185
Friedman, 461, 484
Fundamental cardinal spline function, 294
Fundamental spline function, 292
   for the operator $M_{k,p}$, 298, 300, 302
      estimate, 301, 303
   polynomial case, 294, 296

$\tilde{\Gamma}$, 272
   see fndamental cardinal L-spline
Γ_j, 409
$G(z; x, \lambda)$, 233
Garabedian, 394, 400
Gauss representation of polynomial, 137, 139
General polyspline, 425, 433
Gilbarg–Trudinger, 465
Green formula, 12, 479
Green function ϕ_Z^+
   of the operator $\mathcal{L}_{Z+1}$, 227, 229
Grisvard, 461, 465, 484

$\widetilde{H}^s$, 463
$H^s(D \setminus ST)$, 411
$H^s(\mathbb{M})$, 462
$h\mathbb{Z}$, 239
HHom$_2$, 139
HHom$_k$, 139
Hom$_2$, 139
Hom$_k$, 138
Hölder exponent, 465
Hölder spaces, 465, 482
Hörmander, 484

Harmonic functions
   in annulus, 77, 79
   in spherical domains, 75, 77
   periodic, 43
   radially symmetric, 76, 78
   representation through spherical harmonics, 164, 166
   in strips, 38, 40
Harmonic, 37, 39
Harmonic analysis, 219
Harmonic polynomials
   with separated variables, 150, 152
Helms, 130
$\mathcal{H}_k$, 145, 147
Hobson, 153
Holladay property, 179, 399, 405, 454
   for one-dimensional splines, 21, 23, 25, 27
   for polysplines, 417, 425
Holladay-type theorem, 110
Homogeneous polynomials, 136, 138
   harmonic, 137, 139

Interface, 1
   see beak-surface, 1
   see kot-surfaces, 1
   not C^∞, 453
Interface problems, 459
   smoothness across, 470
Interpolation spline, 19, 21
Inversion, 212

k, 371
$K(z)$, 338
Ker, 481
Kalf, 158
Karlin, 167
Karlin-Studden, 188
Knot set, 20
   in two-dimensional case, 28, 30
   for one-dimensional splines, 19, 21
Knot-surfaces (curves)
   see beak-surfaces
   for polysplines in the plane, 100, 102
   knot-surfaces, 219, 401
Kriging, 8, 75

Λ, 221, 223
$\Lambda + \gamma$, 231
Λ_{Z+1}, *see* nn-ordered vector, 223
$\Lambda_k + \frac{n}{2}$, 371
$\Lambda_k + \gamma$, 276, 278
Λ_k, 265, 267, 276, 278, 367, 371
$\mathcal{L}S_k^{j-1}$, 418

$\mathcal{L}[h\Lambda]$, 325
$\mathcal{L}_s$, 226
$\mathcal{L}_{Z+1}$, 222, 224, 226
$\mathcal{L}^*_{Z+1}$, 234, 236, 246, 248
λ_j, 111, 223
$\tilde{\Lambda}$, 271
$\ell_2^{(3)}$, 375
$[\tilde{L}_{(k)}]^p$, 164
$L(x)$, 225
L, see fndamental cardinal L-spline, 225
L-polynomial, 232
L-polynomials, 224, 238
L-splines, 4, 63, 64, 179
$L = M_{k,p}$, 261
$L = \mathcal{L}[\Lambda_k]$, 371
$L[\Lambda]$, 325
L^*u, 476
$L_{(k)}$, 80, 167
$L^p_{(k)}$, 109, 111
    see oerator $L^p_{(k)}$, 87
L-splines
    cardinal, 219, 221
    compactly supported, 220, 222
    exponential Euler, 252, 254
    null, 223, 225
    uniqueness of interpolation, 62
    see Cebyshev splines
    see ET-systems
L-polynomials, 230, 232
$L^p_{(k)} f(r)$, 91
Lagrange–Hermite interpolation
    for ECT-systems, 196
Laplace operator
    in $\mathbb{R}^2$, 37, 39
    in polar coordinates in $\mathbb{R}^2$, 76, 78
    radialpart, 76, 78
    spherical part, 76, 78
Laplace operator in $\mathbb{R}^n$, 130, 132
Laplace transform $\mathfrak{L}$, 241
Laplace–Beltrami operator, 131, 133, 463, 464
Laurent, 401
Laurent polynomial, 317, 319
Laurent, 21, 23
Lindemann, 291
Linear spline, 17, 19, 20, 22
Lions–Magenes, 406, 484
Lumer, 459
Lyche, T., 64
Lyche-Schumaker, 5

$\mathbb{M}$, 462
$M_{k,1}$, 112
$M_{k,2}$, 112, 278

$M_{k,3}$, 112
    see oerator $M_{k,p}$
$M_{k,p}$, 169, 171
$M^+_{k,p}$, 173, 177
$M_m(x)$, 221
$M_{0,p}$, 107
$M_{k,1}$,
    see oerator $M_{k,p}$, 86
$M_{k,p}$, 125
$M^+_{k,p}$, 126
Madych W, 1
Magnus-Oberhettinger-Soni, 132
Mahler, K, 291
Mallat, 309, 311
Manifold, 462
Markov–Bernstein-type inequality, 159
Marsden identity, 257
Mean-value theorems, 485
Meyer, 309, 311, 368, 379, 465
Micchelli, 217, 219, 221, 223, 225, 238
Micchelli, Ch, 5, 6
Mikhlin, 159, 161, 167, 394, 400
Minimum curvature, 75
$M_{k,p}$, 109, 111
Multiplicity, 223
Multiresolution analysis
    polyharmonic, 369, 373
    standard, 365, 368
    see MA
Multiresolution analysis, 222
    r-regular, 368
Muskhelishvili, 55

N, 481
N^*, 481
$N_{M_{k,p}}$, 171
$N_\Lambda(x)$, 259
Narasimhan, 150
Necas, 484
Neighboring domains, 477
Neural networks, 485
Newton potential in spherical harmonics, 159
Nicaise, 459
Nicaise–Sändig, 459
Nonordered vector Λ, 221, 223
    symmetric, 249, 251, 254, 256
    symmetrized $\tilde{\Lambda}$, 269, 271
Nonordered vector $\Lambda + \gamma$, 229, 231
Nonstationary scaling, 328

Ω_n, 129, 130
ord S^j_k, 418

Odd degree spline, 19, 21, 24, 26
ODEs, 227
Operator $L_{(k)}^p$
  see $_{(k)}$
  solution set, 87, 89
Operator $L_{(k)}^p$ and Operator $M_{k,p}$, 109
Operator $M_{k,p}$
  solution set, 111, 113
Order of polyharmonicity, 179, 183

$\psi(x)$, 317
$\psi^{j,k}(v)$, 385
$\psi_h(x)$, 338
$\phi(x)$, 314
$\phi_1^{k,\ell}(v,\theta)$, 382
$\phi_{2^{-j}}^k(v)$, 374
$\phi_Z(x)$, 229
ϕ_Z^+, 227, 229
$\Pi_Z(\lambda)$, 233, 235, 237
Π^n, 135
$\Pi_Z(\lambda)$, 261
$\Pi_Z(\lambda;x)$, 235
$\mathcal{P}$, 435
$\mathcal{P}^*$, 435
$\Phi_Z(x;\lambda)$, 252, 254
$\widetilde{p}_\sigma$, 372
$\lfloor \phi \rfloor$, 315, 317
$\hat{\phi}(\xi)$, 314
$\hat{\psi}(\xi)$, 317
$P(z)$, 314
PV_0, 376
PV_j, 373, 375
PW_j, 377, 384
$P_h(z)$, 332
Parameter α of cardinal splines, 286, 288, 290
Parameter α of cardinal interpolation, 223, 225
Partial Differential Equations (PDEs), 1
  see PEs
PDEs, 1
  see Prtial Differential Equations
Peano kernel, 256, 257, 485
Pizzetti formula, 12
Pock marks, 70, 75
Polyharmonic functions
  representation in annulus and curcle, 85, 87
Polyharmonic, 38, 40
Polyharmonic functions, 38, 40
  in spherical domains, 84, 86
  radially symmetric, 95, 97
  representation through spherical harmonics
    in ball, 172, 173
    in annulus, 172, 173
Polyharmonic mother wavelet, 385

Polyharmonic MRA, 373, 380
Polyharmonic Multiresolution Analysis, 309, 311
  decomposition and reconstruction, 379, 384
  properties, 374, 379
  Riesz basis, 373, 377
  the sets PV_j, 371, 375
  the spaces PW_j, 373, 377, 379, 384
Polyharmonic Paradigm, 10, 401, 485
  and Gauss representation, 143, 145
  data concept, 27, 29
  objects concept, 37, 39
  smoothness concept, 32, 34
Polyharmonic splines, 1
Polyharmonic wavelets, 367, 371
  zero moments, 391
Polyharmonicity order
  see oder of polyharmonicity
Polynomial spline of degree, 20
Polysplines
  applications in CAGD, 65, 67
  applications in Magnetism, 65, 67
  biharmonic
    on annuli, 101, 103
    on strips, 59, 61
    radially symmetric, 106, 108
  cardinal interpolation, 288
  compactly supported on annulus, 281, 283
  definition for general interfaces, 403, 411
    of Hölder class, 403, 411
    of Sobolev class, 403, 411
  general
    convergence, 446, 454
    existence, 437, 445
  harmonic
    on strips, 57, 59
  interpolation, 404, 412
    uniqueness, 62, 64
  interpolation on annuli in $\mathbb{R}^n$
    existence, 122, 124
  interpolation, on strips in $\mathbb{R}^n$
    with periodic data, existence, 117, 119
  of order p, 403
  on annuli
    computation, 107, 109
    interpolation, 100, 102
    uniqueness, 108, 110
  on strips in $\mathbb{R}^n$
    interpolation, 121, 123
  on annuli
    of class C^{2p-2}, 121, 123
    of Sobolev class, 121, 123
  on annuli in $\mathbb{R}^2$, 99, 101, 120, 123
  on not C^∞ interfaces Γ_j, 450
  on strips, 55, 57

Polysplines (*cont.*):
    interpolation, 56, 58
    on strips in $\mathbb{R}^n$, 116, 118
        of class C^{2p-2}, 116, 118
        of Sobolev class, 116, 118
        with compact data, 119, 121
    radially symmetric, 102, 104
    supported in annulus in $\mathbb{R}^2$, 99, 101, 120, 123
    supported in ball in $\mathbb{R}^2$, 99, 101, 120, 123
    uniqueness of interpolation, 413, 421
Polysplines
    biharmonic, 455
        on strips, 58, 60
    convergence, 454
    general, 433
    numerical analysis of, 416
    odd-order, 415
    of order $2q$, 402
    on annuli
        exixtence, 449
    on strips
        exixtence, 448
    on strips, compact data
        existence, 450
    smoothness, 60
    uniqueness of interpolation, 64
Polysplines of odd-order, 406
Pontryagin, 172, 227
Powell, 19, 21, 200
Prewavelets, 325, 368
Properly elliptic, 429

$\widetilde{Q}$, 283
$q_{Z+1}(z)$, 224
$Q_{Z+1}(x)$, 239
$Q_{Z+1}(x)$, 331
$Q_{Z+1}[\Lambda; h](x)$, 331
$Q^*_{Z+1}(x)$, 248
Q_{Z+1}, 237, 239
q^*_{Z+1}, 234, 236, 246, 248
q_{Z+1}, 221, 223

$r_h(x)$, 249
$\mathbb{R}^n_+$, 429, 466
$\mathcal{R}S^j_k$, 418
$R(z)$, 338
$r(\lambda)$, 233, 235
Radial Basis Functions, 4, 28, 30
    see RFs
Ragozin, 158
RBF, 67, 75

Rectangular domains, 37, 39
Refinement, 259
Refinement equation, 259, 314
    see to-scale relation
Refinement mask, 334
Residuum representation, 235
Riesz basis, 315
Riesz bounds, 267
Riesz–Fisher theorem, 209
Ron, A., 5
Rudin, 150, 211

$\sigma(\cdot)$, 131
σ_{n-1}, 130
$s_h(x)$, 239
$\mathcal{S}_{Z+1}$, 224
$\mathbb{S}^{n-1}$, 130
$\mathcal{S}_{Z+1}[a, b]$, 232
$S(\eta)$, 335
$S(\xi)$, 272
ST, 410, 433
ST_1, 433
$S[\Lambda; h](\xi)$, 276
$S_j(x)$, 238
S^j_k, 418
$\mathbb{S}^{n-1}$, 149
$\mathcal{S}^0_{Z+1}$, 223, 225
$s(\lambda)$, 233, 235
Scaling function, 327
Scaling function and MRA, 220, 222
Schauder's method, 454, 483
Schechter, 484
Schoenberg, 217, 219, 221, 223, 225, 232, 238, 294
Schoenberg, I., 5, 6
Schumaker, 6, 64, 187, 221, 223
Schumker, 167
Seeley, 134, 185
Segment property, 461, 462
Self-adjoint, 133, 177
Separated variables, 152
Shape-preserving property, 73
Shubin, 484
Smirnov, 153, 209
Sobolev, 129, 159, 168
Sobolev embedding
    and polysplines, 452
Sobolev spaces on manifolds, 461
Sobolev, S. L., 4, 6
Sobolev-type embedding, 469
Solution set U_{Z+1}, 222, 224
    variation, 226, 228

Sphere, 128, 130
　area, 128, 130
Spherical coordinates, 129, 131
Spherical harmonics
　basis, 147, 149
　surface, 145, 147
Spherical operator $L_{(k)}$, 78, 80, 166, 168
Spline of Schoenberg, 224
Splines
　Bernoulli, 406
　existence and uniqueness of interpolation, 24
　exponential, 189
　multivariate – box, simplex, 28, 30
　one-dimensional
　　uniqueness and existence, 24, 26
　one-dimensional cubic, 17, 19
　one-dimensional linear, 19, 21
　polynomial, 18
Star-shaped domain, 183
Stein-Weiss, 6, 134, 379
Strip, 30, 32
　definition, 30, 32
Strips, 32, 34
SURFER
　Golden Software, 4
Symmetrized vector $\widetilde{\Lambda}$, 271

$\mathbb{T}^n$, 463
Taylor, 400, 406, 461, 484
TB-spline, 237, 239, 248, 250
　convolution formula, 241, 243
　differentiation, 242, 244
　estimate, 262, 264
　Fourier and Laplace transform, 239, 241
　Hermite–Genocchi formula, 244, 246
　normed, 281, 283
　　refinement equation, 368, 372
　recurrence relation, 244, 246
　refinement equation, 257, 259
　Riesz bounds, 265, 267
　　asymptotic, 279, 281
　　asymptotic on the mesh $h\mathbb{Z}$, 280, 282
　symmetry, 260, 262
　two-scale relation, 257, 259
Timoshenko-Goodier, 55
Tace theorem, 468
Transmission, 459
Treves, 132
Triebel, 131, 465
Two-scale relation, 257, 259, 314, 331
　see rfinement equation, 259

Two-scale sequence, 259
Two-scale symbol, 259, 334

$\widetilde{U}_{0,p}$, 115
$[u]^{(\beta)}$, 465
$\mathcal{U}_{0,p}$, 107, 115
$\widetilde{\mathcal{U}}_{0,p}$, 107
$\widetilde{\mathcal{U}}_{k,p}$, 112, 114
　for plysplines, 434
U_{Z+1}, 226
$U_{Z+1}[\Lambda]$
　variation, 228
U_{Z+1}, 222, 224
$U_{0,p}$, 106
$U_{k,p}$, 89
Uniform, 461
Uniform C^m-regularity, 467

v_n, 130
V_j, 319
V_0, 314
V_j, 314, 326
$V_j^{(k)}$, 373, 384
$v = \log r$, 105, 107, 109, 111, 166, 168
$v = \log r$, 84, 86
Variational property
　see Hlladay property
Vekua, I., 4, 129
Vishik–Eskin, 484

$\widetilde{W}_j$, 318
W_0, 329
W_j, 315, 326, 328
$W_j^{(k)}$, 384
$W_j^{(k)}$, 373
Warner, 463, 484
Wavelet
　mother, 316
Wavelet analysis, 259
　and cardinal L-splines, 222
　non-stationary, 223
Wavelet spaces, 315
　Decomposition and reconstruction algorithms, 321
　decomposition relation, 320
　dual scaling function $\widetilde{\phi}(x)$, 319
　semi-orthogonal, 325
　symmetry, 323
　zero moments, 322
Weinberger, 45, 83
Whittaker-Watson, 209

Winberger, 210
Wronskian determinant, 188

x^α, 130

$Y_{k,\ell}$, 147, 149

$Z_{\theta'}^{(k)}(\theta)$, 153
Zonal harmonics, 152, 154
Zonal harmonics
  and Newton potential, 157, 159